Handbook of Machine Vision

Edited by
Alexander Hornberg

Related Titles

Florczyk, S.
Robot Vision
Video-based Indoor Exploration with Autonomous and Mobile Robots
216 pages with 89 figures and 3 tables
2005
Hardcover
ISBN 3-527-40544-5

Mather, P. M.
Computer Processing of Remotely-Sensed Images
An Introduction
442 pages
2004
Softcover
ISBN 0-470-84919-3

Bose, T., Meyer, F.
Digital Signal and Image Processing
International Edition
approx. 656 pages
2003
Hardcover
ISBN 0-471-45230-0

Mann, S.
Intelligent Image Processing
360 pages
2002
Hardcover
ISBN 0-471-40637-6

Pratt, W. K.
Digital Image Processing
PIKS Inside
756 pages
2001
Hardcover
ISBN 0-471-37407-5

Nikolaidis, N., Pitas, I.
3-D Image Processing Algorithms
190 pages
2001
Hardcover
ISBN 0-471-37736-8

Pitas, I.
Digital Image Processing Algorithms and Applications
432 pages
2000
Hardcover
ISBN 0-471-37739-2

Vincze, M., Hager, G.D. (eds.)
Robust Vision for Vision-Based Control of Motion
approx. 262 pages
2000
Hardcover
ISBN 0-7803-5378-1

Petrou, M., Bosdogianni, P.
Image Processing
The Fundamentals
354 pages
1999
Hardcover
ISBN 0-471-99883-4

Handbook of Machine Vision

Edited by
Alexander Hornberg

WILEY-VCH Verlag GmbH & Co. KGaA

The Editor

Prof. Dr. Alexander Hornberg
Hochschule Esslingen
FB Mechatronik u. Eletrotechnik
alexander.hornberg@fht-esslingen.de

Cover
Individual controllable stack of red and IR ring lights for presence check of circuits and quality check of soldering points.
© Vision & Control GmbH, Suhl/Germany

■ All books published by Wiley-VCH are carefully produced. Nevertheless, editors, authors and publisher do not warrant the information contained in these books to be free of errors. Readers are advised to keep in mind that statements, data, illustrations, procedural details or other items may inadvertently be inaccurate.

Library of Congress Card No.:
applied for

British Library Cataloguing-in-Publication Data:
A catalogue record for this book is available from the British Library.

Bibliographic information published by Die Deutsche Bibliothek
Die Deutsche Bibliothek lists this publication in the Deutsche Nationalbibliografie; detailed bibliographic data is available in the Internet at http://dnb.ddb.de

© 2006 WILEY-VCH Verlag GmbH & Co KGaA, Weinheim

All rights reserved (including those of translation into other languages). No part of this book may be reproduced in any form – by photocopying, microfilm, or any other means – nor transmitted or translated into a machine language without written permission from the publishers. Registered names, trademarks, etc. used in this book, even when not specifically marked as such, are not to be considered unprotected by law.

Printed in the Federal Republic of Germany
Printed on acid-free and chlorine-free paper

Cover Design: aktivcomm, Weinheim
Typesetting: Steingraeber Satztechnik GmbH, Ladenburg
Printing: betz-druck GmbH, Darmstadt
Bookbinding: Litges & Dopf Buchbinderei GmbH, Heppenheim

ISBN-13: 978-3-527-40584-8
ISBN-10: 3-527-40584-4

Contents

1	**Processing of Information in the Human Visual System**	*1*
	Prof. Dr. F. Schaeffel, University of Tübingen	
1.1	Preface *1*	
1.2	Design and Structure of the Eye *1*	
1.3	Optical Aberrations and Consequences for Visual Performance *5*	
1.4	Chromatic Aberration *9*	
1.5	Neural Adaptation to Monochromatic Aberrations *12*	
1.6	Optimizing Retinal Processing with Limited Cell Numbers, Space and Energy *13*	
1.7	Adaptation to Different Light Levels *13*	
1.8	Rod and Cone Responses *15*	
1.9	Spiking and Coding *18*	
1.10	Temporal and Spatial Performance *19*	
1.11	ON/OFF Structure, Division of the Whole Illuminance Amplitude in Two Segments *20*	
1.12	Consequences of the Rod and Cone Diversity on Retinal Wiring *21*	
1.13	Motion Sensitivity in the Retina *21*	
1.14	Visual Information Processing in Higher Centers *23*	
1.14.1	Morphology *23*	
1.14.2	Functional Aspects – Receptive Field Structures and Cortical Modules *24*	
1.15	Effects of Attention *26*	
1.16	Color Vision, Color Constancy, and Color Contrast *27*	
1.17	Depth Perception *28*	
1.18	Adaptation in the Visual System to Color, Spatial, and Temporal Contrast *29*	
1.19	Conclusions *30*	
	References *32*	

Handbook of Machine Vision. Alexander Hornberg (Ed.)
Copyright © 2006 WILEY-VCH Verlag GmbH & Co. KGaA, Weinheim
ISBN: 3-527-40584-7

2	**Introduction to Building a Machine Vision Inspection** *35*	
	Axel Telljohann, Consulting Team Machine Vision (CTMV)	
2.1	Preface *35*	
2.2	Specifying a Machine Vision System *36*	
2.2.1	Task and Benefit *37*	
2.2.2	Parts *37*	
2.2.2.1	Different Part Types *38*	
2.2.3	Part Presentation *38*	
2.2.4	Performance Requirements *39*	
2.2.4.1	Accuracy *39*	
2.2.4.2	Time Performance *39*	
2.2.5	Information Interfaces *39*	
2.2.6	Installation Space *40*	
2.2.7	Environment *41*	
2.2.8	Checklist *41*	
2.3	Designing a Machine Vision System *41*	
2.3.1	Camera Type *41*	
2.3.2	Field of View *42*	
2.3.3	Resolution *43*	
2.3.3.1	Camera Sensor Resolution *44*	
2.3.3.2	Spatial Resolution *44*	
2.3.3.3	Measurement Accuracy *44*	
2.3.3.4	Calculation of Resolution *45*	
2.3.3.5	Resolution for a Line Scan Camera *45*	
2.3.4	Choice of Camera, Frame Grabber, and Hardware Platform *46*	
2.3.4.1	Camera Model *46*	
2.3.4.2	Frame Grabber *46*	
2.3.4.3	Pixel Rate *46*	
2.3.4.4	Hardware Platform *47*	
2.3.5	Lens Design *48*	
2.3.5.1	Focal Length *48*	
2.3.5.2	Lens Flange Focal Distance *49*	
2.3.5.3	Extension Tubes *49*	
2.3.5.4	Lens Diameter and Sensor Size *50*	
2.3.5.5	Sensor Resolution and Lens Quality *50*	
2.3.6	Choice of Illumination *50*	
2.3.6.1	Concept: Maximize Contrast *51*	
2.3.6.2	Illumination Setups *51*	
2.3.6.3	Light Sources *52*	
2.3.6.4	Approach to the Optimum Setup *52*	
2.3.6.5	Interfering Lighting *53*	
2.3.7	Mechanical Design *53*	

2.3.8	Electrical Design 53
2.3.9	Software 54
2.3.9.1	Software Library 54
2.3.9.2	Software Structure 54
2.3.9.3	General Topics 55
2.4	Costs 56
2.5	Words on Project Realization 57
2.5.1	Development and Installation 57
2.5.2	Test Run and Acceptance Test 58
2.5.3	Training and Documentation 58
2.6	Examples 58
2.6.1	Diameter Inspection of Rivets 59
2.6.1.1	Task 59
2.6.1.2	Specification 59
2.6.1.3	Design 60
2.6.2	Tubing Inspection 65
2.6.2.1	Task 65
2.6.2.2	Specification 65
2.6.2.3	Design 66

3 Lighting in Machine Vision 73
I. Jahr, Vision & Control GmbH

3.1	Introduction 73
3.1.1	Prologue 73
3.1.2	The Involvement of Lighting in the Complex Machine Vision Solution 75
3.2	Demands on Machine Vision lighting 78
3.3	Light used in Machine Vision 82
3.3.1	What is Light? Axioms of Light 82
3.3.2	Light and Light Perception 84
3.3.3	Light Sources for Machine Vision 88
3.3.3.1	Incandescent Lamps / Halogen Lamps 89
3.3.3.2	Metal Vapor Lamps 91
3.3.3.3	Xenon Lamps 92
3.3.3.4	Fluorescent Lamps 93
3.3.3.5	LEDs (Light Emitting Diodes) 95
3.3.3.6	Lasers 99
3.3.4	The Light Sources in Comparison 100
3.3.5	Considerations for Light Sources: Lifetime, Aging, Drift 100
3.3.5.1	Lifetime 100
3.3.5.2	Aging and Drift 102

3.4	Interaction of Test Object and Light 105
3.4.1	Risk Factor Test Object 105
3.4.1.1	What Does the Test Object do With the Incoming Light? 106
3.4.1.2	Reflection/reflectance/scattering 107
3.4.1.3	Total Reflection 111
3.4.1.4	Transmission/transmittance 112
3.4.1.5	Absorption/absorbance 113
3.4.1.6	Diffraction 114
3.4.1.7	Refraction 115
3.4.2	Light Color and Part Color 116
3.4.2.1	Visible Light (VIS) – Monochromatic Light 116
3.4.2.2	Visible Light (VIS) – White Light 118
3.4.2.3	Infrared Light (IR) 119
3.4.2.4	Ultraviolet Light (UV) 121
3.4.2.5	Polarized Light 122
3.5	Basic Rules and Laws of Light Distribution 125
3.5.1	Basic Physical Quantities of Light 125
3.5.2	The Photometric Inverse Square Law 127
3.5.3	The Constancy of Luminance 128
3.5.4	What Light Arrives at the Sensor – Light Transmission Through the Lens 129
3.5.5	Light Distribution of Lighting Components 131
3.5.6	Contrast 134
3.5.7	Exposure 136
3.6	Light Filters 138
3.6.1	Characteristic Values of Light Filters 138
3.6.2	Influences of Light Filters on the Optical Path 140
3.6.3	Types of Light Filters 141
3.6.4	Anti-Reflective Coatings (AR) 143
3.6.5	Light Filters for Machine Vision 144
3.6.5.1	UV Blocking Filter 144
3.6.5.2	Daylight Suppression Filter 144
3.6.5.3	IR Suppression Filter 146
3.6.5.4	Neutral Filter/Neutral Density Filter/Gray Filter 146
3.6.5.5	Polarization Filter 147
3.6.5.6	Color Filters 148
3.6.5.7	Filter Combinations 149
3.7	Lighting Techniques and Their Use 150
3.7.1	How to Find a Suitable Lighting? 150
3.7.2	Planning the Lighting Solution – Influence Factors 151
3.7.3	Lighting Systematics 154
3.7.3.1	Directional Properties of the Light 155

3.7.3.2	Arrangement of the Lighting	*157*
3.7.3.3	Properties of the Illuminated Field	*158*
3.7.4	The Lighting Techniques in Detail	*160*
3.7.4.1	Diffuse Bright Field Incident Light	*160*
3.7.4.2	Directed Bright Field Incident Light	*162*
3.7.4.3	Telecentric Bright Field Incident Light	*163*
3.7.4.4	Structured Bright Field Incident Light	*165*
3.7.4.5	Diffuse/Directed Partial Bright Field Incident Light	*169*
3.7.4.6	Diffuse/Directed Dark Field Incident Light	*173*
3.7.4.7	The Limits of the Incident Lighting	*176*
3.7.4.8	Diffuse Bright Field Transmitted Lighting	*177*
3.7.4.9	Directed Bright Field Transmitted Lighting	*178*
3.7.4.10	Telecentric Bright Field Transmitted Lighting	*180*
3.7.4.11	Diffuse/Directed Transmitted Dark Field Lighting	*184*
3.7.5	Combined Lighting Techniques	*184*
3.8	Lighting Control	*186*
3.8.1	Reasons for Light Control – the Environmental Industrial Conditions	*186*
3.8.2	Electrical Control	*186*
3.8.2.1	Stable Operation	*186*
3.8.2.2	Brightness Control	*189*
3.8.2.3	Temporal Control: Static-Pulse-Flash	*189*
3.8.2.4	Some Considerations for the Use of Flash Light	*191*
3.8.2.5	Temporal and Local Control: Adaptive Lighting	*195*
3.8.3	Geometrical Control	*198*
3.8.3.1	Lighting from Large Distances	*198*
3.8.3.2	Light Deflection	*199*
3.8.4	Suppression of Ambient and Extraneous Light – Measures for a Stable Lighting	*200*
3.9	Lighting Perspectives for the Future	*201*
	References	*202*

4	**Optical Systems in Machine Vision**	***205***
	Dr. Karl Lenhardt, Jos. Schneider Optische Werke GmbH	
4.1	A Look on the Foundations of Geometrical Optics	*205*
4.1.1	From Electrodynamics to the Light Rays	*205*
4.1.2	The Basic Laws of Geometrical Optics	*208*
4.2	Gaussian Optics	*210*
4.2.1	Reflection and Refraction at the Boundary between two Media	*210*
4.2.2	Linearizing the Law of Refraction – the Paraxial Approximation	*212*

4.2.3	Basic Optical Conventions	213
4.2.4	The Cardinal Elements of a Lens in Gaussian Optics	216
4.2.5	The Thin Lens Approximation	221
4.2.6	Beam Converging and Beam Diverging Lenses	222
4.2.7	Graphical Image Constructions	223
4.2.7.1	Beam Converging Lenses	223
4.2.7.2	Beam Diverging Lenses	224
4.2.8	Imaging Equations and Their Related Coordinate Systems	224
4.2.8.1	Reciprocity Equation	225
4.2.8.2	Newton's Equations	226
4.2.8.3	General Imaging Equation	227
4.2.8.4	The Axial Magnification Ratio	229
4.2.9	The Overlapping of Object and Image Space	229
4.2.10	Focal Length, Lateral Magnification and the Field of View	229
4.2.11	Systems of Lenses	231
4.2.12	Consequences of the Finite Extension of Ray Pencils	234
4.2.12.1	Effects of Limitations of the Ray Pencils	234
4.2.12.2	Several Limiting Openings	237
4.2.12.3	Characterizing the Limits of Ray Pencils	240
4.2.12.4	The Relation to the Linear Camera Model	243
4.2.13	Geometrical Depth of Field and Depth of Focus	244
4.2.13.1	Depth of Field as a Function of the Object Distance p	246
4.2.13.2	Depth of Field as a Function of β	247
4.2.13.3	The Hyperfocal Distance	248
4.2.13.4	The Permissible Size for the Circle of Confusion d'	249
4.2.14	The Laws of Central Projection–Telecentric System	251
4.2.14.1	An Introduction to the Laws of Perspective	251
4.2.14.2	Central Projection from Infinity – Telecentric Perspective	260
4.3	The Wave Nature of Light	268
4.3.1	Introduction	268
4.3.2	The Rayleigh–Sommerfeld Diffraction Integral	269
4.3.3	Further Approximations to the Huygens–Fresnel Principle	272
4.3.3.1	Fresnel's Approximation	273
4.3.4	The Impulse Response of an Aberration Free Optical System	275
4.3.4.1	The Case of Circular Aperture, Object Point on the Optical Axis	278
4.3.5	The Intensity Distribution in the Neighbourhood of the Geometrical Focus	278
4.3.6	The Extension of the Point Spread Function in a Defocused Image Plane	282
4.3.7	Consequences for the Depth of Field Considerations	285
4.4	Information Theoretical Treatment of Image Transfer and Storage	287

4.4.1	Physical Systems as Linear, Invariant Filters	*288*
4.4.2	The Optical Transfer Function (OTF) and the meaning of spatial frequency	*296*
4.4.3	Extension to the Two-Dimensional Case	*298*
4.4.3.1	The Interpretation of Spatial Frequency Components (r,s)	*298*
4.4.3.2	Reduction to One-Dimensional Representations	*300*
4.4.4	Impulse Response and MTF for Semiconductor Imaging Devices	*302*
4.4.5	The Transmission Chain	*304*
4.4.6	The Aliasing Effect and the Space Variant Nature of Aliasing	*305*
4.4.6.1	The Space Variant Nature of Aliasing	*312*
4.5	Criteria for Image Quality	*316*
4.5.1	Gaussian Data	*316*
4.5.2	Overview on Aberrations of Third Order	*316*
4.5.3	Image Quality in the Space Domain: PSF, LSF, ESF and Distortion	*317*
4.5.4	Image Quality in Spatial Frequency Domain: MTF	*321*
4.5.5	Other Image Quality Parameters	*323*
4.5.5.1	Relative Illumination (Relative Irradiance) [17]	*323*
4.5.5.2	Deviation from Telecentricity (for Telecentric Lenses only)	*325*
4.5.6	Manufacturing Tolerances and Image Quality	*325*
4.6	Practical Aspects	*326*
	References	*331*
5	**Camera Calibration**	***333***
	R. Godding, AICON 3D Systems GmbH	
5.1	Introduction	*333*
5.2	Terminology	*334*
5.2.1	Camera, Camera system	*334*
5.2.2	Coordinate systems	*335*
5.2.3	Interior Orientation and Calibration	*335*
5.2.4	Exterior and Relative Orientation	*335*
5.2.5	System Calibration	*336*
5.3	Physical Effects	*336*
5.3.1	Optical System	*336*
5.3.2	Camera and Sensor Stability	*336*
5.3.3	Signal Processing and Transfer	*337*
5.4	Mathematical Calibration Model	*338*
5.4.1	Central Projection	*338*
5.4.2	Camera Model	*339*
5.4.3	Focal Length and Principal Point	*340*

5.4.4	Distortion and Affinity	340
5.4.5	Radial symmetrical distortion	341
5.4.6	Radial Asymmetrical and Tangential Distortion	342
5.4.7	Affinity and Nonorthogonality	343
5.4.8	Variant Camera Parameters	343
5.4.9	Sensor Flatness	345
5.4.10	Other Parameters	345
5.5	Calibration and Orientation Techniques	346
5.5.1	In the Laboratory	346
5.5.2	Using Bundle Adjustment to Determine Camera Parameters	346
5.5.2.1	Calibration based Exclusively on Image Information	347
5.5.2.2	Calibration and Orientation with Additional Object Information	349
5.5.2.3	Extended System Calibration	351
5.5.3	Other Techniques	352
5.6	Verification of Calibration Results	354
5.7	Applications	355
5.7.1	Applications with Simultaneous Calibration	355
5.7.2	Applications with precalibrated cameras	356
5.7.2.1	Tube Measurement within a Measurement Cell	356
5.7.2.2	Online Measurements in the Field of Car Safety	357
5.7.2.3	Other Applications	358
	References	358

6 Camera Systems in Machine Vision 361
Horst Mattfeldt, Allied Vision Technologies GmbH

6.1	Camera Technology	361
6.1.1	History in Brief	361
6.1.2	Machine Vision versus Closed Circuit Television (CCTV)	362
6.2	Sensor Technologies	363
6.2.1	Spatial Differentiation: 1D and 2D	364
6.2.2	CCD Technology	364
6.2.3	Full Frame Principle	365
6.2.4	Frame Transfer Principle	366
6.2.5	Interline Transfer	366
6.2.5.1	Interlaced Scan Interline Transfer	367
6.2.5.2	Frame Readout	369
6.2.5.3	Progressive Scan Interline Transfer	370
6.3	CCD Image Artifacts	372
6.3.1	Blooming	372
6.3.2	Smear	372

6.4	CMOS Image Sensor	373
6.4.1	Advantages of CMOS Sensors	374
6.4.2	CMOS Sensor Shutter Concepts	376
6.4.2.1	Comparison of CMOS versus CCD	378
6.4.2.2	Integration Complexity of CDD versus CMOS Camera Technology	379
6.4.2.3	Video Standards	379
6.4.2.4	Sensor Sizes and Dimensions	380
6.4.2.5	Sony HAD Technology	381
6.4.2.6	Sony SuperHAD Technology	381
6.4.2.7	Sony EXView HAD Technology	381
6.5	Block Diagrams and their Description	383
6.5.1	Block Diagram of a Progressive Scan Analog Camera	383
6.5.1.1	CCD Read-Out Clocks	383
6.5.1.2	Spectral Sensitivity	386
6.5.1.3	Analog Signal Processing	387
6.5.1.4	Camera and Frame Grabber	388
6.5.2	Block Diagram of Color Camera with Digital Image Processing	389
6.5.2.1	Bayer™ Complementary Color Filter Array	389
6.5.2.2	Complementary Color Filters Spectral Sensitivity	390
6.5.2.3	Generation of Color Signals	390
6.6	Digital Cameras	393
6.6.1	Black and White Digital Cameras	393
6.6.1.1	B/W Sensor and Processing	394
6.6.2	Color Digital Cameras	396
6.6.2.1	Analog Processing	396
6.6.2.2	One-Chip Color Processing	396
6.6.2.3	Analog Front End (AFE)	398
6.6.2.4	One Push White Balance	399
6.6.2.5	Automatic White Balance	400
6.6.2.6	Manual Gain	400
6.6.2.7	Auto Shutter/Gain	400
6.6.2.8	A/D Conversion	400
6.6.2.9	Lookup Table (LUT) and Gamma Function	401
6.6.2.10	Shading Correction	402
6.6.2.11	Horizontal Mirror Function	403
6.6.2.12	Frame Memory and Deferred Image Transport	404
6.6.2.13	Color Interpolation	404
6.6.2.14	Color Correction	406
6.6.2.15	RGB to YUV Conversion	406
6.6.2.16	Binning versus Area of Interest (AOI)	406
6.7	Controlling Image Capture	407

6.7.1	Hardware Trigger Modes	408	
6.7.1.1	Latency (Jitter) Aspects	409	
6.7.2	Pixel Data	410	
6.7.3	Data Transmission	411	
6.7.4	IEEE-1394 Port Pin Assignment	412	
6.7.5	Operating the Camera	414	
6.7.6	HiRose Jack Pin Assignment	414	
6.7.7	Frame Rates and Bandwidth	415	
6.8	Configuration of the Camera	418	
6.8.1	Camera Status Register	418	
6.9	Camera Noise[1]	420	
6.9.1	Photon Noise	421	
6.9.2	Dark Current Noise	421	
6.9.3	Photo Response Nonuniformity (PRNU)	422	
6.9.4	Reset Noise	422	
6.9.5	1/f Noise (Amplifier Noise)	422	
6.9.6	Quantization Noise	423	
6.9.7	Noise Floor	423	
6.9.8	Dynamic Range	423	
6.9.9	Signal to Noise Ratio	424	
6.10	Digital Interfaces	424	
	References	426	

7 Camera Computer Interfaces 427

Tony Iglesias, Anita Salmon, Johann Scholtz, Robert Hedegore, Julianna Borgendale, Brent Runnels, Nathan McKimpson, National Instruments

7.1	Overview	427
7.2	Analog Camera Buses	429
7.2.1	Analog Video Signal	429
7.2.2	Interlaced Video	430
7.2.3	Progressive Scan Video	431
7.2.4	Timing Signals	431
7.2.5	Analog Image Acquisition	432
7.2.6	S-Video	432
7.2.7	RGB	433
7.2.8	Analog Connectors	433
7.3	Parallel Digital Camera Buses	434
7.3.1	Digital Video Transmission	434
7.3.2	Taps	434
7.3.3	Differential Signaling	436
7.3.4	Line Scan	436

7.3.5	Parallel Digital Connectors	437
7.3.6	Camera Link	437
7.3.7	Camera Link Signals	439
7.3.7.1	Camera Link Connectors	440
7.4	Standard PC Buses	440
7.4.1	USB	440
7.4.1.1	USB for Machine Vision	442
7.4.2	IEEE 1394 (FireWire®)	442
7.4.2.1	IEEE 1394 for Machine Vision	445
7.4.3	Gigabit Ethernet (IEEE 802.3z)	450
7.4.3.1	Gigabit Ethernet for Machine Vision	451
7.5	Choosing a Camera Bus	453
7.5.1	Bandwidth	454
7.5.2	Resolution	455
7.5.3	Frame Rate	457
7.5.4	Cables	457
7.5.5	Line Scan	458
7.5.6	Reliability	458
7.5.7	Summary of Camera Bus Specifications	460
7.5.8	Sample Use Cases	460
7.5.8.1	Manufacturing inspection	460
7.5.8.2	LCD inspection	461
7.5.8.3	Security	461
7.6	Computer Buses	462
7.6.1	ISA/EISA	462
7.6.2	PCI/CompactPCI/PXI	464
7.6.3	PCI-X	465
7.6.4	PCI Express/CompactPCI Express/PXI Express	466
7.6.5	Throughput	469
7.6.6	Prevalence and Lifetime	471
7.6.7	Cost	472
7.7	Choosing a Computer Bus	472
7.7.1	Determine Throughput Requirements	472
7.7.2	Applying the Throughput Requirements	474
7.8	Driver Software	475
7.8.1	Application Programming Interface	476
7.8.2	Supported Platforms	479
7.8.3	Performance	479
7.8.4	Utility Functions	481
7.8.5	Acquisition Mode	481
7.8.5.1	Snap	481
7.8.5.2	Grab	482

7.8.5.3   Sequence   *483*
7.8.5.4   Ring   *483*
7.8.6   Image Representation   *485*
7.8.6.1   Image Representation in Memory   *485*
7.8.6.2   Bayer Color Encoding   *489*
7.8.6.3   Image Representation on Disk   *491*
7.8.7   Image Display   *492*
7.8.7.1   Understanding Display Modes   *492*
7.8.7.2   Palettes   *494*
7.8.7.3   Nondestructive Overlays   *494*
7.9   Features of a Machine Vision System   *495*
7.9.1   Image Reconstruction   *496*
7.9.2   Timing and Triggering   *497*
7.9.3   Memory Handling   *500*
7.9.4   Additional Features   *502*
7.9.4.1   Look-up Tables   *502*
7.9.4.2   Region of Interest   *504*
7.9.4.3   Color Space Conversion   *505*
7.9.4.4   Shading Correction   *507*
7.9.5   Summary   *508*

8   **Machine Vision Algorithms**   **511**
Dr. Carsten Steger, MVTec Software GmbH
8.1   Fundamental Data Structures   *511*
8.1.1   Images   *511*
8.1.2   Regions   *513*
8.1.3   Subpixel-Precise Contours   *515*
8.2   Image Enhancement   *516*
8.2.1   Gray Value Transformations   *516*
8.2.2   Radiometric Calibration   *519*
8.2.3   Image Smoothing   *525*
8.3   Geometric Transformations   *536*
8.3.1   Affine Transformations   *537*
8.3.2   Projective Transformations   *538*
8.3.3   Image Transformations   *539*
8.3.4   Polar Transformations   *544*
8.4   Image Segmentation   *544*
8.4.1   Thresholding   *545*
8.4.2   Extraction of Connected Components   *551*
8.4.3   Subpixel-Precise Thresholding   *553*
8.5   Feature Extraction   *555*

8.5.1	Region Features	556
8.5.2	Gray Value Features	560
8.5.3	Contour Features	565
8.6	Morphology	566
8.6.1	Region Morphology	566
8.6.2	Gray Value Morphology	583
8.7	Edge Extraction	587
8.7.1	Definition of Edges in 1D and 2D	587
8.7.2	1D Edge Extraction	591
8.7.3	2D Edge Extraction	597
8.7.4	Accuracy of Edges	605
8.8	Segmentation and Fitting of Geometric Primitives	612
8.8.1	Fitting Lines	613
8.8.2	Fitting Circles	617
8.8.3	Fitting Ellipses	619
8.8.4	Segmentation of Contours into Lines, Circles, and Ellipses	620
8.9	Template Matching	624
8.9.1	Gray-Value-Based Template Matching	626
8.9.2	Matching Using Image Pyramids	632
8.9.3	Subpixel-Accurate Gray-Value-Based Matching	636
8.9.4	Template Matching with Rotations and Scalings	637
8.9.5	Robust Template Matching	638
8.10	Stereo Reconstruction	653
8.10.1	Stereo Geometry	654
8.10.2	Stereo Matching	662
8.11	Optical Character Recognition	666
8.11.1	Character Segmentation	667
8.11.2	Feature Extraction	669
8.11.3	Classification	672
	References	688

9	**Machine Vision in Manufacturing**	**693**
	Dr.-Ing. Peter Waszkewitz, Robert Bosch GmbH	
9.1	Introduction	693
9.2	Application Categories	694
9.2.1	Types of Tasks	694
9.2.2	Types of Production	696
9.2.2.1	Discrete Unit Production Versus Continuous Flow	697
9.2.3	Types of Evaluations	698
9.2.4	Value-Adding Machine Vision	699
9.3	System Categories	700

9.3.1	Common Types of Systems	700
9.3.2	Sensors	701
9.3.3	Vision Sensors	702
9.3.4	Compact Systems	703
9.3.5	Vision Controllers	704
9.3.6	PC-Based Systems	704
9.3.7	Summary	707
9.4	Integration and Interfaces	708
9.5	Mechanical Interfaces	710
9.5.1	Dimensions and Fixation	711
9.5.2	Working Distances	711
9.5.3	Position Tolerances	712
9.5.4	Forced Constraints	713
9.5.5	Additional Sensor Requirements	713
9.5.6	Additional Motion Requirements	714
9.5.7	Environmental Conditions	715
9.5.8	Reproducibility	716
9.5.9	Gauge Capability	718
9.6	Electrical Interfaces	720
9.6.1	Wiring and Movement	721
9.6.2	Power Supply	721
9.6.3	Internal Data Connections	722
9.6.4	External Data Connections	724
9.7	Information Interfaces	725
9.7.1	Interfaces and Standardization	725
9.7.2	Traceability	726
9.7.3	Types of Data and Data Transport	727
9.7.4	Control Signals	727
9.7.5	Result and Parameter Data	728
9.7.6	Mass Data	729
9.7.7	Digital I/O	729
9.7.8	Field Bus	730
9.7.9	Serial Interfaces	730
9.7.10	Network	731
9.7.11	Files	732
9.7.12	Time and Integrity Considerations	732
9.8	Temporal Interfaces	734
9.8.1	Discrete Motion Production	735
9.8.2	Continuous Motion Production	737
9.8.3	Line-Scan Processing	740
9.9	Human–Machine Interfaces	743
9.9.1	Interfaces for Engineering Vision Systems	743

9.9.2	Runtime Interface	745
9.9.3	Remote Maintenance	749
9.9.4	Offline Setup	750
9.10	Industrial Case Studies	752
9.10.1	Glue Check under UV Light	752
9.10.1.1	Solution	752
9.10.2	Completeness Check	754
9.10.3	Multiple Position and Completeness Check	757
9.10.4	Pin Type Verification	760
9.10.5	Robot Guidance	763
9.10.6	Type and Result Data Management	766
9.11	Constraints and Conditions	769
9.11.1	Inspection Task Requirements	769
9.11.2	Circumstantial Requirements	770
9.11.3	Limits and Prospects	773
	References	775
	Index	781

Preface

Why a Further Book on Machine Vision?

The idea to write another book about machine vision (or image processing) seems to be unnecessary. That is only the first impression. The search of a book which describes the whole information processing chain was unsuccessful, because the most books deal predominantly with the algorithms of image processing which is an important part of a machine vision application. They do not take into account that first we have to acquire a digital image of the real part of the world, which have some important properties:

- high contrast
- high resolution

The success of developing a machine vision system depends on the understanding all parts of the imaging chain. The charm and the complexity of machine vision lies in the range of the specialized engineering fields involved in it, namely

- mechanical engineering
- electrical engineering
- optical engineering
- software engineering

each of which struggles for a primary role. The interdisciplinary thought is the base for the successful development of a machine vision applications. Today we have a new term for this field of engineering called *mechatronics*. This fact of the case determine the difficulties and the possibilities of machine vision inspection.

Handbook of Machine Vision. Alexander Hornberg (Ed.)
Copyright © 2006 WILEY-VCH Verlag GmbH & Co. KGaA, Weinheim
ISBN: 3-527-40584-7

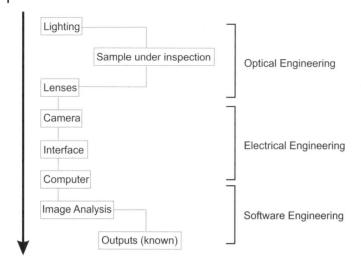

Fig. 1 Information processing chain

The book is written for users and developers who have a basic familiarity with machine vision technology and want to gain further insight into how to develop a machine-vision inspection system in industrial field.

The goal of the book is to present all elements of the information processing chain (see Chapters 3 to 8) in a manner such that even a nonspecialist (e.g., a system integrator or a user) can understand the meaning and function of all these elements.

Chapter 1, "Processing of Information in the Human Visual System," may at first glance seem to have no relevance to the subject of the book. Yet, for understanding the problems and methods of machine vision systems it is useful to know some of the properties of the human eye. There are many similarities between the human eye and a digital camera with its lens. Also the items *color* and *color models* have their roots in the visual optics.

Chapter 2, "Introduction to Building a Machine Vision Inspection," gives a first overview and an introduction of what the user have to do when he wants to build a machine vision inspection. The goal is to give an assistance and a guideline for the first few steps. Of course the individual practice can skip it.

Chapters 3–8 treat different elements of the information processing chain (Fig. 1) in detail.

- Lighting (Chapter 3)
- Lenses (Chapter 4)
- Camera calibration (Chapter 5)
- Camera (Chapter 6)

- Camera computer interfaces (Chapter 7)

- Algorithms (Chapter 8)

The last chapter, "Machine Vision in Manufacturing," concludes the loop to Chapter 2 and demonstrates the application of machine vision inspection in the industrial sector.

Finally, I would like to express my special thanks to the respected authors and their companies for their great engagement in making this book possible and Wiley-VCH for the opportunity to present my understanding of machine vision to an international audience. In particular, I would like to thank Dr. Thoß and Mrs. Werner for their commitment and patience.

Plochingen,
May 2006

Alexander Hornberg

1
Processing of Information in the Human Visual System
Prof. Dr. F. Schaeffel, University of Tübingen

1.1
Preface

To gather as much necessary information as possible of the visual world, and neglect as much unnecessary information as possible, the visual system has undergone an impressive optimization in the course of evolution which is fascinating, in each detail that is examined. A few aspects will be described in this chapter. Similar limitations may exist in machine vision, and comparisons to the solutions developed in the visual system in the course of 5 billion years of evolution might provide some insights.

1.2
Design and Structure of the Eye

As in any camera, the *first step in vision* is the projection of the visual scene on an array of photodetectors. In the vertebrate *camera eye*, this is achieved by the cornea and lens in the eye (Fig. 1.1) which transmit the light in the visible part of the spectrum, 400 nm to 780 nm, by 60 to 70%. Another 20–30% are lost due to scattering in the ocular media. Only about 10% are finally absorbed by the photoreceptor pigment [2]. Due to the content of proteins, both cornea and lens absorb in the ultraviolet and due to the water content, the transmission is blocked in the far infrared. The cornea consists of a *thick* central layer, the stroma, which is sandwiched between two semipermeable membranes (total thickness about 0.5 mm). It is composed of collagen fibrils with mucopolysaccharides filling the space between the fibrils. Water content is tightly regulated to 75–80%, and clouding occurs if it changes beyond these limits. The crystalline lens is built up from proteins, called crystallines, which are characterized on the basis of their water solubility. The proteins in the

Handbook of Machine Vision. Alexander Hornberg (Ed.)
Copyright © 2006 WILEY-VCH Verlag GmbH & Co. KGaA, Weinheim
ISBN: 3-527-40584-7

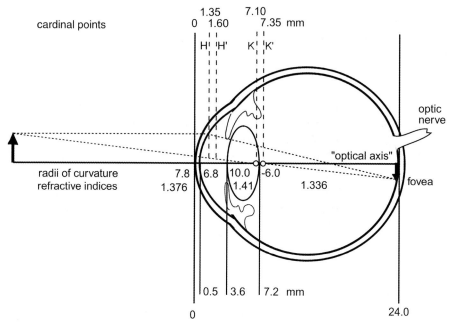

Fig. 1.1 Dimensions and schematic optics of the left human eye, seen from above. The anterior corneal surface is traditionally set to coordinate zero. All positions are given in Millimeters, relative to the anterior corneal surface (drawing not to scale). The refracting surfaces are approximated by spheres so that radii of curvatures can be defined. The cardinal points of the optical system, shown on the top, are valid only for rays close to the optical axis (*Gaussian approximation*). The focal length of the eye in the vitreous (the posterior focal length) is 24.0 mm − H = 22.65 mm. The nodal points K and K′ permit us to calculate the retinal image magnification. In the first approximation, the posterior nodal distance (*PND*, distance K to the focal point at the retina) determines the linear distance on the retina for a given visual angle. In the human eye, this distance is about 24.0 mm − 7.3 mm = 16.7 mm. One degree in the visual field maps on the retina to 16.7 tan(1°) = 290 µm. Given that the foveal photoreceptors are about 2 µm thick, about 140 receptors are sampling 1° in the visual field, which leads to a maximal resolution of about 70 cycles per degree. The schematic eye by Gullstrand represents *an average eye*. The variability in natural eyes is so large that it does not make sense to provide average numbers on dimensions with several digits. Refractive indices, however, are surprisingly similar among different eyes. The index of the lens (here homogenous model, *n* = 1.41) is not real but calculated to produce a lens power that makes the eye emmetropic. In a real eye, the lens has a gradient index (see the text).

periphery have high solubility, but in the center they are largely insoluble. The vertebrate lens is characterized by its continuous growth throughout life, with the older cells residing in the central core, the nucleus. With age, the lens becomes increasingly rigid and immobile and the ability to change its shape and focal length to accommodate for close viewing distances disappears – a disturbing experience for people around 45 who now need reading glasses (presbyopia). Accommodation is an active neuromuscular deformation of the crystalline lens that changes focal length from about 53 mm to about 32 mm in young adults.

Both media have higher refractive index as the water-like solutions in which they are embedded (tear film – on the corneal surface, aqueous – the liquid in the anterior chamber between the cornea and the lens, and vitreous humor – the gelly-like material filling the vitreous chamber between the lens and retina, Fig. 1.1). Due to their almost spherically curved surfaces, the anterior cornea and both surfaces of the lens have positive refractive power with an optical focal length of together about 22.6 mm. This matches almost perfectly (with a tolerance about 1/10 of a Millimeter) the distance from the first principal plane (Fig. 1.1, H) to the photoreceptor layer in the retina. Accordingly, the projected image from a distant object is in focus when accommodation is relaxed. This optimal optical condition is called emmetropia but, in about 30% of the industrialized population, the eye has grown too long so that the image is in front of the retina even with accommodation completely relaxed (myopia).

The projected image is first analyzed by the retina in the back of the eye. For developmental reasons, the retina in all vertebrate eyes is *inverted*. This means that the photoreceptor cells, located at the backside of the retina, are pointing away from the incoming light. Therefore, the light has to pass through the retina (about a fifth of a millimeter thick) before it can be detected. To reduce scatter, the retina is highly translucent, and the nerve fibers that cross on the vitreal side, from where the light comes in, to the optic nerve head are not surrounded by myelin, a fat-containing sheet that normally insulates spiking axons (see below). Scattering in retinal tissue still seems to be a problem since, in the region of highest spatial resolution, the fovea, the cells are pushed to the side. Accordingly, the fovea in the vertebrate eye can be recognized as a pit. However, many vertebrates don not have a fovea [3]; they have then lower visual acuity but their acuity can remain similar over large regions of the visual field, which is then usually either combined with high motion sensitivity (i.e., rabbit) or high light sensitivity at dusk (crepuscular mammals). It is striking that the retina in all vertebrates has a similar three-layered structure (Fig. 1.9), with similar thickness. This makes it likely that the functional constraints were similar. The function of the retina will be described below.

The *optical axis* of the eye is not perfectly defined because the cornea and lens are not perfectly rotationally symmetrical and also are not centered on one axis. Nevertheless, even though one could imagine that the image quality is best close to the *optical axis*, it turns out that the human fovea is not centered in the globe (Fig. 1.2). In fact, it is displaced to the temporal retina by the angle κ, ranging in different subjects from $0°$ to about $11°$ but highly correlated in both eyes. Apparently, a few degree away from the optical axis, the optical image quality is still good enough to not limit visual acuity in the fovea.

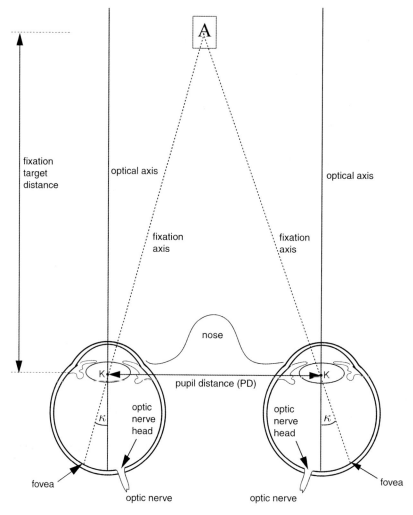

Fig. 1.2 Binocular geometry of human eyes, seen from above. Since the fovea is temporally displaced with regard to the optical axis by the angle κ, the optical axes of the eyes do not reflect the direction of fixation. κ is highly variable among eyes, ranging from zero to 11°, with an average of about 3.5°. In the illustrated case, the fixation target is at a distance for which the optical axes happen to be parallel and straight. The distance of the fixation target, for which this is true, can be easily calculated: for an angle κ of 4°, a pupil distance of 64 mm, this condition would be met if the fixation target is at 32 mm/tan(4°), or 457.6 mm. The optic nerve head (also called *optic disc*, or *blind spot*, the position at which the axons of the retinal ganglion cells leave the eye) is nasally displaced relative to the optical axis. The respective angle is in a similar range as κ. Under natural viewing conditions, the fixation angles must be extremely precise since double images are experienced if the fixation lines do not exactly cross on the fixation target – the tolerance is only a few minutes of arc).

1.3
Optical Aberrations and Consequences for Visual Performance

One could imagine that the optical quality of the cornea and lens must limit the visual acuity since the biological material is mechanically not as stable and the surfaces are much more variable than in technical glass lenses. However, this is not true. At day light pupil sizes < 2.5 mm are the optics of the human eye close to the diffraction limit (further improvement is physically not possible due to the wave properties of light). An eye is said to be diffraction limited when the ratio of the area under its MTF (Fig. 1.3) and the area under the diffraction-limited MTF (Strehl *ratio*) is higher than 0.8 ([4], Rayleigh *criterion*). With a 2 mm pupil, diffraction cuts off all spatial frequencies higher than 62 cyc/° – a limit that is very close to the maximal behavioral resolution achieved by human subjects. By the way, diffraction-limited optics is achieved only in some birds and primates, although it has been recently claimed that also the alert cat is diffraction-limited [5]. A number of tricks are used to reduce the aberrations that are inherent in spherical surfaces: the corneal surface is, in fact, clearly aspheric, flattening out to the periphery, and the vertebrate lens is always a *gradient index structure*, with the refractive index continuously increasing from the periphery to the center. Therefore, peripheral rays are bent less than central rays, which compensates for the steeper angles that rays encounter if they hit a spherical surface in the periphery. The gradient index of the lens reduces its spherical aberration from more than 12 diopters (for an assumed homogenous lens) to less than 1 diopter (gradient index lens). Furthermore, the optical aberrations seem to be under tight control (although it remains uncertain whether this control is visual [6]). The remaining aberrations of cornea and lens tend to cancel each other and this is true, at least, for astigmatism , horizontal coma and spherical aberration [7, 8]. However, aberrations have also advantages: they increase the depth of field by 0.3 D, apparently without reducing visual acuity; there is no strong correlation between the amount of aberrations of subjects and their letter acuity. They also reduce the required precision of accommodation , in particular, during reading [9]. It is questionable whether optical correction of higher order aberrations by refractive surgery or individually designed spectacle lenses would further improve acuity for high contrast letters in young subjects, creating an *eagle's eye*. It is however clear that an extended MTF can enhance the contrast sensitivity at high spatial frequencies. Correcting aberrations might also be useful in older subjects, since it is known that the monochromatic aberrations increase by a factor of 2 with age [10]. Aberrations may also be useful for other reasons; they could provide directionality cues for accommodation [11] and, perhaps, for emmetropization.

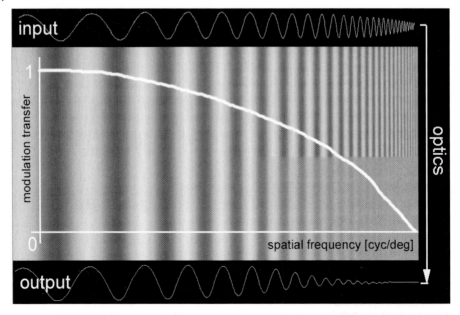

Fig. 1.3 The spatial information in an image can be reconstructed as a linear superposition of sine wave components (spatial frequencies) with different amplitudes and phases (Fourier components). Low spatial frequencies (SF) are generally available with high contrast in the natural visual environment, whereas the contrast declines for higher spatial frequencies, generally proportional to 1/SF (*input*). Due to optical imperfections and diffraction, the image on the retina does not retain the input contrast at high spatial frequencies. The decline of *modulation transfer*, the ratio of output to input contrast, is described by the modulation transfer function (MTF, thick white line). At around 60 cyc/°, the optical modulation transfer of the human eye reaches zero, with small pupil sizes due to diffraction and with larger pupils due to optical imperfections. These factors limit our contrast sensitivity at high spatial frequencies, even though the retina extracts surprisingly much information from the low contrast images of the high spatial frequencies, by building small receptive fields for foveal ganglion cells with antagonistic ON/OFF center/surround organization.

In a healthy emmetropic young eye, the optical modulation transfer function (Fig. 1.3) appears to be adapted to the sampling interval of the photoreceptors.

The contrast modulation reaches zero at spatial frequencies of around 60 cyc/°, and the foveal photoreceptor sampling interval is in the range of 2 µm. Since one degree in the visual field is mapped on a 0.29 mm linear distance on the retina, the highest detectable spatial frequency could be about 290/4 µm or about 70 cyc/°. The modulation transfer function (MTF) shows that the contrast of these high spatial frequencies in the retinal image is approaching zero (Fig. 1.3). With defocus, the MTF drops rapidly. Interestingly, it does not stop at zero modulation transfer, but rather continues to oscillate (although with

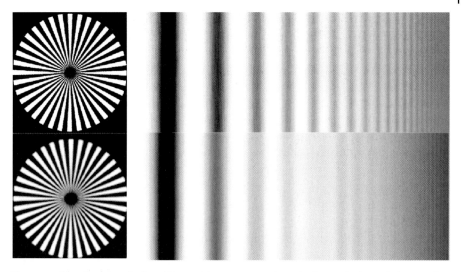

Fig. 1.4 Spurious resolution. The modulation transfer function (Fig. 1.3) shows oscillations beyond the cutoff spatial frequency which up in defocused gratings as contrast reversals. On the left, a circular grating shows the contrast reversals at the higher spatial frequencies in the center (top: in focus, below: defocused). On the right, the grating shown in Fig. 1.3 was defocused. Note the lack of contrast at the first transition to zero contrast, and the repeated subsequent contrast reversals. Note also that defocus has little effect on the low spatial frequencies.

low amplitude) around the abscissa. This gives rise to the so-called *spurious resolution* ; defocused gratings can still be detected beyond the cutoff spatial frequency, although in part with reversed contrast (Fig. 1.4).

The sampling interval of the photoreceptors increases rapidly over the first few degrees away from the fovea and visual acuity declines (Fig. 1.5), both because rods are added to the lattice which increases the distances between individual cones and because their cone diameters increase. In addition, many cones converge on one ganglion cell. Furthermore, only the foveal cones have *private lines* to a single ganglion cell (Fig. 1.9).

Because the optical quality does not decline as fast in the periphery as the spatial resolution of the neural network, the retinal image is undersampled. If the receptor mosaic would be regular, like in the fovea, stripes that are narrower than the resolution limit would cause spatial illusions (*moiré* patterns , Fig. 1.6). Since the receptor mosaic is not so regular in the peripheral retina, this causes just spatial noise. Moiré patterns are, however, visible in the fovea if a grating is imaged which is beyond the resolution limit. This can be done by presenting two laser beams in the pupil, which show interference [13].

Moire patterns are explained from the Shannon´s sampling theorem which states that regularly spaced samples can only be resolved when the sampling rate is equal to or higher than twice the highest spatial frequency – to resolve

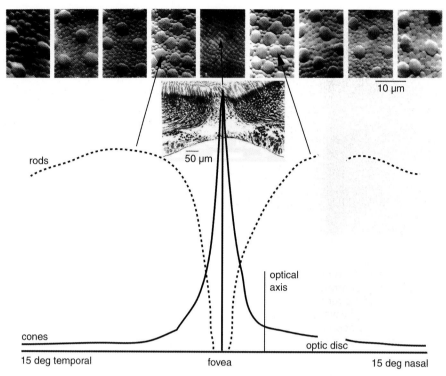

Fig. 1.5 Regional specializations of the retina. The fovea is free from rods and L and M cones are packed as tightly as possible (reaching a density of 200,000 per mm^2 – histology on top replotted after [14]). In the fovea, the retinal layers are pushed to the side to reduce scatter of light that reached the photoreceptors – resulting in the foveal pit. Rods reach a peak density of about 130,000 per mm^2 about 3° away from the fovea. Accordingly, a faint star can only be seen if it is not fixated. As a result of the drop in cone densities and due to increasing convergence of cone signals, visual acuity drops even faster: at 10°, visual acuity is only about 20% of the foveal peak. Angular positions relative to the fovea vary among individuals and are therefore approximate.

the samples, between each receptor that is stimulated there must be one that is not stimulated. The highest spatial frequency that can be resolved by the photoreceptor mosaic (*Nyquist limit*) is half the sampling frequency. In the fovea, the highest possible spatial sampling was achieved. Higher photoreceptor densities are not possible for the following reason: because the inner segments of the photoreceptors have higher refractive indices than their environment, they act as light guides. But if they become very thin, they show properties of waveguides. When their diameter approaches the wavelength of the light, energy starts to *leak out* [15], causing increased optical crosstalk to neighboring photoreceptors. As it is, about 5% of the energy is lost which

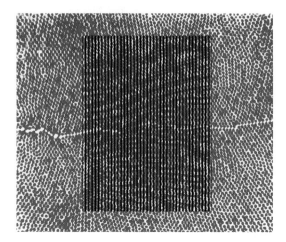

Fig. 1.6 Aliasing (undersampling) and Moiré patterns. If a grating is imaged on the photoreceptor array, and the sampling interval of the receptors is larger than half the spatial wavelength of the grating, patterns appear. The photoreceptor lattice (left) is from a histological section of a monkey retina. If laser interferometry is used to image fine gratings with spatial frequency beyond the resolution limit on the fovea of human subjects, the subjects see Moiré patterns, which they have drawn on the right (redrawn after [13]).

seems acceptable. But if the thickness (and the sampling interval) is further reduced to 1 μm, already 50% is lost.

Since the photoreceptor sampling interval cannot be decreased, the only way to increase visual acuity is then to enlarge the globe and, accordingly, the PND and the retinal image. This solution was adopted by the eagle eye, with an axial length of 36 mm, and the highest spatial acuity in the animal kingdom (grating acuity 135 cyc/° [16]).

1.4
Chromatic Aberration

In addition to the monochromatic aberrations (those aberrations that persist in monochromatic light) there is also chromatic aberration which results from dispersion of the optical media, i.e., the fact that the refractive index is wavelength dependent. In technical optical systems, lenses with different refractive indices are combined in such a way that the focal length does not vary much across the visible spectrum. In natural eyes, no attempt was made to optically balance chromatic aberration. Neural image processing makes it possible that we are not aware of the chromatic image degradation under normal viewing conditions, and there are morphological adaptations in the retina in

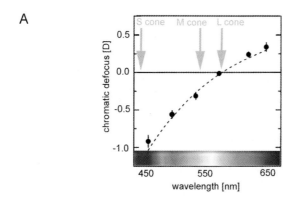

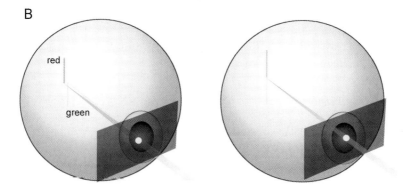

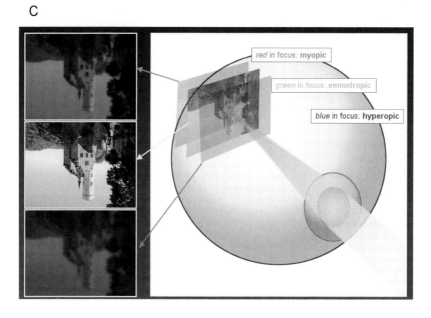

addition. Inspection of the chromatic aberration function in the human eye ([17]; Fig. 1.7A) shows that a large dioptric change occurs in the blue end of the spectrum (about 1 D from 570 to 450 nm) while the change is smaller in the red end (about 0.5 D from 570 to 680 nm). We have three cone photopigments, absorbing either at *long* wavelengths (*L cones*, peak absorption typically at 565 nm), at *middle* wavelengths (*M cones*, typically at 535 nm), or at *short* wavelengths (*S cones*, typically at 440 nm). The dioptric difference between L and M cones is small (about 0.2 D) but the dioptric differences to the S cones are significant (> 1 D). It is, therefore, impossible to see sharply with all three cone types at the same time (Fig. 1.7C). A *white* point of light is, therefore, imaged in the fovea as a circle with a diameter of up to 100 cones diameters, with a 6 mm pupil. Perhaps as a consequence, the S cone system was removed from the high acuity region of the central 0.5° of the fovea (the foveola); one cannot focus them anyway and they would only occupy space that is better used to pack the M and L cones more densely, to achieve best sampling. High acuity tasks are then performed only with the combined L and M cones. The *blue information* is continuously *filled in* because small eye movements make it possible to stimulate the parafoveal S cones. Therefore, this scotoma is normally not visible, similar to the blind spot, where the optic nerve leaves the eye – surprising, given that the blind spot has about five times the diameter of the fovea. Nevertheless, a small blue spot viewed on yellow background from a distance appears black, and a blue field of about 440 nm that is sinusoidally modulated at a few Hertz makes the blue scotoma in the fovea visible as a star-shaped black pattern [18]. Also in the periphery, the S cones sample more coarsely than the L and M cones and reach a spatial resolution of maximally 5-6 cyc/° in the parafoveal region at about 2° away from the fovea. The dispersion of the ocular media does not only produce different focal planes for each wavelength (*longitudinal chromatic aberration*), but also different image magnifications (*transverse chromatic aberration*). Accordingly, a point on an object's line are imaged on the retina through selected parts of the pupil, and the subject can align them via a joystick, the *achromatic axis* can be determined. Light of different wavelengths, entering the eye in the achromatic axis, is imaged at the same retinal position (although with wavelength-dependent focus). (**C**) Due to longitudinal chromatic aberration, light of different wavelengths is focused in different planes. Accordingly, myopic subjects (with too long eyes) see best in the red and hyperopic subjects (with too short eyes) best in the blue.

Fig. 1.7 Chromatic aberration and some of its effects on vision. Due to the increase of the refractive indices of the ocular media with decreasing wavelength, the eyes become more myopic in the blue. (**A**) The chromatic aberration function shows that the chromatic defocus between L and M cones is quite small (about a quarter of a diopter) but close to a diopter for the S cone. (**B**) Due to transverse chromatic aberration, rays of different wavelength that enter the pupil, reach the retina normally not in the same position. If a red and green surface is imaged in the different focal planes along a line only in the *achromatic axis* of the eye (which can be psychophysically determined; Fig. 1.7B). Even a few degrees away from the achromatic axis, the blue light emerging from an object point will be focused more closer to achromatic axis than the red. In particular, since the fovea is usually neither in the optical nor in the achromatic axis (see above), the images for red and blue are also laterally displaced with respect to each other. With a difference in image magnification of 3%, a κ of 3.5°, and a linear image magnification of 290μm per deg, the linear displacement would be $3.5 \times 290 \times 0.03$ μm or about 30 μm, which is about the distance from one S cone to the next. Human subjects are not aware of this difference in magnification and the rescaling of the *blue versus the red image* by neural processing seems to occur without effort.

1.5
Neural Adaptation to Monochromatic Aberrations

The neural image processor in the retina and cortex can relatively easily adapted to the changes of aberrations and field distortions. This can be seen in spectacle wearers. Even though spectacles, in particular progressive addition lenses, cause complex field distortions and additional aberrations, such as astigmatism and coma in the periphery, the wearer is usually not aware of these optical problems, already after a few days. The necessary neural image transformations are impressive. Nevertheless, it is realized that the individual visual system is best trained to the natural aberrations of the eye, even though it can learn to achieve a similar acuity if the same aberrations are experimentally rotated [19]. One of the underlying mechanisms is *contrast adaptation*. If all visible spatial frequencies are imaged on the retina with similar contrast (Fig. 1.3), it can be seen that the contrast sensitivity varies with spatial frequency. The highest sensitivity is achieved around 5 cyc/°. However, the contrast sensitivity at each spatial frequency is continuously re-adjusted, depending on how much contrast is available at a given spatial frequency. If the contrast is low, the sensitivity is increased, and if it is high, it is reduced. This way, the maximum information can be transmitted with limited total channel capacity. Contrast adaptation also accounts for the striking observation that a defocused image (lacking high contrast at higher spatial frequencies) appears sharper after a while ([20]; see movie, Nature Neuroscience[1]). If myopic subjects take off their glasses, they initially experience very poor visual acuity, but some improvement occurs over the first minutes without glasses. These changes are not in the optics, but neuronal. Contrast adaptation occurs both in retina and cortex [21].

1) http://www.nature.com/neuro/journal/v5/n9/suppinfo/nn906_S1.html

1.6
Optimizing Retinal Processing with Limited Cell Numbers, Space and Energy

A striking observation is that only the visual system has extensive peripheral neural preprocessing, which starts already at the photoreceptors. Although the retina is a part of the brain, it is not immediately obvious why the nerves from the photoreceptors do not project directly to the central nervous system, as in other sensory organs. The reason is probably that the amount of information provided by the photoreceptors is just too much to be transmitted through the optic nerve (about 1 million fibers) without previous filtering [22].

There are about 125 million photoreceptors, and their information supply converges into about 1 million ganglion cells which send their axons through the optic nerve to the brain (Fig. 1.11). In the optic nerve, the visual information is more compressed than ever before, or after. It follows then that why the optic nerve cannot be made any thicker, with more fibers so that extensive information compression would be unnecessary. The reason appears to be that the visual cortex can process high spatial acuity information only from a small part of the visual field. As it is, the foveal region occupies already about 50% of the cortical area, and processing the whole visual field of about 180° would require a cortex perhaps *as large as a classroom*. On the other hand, confining high acuity to a small part of the visual field requires extensive scanning eye (or head) movements. A thicker optic nerve would impair such eye movements and it would also increase the size of the blind spot. Eye movements occupy considerable processing capacity in the brain since they are extremely fast and precisely programmed (for instance, the tolerance for errors in the angular position of the eye with binocular foveal fixation is only a few minutes of arc – otherwise, one sees double images). But this seems to require still less capacity than the one required for extending the foveal area.

1.7
Adaptation to Different Light Levels

The first challenge that the retina has to deal with is the extreme range of ambient illuminances. Between a cloudy night and a sunny day at the beach, illuminances vary by a factor of about 8 log units. Without adaptation, a receptor (and also a CCD photodetector or a film) can respond to 1.5 or 2 log units of brightness differences. This is usually sufficient because natural contrasts in a visual scene are rarely higher. But if the receptors are not able to shift and flatten their response curves during light adaptation (Fig. 1.8), they would saturate very soon if the ambient illuminance increases, and the con-

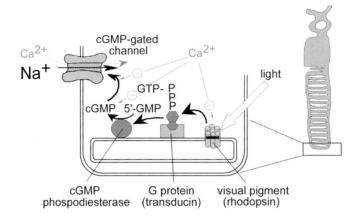

Fig. 1.8 Principle of the phototransduction cascade and the role of calcium in light/dark adaptation in a rod photoreceptor. The pigment molecule embedded in the photoreceptor outer segment disc membrane, consisting of a protein (opsin) and retinal (an aldehyde), absorbs a photon and converts into an activated state which can stimulate a G protein (transducin in rods). Transducin, in turn, activates an enzyme, cGMP phosphodiesterase, which catalyzes the breakdown of cGMP to 5'-GMP. cGMP has a key role in phototransduction: to open the *cGMP-gated cation channels*, 3 cGMP molecules have to bind to the channel protein. Therefore, if cGMP is removed by cGMP phosphodiesterase, the channels cannot be kept open. The Na^+ influx stops (which depolarizes the cell against its normal resting potential), and the membrane potential moves to its resting potential; this means hyperpolarization. When the channels are closed during illumination, the intracellular calcium levels decline. This removes the inhibitory effects of calcium on (1) the cGMP binding on the channel, (2) the resynthesis pathway of cGMP, and (3) the resynthesis of rhodopsin. All three steps reduce the gain of the phototranduction cascade (*light adaptation*). It should be noted that complete dark or light adaptation is slow: it takes up to 1 h.

trast of the image would decline to zero. So, the major role of adaptation is to prevent saturation.

It is clear that light/dark adaptation has to occur either prereceptoral or in the photoreceptors. It is not possible to generate an image with the spatial contrasts if the photoreceptors are saturated. In the visual system, some adjustment occurs through the size of the pupil which can vary from 2 to 8 mm in young subjects (a factor of 16, or little more than 1 log unit). It is clear that the remaining 8 log units have to be covered. In vertebrates this is done by dividing the illuminance range into two parts, the scotopic part, where rod photoreceptors determine our vision, and the photopic part, where cones take over. There is a range in between where both rods and cones respond, and this is called the mesopic range. Both rods and cones can shift their response curves from higher to lower sensitivity. This is done by changing the gain of the biochemical phototransduction cascade in the photoreceptor cells that converts the energy of a photon of light that is caught by the photopigment into an electrical signal at the photoreceptor membrane. Adaptation occurs by changes in the intracellular calcium concentration which, in turn, affects the gain of at least three steps in the cascade (Fig. 1.8). Strikingly, photoreceptors hyperpolarize in response to a light excitation and basically show an inverted response, compared to other neurons. This inverted response is energetically costly since it means that *dark* represents the adequate stimulus, with a constant high release of the transmitter, glutamate, from the presynaptic terminals, with a high rate of resynthesis. Glutamate release is controlled by intracellular calcium, with high release at high calcium levels and vice versa. If the light is switched on, the cation channels in the outer segment cell membrane are closed which causes a constant influx of positive ions into the outer segment and, thereby, its depolarization. Why the photoreceptors respond to light, the adequate stimulus, in an inverted fashion, has not been convincingly explained.

The inverted response makes it necessary to reverse the voltage changes at a number of synapses in the retina. Normally, the signal for excitation of a spiking neuron is depolarization, not hyperpolarization. In fact, already at the first synapse in the retina, the photoreceptor terminals, the transmitter glutamate can either induce depolarization of the postsynaptic membrane (OFF bipolar cells) or hyperpolarization (ON bipolar cells), depending on the type of receptors that bind glutamate (OFF: ionotropic AMPA/Kainate receptor; ON: metabotropic mGluR6 receptor).

1.8
Rod and Cone Responses

Rod photoreceptors can reach the maximum possible sensitivity: they show a significant transient hyperoparization in response to the absorption of a single photon. Photons reach the retina in a star light like rain hits a paved road. Not each rod receives a photon and not each paved stone is hit by a rain drop. Accordingly, the image is *noisy* (Fig. 1.9, top).

The probability of catching a certain number of photons during the integration time is described by the Poisson statistics , and the standard deviation is the square root of the number of photons caught per unit time. Therefore, the signal to noise ratio can be calculated as the square root of the number of photons divided by the number of photons. If 100 photons are absorbed during integration time of the rod (about 200 ms), only changes in contrast that are larger than 10% can be detected. Low contrast detection requires many photons: to distinguish contrasts of 1%, at least 10,000 photons must be absorbed during the integration time [22].

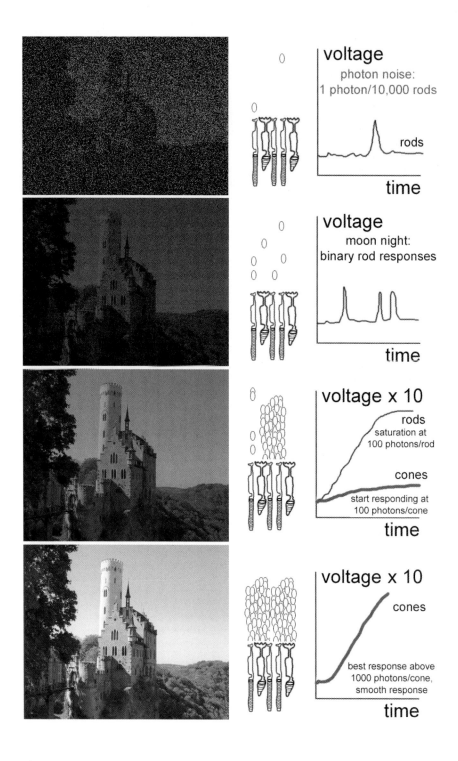

Another limiting factor is thermal noise. The rhodopsin molecule has an average lifetime of 300 years at 37°C but when it decays, rod photoreceptors cannot distinguish between a photon absorption and spontaneous decay. Due to the abundancy of rhodopsin in the photoreceptors, 10^6 decays occur spontaneously each second. Thermal noise matches photon noise in a clear star night. If rhodopsin decays only 1 log unit faster, our threshold sensitivity would rise by 3 log units.

With increasing ambient illuminance, rods continue to give binary responses (either hyperpolarization or not) over the first 3 log units [22]. If brightness further increases, in the mesopic range, the degree of hyperpolarization of the rod membrane increases linearly with the number of photons caught during the integration time, up to about 20 photons. At about 100 photons, the rods saturate, although adaptation can reduce their sensitivity so that a graded response is possible up to 1000 photons. Cones take over at 100 photons per integration time (here only a few ms), and only at this number their response rises above the dark noise. They work best at 1000 photons or more. In summary, rods' responses are always corrupted by photon noise, whereas cones respond in a smoothly graded fashion, and their responses contain more bits (Fig. 1.9, bottom). Generally, the response function of sensory organs is logarithmic, which means that for a weak stimulus, a small change in stimulus strength is detectable, whereas for a strong stimulus, the change must be larger to be detected (Weber's law: detectable stimulus strength difference proportional to stimulus strength).

Since rods are hunting *each photon*, they occupy, in most mammals, as much territory of the retinal area as possible ($> 95\%$) whereas cones which are not limited by photon noise are densely packed only in the fovea where they per-

Fig. 1.9 Photon responses of rods and cones. From complete darkness to moon night, rods respond to single photons: their signals are binary (either *yes* or *no*). Because not each rod can catch a photon (here illustrated as little white ellipses), and because photons come in randomly as predicted by a Poisson distribution, the image appears noisy, and has low spatial resolution and little contrast. Even if it is 1000 times brighter (a bright moon night), rods do not catch several photons during their integration time of 100–200 ms and they cannot summate responses. Until up to 100 photons per integration time, they show linear summation but their response curve is still corrupted by single photon events. Beyond 100 photons per integration time, rods show light adaptation (see Fig. 1.8). At 1000 photons/integration time, they are saturated and silent. Cones take over, and they work best above 1000 photons per integration time. Because cone gather their signal from so many photons, photon noise is not important and their response to brightness changes is smooth and gradual. If the number of photon rises further, the sensitivity of the cone phototransduction cascade is reduced by light adaptation, and their response curve is shifted to higher light levels. Similar to rods, they can respond over a range of about 4 log units of ambient illuminance change.

mit high spatial sampling. Rods were left out because they would increase the foveal sampling intervals of the M and L cones.

To match the *information channel capacity* of rods and cones to the available information, the rod axons are thin and the terminal synapses small, with only about 80 transmitter vesicles released per second, and with only one region of transmitter release (*active zone*), whereas the cone axon is thick, with up to 1500 vesicles released per second, and a synapse (*cone pedicle*) which is perhaps the most complicated synapse in the whole nervous system. It makes contacts to several hundred other retinal neurons (horizontal and bipolar cells) and has a typical appearance (the *cone pedicle*). The signal is diverged into 10 parallel channels [22], 5 ON- and 5 OFF-type bipolar cells. Rather than using one broadband *super channel*, 10 parallel channels are used, each with different bandwidth. The separation into different channels is thought to occur (1) because different aspects of visual processing can be separated into different pathways already at an early level and (2) because such a *broadband superchannel* cannot be made due to the limited range of possible spike frequencies (see the next section).

1.9
Spiking and Coding

In electronic devices, electrical signals can be transmitted either analog or binary. In fact, both principles are also realized in the nervous system. Short distance signal transmission, like through the dendrites to the cell body of neurons, occurs via an electronically propagating depolarization of the membrane. For long distance traveling, electronic signals become too degraded and lose their reliability. Therefore, binary coding (*0 or 1*), as used in action potentials, is used which is much more resistant to degradation. At the root of the neuron's axon, and along the axon, voltage-dependent sodium channels are expressed which are necessary for the generation of action potentials (*spikes*). Action potentials are rapid and transient depolarizations of the membrane which travels with high speed (up to 120 m/ s) along the fibers. The level of excitation is now encoded in the frequency of the spikes. Due to a recovery phase after each spike of about 2 ms, the maximal frequency is limited to about 500 Hz. This means that the dynamic range is limited. Because, even with a constant stimulus, the spike frequency displays some noise, the response functions of the neurons have no *steps*. It turns out that noisy signals are a common principle in the nervous system, and high precision is achieved by parallel channels, if necessary (probability summation) or by temporal summation. The signal to noise ratio, when a difference in stimulus strength should be detected, is determined by the standard deviation of the firing rate. If the firing rate varies little with constant stimulation (small standard deviation), a small change would be detectable, but if it varies much (large standard deviation), only large differences are detected. The signal to noise ratio can be calculated from the differences in spike rates at both stimulus strengths, divided by the standard deviation of spike rates (assuming that the standard deviation is similar in both cases). Summation of several channels is not always possible: sometimes decisions need to be made based on the signals from only two cells, for instance when two spots are resolved that are at the resolution limit of the foveal cones and, accordingly, two retinal ganglion cells.

In the retina, most neuron are nonspiking. This is possible because the distances are short and the signals can be finely graded – *the retina is a tonic machine*. Only at the output side, mostly in ganglion cells (but also in a few amacrine cells), the signals are converted into spikes which can then travel down the long axons, a few centimeters, to the first relay station, the lateral geniculus (LGN, Fig. 1.11). Since ganglion cells can be excited and inhibited (since they can generate ON or OFF responses), it is necessary that they have a baseline spontaneous spike activity (ranging from 5 to about 200 Hz). It is clear that those ganglion cells with high spontaneous activity can encode smaller changes in stimulus strength than those with low activity.

1.10
Temporal and Spatial Performance

The temporal resolution of the retina is ultimately limited by the integration time of the photoreceptors. The integration time, in turn, determines the light sensitivity of the receptors. Rods have longer integration time (about 200 ms) and, accordingly, have lower flicker fusion frequencies (up to about 10 Hz). Cones, with integrations times of about 20 ms, can resolve stroboscopic flicker light of up to 55 Hz. But under normal viewing conditions, the flicker fusion frequency is much lower, around 16–20 Hz. If it were 55 Hz, watching TV would be disturbing. The European TV or video format has a frame rate of 25 Hz, but two frames with half vertical resolution, presented alternatingly at 50 Hz, are interlaced to prevent that the flicker is seen.

The complete description of the eye's spatial performance is the contrast sensitivity function. This function describes contrast sensitivity (1/contrast threshold) as a function of spatial frequency. It is clear that contrast sensitivity must decline with increasing spatial frequency, just based on the optics of the eye: the higher the spatial frequency, the less contrast is preserved in its retinal image, due to aberrations and diffraction (the modulation transfer function, Fig. 1.3). Even if the neural processor in the retina has the same contrast sensitivity, its sensitivity would decline in a psychophysical measurement. On the other hand, it is not trivial that contrast sensitivity also declines in the low spatial frequency range. This decline is determined by neural processing: there are no such large ON/OFF receptive fields to provide high sensitivity to low spatial frequencies. This represents probably an adaptation on the abundancy of low spatial frequencies in natural scenes. It has been shown that the energy at spatial frequencies (SFs) falls off with about 1/SF [12]. The peak of the contrast sensitivity function moves to lower SFs with declining retinal illuminance. At daylight, the peak contrast sensitivity is at about 5 cyc/°. Here, brightness differences of only 1/200 (0.5%) can be detected. The contrast sensitivity is down to zero at spatial frequencies of 50–60 cyc/°. However, due to the striking feature of the optical transfer function (the first Bessel function), the MTF shows a number of phase reversals beyond the *cutoff frequency* at which the contrast first declines to zero (Fig. 1.4). Therefore, it may be possible to detect a grating even though its spatial frequency is higher than the cutoff frequency, even though it may have reversed contrast. For this reason, grating acuity is not the best measure of spatial vision, in particular with defocus [23].

1.11
ON/OFF Structure, Division of the Whole Illuminance Amplitude in Two Segments

Perhaps because no important information is represented in absolute brightness values in the visual environment, the visual system has confined its processing almost exclusively to differences – spatial and temporal contrasts. Recording from neurons along the visual pathways shows that the cells have structured *receptive fields* – defined angular positions in the visual field where they respond to the stimulation. The receptive fields are initially circular (retina, lateral geniculate (LGN), striated cortex) but become later elongated at higher cortical areas, finally larger and may even cover the whole visual field (for example, neurons that recognize highly specific features, like a face). Receptive fields in retina and LGN are organized in an ON center–OFF surround structure, or vice versa. If a small spot of light is projected on the center of an ON center ganglion cell, the cell fires vigorously, but if also the surround is illuminated, the response returns to baseline activity. If only the surround is illuminated the cell's activity is reduced below the resting level. From the structure of the receptive fields, it is already clear that homogenously illuminated surfaces without structure are poor stimuli for our visual system; *the inside of a form does not excite our brain* (David Hubel).

Dividing the processing into ON and OFF channels, starting from an intermediate activity level, has also the advantage that the dynamic range of the responses can be expanded. It is surprising that OFF ganglion cells have smaller dendritic fields and denser sampling arrays than the ON cells. This asymmetry seems to correspond to an asymmetric distribution of negative and positive contrasts in natural images [24].

1.12
Consequences of the Rod and Cone Diversity on Retinal Wiring

Since the illuminance range is divided by the visual system into a scotopic and a photopic range, and both are not used at the same time, it would be a waste to use separate lines in the rod and cone pathway. In fact, there is no rod OFF biploar cell at all, and the rod ON pathway has no individual ganglion cells. Rather, the rods are piggy-packed on the cone circuitry at low light levels, via the A2 amacrine cell (Fig. 1.10) (at least 40 amacrine cell types were classified on the basis of their morphological appearance, their transmitters, and their electrical responses). Since the information contained in the cone signals is much richer (since it is not limited by photon noise), the neurons that carry their information have thicker axons and much more synapses. They also have higher spike rates and contain more mitochondria – the energy sources of the cell.

1.13
Motion Sensitivity in the Retina

The ability to detect motion and, in particular, its direction is present already not only at the amacrine cell level in the retina, but also in many neurons of the cerebral cortex. Typically, motion selective cells fire strongly when an edge moves across their receptive field in the *preferred direction*, and they are inhibited or even silent when the motion is in the *null direction*. The illusion of motion can be provoked by stimulating briefly at two different positions in the receptive field, with a short time delay, and this illusion appears to work both during electrophysiological recordings at the cellular level, and psychophysically. It was concluded that excitation evoked by motion in the preferred direction must reach the ganglion cell before inhibition can cancel it; and inhibition evoked by motion in the null direction must arrive before excitation can cancel it. The asymmetric signal transmission speed was assumed to result from morphologically asymmetric input of the so-called starburst amacrine cells to the directionally sensitive ganglion cells, although neither the developmen-

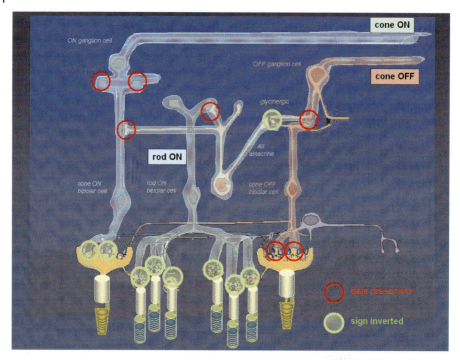

Fig. 1.10 Rod and cone pathways and ON/OFF channels. To make the system more sensitive to differences, rather than to absolute brightness, the image on the retina is analyzed by an ON/OFF system, i.e., cells that respond preferably by changes in brightness in either direction. The division into these two major channels occurs already at the first synapse, the photoreceptor *endfoot*. Because the photoreceptors can only hyperpolarize in response to illumination, the subsequent cells must be either depolarized (*excited*) or hyperpolarized (*inhibited*). This means that the signal must either be inverted (*ON channel*) or conserved (*OFF channel*). It is shown how the signals change their signs along the processing pathway. Since rods and cones respond to different illuminance ranges, it would be a waste of space and energy, to give both of them separate lines. In fact, the rods have only the first cells, the rod ON bipolar cell which then *jumps* on the cone pathways via the *AII amacrine cell*; they are not used by the cones at low light. Rods do not have an own OFF bipolar cell. Cones, with the need of coding small differences in brightness with high resolution and with large information content, have two separate lines (*ON* and *OFF*) to increase information capacity.

tal mechanism for the assymmetric connections nor the role of the involved transmitters, acetylcholine, and GABA is completely understood [25].

1.14
Visual Information Processing in Higher Centers

About 15 types of ganglion cells can be classified in the retina, and they are characterized by different morphologies, band widths, and response characteristics. The underlying hypothesis is that each part of the visual field is sampled by a group of ganglion cells that process different aspects of the visual information – how many aspects are not exactly known but the number should be between 3 and 20 [26].

1.14.1
Morphology

Researchers have always attempted to divide the visual pathways into functionally different channels. This is more successful in the initial steps, up to the primary cortex, but the separation of the pathways becomes more diffuse in higher centers. Three pathways were identified: the magnocellular, parvocellular, and koniocellular pathways. The magnocellular pathway is basically a luminance channel with low spatial acuity, large receptive fields of ganglion cells, high motion sensitivity, and high contrast sensitivity under scotopic conditions but with little or no spectral oppenency. It is relayed in the LGN in the two basal layers 1 and 2 (1: contralateral eye, 2: ipsilateral eye (Fig. 1.11)), and makes about 10% of the LGN population. The parvocellular pathway is the high spatial acuity channel, low temporal resolution, with smaller receptive fields of ganglion cells, low contrast sensitivity under scotopic conditions, but with color opponency. It is relayed in layers 3–6 (layers 4 and 6 contralateral, 3 and 5 ipsilateral) and makes about 80% of the LGN population. The koniocellular pathway is specific for blue-yellow oppenency, large receptive fields of ganglion cells, but no distinct projection to the LGN layers (*intercalated projection*), making about 10% of the optic nerve fibers. The separation of these pathways is preserved to the primary visual cortex, also called V1 or area A17. Here, the M cells project onto layer $4C\alpha$ and the P cells onto layer $4C\beta$. The koniocellular pathway feeds into the upper layers (1–3), and there into the cytochrome oxidase-rich regions, the *blobs* (Fig. 1.11).

Collaterals of M and P cells also project onto layer 6 from where they project back to the LGN; feedback is one of the common principles in the visual system: apparently, the selected input can be shaped, depending on the demand of the unit that sends the feedback. For instance, the feedback seems to enhance the resolution of an earlier level to the pattern isolated by the higher level [27], for instance by enhancing the inhibitory surround in the receptive field of a cell. The feedback seems to be extremely well developed: only about

10% of the input in the LGN comes from the periphery, the retina, while the corticofugal feedback connections make up to 30% of its input.

1.14.2
Functional Aspects – Receptive Field Structures and Cortical Modules

The receptive fields of retinal ganglion cells and cells in the LGN are concentrical with a typical ON-center/OFF surround structure, or vice versa. At the level of V1, receptive fields become elongated, and they respond best to elongated moving slits, bars, or edges at a particular orientation but the excitatory and inhibitory regions are no longer concentrical (*simple cells*). Different cells require different orientations and the response can be improved if the stimulus moves in the preferred direction. These cells can now be stimulated through

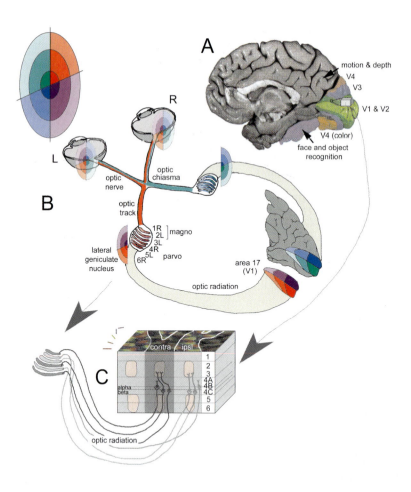

either eye; in fact, the primary visual cortex is the first level at which binocularly driven cells are found. There are also *complex cells* which are selective for the position and orientation of an edge but they have no longer excitatory or inhibitory regions. In the topographic representations of the visual field in the visual system, the receptive field sizes generally increase with the distance from the fovea. They also increase with hierarchic level of the brain area, and the stimuli that are necessary to excite the cells become increasingly specific. A common view is (Fig. 1.10) that there are two major *streams* of information processing from the visual cortex: a dorsal stream that processes predominantly the *where* aspects of an object (location in space, depth, movement) and a ventral stream that processes the *what* aspects (shape, color, details). The P-stream is assumed to feed predominantly into the *what* stream and the M-stream more into the *where* pathway. A current view is that the ventral *what* stream is actually responsible for *seeing*, whereas the dorsal *where* stream is only necessary for the direction of attention and for the control of visually guided movements [28]. Along the ventral *what* stream, cells that are extremely specific for certain features of the visual stimuli are found: for instance, they may respond only to faces or even to facial expressions, and this happens largely independently

Fig. 1.11 Feed-forward projections from the eyes to the brain and topographic mapping. In each eye the visual field on the left and right of the fovea (the cut goes right through the fovea!) projects to different cortical hemispheres: the ipsilateral retina projects to the ipsilateral visual cortex, and the contralateral retina crosses the contralateral cortex (*hemifield crossing* in the optic chiasma). The first synapse of the retinal ganglion cells is in the lateral geniculate nucleus (LGN), but information from the left (L) and right (R) eye remains strictly separated. The LGN consists of six layers, layers 1 and 2 are primarily occupied by the magnocellular pathway, and 3–6 by the parvocellular. Information from both eyes comes first together in the visual cortex, area 17, layers 2 and 3, and a strict topographic projection is preserved (follow the color-coded maps of the visual field areas (B)). The wiring in A17 (C) has been extensively studied, in particular by Nobel prize winners Hubel and Wiesel (1981). The input from the LGN ends in layer 4C alpha (magnocellular) and 4C beta (parvocellular) and layers 1–3 (koniocellular). These cells project further into the *blobs*, cytochromoxidase-rich peg-shaped regions (pink spots in (C)). The innervation has a remarkable repetitive pattern: parallel to the cortical surface, the preferred orientation for bars presented in the receptive field of the cells shifts continuously in angle (illustrated by color-coded orientation angles on top of the tissue segment shown in (C)). Furthermore, the regions where the contra or ipsilateral eye has input into layer 4 interchange in a striking pattern. A17 is the cortical input layer with mostly *simple cells*, i.e., cells that respond to bars and edges with defined directions of movements. At higher centers (A), two streams can be identified on the basis of single cell recordings and functional imaging with new imaging techniques (functional magnetic resonance imaging, fMRI): a *dorsal stream*, concerned about motion and depth (*where?* stream) and a ventral stream concerned about object features, shape color, structure (*what? stream*). Feedback projections are not shown, and only the major projections are shown.

from shading and the visual angle of presentation – a demanding task also in machine vision. The receptive fields of cells that respond to faces may cover the entire visual field, but already in area MT, the major motion processing center in the *dorsal stream*, the receptive field sizes are 10 times as large as in V1.

Hubel and Wiesel won their Nobel prize (1981) also because they discovered the modular organization of the striate cortex (Fig. 1.11C). There are topographically arranged units with about 0.5 mm diameter which contain the following subunits: (1) a column of cells with a defined orientation selectivity (*orientation column*), (2) peg-shaped structures that extend through layers 2 and 3 and stain heavily for cytochrome oxidase, a mitochondrial enzyme linked to metabolic activity (*blobs*, assumed to be involved in color processing, with predominant input from the koniocellular stream), and (3) two ocular dominance columns, with preferential input from either eye (*ocular dominance column*). These units repeat each other and the preferred orientation (see Fig. 1.11C) smoothly rotates until a 180° reversal is achieved after about 3/4 mm. A unit with a complete set of orientations has been termed *hypercolumn*.

Topographic representation of the visual world occurs in the visual system at many levels: first certainly, in the retina, but then also in the LGN, in the different cortical layers, the superior colliculus and the motion processing center area MT (V4), and others.

1.15
Effects of Attention

Most observations suggest that attention alters the sensitivity of neurons without affecting their stimulus preferences [29]. Only some neurons in V1 are modulated by attention; others ignore it and some respond exclusively when the stimulus is attended. The influence of attention increases with cortical hierarchy, perhaps also with increasing feature specificity of the cells. Effects of attention can be measured, for instance, by training a monkey to fixate a red or green spot. A neuron, for instance in area V4, is recorded, and the receptive field is mapped. A red or green stimulus, or nothing (for instance, a bar) appears in the receptive field. The monkey is trained to respond if the stimulus color matches the color of the fixation spot or not. In this kind of experiment, the responses of the V4 cell can be compared with different levels of attention and different stimuli.

Attention may also change the synchrony of neuronal signals in the visual cortex, although it is not yet studied how synchrony affects the strength of the neural responses. If attention can shape the responses of neurons to better performance, the question arises why not all neurons are at maximal sensitivity at all times. Presumably, the sensitivity is set to produce an appropriate balance between false alarms about the presence of a stimulus and failure to detect it [29].

1.16
Color Vision, Color Constancy, and Color Contrast

Both the location of the absorption peak of the photopigment and the number of photons that arrive determine the probability that a photon is caught by the photoreceptor. These two variables cannot be separated because the photoreceptor is only a *photon counter* . For the same reason, a single photoreceptor type also cannot provide information on light intensity. Only the sum of the responses from several receptor types can provide this information, whereas the differences between their responses provide information on the spectral composition of the light reflected from an object and its *color*. For this purpose, photoreceptors have photopigments with different spectral absorption. Most mammals, including male new world primates, are dichromatic, indicating that they have only two cone pigments. Only old world monkeys and humans are regularly trichromatic, but fish, reptiles, and birds may even be tetrachromatic.

Perhaps because the spectral absorbance curves of the photopigments are wide, and there is particularly much overlap in the spectra of the L and M cones with high correlations in the signals, fine wavelength discrimination (as present in our visual system, best performance at about 550 and 470 nm with a detection of only 1 or 2 nm difference) can be achieved only by antagonistic circuitry. Already at the level of the ganglion cells, the initial three spectral sensitivities of the cones are recombined into three mechanisms: (1) a luminance channel which consists of the added L+M signals, (2) a L−M color opponent channel where the signals from L and M cones are subtracted from each other to compute the red-green component of a stimulus; and (3) an S − (L+M) channel where the sum of the L and M cone signals is subtracted from the S cone signal to compute the blue-yellow variation of the stimulus. These three channels represent the *cardinal directions in color space*. They are also anatomically different in the retino-geniculo-cortical pathways.

The so-called magnocellular pathway (large *magnocellular* ganglion cells, projecting to the bottom two layers in the LGN) is most sensitive to luminance information, with high contrast sensitivity, low spatial resolution, and no color sensitivity. In the parvocellular pathway (small *parvocellular* ganglion cells, projecting to the upper four layers of the LGN), the red-green information is transmitted and in the koniocellular pathway (intermediate ganglien cell sizes to all LGH layers), the blue-yellow information is transmitted.

Up to the LGN, it seems to be possible to define cell classes with different functions, but later, in the cortex, this becomes increasingly diffuse. Considerable efforts were devoted to link structural differences to functional differences, looking for *the cell type* that processes *color or form, or motion*, or locating *the brain area* that processes a certain aspect of a stimulus. However, the more experiments are done, the less clear becomes the link between function, position, and structure, and it seems as if most cells in the central visual system have access to most features of a stimulus and that there is no complete segregation of processing aspects. These so-called multiplexing properties are found in cortical neurons as early as in V1.

Also color processing does not seem to occur independently from form [30]. For instance, patients with normal photopigments but with loss of color vision due to accidental damage of the cortex (*acquired achromatopsia*) may see objects without color but may still have near normal sensitivity to chromatic gratings [31].

The color of an object is determined by the proportion of light that it reflects at a given wavelength, which is described by the spectral reflectance function. Color vision is confounded by the spectral composition of a light source. For instance, if the light source includes more light of long wavelengths, the L cones absorb more photons and this should cause a reddish impression of the scene. However, this does not usually happen because the effect of illumination is successfully compensated by the visual system (*color constancy*). Color constancy is not only locally controlled at the receptoral level, since the spectral composition of the light at distant regions in the visual field changes the local spectral sensitivity function.

Analogous to luminance contrasts, which were best detected by ON center/OFF surround cells (or vice versa), color contrasts would best be detected by cells that have antagonistic red-green mechanisms (for example, $+L-M$) both in the center and surround, and measure the differences between center and surround (*double opponent cells*). Such cells, with concentric receptive fields, were found in the primary visual cortex of monkeys. If their receptive fields are larger they could also mediate color constancy over extended regions in the visual field.

1.17
Depth Perception

The visual system uses several independent cues to determine the distance of objects in depth. These can be divided into monocular and binocular cues.

Monocular depth estimations are typically possible from motion parallax, familiar size, shading, perspective and, to a minor extent and only for close distances, from the level of accommodation necessary to focus an object. The major depth cue, however, is binocular and results from the fact that both eyes see an object under a different angle. Accordingly, the retinal images best match at the fixation point, i.e., the images in the fovea. The peripheral parts of the images do not match; they show *disparity* (Fig. 1.3). Rather than producing the impression of double images (*diplopia*), the cortex has some tolerance to these noncorresponding images and can still put them together. But the differences are recognized and provide a highly sensitive mechanism for depth detection, called *stereopsis*. The two most striking features of stereopsis are that (1) it does not require object recognition but rather works also on Julesz´s random dot patterns and (2) the difference between the images in both retinas is detected that are smaller than a photoreceptor diameter: stereopsis involves a *hyperacuity*. This is necessary, for instance, to place a thread through a needle's eye. Considering that the visual processing may be based on displacements of less than a photoreceptor's diameter, it is even more striking that displacement of *blue versus red images*, equivalent to about 15 photoreceptor diameters (Fig. 1.7), remains undetected.

Disparity sensitive neurons have been extensively recorded not only in the visual cortex in V1, but also in V2 and the motion areas in the dorsal stream. Cells can be classified as *near cells* which respond best to crossed disparities and *far cells* that respond best to uncrossed disparities. These both types are most frequent in the motion areas, whereas cells that respond best to zero disparity (*zero disparity cells*) are abundant in areas V1 and V2 (Fig. 1.11).

1.18
Adaptation in the Visual System to Color, Spatial, and Temporal Contrast

One of the most striking features of neural processing in our visual system is that the gains for all aspects of vision are continuously adapted. The most immediate adaptation is light/dark adaptation, independently in each photoreceptor, but this adaptation in cones modifies their relative weight and, therefore, also color vision. Not only the receptors adapt but also higher processing steps in the visual system (review [32]). A few examples are that there are (1) motion adaptation (after the train stops, the environment appears to move in the opposite direction); (2) tilt adaptation (after one looks at tilted bars, vertical bars appear to be tilted in the opposite direction); (3) contrast adaptation (after prolonged viewing of a high contrast grating, the sensitivity for detection of similar gratings is severely reduced – this also affects the impression of *sharpness* of an image); (4) adaptation to scaling (if one looks at a face in which the distance between the eyes is artificially reduced, the *control face* appears to have a larger interocular distance, the *face-distortion aftereffect*; and (5) adaptation to optical aberrations and field distortions of the image on the retina which are well known by spectacle wearers. Recovery in most of these adaptations is generally in the range of seconds, but extended exposure may also cause extended changes in perception. A most striking example here is that wearing red or green spectacles for 3 or 4 days will shift the ratio of the red-green (L–M cone) weighting also for several days [33]. That the weighting of the different cone inputs can be extremely shifted (perhaps also adapted) could explain why subjects with very different L/M cone ratios (naturally occurring variability: 0.25:1 to 9:1) can all have normal color vision. Selective adaptation of color channels can also nicely be seen at the homepage of Professor Michael Bach, http://www.michaelbach.de/ot/col_rapidAfterimage/.

1.19
Conclusions

The most striking features of the natural visual system are the extreme plasticity of image processing, and the apparently optimal use of energy, space, and cell numbers.

1. To prevent saturation of photoreceptors over a range of possible ambient illuminances of, at least, 8 log units, in the presence of only about 2 log units of simultaneous contrast in natural scenes, their response curves are shifted by altering the gain of the phototransduction cascade. Furthermore, the retina divides the entire illuminance range into two, the low illuminance range, where rods respond, and the high illuminance range, where cones take over. To save energy and wire volume in the retina, both photoreceptors use the same circuitry, and their inputs to these circuits are automatically switched over by synaptic plasticity, although, as in most cases in nature, with a smooth transition.

2. Cable capacity is matched to information content, and multiple parallel channels with different bandwidth are used if necessary in the case of cones to make it possible to transmit all relevant information.

3. There is extensive image preprocessing directly in the initial light sensor, the retina. This is necessary because 125 million photoreceptors converge into an optic nerve with only about 1 million lines, and also, because the information from the cones is too abundant to be fully processed by the cortex. For the same reason, high acuity information is processed only in the central one deg of the visual field, and this limitation is

compensated by eye movements. The spatial resolution here is as good as physically possible for the given eye size of 24 mm, since diffraction at the pupil and the waveguide properties of the cones preclude denser spatial sampling. The preprocessing in the retina focuses on temporal and spatial brightness differences, spatial and temporal bandpass filtering, by building antagonistic receptive field structures of the output cells of the retina (ON/OFF, color opponency, motion selectivity with preferred direction), to enhance sensitivity to small changes. By matching the sensitivity to the available stimulus strength, maximal information is extracted; for example, the contrast sensitivity is continuously adapted at each spatial frequency, to make optimal use of the stimulus contrast.

4. Rescaling the retinal image, or compensating for local distortions or optical aberrations seem to occur largely without effort – a most impressive performance.

5. Depth is determined from several monocular cues but the most powerful mechanism is stereopsis, derived from binocular disparity. Stereopsis does not require object recognition and works on random dot stereograms; it is impressive how corresponding dots in the two retinal images are identified, and it demonstrates extensive parallel comparisons over a wide range of the visual field.

6. In the cortex, different aspects of the image are initially analyzed at each position in the visual field (small receptive fields of the respective neurons). Later, the trigger features of the cells become more and more specific and the receptive field sizes increase. Cells can be recorded in higher cortical centers that are selective for faces and expressions, and that retain their selectivity for these stimuli at different illuminations. How this information is extracted is not yet clear.

7. An unsolved problem is how and where all the separately processed features finally converge to provide the complete picture of a visual object (the *binding problem*).

8. It is interesting to note that there is an extensive description of the neural responses in the visual system (i.e., retinal ganglion cells) to stimulation with light spots and simple patterns but no one would dare to describe the visual scene given only a recording of the optic nerve trains. *Natural vision has not been the focus of much research* [34]. It seems that there is need for more research on the responses of the cells in the visual system with natural stimulation.

Acknowledgements I am grateful to Annette Werner, Howard C. Howland, Marita Feldkaemper and Mahmound Youness for reading and commenting on an earlier version of this manuscript.

References

1 Material that is covered by most common text books on the visual system is not referenced.
2 Rodieck RW (1973) "The Vertebrate Retina", Freeman, San Francisco.
3 Hughes A (1977) The topography of vision in mammals of contrasting life styles: comparative optics and retinal organization. In: "Handbook of Sensory Physiology", Vol VII/5, part A (Crecitelli F, ed.), Springer, Berlin, pp 615–637.
4 Marcos S (2003) Image quality of the human eye. Int Ophthalmol 43, 43–62.
5 Huxlin KR, Yoon G, Nagy J, Poster J, Williams D (2004) Monochromatic ocular wavefront aberrations in the awake-behaving cat. Vis Res 44, 2159–2169.
6 Howland HC (2005) Allometry and scaling of wave aberrations of eyes. Vis Res 45, 1091–1093
7 Artal P, Guirao A, Berrio E, Williams DR (2001) Compensation of corneal aberrations by the internal optics of the human eye. J Vis 1, 1–8.
8 Kelly JC, Mihahsi T, Howland HC (2004) Compensation of corneal horizontal/vertical astigmatism, lateral coma, and spherical aberration by the internal optics of the eye. J Vis 4, 262–271.
9 Collins MJ, Buehren T, Iskander DR (2005) Retinal image quality, reading and myopia. Vis Res 45, in press.[**Author: Please provide page range in reference [9].**]
10 Guirao A, Gonzalez C, Redono M, Geraghty E, Norrby S, Artal P (1999) Average optical performance of the human eye as a function of age in a normal population. Invest Ophthalmol Vis Sci 40, 203–213.
11 Wilson BJ, Decker KE, Roorda A (1992) Monochromatic aberrations provide an odd-error cue to focus direction. J Opt Soc Am A 19, 833–839.
12 Field DJ, Brady N (1997) Visual sensitivity, blur and sources of variability in the amplitude spectra of antural scenes. Vis Res 37, 3367–3383.
13 Williams DR (1985) Aliasing in human foveal vision. Vis Res 25, 195–205.
14 Curcio CA, Sloan KR, Kalina RE, Hendrikson AE (1990) Human photoreceptor topography. J Comp Neurol 292, 497–523.
15 Kirschfeld K (1983) Are photoreceptors optimal? Trends Neurosci 6, 97–101.
16 Reymond L (1985) Spatial visual acuity of the eagle Aquila audax: a behavioral, optical and anatomical investigation. Vis Res 25, 1477–1491.
17 Marcos S, Burns SA, Moreno-Barriusop E, Navarro R (1999) A new approach to the study of ocular chromatic aberrations. Vis Res 39, 4309–4323.
18 Magnussen S, Spillmann L, Sturzel F, Werner JS (2001) Filling-in of the foveal blue scotoma. Vis Res 41, 2961–2971.
19 Artal P, Fernandez EJ, Singer B, Manzamera S, Williams DR (2004) Neural compensation for the eye's optical aberrations. J Vis 4, 281–287.
20 Webster MA, Georgeson MA, Webster SM (2002) Neural adjustments to image blur. Nat Neurosci 5, 839–840.
21 Heinrichs TS, Bach M (2001) Contrast adaptation in human retina and cortex. Invest Ophthalmol Vis Sci 42, 2721–2727.
22 Sterling P (2003) How retinal circuits optimize the transfer of visual information. In: "The Visual Neurosciences", Vol 1 (Chalupa LM, Werner JS, eds.), MIT Press, Cambridge, MA, pp 234–259.
23 Gislen A, Gislen L (2004) On the optical theory of underwater vision in humans. J Opt Soc Am A 21, 2061–2064.
24 Ratliff L, Sterling P, Balasubramanian V (2005) Negative contrasts predominate in natural images. Invest Ophthalmol Vis Sci 46 (suppl.), 4685 (ARVO abstract).
25 Sterling P (2002) How neurons compute direction. Nature 420, 375–376.

26 Kaplan E (2003) The M, P, and K pathways of the primate visual system. In: "The Visual Neurosciences", Vol 1 (Chalupa LM, Werner JS, eds.), MIT Press, Cambridge, MA, pp 481–493.
27 Silito AM, Jones HE (2003) Feedback systems in visual processing. In: "The Visual Neurosciences", Vol 1 (Chalupa LM, Werner JS, eds.), MIT Press, Cambridge, MA, pp 609–624.
28 Movshon JA (2005) Parallel visual cortical processing in primates. Invest Ophthalmol Vis Sci 46 (suppl.), 3584 (ARVO abstract).
29 Maunsell JHR (2003) The role of attention in visual cerebral cortex. In: "The Visual Neurosciences", Vol 2 (Chalupa LM, Werner JS, eds.), MIT Press, Cambridge, MA, pp 1538–1545.
30 Werner A (2003) The spatial tuning of chromatic adaptation. Vision Res 43, 1611–1623.
31 Gegenfurtner K (2003) The processing of color in extrastriate cortex. In: "The Visual Neurosciences", Vol 2 (Chalupa LM, Werner JS, eds.), MIT Press, Cambridge, MA, pp 1017–1028.
32 Webster MA (2003) Pattern-selective adaptation in color and form perception. In: "The Visual Neurosciences", Vol 2 (Chalupa LM, Werner JS, eds.), MIT Press, Cambridge, MA, pp 936–947.
33 Neitz J, Carroll J, Yamauchi Y, Neitz M, Williams DR (2002) Color perception is mediated by a plastic neural mechanisms that is adjustable in adults. Neuron 35, 783–792.
34 Meister M, Berry MJ II (1999) The neural code of the retina. Neuron 22, 435–450.

2
Introduction to Building a Machine Vision Inspection

Axel Telljohann, Consulting Team Machine Vision (CTMV)

2.1
Preface

This chapter is the introduction to the basics for designing a machine vision inspection. The next chapters give a detailed description of different terms such as lighting, optics, cameras, interfaces, algorithms, and the application of these components to the machine vision system in the environment of manufacturing.

Now, the following sections focus on how to solve a concrete vision task in practice and which steps have to be taken on the road to a successful industrial application.

This roadmap provides a basis for the major decisions that have to be made for a design. It also calls attention to the optimum sequence in realizing a system. It is based on CTMV's long-time experience in designing machine vision systems and displays their method of approaching a vision task.

The sequence of a project realization can be seen as follows:

1. specification of the task
2. design of the system
3. calculation of costs
4. development and installation of the system

A successful design is based on a detailed *specification*. The task and the environment need to be described. Often, ambient influences such as mechanical tolerances and sometimes even the task are not specified precisely. This might be caused by a lack of knowledge about these factors or by the estimation that image processing is mainly done by software and thus can be changed easily. Even though software is easy to modify, the consequence of an insufficient

Handbook of Machine Vision. Alexander Hornberg (Ed.)
Copyright © 2006 WILEY-VCH Verlag GmbH & Co. KGaA, Weinheim
ISBN: 3-527-40584-7

specification is a hazard to an effective project workflow. Section 2.2 briefly describes the information that is necessary for a design.

Besides the specification, it is essential to provide a set of sample parts that covers error-free and error parts as well as such being in the range between just good and already inaccurate. These parts are required to design the illumination, to determine the required resolution and to get an impression of the diversity of feature changes.

As for the *design*, Section 2.3 provides the guideline for the crucial steps, such as

- choosing the camera scan type
- determining the field of view
- calculating resolution
- choosing a lens
- selecting a camera model, frame grabber, and hardware platform
- selecting the illumination
- addressing aspects of mechanical and electrical interfaces
- designing and choosing software

If the system is designed, *costs* can be evaluated for hardware and software. Furthermore, the required effort for development and installation can be estimated so that a *project plan* including costs can be created. Section 2.4 briefly focuses on the costs for a vision system.

Finally, the project development can be launched; Section 2.5 addresses issues for a successful development and installation of vision systems.

This chapter concludes with the presentation of two examples as realized by CTMV. The steps of specification and design are displayed on the basis of these projects.

2.2
Specifying a Machine Vision System

Before launching a vision project, the task and conditions need to be evaluated. As for the conditions, a part description as well as requirements of speed and accuracy needs to be defined.

This section describes the essential subjects for designing a machine vision system. These topics should be summarized to a system specification in written form. In combination with sample parts this document displays the initial situation. Change requests, that might occur during project realization, then can be checked with the initial setup and can be added to the requirements list. Thus, this outline provides an overview of the system's requirements and potentials to both the user and the developer. Furthermore, this specification provides the basis for the acceptance test.

2.2.1 Task and Benefit

Certainly, the task and the benefit are the most important topics of the specification. This part needs to cover the requirements of the system. Any operation performed and result generated by the system needs to be defined, including the expected accuracy. What is the inspection about? Which measurements have to be performed? As described in Chapter 9, the task type can be categorized, which gives the system a brief title.

The present method of the operation is a key to gather more information about the task and to estimate the benefit. The advantages of a machine vision system can be multiple, for instance: the task is performed with a higher precision or a 100% inline inspection of every part might be possible where random examination used to be the predominant method.

The sum of these benefits justifies the expenses for a vision system. If a cost justification is done by an ROI calculation, these benefits are fundamental to evaluate the budget or the point of time when the system is profitable.

2.2.2 Parts

As mentioned above, a precise description of the parts and a sufficient set of samples are necessary for the outline. The following characteristics and their range of diversity need to be specified:

- discrete parts or endless material (i.e., paper or woven goods)
- minimum and maximum dimensions
- changes in shape
- description of the features that have to be extracted
- changes of these features concerning error parts and common product variation
- surface finish
- color

- corrosion, oil films, or adhesives
- changes due to part handling, i.e., labels, fingerprints

A key feature of vision systems is their ability to operate without the need of touching parts. Nevertheless, for sensitive test parts a damage caused by lamp heat or radiation should be checked as well as the compatibility to the part handling.

2.2.2.1 Different Part Types

The difference of the part types needs a detailed description. What features differ in which way, for instance, dimensions, shape, or color?

With a greater type variety, type handling becomes more important. The question of how to enable the software to deal with new types will find different answers for a small range of types in comparison to larger ranges. For a vision system that frequently has to cope with new types, it can be essential to enable the user to learn new parts instead of requiring a vision specialist. Besides the difference of the parts it is important to know whether the production is organized as a mixed type or as a batch production. For the latter, the system might not need to identify the part type before inspection, which in most cases will save computation time.

2.2.3
Part Presentation

As for part presentation, the crucial factors are part motion, positioning Tolerances, and the number of parts in view.

Regarding part motion, the following options are possible:

- indexed positioning
- continuous movement

For indexed positioning the time when the part stops needs to be defined, as it influences the image acquisition time. For a continuous movement, speed and acceleration are the key features for image acquisition.

As for positioning, the tolerances need to be known in translation and rotation. The sum of these tolerances will affect the field of view and depth of view.

If there is more than one part in view, the following topics are important:

- number of parts in view
- overlapping parts

- touching parts

The main concern with overlapping or touching parts is that features are not fully visible. For contour-based algorithms used for the outer part's shape touching and overlapping parts can be a serious problem.

2.2.4 Performance Requirements

The performance requirements can be seen in the aspects of

- accuracy and
- time performance

2.2.4.1 Accuracy

The necessary accuracy needs to be defined, as it influences the required resolution.

2.2.4.2 Time Performance

Since a vision system usually is one link in the production chain, its task has to be finished within a specified time. The requirements regarding processing time will influence the choice of the hardware platform and will eventually limit the possibility of using certain algorithms.

For the specification, the following times have to be defined:

- cycle time
- start of acquisition
- maximum processing time
- number of production cycles from inspection to result using (for result buffering)

The last topic can be an issue if the part is handled on a conveyor belt and the inspection result is not used straight away, but at a certain distance, as depicted in Fig. 2.1.

For the first case, the processing must be finished within one cycle. For the second case, the results have to be latched; then, only the mean computation time must be less than the cycle time.

2.2.5 Information Interfaces

As a vision system usually is not a stand-alone system, it will use interfaces to its environment. These can be human machine interfaces to handle the system

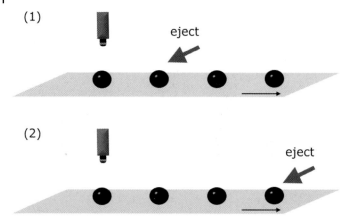

Fig. 2.1 Requirements on the processing time.

by an operator such as TCP/IP, fieldbus, serial or digital-I/O interfaces for machine to machine communication. Additionally, databases, protocols, or log files are common methods of saving and passing information.

Interfaces are most commonly used for

- user interface for handling and visualizing results
- declaration of the current part type
- start of the inspection
- setting results
- storage of results or inspection data in log files or databases
- generation of inspection protocols for storage or printout

2.2.6
Installation Space

For installing the equipment, the installation space needs to be evaluated. The possibility of aligning the illumination and the camera has to be checked. Is an insight into the inspection scene possible? What variations are possible for minimum and maximum distances between the part and the camera? Furthermore, the distance between the camera and the processing unit needs to be checked for the required cable length.

2.2.7
Environment

Besides the space, the environment needs to be checked for

- ambient light
- dirt or dust that the equipment needs to be protected from
- shock or vibration that affects the part of the equipment
- heat or cold
- necessity of a certain protection class
- availability of power supply

2.2.8
Checklist

To gather the information required for designing a machine vision system, a checklist can be used (with kind permission from CTMV. This document can be found in the appendix. It can also be downloaded from *www.ctmv.de*.

2.3
Designing a Machine Vision System

At this point, the information about the task, the parts, and the miscellaneous topics is available. On this basis the project can be designed. This section provides a guideline for designing a vision project. In general, this will provide a reasonable method for the procedure, although there will be exceptions.

2.3.1
Camera Type

Choosing an area or line scan camera is a fundamental decision for the design. It influences the choice of hardware and the image acquisition. 3D techniques are not covered by this guideline as their range and variety are too great.

For line scan cameras, the following sections will use the labeling of directions as shown in Fig. 2.2.

Area cameras are more common in automation and provide advantages in comparison to line scan cameras. Setting up an area camera usually is easier as a movement of the part or the camera is not required. Adjusting a line scan camera in a nonmoving arrangement by using a line profile though can

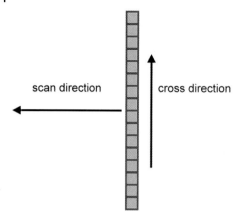

Fig. 2.2 Directions for a line scan camera.

be a challenging task. Besides the setup, the triggering of a line scan camera requires detailed attention. In general, line scan cameras and frame grabbers are more expensive than area cameras.

Why choosing line scan cameras then? Using line scan techniques offers higher resolutions in both cross direction and scan direction, where the resolution is defined by the scan rate. Depending on the frame grabber, the use of line scan cameras allows the processing of a continuous image data stream in contrast to single frames captured by area cameras. For applications, such as web inspection, processing a continuous stream offers the advantage that the inspected endless material is not separated into single frames for the image processing. Using area scan cameras instead would implicate the necessity of composing frames or merging defects that are partly visible in two frames.

For moving parts it might be consequential to use line scan cameras – a classical application used for inspecting the surface of a rotating cylinder.

According to this, the choice between an area and a line scan camera can be made. As for verifying whether the necessary resolution can be achieved by area cameras, the resolution calculation that is addressed later on is required in this step already. Besides the selection of the camera technique, the choice of the camera model is done afterward.

A detailed description of camera technologies can be found in Chapter 6.

2.3.2
Field of View

The field of view is determined by the following factors:

- maximum part size

- maximum variation of part presentation in translation and orientation
- margin as an offset to part size
- aspect ratio of the camera sensor

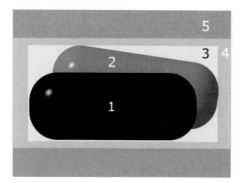

Fig. 2.3 Field of view.

As referred to Fig. 2.3, the black part labeled 1 displays the maximum part size. Due to positioning the part can exceed the maximum variation as shown with the gray part, labeled 2. Frame 3 leads to the size determined by the maximum part size plus the maximum positioning tolerance.

Frame 4 is defined by the additional needs of a margin between the part and the image. For the image processing it might be necessary to provide space between the part and the image edges. Furthermore, for maintenance and installing the camera it is convenient to admit a certain tolerance in positioning.

Frame 4 is the desired field of view. However, the calculated field of view needs to be adapted regarding the camera's sensor resolution. Most area cameras provide an aspect ratio of 4:3.

Thus, for every direction the field of view can be calculated as

$$\text{FOV} = \text{maximum part size} + \text{tolerance in positioning} + \text{margin} \quad (2.1)$$
$$+ \text{adaption to the aspect ratio of the camera sensor}$$

2.3.3 Resolution

If it comes to resolution, the following distinction is necessary:

- camera sensor resolution

- spatial resolution

- measurement accuracy

2.3.3.1 Camera Sensor Resolution

The number of columns and rows that a camera provides is specified by the internal sensor. It is measured in pixels. For line scan cameras, the resolution is defined for one dimension only. Besides the number of pixels, the size of one pixel – the cell size – is required for the lens design.

2.3.3.2 Spatial Resolution

This is a matter of direct mapping of real-world objects to the image sensor. It can be measured in mm/pixel. The resolution is dependent on the camera sensor and the field of view; the mapping is done by the lens.

It has to be considered that some area cameras do not provide square pixels, so that the resulting spatial resolution is not equal in horizontal and vertical direction. For line scan cameras, the lens determines the resolution in cross directions. In scan direction, the resolution is dependent on the scan rate and speed.

2.3.3.3 Measurement Accuracy

This is the overall performance of the system – the smallest feature that can be measured. Dependent on the software algorithm, the measurement accuracy can be different from the spatial resolution.

For the image processing, the contrast of the feature as well as the software algorithms decide whether a feature is measurable. If the contrast of small defects is poor, the software might not be able to detect a single pixel defect; four or five pixels might be necessary then. On the other hand, model algorithms, such as circle fitting, and subpixeling allow higher measurement accuracies than spatial resolution. The following table presents an overview of algorithms and the accuracy that can be expected. Certainly, these values are dependent on the algorithm used and the image quality.

Algorithm	Accuracy in pixel
Edge detection	1/3
Blob	3
Pattern matching	1

Thus, the spatial resolution that is necessary to achieve a measurement accuracy depends on the feature contrast and the software algorithms.

2.3.3.4 Calculation of Resolution

For choosing a camera model, the required resolution has to be determined. Therefore, the size of the smallest feature that has to be inspected and the number of pixels to map this feature are crucial. Evaluate the necessary spatial resolution as follows:

$$Rs = FOV/Rc \qquad (2.2)$$
$$Rc = FOV/Rs \qquad (2.3)$$

Name	Variable	Unit
Camera resolution	Rc	pixel
Spatial resolution	Rs	mm/pixel
Field of view	FOV	mm
Size of the smallest feature	Sf	mm
Number of pixels to map the smallest feature	Nf	pixel

The necessary spatial resolution can be calculated as follows:

$$Rs = \frac{Sf}{Nf} \qquad (2.4)$$

If the field of view is known, the camera resolution can be evaluated as

$$Rc = \frac{FOV}{Rs} = FOV \cdot \frac{Nf}{Sf} \qquad (2.5)$$

This calculation has to be performed for both horizontal and vertical directions. For an area camera this is done straightforwardly. However, the aspect ratio of the camera sensor has to be considered.

2.3.3.5 Resolution for a Line Scan Camera

For a line scan camera, the resolution in cross direction can be calculated as above. In scan direction the resolution defines the necessary scan rate – dependent on the speed – as follows:

$$fs = v/Rs \qquad (2.6)$$
$$ts = 1/fs \qquad (2.7)$$

Name	Variable	Unit
Spatial resolution	Rs	mm/pixel
Relative speed	v	mm/s
Line frequency	fs	Hz
Scan time	ts	s/scan

2.3.4
Choice of Camera, Frame Grabber, and Hardware Platform

At this point, the camera scan type and the required resolution are known so that an adequate camera model can be chosen. The decisions about the camera model, the frame grabber, and the hardware platform are interacting and basically done in one step.

2.3.4.1 Camera Model
As the camera scan type is defined, further requirements can be checked, such as

- color sensor
- interface technology
- progressive scan for area cameras
- packaging size
- price and availability

2.3.4.2 Frame Grabber
The camera model affects the frame grabber choice and vice versa. Obviously, the interfaces need to be compatible. Furthermore, the following topics should be considered:

- compatibility with the pixel rate
- compatibility with the software library
- number of cameras that can be addressed
- utilities to control the camera via the frame grabber
- timing and triggering of the camera
- availability of on-board processing
- availability of general purpose I/O
- price and availability

2.3.4.3 Pixel Rate
The topic addressed above is the pixel rate. This is the speed of imaging in terms of pixels per second. For an area camera, the pixel rate can be determined as

$$PR = Rc_{\text{hor}} \cdot Rc_{\text{ver}} \cdot fr + \text{overhead} \tag{2.8}$$

Name	Variable	Unit
Pixel rate	PR	pixel/s
Camera resolution horizontal	Rc_{hor}	pixel
Camera resolution vertical	Rc_{ver}	pixel
Frame rate	fr	Hz
Camera resolution	Rc	pixel
Line frequency	fs	Hz

An overhead of 10% to 20% should be considered due to additional bus transfer.

For a line scan camera, the calculation is similar:

$$\text{PR} = Rc \cdot fs + \text{overhead} \qquad (2.9)$$

The pixel rate has to be handled by the camera, the frame grabber, and the processing platform. For the grabber and the computer, the sum of the pixel rates of all cameras is essential. As a guideline, the following figures can be used:

Bus technology	Maximum bandwidth for application in MB/s
PCI	96
PCI-Express	250 per lane (lanes can be combined to increase bandwidth)
IEEE 1394	32
CameraLink	max 680 for full frame CameraLink

2.3.4.4 Hardware Platform

As for the hardware platform a decision can be made between smart cameras, compact vision systems, or PC-based systems. Costs and performance are diverging. Essential topics for choosing the hardware platform are as follows:

- *Compatibility with frame grabber*

- *Operating system.* Obviously, the software library has to be supported. A crucial factor is the decision whether a real-time operating system is mandatory. This might be the case for high-speed applications.

- As for the operating system and software, also the *development process* has to be considered. The hardware platform should provide an easy handling for development and maintenance

- If an operator needs to set up the system frequently, the platform should provide *means for a user-friendly human machine interface.*

- *Processing load.* The hardware platform has to handle the pixel rate and the processing load. For high-speed or multiple camera applications, compact systems might be overstrained.

- *Miscellaneous points*, such as available interfaces, memory, packaging size, price, and availability

2.3.5 Lens Design

As the field of view and the camera resolution are known, the lens can be chosen. An important parameter for the lens design is the standoff distance. In general, using greater distances will increase the image quality. The available space should be used to obtain an appropriate standoff distance. This distance is used to calculate the focal length. A detailed discussion can be found in Chapter 4.

2.3.5.1 Focal Length

On the basis of thin lenses, the focal length f' can be determined. Even though this formula is not correct for a set of thick lenses that camera lenses are made of, it provides a reasonable indication for the focal length to choose. Figure 2.4 displays a thin lens model (see also Section 4.2.5) used for calculation. The optical convention is the same as explained in Section 4.2.3. A real-world object with the size y is mapped by the lens to an image object of the size y'. The

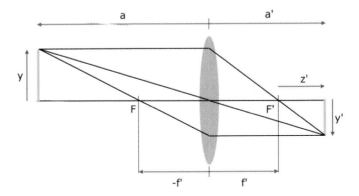

Fig. 2.4 Model of a thin lens.

standoff distance $a < 0$ is a function of the focal length $f' > 0$ and the distance $a' > 0$ between the lens and the image sensor (4.20):

$$\frac{1}{f'} = \frac{1}{a'} - \frac{1}{a} \qquad (2.10)$$

The magnification β is determined by (4.23)

$$\beta = \frac{y'}{y} = \frac{a'}{a} \quad . \qquad (2.11)$$

Considering that the field of view is mapped to the size of the image sensor, the magnification can also be evaluated by

$$\beta = -\frac{\text{sensor size}}{\text{FOV}} \quad . \tag{2.12}$$

So

$$f' = a \cdot \frac{\beta}{1-\beta} \tag{2.13}$$

Hence, for calculating the focal length f', the magnification β and the standoff distance a are necessary. Using an appropriate standoff distance a, formula (2.13) provides a lens focal length. Standard lenses are available with focal lengths such as 8 mm, 16 mm, 25 mm, 35 mm, 50 mm, or greater. Thus, a selection has to be done within this range. After choosing a lens with the focal length closest to the calculated value, the resulting standoff distance a can be evaluated by (4.20)

$$a = f' \cdot \frac{1-\beta}{\beta} \quad . \tag{2.14}$$

Referring to Fig. 2.4, the lens extension z' – the distance between the focal point of the lens and the sensor plane (see Section 4.2.8.2) – can be calculated as (4.24)

$$z' = a' - f' = -f' \cdot \beta \quad . \tag{2.15}$$

In addition to the focal length, the following characteristics have to be considered.

2.3.5.2 Lens Flange Focal Distance
This is the distance between the lens mount face and the image plane. There are standardized dimensions; the most common are as follows:

Mount	Size (mm)
C-Mount	17.526
CS-Mount	12.526
Nikon F-Mount	46.5

As seen from the camera perspective the mount size determines the shortest possible distance between the lens and the image sensor. For a CS-Mount camera any C-Mount lens can be used by inserting a 5 mm extension tube. However, a C-Mount camera cannot be used with a CS-Mount lens. F-Mount is commonly used for line scan cameras.

2.3.5.3 Extension Tubes
For a focused image, decreasing the standoff distance a between the object and the lens results in an increasing focus distance a'. The magnification is also

increased. The lens extension l can be increased using the focus adjustment of the lens. If the distance cannot be increased, extension tubes can be used to focus close objects. As a result, the depth of view is decreased. For higher magnifications, such as from 0.4 to 4, macro lenses offer better image quality.

2.3.5.4 Lens Diameter and Sensor Size

Sensor sizes vary; they are categorized in sizes of 1/3", 1/2", 2/3", and 1". The size is not a precise dimension but determines that a the sensor lies within a circle of the named diameter.

Also for lenses a maximum sensor format is specified. The choice of lens and camera sensor must be suitable. Using a 1" sensor with a 2/3" lens results in a poorly illuminated sensor; low light and aberration will be an issue. On the other hand, using a lens that is specified for a 1" sensor in combination with a 2/3" sensor is possible.

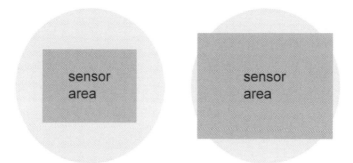

Fig. 2.5 Areas illuminated by the lens and camera; the left side displays an appropriate choice.

2.3.5.5 Sensor Resolution and Lens Quality

As for high resolution cameras, the requirements on the lens are higher than those for standard cameras. Using a low-budget lens might lead to poor image quality for high resolution sensors, whereas the quality is acceptable for lower resolutions.

2.3.6
Choice of Illumination

Illumination for a machine vision system is an individual selection of the optimum concept. Finding the best setup usually is a result of experiments based on a theoretical approach. This section provides an overview of the factors that can be changed and the general illumination concepts, such as back–front lighting. A detailed description of lighting can be found in Chapter 3.

2.3.6.1 Concept: Maximize Contrast

The illumination concept determines the quality of the feature signals in the image. The features need to be presented with a maximum of contrast. The challenge of illumination is to increase the signal to noise ratio, and to emphasize and expose these features to maximize the contrast.

Any effort invested in the optimum illumination concept will increase the system's inspection performance and reliability; it will also decrease the complexity of the software. What means are available to increase the contrast?

- An essential factor is the *direction* of light. It can diffuse from all directions or directed from a range of angles.

- The *light spectrum* also influences the contrast. Effects such as fluorescence or the influence of infrared or ultraviolet light should be checked as well as optical filters. For color applications, the light spectrum needs to be verified for usability; white LEDs for instance usually do not provide a homogeneous spectrum.

- *Polarization*. The effect of polarization increases the contrast between object areas that directly reflect light in comparison to diffuse reflection. Polarization will show effect on surfaces, such as metal or glass.

2.3.6.2 Illumination Setups

The main setups in illumination are

- backlight and
- frontlight.

Backlight usually is realized in a back-illuminated panel. Common light sources are LEDs and fluorescent tubes. They are available in a wide range of sizes and intensities. This backlight can diffuse; it provides light from a wide range of angles.

A further technique is the condenser illumination; a condenser lens is used to focus backlight in the direction of the camera. This lighting provides a telecentric path of rays and therefore is used for measurement tasks.

For *frontlight*, different techniques can be listed:

- *Diffused light*. This light is provided from all angles. The commonly used diffuse lights are dome lights with different diameters. They often come with an LED illumination inside the dome.

- *Directed light*. This light is provided from a range of angles. This can be ring lights or line lights.

- *Confocal frontlight.* Beam splitters make the light come from the direction of the camera's optical axis.

- *Bright field.* This is a variety of directed light. This light is supplied in a way that it is reflected by the part's surface into the camera. The surface appears bright in the image; part regions that do not reflect light appear dark.

- *Dark field.* This is another variety of directed light where the reflected light of the part is directed away from the camera. The surface then appears dark in the image; irregular part regions reflect light and appear bright.

2.3.6.3 Light Sources

Common light sources are as follows:

- *Fluorescent tubes.* These tubes are available in ring and straight forms; the ring form does not provide a continuous light ring, as the tube needs to be mounted. Essential for machine vision applications is an electrical high frequency setup. If the tube is used with a 50 Hz setup, intensity oscillation will be visible in the images.

- *Halogen and xenon lamps.* These lights run on direct-current voltage; thus, a light oscillation is not an issue. Halogen lights are often used in combination with fiber glass and different attachment caps, such as lines or rings. Xenon lamps are used for flash-light applications.

- *LED.* LED lights are becoming more and more important in machine vision. Advantages of LEDs are the use of direct-current voltage, a long life that exceeds the durability of halogen and fluorescent tubes by multiples. Due to their size, they come in smaller packaging sizes and usually do not need further electrical components. Flashing can easily be realized with LEDs; the light intensity can be increased compared to continuous operation.

- *Laser.* Laser light is used for special applications, such as triangulation.

2.3.6.4 Approach to the Optimum Setup

For finding the optimum setup, a theoretical idea is adjuvant rather than trying. Nevertheless, a confirmation of the setup based on experiments with sample parts is mandatory. Using a camera is not obligatory as for the first step; it can be replaced by the human eye. If at all a camera is required, an area scan camera is recommended due to an easier handling. For finding illumination angles, it might be helpful to use a single discrete lamp and then proceed with a choice of suitable lamps.

The alignment of light, the part and the camera needs to be documented. To balance between similar setups images have to be captured and compared for the maximum contrast. Furthermore, a compatibility with the mechanical environment needs to be checked before proceeding.

2.3.6.5 Interfering Lighting

When inspecting a number of features, different illumination setups might be required. The influences of different lamps on the images have to be checked. To avoid interfering, a spatial separation can be achieved by using different camera stations. Then, the part is imaged with different sets of cameras and illuminations. Furthermore, a separation in time is possible; images of different cameras are taken sequentially, whereas only the lamp required for the imaging camera is switched on. Another solution is the use of different colors for the cameras. This can be achieved by colored lamps in combination with color filters for the camera that belongs to the lamp.

2.3.7
Mechanical Design

As the cameras, lenses, standoff distances, and illumination devices are determined, the mechanical conditions can be defined. As for mounting of cameras and lights the adjustment is important for installation, operation, and maintenance. The devices have to be protected against vibration or shock. In some cases a mechanical decoupling might be necessary.

The position of cameras and lights should be changed easily. However, after alignment the devices must not be moved by operators. An easy positioning is achieved by a setup that allows the crucial degrees of freedom to be adjusted separately from each other. If the camera has to be adopted to different standoff distances, a linear stage might be easier in handling than changing the lens focus.

2.3.8
Electrical Design

For the electrical design the power supply is specified. If a certain protection class is necessary, the housing of cameras and illumination need to be adequate. The length of cables as well as their laying, including the minimum tolerable bend radius, need to be considered. The following table provides an overview of the specified cable lengths. Using repeaters or optical links can increase the lengths. For digital busses, such as Camera Link or IEEE 1394, the length is also dependent on the bandwidth.

System	Specified cable length
Camera Link	10 m
IEEE 1394	4.5 m
Analog	up to 15 m

2.3.9 Software

As for software, two steps have to be performed:

- selection of a software library
- design and implementation of the application-specific software.

In most cases, not all software functions are programmed by the developer; software libraries or packages are used, which provide image processing algorithms.

2.3.9.1 Software Library

When selecting a software library, the functionality should be considered; therefore, it is necessary to have a basic concept of the crucial algorithms in mind, which have to be used. Obviously, the software needs to be compatible with the hardware used for imaging as well as with the operating system.

Furthermore, the level of development is important. The machine vision market offers software packages that are configurable; without the need of programming an application can be realized. Often these packages are combined with the required hardware.

For more complex tasks, the means of changing algorithms and procedures might not be sufficient. In this case, a programmable software package provides the options to adapt the software to the application needs. The integration is more complex as more programming is involved; the software structure has to be programmed.

The highest level of programming and flexibility offers an application programming interface (API). This is a set of functions that have to be combined to an application software.

2.3.9.2 Software Structure

The software structure and the algorithms used are highly dependent on the vision task. A general guideline cannot be provided. However, for most applications, the software for the image processing routines follows a structure as follows:

- image acquisition
- preprocessing

- feature localization
- feature extraction
- feature interpretation
- generation of results
- handling interfaces

Obviously, the sequence starts with the acquisition of images. Certain requirements, such as triggering or addressing flash units, might have to be met.

If the image or the images are acquired, they might have to be preprocessed. Commonly used are filters, such as mean filters or shading. Shading is an important issue to overcome nonhomogeneous illumination scenes. Depending on the hardware, shading can be realized in a camera, respectively, the grabber, or in software. Preprocessing usually consumes plenty of computation time and should only be used if mandatory.

Due to the tolerances in positioning, the position of the features needs to be localized. Based on the position, regions of interest (ROIs) can be defined. In contrast to processing the entire image, ROIs offer the possibility of processing local image areas; computation time can be economically used.

Feature extraction addresses the basic algorithms to present features for an interpretation. Basic algorithms are blob analysis, texture analysis, pattern match, or edge detection – often in combination with geometric fitting.

Feature interpretation basically accomplishes the vision task – a measurement is gauged, a verification is done, and a code is read.

The generation of results is the next step. Depending on the feature interpretation, the results can be compared to tolerances; the part can be an error or error-free part.

Handling of interfaces addresses means, such as digital-I/O, data logging, or visualization.

2.3.9.3 General Topics

The software should cover requirements for an easy handling of the vision system:

- *Visualization* of live images for all cameras
- Possibility of *image saving*
- *Maintenance modus*. Camera and illumination need to be aligned, image contrast needs to be tested, and a camera calibration might be necessary. The software needs to meet these requirements for easy handling and maintenance.

- *Log files* for the system state. The system state and any error occurred while processing parts should be logged in a file for the developer. Software errors sometimes are hard to reproduce; in this case a log file might lead to the reasons.

- Detailed *visualization* of the image processing. Subsequent processing steps should be displayed in the user interface. This provides the possibility of estimating the processing reliability and reasons for failures.

- Crucial *processing parameters*, such as thresholds, should be accessible from the user interface for a comfortable system adaption.

2.4 Costs

Before launching the project, the costs have to be evaluated. They can be divided into the initial development costs and the operating costs. The development costs consist of expenses for

- project management
- base design
- hardware components
- software licenses
- software development
- installation
- test runs, feasibility tests, and acceptance test
- training
- documentation

If more than one system is manufactured, the costs for the subsequent systems will be less than that for the prototype; base design, software development, feasibility test, and documentation are not required.

As for the operating costs, the following factors should be considered:

- maintenance, such as cleaning of the optical equipment
- change of equipment, such as lamps
- utility, for instance electrical power or compressed air if needed
- costs for system modification due to product changes.

2.5 Words on Project Realization

The sequence of a project realization usually is as follows:

1. specification
2. design
3. purchase of hardware and software
4. development
5. installation
6. test runs
7. acceptance test
8. Training and documentation

Specification and design are addressed above.

2.5.1 Development and Installation

As known from experience, it is advisable to split the installation into two parts. The first part focuses on the setup of the components. As soon as they are available, the installation can be done at the designated location. Even though the software is not fully developed, this early project step is required to check for reliability and image quality

The image quality can be tested directly in the production environment. Therefore, the system needs a basic software for image grabbing, triggering and image saving. The reliability of the system in terms of triggering, and imaging can be checked. At this point, the system comes across influences that have not been or could not be specified. This might be part vibration or electrical interference.

In general, this is a fundamental step in a vision project:

- getting familiar with the imaging routine,
- getting to know the influences of the production process and
- approving the desired image quality.

Before proceeding, problems with these factors need to be resolved. For the software development, sample images should be gathered as they are imaged with the chosen equipment.

2.5.2
Test Run and Acceptance Test

The software needs to be tested and optimized until the specifications are met. Depending on the complexity of the system, a couple of test and optimization sequences should be expected

Due to a successful project workflow, an acceptance test states that the system works regarding the requirements.

2.5.3
Training and Documentation

Finally, a documentation and detailed training of operators are mandatory to accomplish the project. A documentation should cover the following points:

- system specification

- handling and usage of the system, and hardware and software handbook

- maintenance

- spare-part list as well as a recommendation of anticipated parts to hold on stack (i.e., lamps)

- mechanical drawings

- circuit diagram

- handbooks of the used components, i.e., cameras

2.6
Examples

Two applications, as realized by CTMV, are displayed as practical examples.

2.6.1
Diameter Inspection of Rivets

2.6.1.1 Task

In the production of floating bearings, the bearing and the shafts are riveted. Due to material and processing influences, the diameter of the rivet needs to be inspected. Every rivet has to be checked.

Fig. 2.6 Bearing with rivet and disk.

2.6.1.2 Specification

- *Task and benefit*. The diameters of two similar rivets have to be inspected; the task can be categorized into a measurement application. The 100% inspection of every part has to be performed inline. The nominal size of the rivet is 3.5 mm, and the required accuracy is 0.1 mm. The inspection used to be performed manually.

- *Parts*. The nominal size of the rivets lies in a range of 3 mm to 4 mm; it is placed in front of a disk. The surface color of the disk might change due to material changes. The rivet material does not change. The bearings can be covered with an oil film.

- *Part positioning*. The positioning is indexed by the use of an automated belt. The parts are presented without overlap. The tolerance of part positioning is less than ± 1 mm across the optical axis and ± 0.1 mm in the direction of the optical axis. The belt stops for 1.5 s. Part vibration might be an issue. The belt control can provide a 24 V signal for triggering the cameras. There is only one part type to inspect.

- *Performance requirements.* The diameter of the rivet needs to be measured with an accuracy of 0.1 mm. The processing result has to be presented immediately. The maximum processing time is 2 s; the cycle time is 2.5 s.
- *Information interfaces.* The inspection result is passed using a 24 V signal, indicating an error part. The measurement results need to be visualized.
- *Installation space.* A direct insight into the rivet is possible. The maximum space for installing equipment is 500 mm. The distance between the cameras and the computer is 5 m. A certain protection class is not necessary.

2.6.1.3 Design

(1) Camera type. As the part positioning is indexed and the rivet can be imaged with one frame, area cameras are used.

(2) Field of view. Referring to Eq. (2.1) the field of view is calculated as

$$\text{FOV} = \text{maximum part size} + \text{tolerance in positioning} \\ + \text{margin} + \text{adaption to aspect ratio of camera sensor}$$

In this case, the following values are specified:

maximum part size	4 mm
tolerance in positioning	1 mm
margin	2 mm
aspect ratio	4:3

Hence, the field of view is calculated as

$$\text{FOV}_{\text{ver}} = 4\,\text{mm} + 1\,\text{mm} + 2\,\text{mm} = 7\,\text{mm}$$

As the aspect ratio of the camera sensor is 4:3, the field of view in the horizontal direction is adapted to

$$\text{FOV}_{\text{hor}} = 7\,\text{mm} \cdot \frac{4}{3} = 9.33\,\text{mm}$$

Thus, the field of view is determined to be 9.33 mm × 7 mm.

(3) Resolution. As the field of view and the accuracy of the measurement are known, the necessary sensor resolution can be calculated as follows (referring to (2.5)):

$$Rc = \frac{FOV}{Rs} = FOV\frac{Nf}{Sf}$$

The diameter will be measured using an edge detection and a subsequent circle fitting. Hence, the number of pixels for the smallest feature can be 1/3 pixel. Due to changing disk material the edges of the rivet can be low in contrast. Hence, the number of pixels for the smallest feature is set to 1 pixel. The size of the smallest feature is 0.1 mm.

The horizontal and vertical resolutions can be evaluated as:

$$Rc_{hor} = 7\,\text{mm} \cdot \frac{1\,\text{pixel}}{0.1\,\text{mm}} = 70\,\text{pixels}$$

$$Rc_{ver} = 9.33\,\text{mm} \cdot \frac{1\,\text{pixel}}{0.1\,\text{mm}} = 93.3\,\text{pixels}$$

(4) Choice of camera, frame grabber, and hardware platform. Due to these values, a standard VGA camera can be chosen. As described in Section 2.3.4, the choices of camera, frame grabber, and hardware platform are dependent on each other and basically done within one step. As hardware platform, a National Instruments Compact Vision System NI CVS 1454 is chosen, since it combines an embedded high-performance processor, direct IEEE 1394 connection, TCP/IP interface, and the use of a powerful software library. For visualization, a computer monitor can be connected. As for the camera, an AVT Marlin F-033B IEEE 1394 CCD camera is used. It provides a camera resolution of 656 × 494 pixels. The choices of IEEE 1394 and the compact vision system save costs for a frame grabber and additional hardware.

(5) Lens design. As a precise mass is to be measured and the part moves in the direction of the optical axis, a telecentric lens is chosen. For choosing a telecentric lens, the key factors are field of view and magnification rather than the focal length. The field of view was determined to be 9.33 mm. Hence, a lens with a field of view of 10.7 mm was chosen. The standoff distance is specified to be 100 mm.

Thus, the spatial resolution is

$$Rs = \frac{FOV}{Rc} = \frac{10.7\,\text{mm}}{656\,\text{pixels}} = 0.016\,\frac{\text{mm}}{\text{pixel}}.$$

(6) Choice of illumination. The rivet has a convex shape, as depicted in Fig. 2.7. To illuminate the outer edges of the shape, a dome is chosen as diffuse frontlight. The setup of a camera, its illumination, and the part are displayed in Fig. 2.8. Regarding the longevity, a LED illumination was preferred. As depicted in Fig. 2.8, the diameter needs to have a certain size to illuminate the rivet properly. A dome was chosen with a diameter of 100 mm.

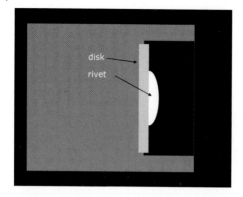

Fig. 2.7 Bearing, rivet, and disk in lateral view.

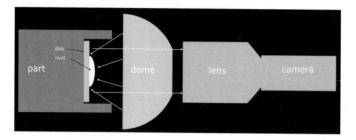

Fig. 2.8 Setup of part, illumination, and camera.

(7) Mechanical design. For the mechanical design, no particularities have to be considered. The cameras and light domes are mounted on aluminum profiles for easy adjustment.

(8) Electrical design. As power supply, only 24 V is necessary for the compact vision system and the illumination. The cameras are supplied via the IEEE 1394 connection.

(9) Software. For the software library, National Instruments LabVIEW and Imaq Vision are chosen as they provide a powerful image processing functionality and short implementation efforts.

As mentioned in section 2.3.9, the software is structured in

- image acquisition
- preprocessing
- feature localization
- feature extraction
- feature interpretation
- generation of results
- handling interfaces

Fig. 2.9 Rivet and disk as imaged by the system.

The image acquisition is done by a simultaneous external triggering of the cameras. No preprocessing is used.

Figure 2.9 shows the rivet as depicted by the system.

The feature localization is done by thresholding the image for a blob analysis. Figure 2.10 displays the thresholded image. The center of the greatest blob is used as a feature center position.

For the feature extraction, a circular edge detection and circle fitting are performed as displayed in Fig. 2.11.

The feature interpretation and generation of results are performed by comparing the diameter to the defined tolerances. Depending on the result, a digital output is set.

Fig. 2.10 Feature localization by thresholding and blob analysis.

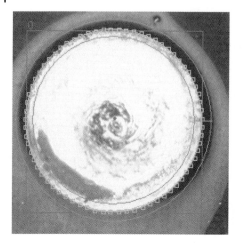

Fig. 2.11 Circular edge detection and subsequent circle fitting.

The user interface is designed to display the images of both cameras. Furthermore, the processing results, such as feature localization and Interpretation, are displayed.

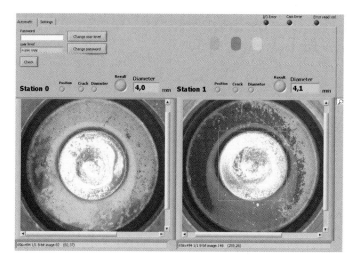

Fig. 2.12 User interface for the rivet inspection.

2.6.2
Tubing Inspection

2.6.2.1 Task

Producing tubings for sophisticated applications, defects, such as particles and material drops, can be a quality issue. A vision system for an inline tubing inspection was realized by CTMV.

2.6.2.2 Specification

- *Task and benefit*. As mentioned above, an inline inspection for the named defects has to be realized. The smallest defect that has to be detected has a size of 0.08 mm. The defects are classified into different classes regarding defect type and size. For each class, tolerances can be defined as for the size and frequency of occurrence. For instance, particles are tolerable if their size is between 0.1 mm and 0.2 mm and not more than 5 defects per 1 m of tube are detected.
  An inspection protocol is necessary, which displays the defects, their ongoing meter from the inspection start, size, and image. Additionally, these data have to be provided for an online access from remote computers over the TCP/IP protocol. The inspection is performed manually.

- *Parts*. The tube diameter varies between 5 mm and 32 mm. The tubes are transparent. A diameter change can be addressed to the system. The tube's surface is free of dirt or adhesives; color changes are not expected.

- *Part positioning*. The tubes are produced in an horizontal movement with a maximum speed of 3 m/min. The position tolerance in the cross direction is 0.5 mm.

- *Performance requirements*. The smallest defect size that has to be detected is 0.08 mm. The processing time is defined as a function of processing speed. An image needs to be processed before the next acquisition is accomplished.

- *Information interfaces*. As mentioned above, a user interface for controlling and setting the tube diameter, an inspection protocol for printout and storage, and an online access to the defect data over a TCP/IP connection is required.

- *Installation space*. A direct insight into the tube is possible. The maximum distance from the tube center is 400 mm. In the direction of movement, a distance of 700 mm can be used for the system. The distance between the cameras and the computer is 3 m. The components should be covered from dripping water.

2.6.2.3 Design

(1) Camera type. As the tube is moving and a rather high resolution will be mandatory, a line scan setup is preferred. To cover 360° of the perimeter, at least six cameras have to be used. At this point, a cost calculation of six line scan cameras, an adequate number of frame grabbers, and processing hardware are displayed; the costs exceed the budget.

Hence, area cameras have to be used. For acquiring single frames, camera triggering and merging of defects, which are partly visible in two or more images, will be an issue.

(2) Field of view. When using six cameras, each camera needs to cover a field of view of the radius' size, as depicted in Fig. 2.13.

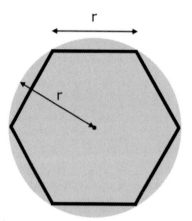

Fig. 2.13 Required field of view when using six cameras.

The maximum diameter is specified to be 32 mm and the radius to be 16 mm. The positioning tolerance is less than 0.5 mm. Hence, the required field of view for one camera can be calculated as

$$\text{FOV} = \text{maximum part size} + \text{tolerance in positioning} + \text{margin}$$
$$+ \text{adaption to aspect ratio of camera sensor}$$

$$\text{FOV}_{\text{hor}} = 16\,\text{mm} + 0.5\,\text{mm} + 1\,\text{mm} = 17.5\,\text{mm}$$

Using an area camera with a sensor ratio of 4:3, the vertical field of view is determined as

$$\text{FOV}_{\text{vert}} = \text{FOV}_{\text{hor}} \cdot \frac{3}{4} = 17.5\,\text{mm} \cdot \frac{3}{4} = 13.125\,\text{mm}$$

Hence, the field of view is calculated to be 17.5 mm × 13.125 mm. The camera is mounted as depicted in Fig. 2.14.

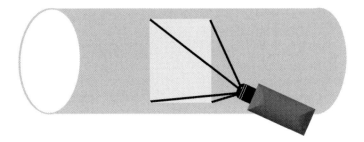

Fig. 2.14 Positioning of the camera.

(3) Resolution. The size of the smallest defect is defined as 0.08 mm. As the processing routine will be based on blob analysis, at least 3 pixels should be used to map the smallest defect. Hence, a spatial resolution of 0.027 mm/pixel is required.

With the field of view, the camera resolution can be calculated to be

$$Rc = \frac{FOV}{Rs} = \frac{17.5 \text{ mm}}{0.027 \text{ mm/pixel}} = 656 \text{ pixels}$$

(4) Choice of camera, frame grabber and hardware platform. Due to these values, a standard VGA camera can be chosen. A camera interface technique, IEEE 1394, is chosen due to an easy integration and low costs in comparison to systems such as Camera Link. The choice is made for a Basler 601f CMOS camera with a sensor resolution of 656 × 491 pixels.

Using 656 pixels to map the 17.5 mm FOV, the resulting spatial resolution is

$$Rs = \frac{FOV}{Rc} = \frac{17.5 \text{ mm}}{656 \text{ pixels}} = 0.027 \text{ mm/pixel}$$

The smallest defect of 0.08 mm then is mapped with 3 pixels.

A 19" Windows XP-based computer is used as hardware platform. The cameras are connected to two National Instruments PCI-8254R boards with a reconfigurable I/O and IEEE 1394 connectivity.

(5) Lens design. The maximum distance from the tube center is defined as 400 mm. The magnification can be calculated as

$$\beta = -\frac{\text{sensor size}}{\text{FOV}} = -\frac{6.49 \text{ mm}}{17.5 \text{ mm}} = -0.371.$$

The sensor size results from the multiplication of the cell size of 9.9 μm/pixel and the sensor resolution of 656 pixels.

Using the magnification and the maximum distance from the tube center of 400 mm subtracted by a value of 200 mm for the camera and lens, the focal length can be calculated as

$$f' = a \cdot \frac{\beta}{1-\beta} = 200\,\text{mm} \frac{0.371}{1+0.371} = 54.1\,\text{mm}.$$

The choice is made for a 50 mm lens.

The resulting standoff distance d is

$$a = f' \cdot \frac{1-\beta}{\beta} = 50\,\text{mm} \cdot \frac{1+0.371}{-0.371} = -184.8\,\text{mm}$$

As referred to 2.15, the lens extension l can be evaluated to

$$l = a' - f' = -f \cdot \beta = 50\,\text{mm} \cdot 0.371 = 18.55\,\text{mm}$$

As this distance cannot be realized by focus adjustment, an extension tube of 15 mm is used.

(6) *Choice of illumination.* As the tube is translucent, a diffuse backlight is used. Defects appear dark then. As the shutter time needs to be set to a low value, high intensities are required. The time the tube takes to move a distance of 1 pixel in the image is calculated as

$$t = \frac{Rs}{v}$$

where v is the speed (3 m/min = 50 mm/s) and Rs is the spatial resolution scan direction.

Hence

$$t = \frac{0.027\,\text{mm/pixel}}{50\,\text{mm/s}} = 540\,\mu s$$

The choice is made for a high power LED backlight with a size of 50 mm × 50 mm. Due to the intensity, a flash operation is not necessary.

(7) *Mechanical design.* For the mechanical design, the mounting of the cameras and lights needs to be considered. Since different illuminations could interfere with each other, the sets of camera and light are positioned in a row. The setup for one camera is depicted in Fig. 2.15.

As the equipment has to be covered from dripping water, the lights and cameras are mounted in housings, so is the computer.

(8) *Electrical design.* The cable length is below 4.5 m and within the IEEE 1394 specification.

(9) *Software.* For the software library, a CTMV software package was programmed using Microsoft Visual C#. For image acquisition the API of National Instruments Imaq for IEEE 1394 was chosen.

For *image acquisition*, the cameras have to be triggered to capture the frames with a defined overlap of 2 mm. Figure 2.16 displays four subsequent frames as imaged by one camera.

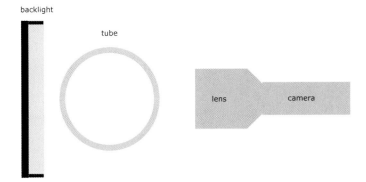

Fig. 2.15 lateral view of one set of camera and light.

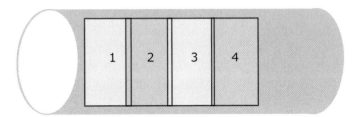

Fig. 2.16 Frames, as captured by one camera.

For triggering, a rotary encoder is used, which indicates the tube movement (Fig. 2.17). The encoder signals are connected to a specially designed input of the frame grabber. Using an FPGA counter, the trigger signal is created by a card and set to the camera. The application software on the host computer does not handle the triggering; it is done by the FPGA. This saves computation time and guarantees high reliability of the triggering process.

Since the tubes are curved, a homogeneous light uniformity in the image is not present. Figure 2.18 displays an image of a tube. To achieve uniformity

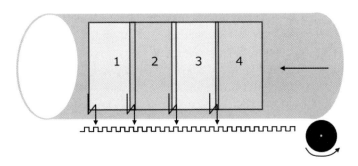

Fig. 2.17 Generating trigger signals using a rotary encoder.

Fig. 2.18 Tube, as imaged by the system.

for the latter inspection, a shading is used. The teach-in is done at the start of the inspection; the reference is a mean computation of several images.

The feature localization and segmentation is done by thresholding. Since shading is used, an adaption of the threshold for different tubes is not mandatory. Figures 2.19(a) and (b) display a defect in the original image and as segmented by thresholding.

(a)　　　　　　　　(b)

Fig. 2.19 (a) Defect as imaged. (b) Defect as thresholded by the system.

After segmentation, feature interpretation is done by blob analysis. Every blob is measured in height, width, and area. Furthermore, it is classified into the defect classes, such as particles and drops. For measurement, it has to be checked whether the defect is visible in more than one frame and therefore has to be merged due to a correct measurement. Figure 2.20 displays the situation.

After measurement and classification, the defect is added to the appropriate defect class. If the number of tolerated defects exceeds the defined tolerance, an error signal is set.

Furthermore, an entry in the defect logging database including width, height, and an image of the defect is performed.

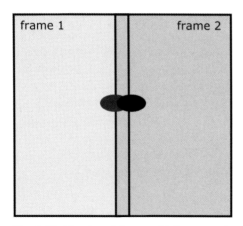

Fig. 2.20 Merging defects, which are partly visible in two frames.

3
Lighting in Machine Vision

I. Jahr, Vision & Control GmbH

3.1
Introduction

3.1.1
Prologue

One simple part and a hand full of lighting techniques (examples: Fig. 3.1 and Fig. 3.2). And everyone looks different. However, there still remain many questions, some of which are

- What do you prefer to see?
- What shall the vision system see?
- What do you need to see? What does the vision system need to see?
- How do you emphasize this? How does the vision system emphasize this?
- Where are the limits?
- Does it work stable in practice?
- What are the components used?
- What light sources are in use?
- What are their advantages and disadvantages?

Questions over questions. But on the first view everything seemed to be very simple: it is all only made by light, basically caused by the presence of light. But many people do not know how to do that. For them light is a closed book.

Handbook of Machine Vision. Alexander Hornberg (Ed.)
Copyright © 2006 WILEY-VCH Verlag GmbH & Co. KGaA, Weinheim
ISBN: 3-527-40584-7

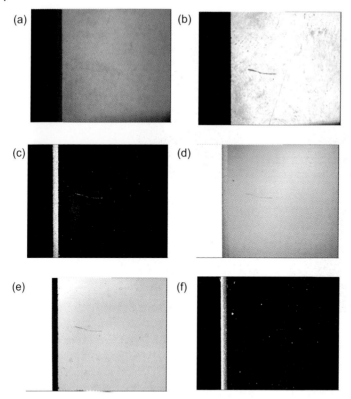

Fig. 3.1 Different lighting techniques applied on a glass plate with a chamfer: (a) diffuse incident bright field lighting, (b) telecentric incident bright field lighting, (c) directed incident dark field lighting, (d) diffuse transmitted bright field lighting, (e) telecentric transmitted bright field, lighting, (f) directed transmitted dark field lighting.

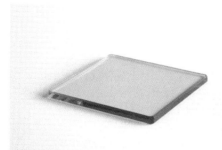

Fig. 3.2 Glass plate with a chamfer in the familiar view of the human eye.

The reader's mission is to learn, how to illuminate, that some features appear dark and others bright. But everything as desired (see Fig. 3.3)!

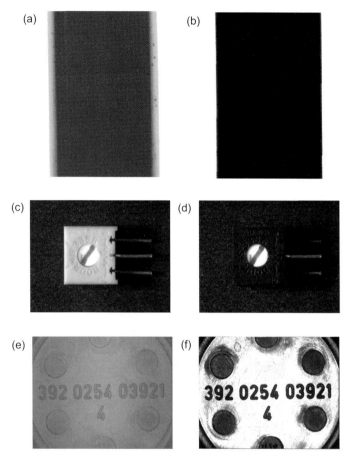

Fig. 3.3 Different parts under different lighting conditions. (a) Metal bolt with diffuse backlight, (b) metal bolt with telecentric backlight, (c) blue potentiometer under blue light, (d) blue potentiometer under yellow light, (e) cap with diffuse lighting, (f) cap with directed lighting.

The aim of this chapter is to teach the reader to recognize, for example, why the background is dark, why the lettering is dark or how can I avoid from hot spots. They shall understand how light works.

3.1.2
The Involvement of Lighting in the Complex Machine Vision Solution

In the beginning there was light.

Not only in the historical view, but also in the view of Machine Vision people this is one of the most important proverbs. Let us consider this: photography means *writing with light*. And one can consider Machine Vision as an (extended) contemporary further development of photography. So the light is the base of Machine Vision. It does not only mean *writing with light*, but also *working with light* as their information carrier.

You may ask: why the light? A vision system consists of much more parts than a *lamp*! True, but all information that is processed comes from the light. The light information is the origin. The lighting is as important as the optics, because it carries the primary information. To mathematically do image processing, the brightness values of the object and the background have to differ. Contrast, brightness and darkness, shadows, textures, reflexes, streaks are necessary. And all this is done by light. That is why the experts know that two-third of a robust Machine Vision solution is lighting.

Let me express in this way: garbage in – garbage out. Where none takes care of the lighting design, none should be surprised about worse results delivered by the vision system. No brightness and contrast – no algorithm will find the edges. What are a few hours to plan the lighting solution and a few hundreds of Dollars for a professional lighting against a few man-weeks of software engineering to save the consequences of bad lighting. And after that there still remains a lot of unsteadiness.

To systematically search for a matching lighting saves real money, time and nerves!

Knowing the main parts of a vision system one fact is obvious: Machine Vision is a complex teamwork of totally different technical disciplines that are involved (see Fig. 3.4). And depending on the discipline you are qualified the sight can be a totally different one. If you are right in the middle the view for the entirety can be lost. The same applies to Machine Vision. It is a synthesis technology, consisting of

- lighting

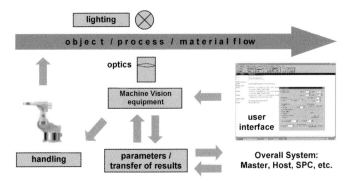

Fig. 3.4 Basic structure of a Machine Vision solution and main parts of a vision system.

- optics
- hardware (electronics/photonics)
- software engineering

These are the core disciplines. But Machine Vision is always embedded in

- automation
- mechanical engineering
- connected with the information technology
- electrical engineering

In this delicate mixture the lighting is the key. However, on the other hand, to find one lighting solution we cannot divide the lighting from the rest of the vision system. All parts are in interaction (see Fig. 3.5). If you change something in the lighting design, other parameters of the vision system also change. There are many feedbacks. The lighting determines other parts of the vision system (optics, sensor, hardware, software).

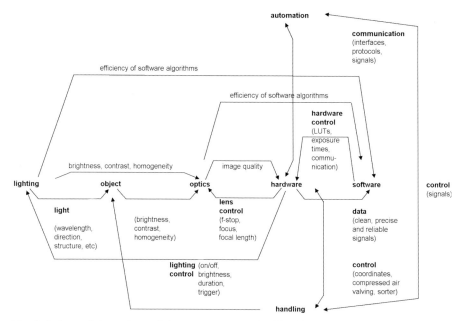

Fig. 3.5 Some interactions of the components of a Machine Vision system (selection, not complete).

The intention of this chapter is to give you an approach to solve lighting problems. It was made with the view to the practical everyday life in the factory floor of Machine Vision.

3.2
Demands on Machine Vision lighting

A Machine Vision Lighting for industrial applications is not only a *lamp* but also more than only a cluster of LEDs. Against the expectations of an uninitiated user there are a few important aspects to pay attention (see Fig. 3.6).

State-of-the-art Machine Vision illumination components for industrial use are complex technical systems consisting of

- light sources
- mechanical adjustment elements
- light guiding optical elements
- stabilizing, controlling and interface electronics
- if necessary software (firmware)
- stable and mountable housing
- robust cabling.

All these features are necessary to form a device that resists the adverse environmental conditions of the industrial floor.

The demands of industrial lighting components are manifold. Partly they are opposing and challenge the developers. Not all criteria are always necessary:

Demands of optical features:

- wavelength (light color):
  - defined and constant distribution, especially for color image processing
  - no (low) aging
  - no (low) drift caused by temperature
  - no differences from device to device
- brightness:

3.2 Demands on Machine Vision lighting | 79

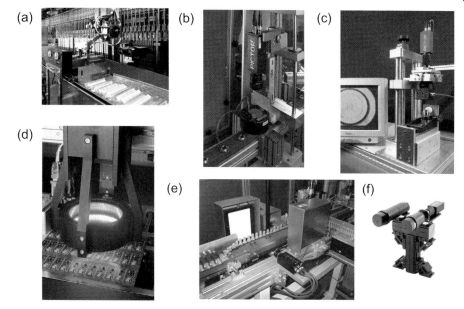

Fig. 3.6 Examples for the variance of Machine Vision lighting. (a) Illumination of the calibration process for medical thermometers. Demands: robustness, brightness, homogeneity, protection against vibrations and splashes of water. (b) Illumination of mechanical parts, running on a dirty and dark conveyor belt. Demands: robustness, tough mounting points, brightness, protection against dust, protection against voltage peaks. (c) Precise illumination for measurement of milled parts. Demands: obvious mounting areas for adjustment, brightness, homogeneity. (d) Lighting stack of a free adaptable combination of red and IR light in an automat for circuit board inspection. Demands: brightness control of different parts of the illuminations using standard interfaces, brightness, shock and vibration protection. (e) Lighting plates for inspection in the food industry. Demands: homogeneity, wide range voltage input, defined temperature management. (f) Telecentric lighting and optics' components for highly precise measurements of optically unfavorable (shiny) parts. Demands: stable assembly with option to adjust, homogeneity, stabilization, possibility to flash (source: www.visioncontrol.com).

- defined and constant
- no (low) aging
- no (low) drift caused by temperature
- no differences from device to device

• homogenous or defined, known and repeatable brightness profile

- only one component for static, pulsed and flash light
- possibility of flashing independent of the illumination wavelength
- bright and powerful, but not dangerous (lasers and in some cases LEDs can also cause damages at the human eye retina)

Demands of electrical features:

- operation with no dangerous low voltage
- wide range voltage input with stabilization for typical industrial voltages between 10 and 30 V dc
- protection against wrong connection
- different controlling/operation modes (static, pulsed, flash, programmable)
- process interfacing: standard interfaces such as PLC inputs/outputs, data interfaces with USB or Ethernet (if necessary).
- simple storage and adjustment of operating parameters (see Fig. 3.7)
- all controlling circuitry included in the housing
- temperature management to avoid overheating
- flexible and tough cables for operation in robotics with a high resistance against bending stress

Fig. 3.7 Elements to adjust lighting characteristics (brightness and flash duration) on a telecentric backlight directly on the lighting component. All electronics are included (source: vision-control.com).

Demands of mechanical features:

- mechanically robust, packed in black anodized aluminum housing or industrial polyamide

- diverse and solid mounting points for a firm fixing and adjustment (see Fig. 3.8)

- protection of all elements against vibrations and strong acceleration (lighting in robotics can be stressed up to 10 g or > 100 m s^{-2})!)

- dust and splash water protection

Fig. 3.8 Robust mounting threads for a tough fixing and adjustment (source: vision-control.com).

Above all stands the demand for an easy and fast installation without additional components, such as mechanical adaptors or holders, electrical convertors or boxes. Further, the compliance with national and international standards such as CE, IP, radiation protection decree and so forth is needed.

Industrial users usually accept a 10,000 h (longer than a year) minimum life and operation time of the light sources. Only this is the base for short downtimes and low maintenance. This fact often limits the choice of a matching lighting components.

It is expected from the users that standard components are in use. This is not only meant for the light source but also for the complete lighting. These components should be – if sometimes necessary – replaced very fast. A supplier with an ISO9001:2000 certificate and a wide distributor network can guarantee a fast spare part delivery even after many years.

3.3
Light used in Machine Vision

3.3.1
What is Light? Axioms of Light

Light as the information carrier of all visual information and Machine Vision is based on electromagnetic waves. Light means a limited sector of the electromagnetic spectrum. The range of light waves extends from wavelengths of 15 nm to 1 mm (see Tab. 3.1). It can be divided into three general ranges:

UV (ultraviolet light): from 15 nm to 380 nm
VIS (visible light): from 380 nm to 780 nm
IR (infrared light): from 780 nm to 1 mm

At the lower limit it is related to the x-rays and at the upper limit it is related to the micro waves. Most Machine Vision application use the range of VIS and the near IR. Some seldom applications extend the use of light to the near UV. But this needs special image sensors.

Tab. 3.1 The spectrum of light according to DIN 5031.

Range			Wavelength λ (nm)	Frequency f (Hz)
UV	UV-C	VUV/vacuum UV	100 to 200	3×10^{15}
		FUV/far UV	200 to 280	
	UV-B	Middle UV	280 to 315	
	UV-A	Near UV	315 to 380	7.9×10^{14}
VIS		Violet	380 to 424	7.9×10^{14}
		Blue	424 to 486	
		Blue green	486 to 517	
		Green	517 to 527	
		Yellow green	527 to 575	
		Yellow	575 to 585	
		Orange	585 to 647	
		Red	647 to 780	3.85×10^{14}
IR	IR-A	NIR/near IR	780 to 1,400	
	IR-B		1,400 to 3,000	
	IR-C	MIR/middle IR	3,000 to 50,000	
		FIR/far IR	50,000 to 1,000,000	$3.0 \cdot 10^{11}$

The range of visible light that is accessible for the man contains the whole spectra of colors from blue to red. A more or less mixture of all colors (with changing quantities) appears as white for the man.

It is a philosophical question what light is – waves or particles? Using the property of waves it can be handled with the theory of electrical engineering. The physical optics does it. This allows us to explain the effects of diffraction, polarization and interference.

Using the property of particles, the photons act as carriers of energy. To illustrate the energy of photons it is mentioned that a laser of 1 mW power emits approximately 10^{16} photons s^{-1}.

The direction of propagation of the photons (this is the direction of the normals of the wave fronts too) is symbolized by the *light rays* that use the geometrical optics. Single light rays practically do not occur. That is why usually it is spoken of light ray bundles. The ray model is a clear and simple one for the explanation of lighting in Machine Vision.

So light is always in the form of both wave and particle. This is expressed as the so-called *wave–particle dualism*. Depending on the effect that is to be explained the property of wave or particle can be used.

To explain the effects of lighting and optics, some axioms (established hypothesises) for the light are made:

- light is an electromagnetic wave (with an electric and a magnetic component)
- light propagates in straight lines in homogenous and isotrope media
- there is no interaction between light of different light sources
- refraction and reflection occur at boundaries
- the velocity of propagation is dependent on the crossed medium
- the propagation velocity of light is $c = \lambda \cdot f$ with λ being the wavelength and f the frequency of light.

The velocity of light propagation depends on the medium, where the light passes through. The *absolute speed* of 299,792 km s^{-1} reaches the light only in the vacuum. In all other medium light propagates slowlier. This should be considered when other materials are inserted in the optical path. Changed distances are the consequence. The propagation velocity is connected with the refraction law of the optics (see Fig. 3.9):

$$n \sin \alpha_i = n' \sin \alpha_r$$

and

$$\frac{\sin \alpha_i}{\sin \alpha_r} = \frac{c_0}{c'}$$

where

n, n' is the refraction index of the medium,
n_0 is the refraction index in the vacuum = 1,
c, c' is the velocity of light in the medium,
c_0 is the velocity of light in the vacuum,
α_i is the entrance angle,
α_r is the exit angle.

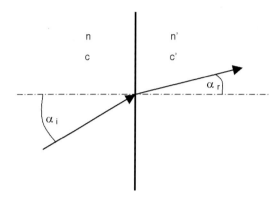

Fig. 3.9 Refraction: refractive indices, angles and velocities.

According to $n = c_0/c$, Table 3.2 of refraction indices and velocity speeds can be made.

Tab. 3.2 Typical refraction indices of optical materials and accompanying light propagation velocity.

Refraction index n	Velocity c (km s^{-1})
1.0	299,792
1.3	230,609
1.5	199,861
1.7	176,348
1.9	157,785

3.3.2
Light and Light Perception

For simplification, the following viewings disregard that the imaging objective influences the spectral composition of the passing light to the image sensor. Each objective has its specific spectral transmission that changes the composition of light (see Chapter 4).

Starting from the spectral perception of light, solid state image sensor materials of Machine Vision cameras and the human eye have different perceptions of light. The spectral response of the human eye covers the range of light between 380 nm and 780 nm and is described by the so-called $V(\lambda)$ curve (for daylight perception). This curve is the base for the calculation of all photometric units such as luminous intensity, luminous flux and so forth (see Section 3.5).

The light perception of Machine Vision cameras differs from the human eye. Machine Vision uses in most cases light with wavelengths between 380 nm and 1100 nm (from blue light to near infrared). This is caused by the reception and spectral response of the imagers. The precise dependence is caused by the donation of the solid state material of the imager. The data sheet of the image sensor used in the camera informs about this.

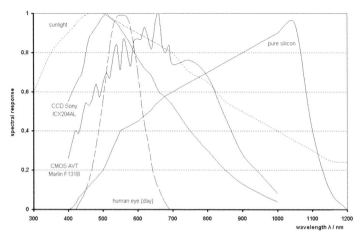

Fig. 3.10 Spectral response of the human eye, typical monochrome CCD image sensor (Sony ICX204AL), typical monochrome CMOS image sensor (camera AVT Marlin F131B). For demonstration, the spectral emission of the sun is also presented.

A solid state imager with a sensitivity as shown in Fig. 3.10 can only output a weighted brightness information per pixel. One single pixel is not able to determine colors.

To interpret and determine colors, it is necessary to combine the information of three (or four) pixels with different color perceptions. Red, green and blue (RGB) or the complementary colors cyan yellow and magenta are typical triples to determine the color values.

In practice this is realized with micro-optical mosaic filters for single chip color cameras (see Fig. 3.11). These color filters cover the single pixels inside the image sensor with a special pattern (for example the Bayer pattern) and make them spectrally sensitive (one-chip-color camera).

Another construction uses three separate sensors. Each sensor is made sensitive for only one color (RGB). The information of all three sensors is correlated and delivers the color information (3-CCD-cameras).

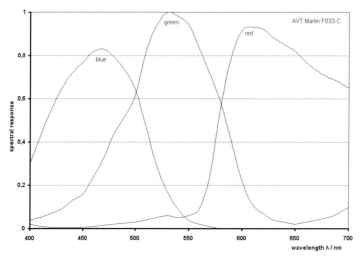

Fig. 3.11 Spectral response of the three color channels (caused by the mosaic filter) of a typical one chip color CCD image sensor (camera AVT Marlin F033C).

The brightness perception of the sensor is one side. The other side is the spectral supply that the light sources emit. Figure 3.12 shows typical spectral supply of different LEDs.

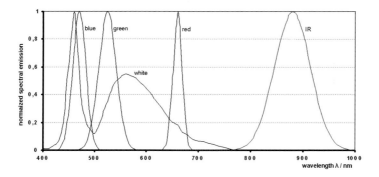

Fig. 3.12 Normalized light emission of different colored LEDs.

The interaction between lighting and receiver means that only a useful coordination of the light source and the sensor gives the base for a contrastful image with a large gray level difference. This makes sure that the following components in the signal chain of the vision system can operate optimally and work reliably in the machine.

The following example shows the brightness perception of a CCD imager combined with different typical colored LED light sources (see Fig. 3.12). The approximation of a Gaussian spectral light distribution is only valid for single colored LEDs. For white LEDs there is no simple approximation.

The power of light sources is usually given in data sheets by their luminous intensity/radiant intensity. In simplification, it represents the area below the relevant curve, multiplied with a constant, that represents the power of the light source without considering the sensitivity of the sensor. It is pointed out that the brightness reception of the sensor and the human eye is different (see Fig. 3.10).

The result of the evaluation of the brightness by the CCD imager (sensitivity see Fig. 3.10) is obtained by some calculations and conversions in [7] and appears as in Table 3.3.

Tab. 3.3 Brightness perception of typical colored 5 mm LEDs. Transformation of the catalog data into sensor relevant values.

LED color	Values from catalog			Calculated sensor relevant I_{CCD} (mW sr^{-1}) from the CCD sensor	Measured gray values
	Luminous radiant intensity (values from data sheet)	Peak wave length λ_p (nm) (see Fig. 3.12)	Half width wavelength radiant intensity $\Delta\lambda_{1/2}$ (nm)		
Blue	$I_v = 760$ mcd	470	30	10.2	232
Green	$I_v = 2400$ mcd	525	40	4.77	129
Red	$I_v = 2700$ mcd	660	20	33.5	378
White	$I_v = 5100$ mcd	$x = 0.31, y = 0.32$		13.8	246
IR	$I_e = 25$ mW sr^{-1}	880	80	3.3	120

It is realized that a difference occurs between the catalog values of luminous intensity (made for the human eye)/radiant intensity and the evaluated radiation by the sensor.

The red LED produces the largest gray value on a CCD sensor, although the luminous intensity (for the human eye) is not the greatest.

On the other hand, the white LED with the largest luminous intensity of 5100 mcd, produces only a medium gray value. This is caused by the difference of the value of luminous intensity with the evaluation of the $V(\lambda)$ curve of the human eye and the perception by a CCD sensor.

Even the infrared LED with its low sensor relevant radiant intensity produces a considerable gray value due to its wide half width wavelength.

These considerations must be made for each sensor – LED combination for finding a new effective way to illuminate most brightness. However, still a few more interactions of light and test object have an effect on the lighting component to choose.

3.3.3
Light Sources for Machine Vision

The basic element of lighting components are the light sources. The built-in light source influences very much the whole lighting set-up.

A practical classification of light sources can be made by the kind of conversion of light. In this publication are only mentioned these kinds of light sources that are typical for the use in Machine Vision.

(a) Temperature radiators

The incandescent emission of these radiators (lamps) produces a mixture of wavelengths and continuous spectrum. The efficiency for a solid state imager is low, because only a part of the emitted spectrum is used. Temperature radiators need a high operation temperature, the higher the more effective (T^4) and whiter they radiate. The radiation maximum shifts to shorter wavelength if it is hot (*Wien's shifting law*). These radiators provide a simple mode of operation.

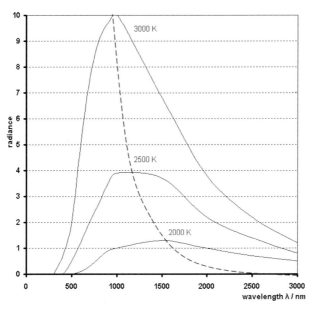

Fig. 3.13 Wavelength composition of emitted light in dependence of the temperature of the radiator (radiance over wavelength) [6].

(b) Luminescence radiators

These are LEDs, lasers. They emit light in a selective and limited spectral band. The efficiency is high (with a adapted receiver) due to the selective emission. At low temperatures these light sources work most efficiently. Some of them have complex modes of operation.

Concerning the converted light power all light sources are comparable by their efficiency, their light output. The theoretical maximum is 683 lumen per Watt from monochromatic green light (555 nm). For white light the theoretical maximum is 225 lm W^{-1}. The efficiency of real light sources is shown in Table 3.4.

Tab. 3.4 Real light sources and their *luminous efficiency* (as of of the 2003) [16].

Lamp type	Luminous efficiency (lm W^{-1})
Na-vapor lamp	to 200
Metal halide lamps	to 100
Xenon lamps	to 60
Fluorescent lamps	40–100
Halogen lamps	to 35
Incandescent lamps	10–20
Colored LED	to 55 (red–range)
White LED	to 25

Within the next few years, it is targeted to increase the luminous efficiency of colored LEDs to 300 lm W^{-1}. Thus the LED will be the most efficient known light source.

3.3.3.1 Incandescent Lamps / Halogen Lamps

These lamps in their classical form use tungsten filament for the incandescent emission. The wide band emission of radiation (from UV to IR, see Fig. 3.13) causes that only about 7% of the energy are converted into visible light. The consequence is a bad efficiency.

To achieve a high luminous flux, the filament must be as hot as possible, i.e. must be driven with a higher voltage than recommendable. But heat is a contraproductive factor for the lifetime of an incandescent lamp. So they always work in the compromise between intensity and lifetime. The dependences of a typical halogen lamp [17] are as follows:

$$\text{lifetime:} \quad t = t_0 (U/U_0)^{-1.2}$$
$$\text{luminous flux:} \quad \Phi = \Phi_0 (U/U_0)^{3.2}$$
$$\text{color temperature:} \quad T = T_0 (U/U_0)^{0.37}$$

with

t_0 being the rated lifetime,
L_0 the rated luminous flux,
T_0 the rated color temperature,
U_0 the rated voltage,
U the operating voltage.

The exponential dependence of the luminous flux makes clear that halogen light sources can only be driven with a stabilized power supply. Unstabilized voltage sources pass through an increased change of the luminous flux.

Halogen lamps have filled the lamp bulb with halogen gas. This takes care of a limited darkening process along the lifetime and also doubles the luminance. The tungsten vapor during the emission again settles at the filament and not at the glass bulb. But the halogen cycle works only between 70 and 105% of the rated voltage. Halogen lamps are usually built in cold light sources. To improve the light output, these lamps are equipped with a cold mirror on the backside to reflect the visible light to the fiber bundle coupling. The IR radiation passes the cold light mirror backwards.

Advantages

- bright light sources

- continuous spectrum (VIS: white light with color temperature 3,000–3,400 K)

- operation with low-voltage

- works in hot environment too (up to 300 °C)

Disadvantages

- typical lifetimes: 300–2,000 h (very short!)

- high power halogen lamps have much shorter lifetimes

- large fall off in brightness (drift)

- sensitivity to vibration

- delay on switching on and off makes it applicable only for statical light

- operating voltage fluctuations are directly passed through to brightness changes – need of a powerful stabilization

- in some cases protection measures are necessary – depending on the glass bulb (quartz glass or hard glass) UV light is emitted (with quartz glass)

- large and heavy when built in a tough casing
- needs additionally fiber optics to form the light (approximately 40% loss of light output (incoupling, transfer of light))
- produce much heat

Considerations for Machine Vision

- useful for color applications, good color rendering
- does not meet industrial demands for a lighting

3.3.3.2 Metal Vapor Lamps

These kinds of gas discharging lamps work very efficiently with a high yield of light. The metal vapor inside is used to produce a vapor pressure that is necessary for the gas discharge.

The application in Machine Vision is also rare because of the large amount of heat that is to lead away. Metal vapor lamps are usually built in cold light sources.

Advantages

- very bright
- cold light (high color temperature)

Disadvantages

- lifetime limited, approximately 10,000 h
- brightness adaptation only optically feasible
- working with high voltage up to 30 kV – protective measures are necessary
- lamp bulbs are under high pressure
- UV quantity in the spectra
- single spectral bands used (see Fig. 3.14) can cause problems for color rendering
- very expensive
- limited temperature range (typically 10–40 °C)
- produces heat

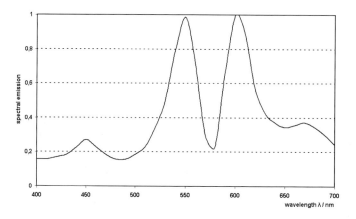

Fig. 3.14 Spectral emission of a metal vapor lamp.

- warm running necessary

Considerations for Machine Vision

- seldom used
- restricted suitability for industrial applications

3.3.3.3 Xenon Lamps

These discharging lamps are available in both, for continuous and flash operation. Xenon flash lamp generators are preferred for Machine Vision. They are usually for the illumination of fast running processes because of their high intensity. To *freeze* motion they are able to produce very short flashes with up to 250,000 candela intensity that can still be seen in 20 km distance.

The wavelength spectrum covers a range from below 150 nm to greater than 6 μm (see Fig. 3.15). Xenon flash lamps provide an almost continuous spectrum of wave lengths in the visible region. This leads to a balanced white light that makes additional color balancing filters generally not necessary.

In contrast to a tungsten halogen lamp that produces a temporal continuous energy, the energy of the xenon lamp can be concentrated in high power flashes. The differences are considerable. In contrast to a 150 W tungsten halogen lamp with 1.2×10^{-2} lm luminous flux produces a 43 W xenon flash light 4 lumens luminous flux [5].

Xenon flash lamps use sensible glass tubes that need housing and further an expensive control circuitry that is built in unwieldy boxes. Some little application uses the direct light from the flash tube; usually the emitted light is coupled into fiber optics to form the light.

Advantages

- extremely bright
- high color temperature (5,500 to 12,000 K), very white, best color rendering
- flashable with flash duration of 1–20 µs (short arc models), 30 µs to several milliseconds (long arc models)
- high flash rate with up to 200 (1,000 with decreasing flash energy) flashes per second
- lifetime of up to 10^8 flashes

Disadvantages

- protection measures needed, because of the use of high voltage
- EMC problems caused by strong electrically pulses to control the light source
- expensive, costly electronics
- bulky, needs a light generator box
- unflexible shape (flash bulb with a reflector) or requirement of additional fiber optics (approximately 40% loss of light caused by incoupling and transfer of light)
- flash to flash variation of intensity (<10%)
- aging : intensity after a few million flashes can be down to 50%

Considerations for Machine Vision

- useful for color image processing where much light is needed
- useful for fast running processes
- reservations for industrial use because of operation conditions

3.3.3.4 Fluorescent Lamps

Fluorescent lamps are discharging lamps that are known from lighting of rooms. Their efficiency is better than that from incandescent lamps. A different distribution of spectra can be chosen due to different coatings of the lamp tubes inside. Customary catalog of fluorescent lamp suppliers inform

3 Lighting in Machine Vision

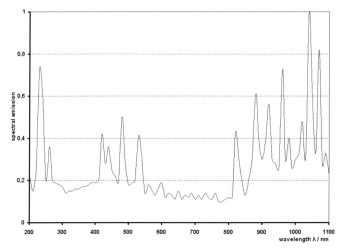

Fig. 3.15 Typical spectral emission of a xenon flash lamp. The wide spectral distribution brings good color balance.

about the widespread possibilities. These coatings ensure that the UV radiation that is produced by the evaporated mercury inside the tube converts into visible light (see Fig. 3.16).

Fluorescent lamps are driven with alternating current (AC). The change of the current direction is converted into a flickering of light with the double frequency of the supplying voltage, because the luminants inside the tubes continue to glow for a maximal time of 1/1,000 s. To avoid brightness interferences between fluorescent light and image acquisition frequency, it is necessary to use HF-ballasts (22 kHz or more are advisable). Without this measure variations of brightness from image to image up to total darkness can occur for the worst case.

Fluorescent lamps are popular for low cost solutions and for achieving a homogenous lighting from a distance.

Advantages

- cheap
- different color temperatures for selection (3,000 to 6,000 K)
- able to illuminate large areas

Disadvantages

- inflexible, limited fixed shapes of the light source (line, ring)
- reduced working lifetime 5,000–12,000 h, small ring lights only 2,500 h

- HF-ballast necessary (possible EMC problems)
- single spectral bands
- large temperature drift
- only applicable for static light, no flash
- rapid aging (after 12,000 h approximately 50% of the brightness from beginning)
- warm running necessary

Considerations for Machine Vision

- use for illumination of large areas
- low cost solutions (with all the disadvantages)
- use of electronic HF-ballast is necessary

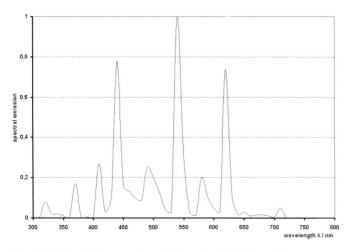

Fig. 3.16 Typical spectral emission of an HF driven fluorescent ring light.

3.3.3.5 LEDs (Light Emitting Diodes)

The progress in materials and electronics technology pushes the triumphant advances of the LED technology. One can say that nowadays LED lighting is the standard lighting of Machine Vision. This development was driven by the demands of the automotive industry for tough, cheap, reliable and powerful light sources.

LEDs are small and very robust light sources and emit cold light with a narrow half-width of wavelength of approximately 30 nm (see Fig. 3.12), which means almost monochromatic light. They are powerful with an efficiency of optical power of up to 55 lm W^{-1}, tendency strongly increasing.

The lifetime for LEDs differs strongly depending on the color, measurement conditions, environmental conditions, design, manufacturer and so forth. For monochromatic LEDs it is supposed that they achieve average lifetimes under optimal conditions between 100,000 and 200,000 h, while white LEDs achieve partially much less. These data of single LEDs do not have to do with the lifetime of an industrial lighting component [15] (see Section 3.5.3.1).

The color of the emitted light depends on the substrate of the p–n transition inside the LED. Many colors are possible (see Table 3.5).

Tab. 3.5 Some solid state materials and their emitted center wavelength in LEDs.

Color	λ_c (nm)	Material
White	$x = 0.32$, $y = 0.31$	InGaN
Blue	470	InGaN
Blue–green	505	InGaN
Green	528	InGaN
Green	570	GaAlP
Yellow	587	InGaAlP
Amber	615	InGaAlP
Orange	606	InGaAlP
Red	633	InGaAlP
Hyper-red	645	GaAlAs
Red	645	GaAlP
IR	950	GaAlAs

Since a few years direct white LEDs are available. They compete more and more with traditional white light sources. What was possible before only with a triple of closely mounted red, green and blue LEDs (Multiled) is now possible in one chip. From their nature are white LEDs blue light emitting LEDs covered with a yellow illuminant. The yellow pigments that are embedded in a transparent synthetic resin result in emitted white light. Depending on the stability of the production of the yellow pigments different hues are achievable or disturb application that need stable color conditions (see Fig. 3.17).

LEDs can operate with up to a ten-fold current overload if they are pulsed for a short time. The result can be an up to ten-fold higher intensity of light emission (infrared). White LEDs can achieve an up to 6 times higher intensity in this operation mode (see Fig. 3.18). The real increase of light emission depends on the used substrate (color) of the LED.

Note for this operation that the allowed pulse-duty factor to avoid the thermal destruction of the LED has to be kept. For the operation of LEDs in the pulse mode no aging is noted even after a few 10 millions of flashes!

The different sizes and small designs (5 mm, 3 mm, SMD types, LED on chip etc.) inspire developers of Machine Vision lighting to various lighting components.

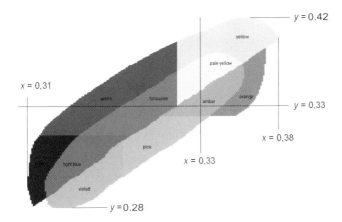

Fig. 3.17 Distribution and span of hues of white LEDs. Use for color classification and sorting of white LEDs [21]. x and y are color coordinates.

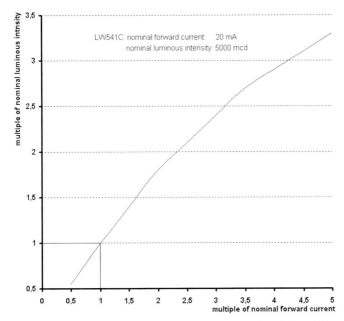

Fig. 3.18 Example for a current–luminous intensity relationship of a white LED (source: data sheet LW541C OSRAM Semiconductors).

Advantages

- particularly suitable for industry
- lifetime > 20,000 to > 100,000 h. Strongly depending on operation conditions.
- vibration insensitive/shock resistant, survive higher G-forces
- nearly monochromatic or white light
- small temperature drift (time, temperature) of brightness
- ideal electrical controllability
- fast reaction down to <1 µs
- flash light is possible in all colors and IR
- free design of lighting shapes, all lighting techniques are possible
- lighting with mixed colors possible
- low power consumption and low emergence of heat
- no maintenance necessary
- low voltage means low danger
- low space necessary

Disadvantages

- aging of white LEDs changes color coordinates (hue and color temperature)
- white LEDs not homogenous to manufacture (sorting LEDs required regarding center wavelength, optical power, color coordinates)
- not yet as bright as halogen lamps (state 2004)
- maximum operation temperature of 60 °C, otherwise strongly increased aging

Considerations for Machine Vision

- equipped with those features LEDs are ideal light sources for Machine Vision
- the progress in LED technology will also inspire Machine Vision

3.3.3.6 Lasers

Laser light sources are relatively seldom used in Machine Vision. If so, only lasers based on laser diode modules are applied, no solid state or gas lasers. Characterizing facts are

- highly concentrated energy
- emission of coherent light
- emission of real monochromatic light
- point-shaped origin of light

The highly concentrated energy makes it possible to work with an optical power of only a few milliwatts to achieve considerable brightness on the camera sensor. On the other hand, the concentrated energy is an obstacle for the use. Many safety measures are taken to protect the human eye and body from the danger of concentrated radiation.

Laser light sources are classified into laser protection classes (see DIN EN 60825-1:1994). These classes describe the protection measures in dependence from the power and wavelength used. This explains the aversion of many companies to forbid the use of lasers in any kind in their production lines.

The emission of coherent light is an optical phenomenon based on the emission of light waves of only one wavelength (monochromatic) in phase. The superposition of many of these waves leads to the appearance of interference and with that to the characteristic speckle patterns. The speckle patterns appear as local and temporal unregular and changing patterns of brighter and darker points. This appears even with a defocused imaging optics or with a beam shaping optics in front of the laser light source (see Fig. 3.19). The speckle patterns prevent the correct function of most image processing algorithms.

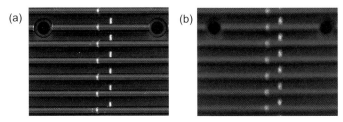

Fig. 3.19 Illumination of a structured surface with a laser line: (a) with focused imaging objective, (b) with unfocused imaging objective. Clearly perceptible in both cases is the pepper–salt pattern of the light line caused by speckles.

To avoid this, one can destroy the coherence of the laser using a fiber coupling into a multimode fiber. This is expensive and needs high precision for manufacturing.

Using a beam shaping optics, laser lighting can produce the best possible parallel light (beam expanders). Using diffractive gratings manifold light structures for projection can be produced (see Section 3.7.4.4).

Electrically seen are laser diodes easy to handle but they need a specialized drive. They can be driven in cw (continuous wave) mode or can be modulated up to a few MHz. They are very sensitive against electrostatic discharging and optical overload. That is why they need an integrated monitor (photo) diode to control the optical output.

Advantages

- special and different light shapes achievable
- emission of almost perfect parallel light possible
- highly intense

Disadvantages

- protection measure needed
- nonhomogenous illumination because of speckles
- used only for special procedure in Machine Vision

Considerations for Machine Vision

- reservation because of danger for man
- useful for some application in 3D

3.3.4
The Light Sources in Comparison

The mesh diagram (see Fig. 3.20) shows the suitability of different light sources for Machine Vision. A larger area typifies a better matching.

3.3.5
Considerations for Light Sources: Lifetime, Aging, Drift

3.3.5.1 Lifetime
The industrial use of light sources implies that they work built in a complete lighting device. Thus the lifetime does not only mean the lifetime of the light

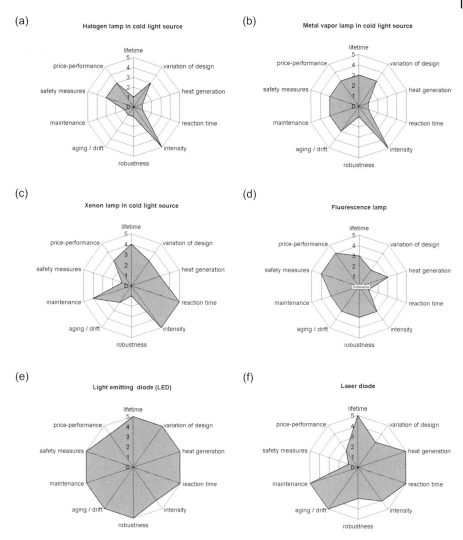

Fig. 3.20 (a) to (f) How much meet a lighting the needs for a use in Machine Vision? Assessments are made from 1 = bad to 5 = very good.

source but also the lifetime of all cooperating electronic and thermal effective components under harsh industrial conditions. And this is usually much less than only the lifetime of the light source that is measured under optimal conditions. That is why diverge the lifetime data from only the light source and from the complete lighting component.

If we keep in mind that all data of lifetime are statistical values (MTBF – mean time between failure), we feel that the nominal values are a rather vague information of how long an illumination device will work (see Table 3.6).

Tab. 3.6 Lifetime in hours, days, month, years.

Life time in hours	Days (approx.)	Month (approx.)	Years (approx.)
500	21	0.7	0.06
1,000	42	1.5	0.10
5,000	208	6.7	0.60
10,000	416	31.4	1.10
20,000	833	26.9	2.20
50,000	2,083	67.2	5.60
100,000	4,166	134.4	11.2

To talk about lifetime of lighting also means to talk about the reliability and availability of the complete vision system. Lighting as the core component and information carrier of the vision system decides fundamentally on the function of the whole machine. False economy for lighting components can cause high costs:

(a) preventive maintenance to replace short-lived light sources cyclically

(b) unforeseen failure of the machine/system with all possible consequences.

Both include

- costs for replaced components
- costs for downtime of the machine (system)
- labor costs for maintenance
- delay time costs to maintenance, delay time costs to delivery of the replacement part (for b)

This should also be considered while selecting a lighting component.

3.3.5.2 Aging and Drift

The light emission of light sources is based on chemical compounds that have characteristic aging behavior in their substantial structure. This temporal influences on the emission of light is called aging. Depending on the kind of light source this aging can be totally different. It is important to know this behavior to limit or to compensate it.

Almost all image processing algorithms are based on brightness and contrast in the image with relatively constant values. Aging light sources are

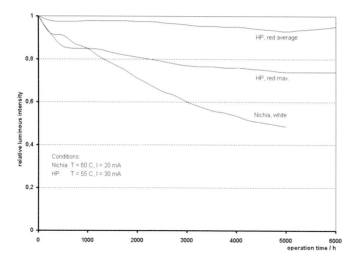

Fig. 3.21 Aging of different LEDs [18].

changing the brightness and contrast. The allowed value of this change depends on the used algorithms. Some algorithms are very sensitive, while others are more tolerant. Measurements, tests or calculations by the software engineers can give information about these dependences related to the software.

Especially measurement application with Machine Vision needs stable lighting for stable images. This is an indispensable condition to maintain the accuracy of detection.

Aging is usually sped up by temperature. For the most light sources a long-term operation with overtemperature will shorten the lifetime and/or the brightness. Thermal processes or even passed time change the chemical structure of the radiant materials irreversibly. The consequences are lower or changed emission of light.

For LEDs this have an effect on the maximum of brightness (monochromatic LEDs) or on the brightness, color distribution and color temperature (white LEDs). Especially the conversion layer of white LEDs tends to strong aging at higher temperatures (see Fig. 3.21).

A typical temporal course of aging of a LED related to the luminous intensity is given by

$$I(t) = I_0 \exp(-t/t_s)$$

with

I_0 being the luminous intensity at beginning of operation,
t the operation time and
t_s the aging constant [21]
(144,270 h for 50% brightness after 100,000 h operation).

To avoid this lighting components need a temperature compensation to bring down the temperature of the sensitive components as much as possible (see Fig. 3.22). The colder the chips the better the stability will be. Well planned constructions of lighting components bear that in mind. The electrical and mechanical-thermical construction ensures that all temperature sensitive elements work at critical temperatures, particularly the light sources. In some cases airflows can help to reduce their warming.

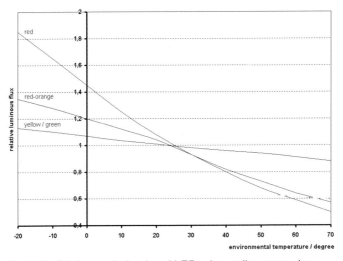

Fig. 3.22 Brightness behavior of LEDs depending on environmental temperature.

LED light sources work most effectively if they are cold, which means below a temperature of approximately 60°C. The concrete temperature depends on the substrate and LED type. If LEDs warm up more, it is observed that the brightness decreases by 1% for each degree increase of temperature [15]. That is why the thermal compensation of lighting is very important and is a distinguishing feature for robust and long-lived lighting components.

A very simple possibility of avoiding overheating is to switch on the light source only, when the camera acquires an image. Rest of time the lighting is switched off.

Working with LED in flash operation mode means to avoid from aging too. A flashed LED does not age over a few million flashes in compliance with a pulse-duty factor of 1:10 minimum (flash duration: rest) and a maximum flash duration of $< 10\mu s$ (average value). This operation mode prolongs the lifetime of the LED light source to the maximum.

Rule of the thumb:
A recommended setting of brightness of a light source is 50%. This optimizes the lifetime and minimizes aging.

Even in short time periods the intensity of lighting changes. This drift is caused by warming up the light source/controlling electronics. Depending on the light source this behavior totally differs. For temperature radiators it is known that they increase their power of radiation with increased temperature.

Fluorescent lamps need a warm operation temperature. That is why they deliver lower light output in cold environment. On the other hand they need a lead time to warm up to achieve the rated light intensity. After further warming they decrease in intensity (see Fig. 3.23).

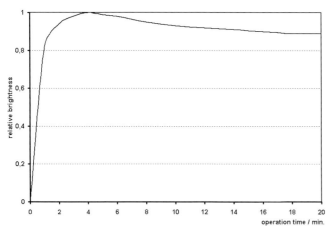

Fig. 3.23 Time dependence of brightness of a typical fluorescent lamp PL-S11W/840/4P [Philips].

3.4
Interaction of Test Object and Light

3.4.1
Risk Factor Test Object

Light begins to work for Machine Vision when it starts interacting with the test object. And instantly the test object becomes the decisive but not the predictable component. What are the problems that the test object brings along?

Above all there are no constant optical properties. We should consider that the optical inspection with Machine Vision is not limited only on parts with stable optical characteristics. Parts with totally different or changing properties can occur such as plastic molding parts of different color. For the functions of the parts may mean nothing, but for the function of the vision system that can mean everything.

The classical dilemma of Machine Vision is to know only a little or almost nothing about the optical properties of the test object. Most of the parts are not even specified for those properties. This means almost to work with a *black box* part.

A few properties of test objects that influences on the possibilities of optical inspection are shown in Table 3.7.

Tab. 3.7 Properties of test objects that influence the vision system inspection.

Factor group	Influence factor	Possible reasons for the change in properties
Optical factors	Part color	Changed material or material mixture
	Pattern	Changed tool quality
	Reflection	Changed material, material mixture, manufacturing method
	Scattering	Changed material, material mixture, surface finish
	Transmission	changed material or material mixture
	Absorption	Changed material or material mixture
Mechanical factors	Shape of edge	new and worn tools (i.e. cutting tool)
	Surface geometry	new and worn tools (i.e. cutting tool)
	Surface imperfections	new and worn tools (i.e. cutting tool), changed contractor
	Surface roughness	Changed tool quality
	Chatter marks	Worn tool
	Surface finish	Changed contractor
Chemical factors	Corrosion	Different reflection
	Oil film	Protection from corrosion needed
	Release agent	Changed manufacturing method

Often even the customer cannot foresee the changes of these properties, because he is dependent on his contractors and usually he does not know the technological processes behind. So the test object remains the no. 1 risk factor. The risk factors can be limited by using a matching lighting.

3.4.1.1 What Does the Test Object do With the Incoming Light?

Light that enters the test object is divided into three fundamental parts (see Fig. 3.24):

- the light quantity that is reflected by the test object – the reflection R
- the light quantity that passes the test object – the transmission T
- the light quantity that is absorbed by the test object – the absorption A

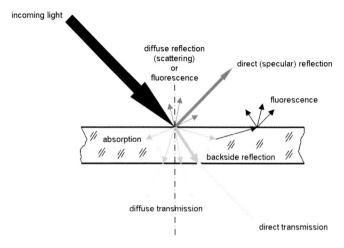

Fig. 3.24 Quantities of the incoming light at the test object and their distribution.

In accordance with the energy conservation law is the sum of all three parts:

$$R + T + A = 100\%.$$

Above all these material specific characteristic values of the test object depend on the wavelength and incident angle but a few more influencing factors are known. A real test object is always an unknown mixture of reflection, absorption and transmission. The effect of fluorescence is not considered in this place.

3.4.1.2 Reflection/reflectance/scattering

The reflected light is essentially optically shaping the part in Machine Vision. Reflection means the deviation of light at surfaces and interfaces of materials with different refraction indices. It forms the basis for imaging in mirrors and prisms. According to the reflection law (see Section 3.1) the light is basically reflected from the part where the incident and reflected light rays are in the same plane. After reflection, the reflected light has changed the direction of oscillation. It is polarized (see Section 3.4.2.5).

Usually the parts from Machine Vision are located in air that the angle of incidence is equal to the angle of reflection. The geometrical macro structure of the test object generally determines where the light is reflected (see Fig. 3.25). But the geometrical micro structure (surface quality and roughness) influences on how the light is reflected. The surface will diffuse or scatter the light and determines the light distribution after reflection. Different reflecting characteristics of the test object occur in different gray values in the image. This fact is the main reason for problems of malfunctioning of vision systems.

Condition for a directed reflection is a smooth surface with a peak-to-valley-height smaller than $\lambda/4$ (The Rayleigh condition). If the peak-to-valley-height is larger than this, the light is diffusely reflected, it is scattered. The directional properties of a reflecting surface depend on the material and above all from the surface treatment. A polished one will have other reflecting properties than a milled surface, a grinded surface others than a painted.

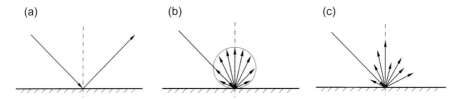

Fig. 3.25 Different qualities of surfaces and their influence on the distribution of the reflected light. (a) Directed reflection, (b) regular diffuse reflection, (c) irregular diffuse (mixed) reflection.

Depending on these different possible light distributions only a smaller amount of light energy is reflected backward to the cameras lens. The rest is distributed into the whole area surrounding the reflecting surface. The reflected light energy can be strongly reduced. Details have to be checked from part to part and from application to application.

Information about the general degree of reflection gives the material characteristic value reflectance R that expresses the relation between reflected and incident luminous flux:

$$R = \Phi_r/\Phi_0$$

where Φ_r is the reflected luminous flux and Φ_0 is the incident luminous flux (see Fig. 3.8).

The reflectance depends on factors such as material, wavelength/color temperature, polarization of light, angle of incidence of light and so forth (see Fig. 3.26).

An extended view to the reflection properties gives the characteristic value *glossfactor* that expresses the ratio of directed to diffuse reflection. A high gloss factor means a large amount of direct reflection.

These values can be taken for an approximate calculation of the light reflected toward the lens and the sensor.

3.1 Example From an incoming illuminance of 100 lx (from the light source) to a nickel surface only 45 to 63 lx are reflected toward the camera. The efficiency of the lighting is drastically reduced.

Transparent materials do not reflect only on the surface. They tend to reflect at the backside too (see Fig. 3.24). The consequence are backside reflections in

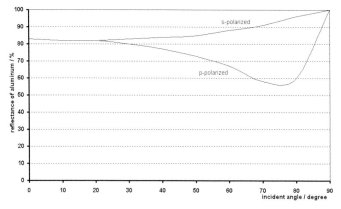

Fig. 3.26 Dependence of the incident angle of light – reflectance for a polished aluminum mirror surface.

the form of double images or ghost images. For targeted reflection use only thin film materials to avoid this or reduce it with anti-reflection coatings if possible.

That can be useful for special measurement methods with Machine Vision for example the measurement of thickness of glass plates using the double images of a laser point. With a known angle of incidence into the glass plate, the thickness evaluating the distance between the incoming spot and the reflected one can be measured.

Last but not least it affects the shape of the test object on the reflected light. A sharp edged and a rounded edged part possess totally different reflection properties. So the shape of the material edges essentially determines the light that reaches the camera lens and the sensor (see Fig. 3.27).

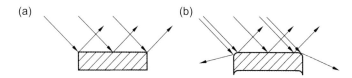

Fig. 3.27 Light reflection from a stamped metal part: (a) stamped with a new cutting tool and (b) stamped with a wear out cutting tool. The light distribution changes totally due to the different shape.

Even chemical processes (surface treatment, corrosion, corrosion protection) influence the reflection. They change the reflectance.

Note this if getting parts from different suppliers that use different manufacturing and surface treatment procedures. Parts that are stored in a humid

Tab. 3.8 Average reflectance, transmission, absorbance of materials for white light (color temperature 3,000 K). Perpendicular incidence of light[8].

Material	Reflectance	Absorbance	Transmission
Aluminum, polished	0.6 to 0.72	0.28 to 0.4	
Aluminum, polished and anodized (nature)	0.75 to 0.9	0.1 to 0.25	
Aluminum, frosted	0.55 to 0.6	0.4 to 0.45	
Aluminum foil	0.8 to 0.87	0.13 to 0.2	
Brass, polished	0.61 to 0.62	0.39 to 0.38	
Brass, frosted	0.52 to 0.55	0.45 to 0.48	
Brick	0.15 to 0.4		
Chrome, polished	0.6 to 0.7	0.3 to 0.4	
Chrome, frosted	0.4 to 0.45	0.55 to 0.6	
Color paint			
White	0.7 to 0.85	0.15 to 0.3	
Cream	0.56 to 0.72	0.28 to 0.44	
Yellow	0.48 to 0.52	0.48 to 0.52	
Brown	0.27 to 0.41	0.59 to 0.73	
Green	0.12 to 0.2	0.8 to 0.88	
Blue	0.07 to 0.1	0.9 to 0.93	
Red	0.1 to 0.27	0.73 to 0.9	
Copper, polished	0.48 to 0.6	0.4 to 0.52	
Cotton, white	0.3 to 0.4	0.23 to 0.4	0.25 to 0.4
Earth, damp	approx. 0.07		
Enamel, white	0.65 to 0.8	0.2 to 0.35	
Glass aluminum mirror	0.9 to 0.94	0.06 to 0.1	
Glass, clear (1 to 4 mm thick)	0.06 to 0.08	0.02 to 0.04	0.9 to 0.92
Glass, frosted (2 to 3 mm thick)	0.07 to 0.2	0.05 to 0.17	0.63 to 0.88
Glass, opal (2 to 4 mm thick)	0.31 to 0.57	0.03 to 0.31	0.12 to 0.66
Grass	approx. 0.06		
Granite	0.1 to 0.2		
Marble	0.6 to 0.65		
Nickel, polished	0.53 to 0.63	0.37 to 0.47	
Nickel, frosted	0.45 to 0.55	0.45 to 0.55	
Paper	0.6 to 0.85	0.05 to 0.39	0.05 to 0.41
Plaster	0.2 to 0.55		
Plastics, white (2 to 3 mm thick)	0.2 to 0.4	0.1 to 0.2	0.4 to 0.6
Sandstone	0.15 to 0.4		
Silk, white	0.25 to 0.38	0.01 to 0.06	
Silver, polished	0.85 to 0.94	0.06 to 0.15	
Skin, untanned	approx. 0.45		
Snow	0.65 to 0.75		
Steel, polished	0.55 to 0.6	0.4 to 0.45	
Tin plate	0.67 to 0.69	0.31 to 0.33	
Velvet, black	0.02 to 0.1	0.9 to 0.98	
Wood bright	to 0.4	to 0.6	
Wood dark	from 0.07	from 0.93	

or chemical aggressive environment can corrode. They will look principally different to parts from the (perfect) pilot production.

Parts that are treated with anti-corrosive agents change their reflective properties too. The difference among a clean metal part, a metal part with an oil film or a wax covered metal part is only mentioned.

Tip

> Take two sample test objects as (visually) different in reflecting as possible. Determine the difference in the reflected light of both parts using the vision system and compare gray values. How do the parameters vary? Do they both fulfill the needs for safe function of the software? If not, change the lighting and/or lighting method!

Tip

> Take the reflections at the test object into consideration according to the reflection law. Think of the micro and/or macro structure of the part. Position the light source(s) in such a way that the light is reflected into the camera. Try it from different positions so long as you get the reflexes in the desired places.

3.4.1.3 Total Reflection

A special phenomenon on reflecting parts (for example transparent glass and plastics) is the total reflection. Above a marginal angle α_l some materials reflect completely. This occur for example on surfaces of water, glass, plastics and can be profitable when used for the reflection of light inside prisms and fiberglass.

The condition for this is an angle of incident light being larger than the marginal angle α_l and a transition of the refraction indices from a higher to a lower value (see Fig. 3.28).

$$\alpha_l = \arcsin(n'/n)$$

where n is the refraction index before reflection and n' is the refraction index of the adjacent medium.

Typical values of the marginal angles are

48.7° for a transition of water ($n = 1.33$) against air ($n' = 1$)
41.8° for a transition of an average optical glass ($n = 1.5$) against air ($n' = 1$)
62.5° for a transition of an average optical glass ($n = 1.5$) against water ($n' = 1.33$)

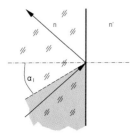

Fig. 3.28 Total reflection can be found only in the gray area. At smaller angles the transparent material is only refracting the light.

Tip

> Note the total reflection when inspecting transparent parts. Changed viewing angle or lighting angle can abolish or let appear the total reflection.

3.4.1.4 Transmission/transmittance

Another part of light can pass the test object. The material specific value to characterize this is the transmission (see Fig. 3.24 and Fig. 3.29).

For backlight applications mean the transmission the amount of light passing through the object that produces brightness for the test object at the imager (see Fig. 3.30). The higher the transmission, the brighter the part occurs. Possibly a high transmission of the part is not desirable, because it prevents a necessary contrast between the bright background (from lighting) and dark test object.

For front light applications mean the transmission a loss of brightness. The incoming light is not reflected to the camera sites as desired. It is not usable for the camera because it passes the object.

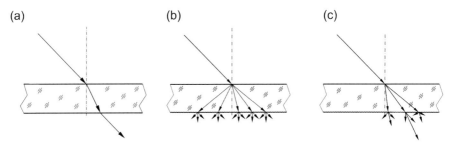

Fig. 3.29 Different qualities of transparent materials and their influence on the distribution of the transmitted light, (a) directed transmission, (b) diffuse transmission, (c) irregular diffuse (mixed) transmission.

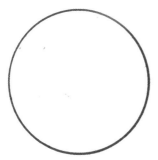

Fig. 3.30 Image of directed and diffuse transmission: round glass block with chamfer. In the middle the light is directly transmitted. An annulus around shows diffuse and strongly reduced transmission caused by the rough surface of the chamfer. Lighting component: telecentric lighting.

Information about the degree of transmission gives the material characteristic value transmittance T:

$$T = \Phi_t/\Phi_0$$

where Φ_t is the transmitted luminous flux and Φ_0 is the incident luminous flux.

The transmittance depends above all on the illuminated material, wavelength/color temperature, polarization of light, angle of incidence of light an so forth. Some average values for transmittance are shown in Table 3.8.

These values can be taken for approximately calculating the transmitted light through the test object.

3.2 Example From a generated luminance in a diffuse light source of 4,000 cd m^{-2} still remain approximately 2,000 cd m^{-2} after passing a part of white plastics with 3 mm thickness.

But transmission does not only occur from the test objects but also from inserted optical filters. Apart from the spectral influence on the light information they reduce the transmission of light (see Section 3.6.1).

3.4.1.5 Absorption/absorbance

The third part of light energy that influences the light distribution of the test object is the absorption. Light is absorbed by opaque and transparent objects too. This portion of light is changed into warmth in the test object. To foresee the amount of absorbed light energy is difficult, because it depends on many imponderable and material factors that are not to determine easily.

To determine these factors, an approximate value can be derived from converting the starting formula

$$R + T + A = 100\% \quad \text{to absorbance} \quad A = 100\% - R - T.$$

Some guide numbers for the absorbance are given in Table 3.8. Note that this part of light is not at all usable for lighting purposes.

3.4.1.6 Diffraction

One axiom of light is the straight propagation within a homogenous medium. But light as an electromagnetic phenomenon can look behind the test object into the geometrical shadow area. If a light wave touches the edge of an object, the light wave is diffracted. That means it changes the direction of propagation. Because light always consists of bundles of light rays interference happens. This is known from specific diffraction patterns (see Fig. 3.31).

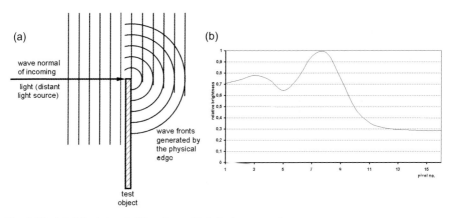

Fig. 3.31 (a) Principle of diffraction with interference of waves touching a body. The incoming wave interferes with the new created wave from the physical edge of the test object. This results in typical diffraction patterns, (b) Gray value distribution of a real edge with notable diffraction. Image scale of 5:1 in combination with a telecentric backlighting of 450 nm wavelength. Pixel resolution is 1.5 µm/pixel.

The size of diffraction effects proportionally depends on the used illumination wavelength. The smaller the used wavelength is, the smaller is the width of the diffraction patterns. This means that a reduction of the illumination wavelength from IR (typically 880 nm) to blue light (typically 450 nm) can reduce the diffraction effects to almost half.

Diffraction will be notable for Machine Vision applications with monochromatic light and image scales larger than 3:1. For smaller image scales it only distorts the width and shape of the gray value/brightness transition of the edge. If diffraction is notable it is necessary to check the image processing software. How the edge detection is done and how does the software treat the diffraction patterns? Will the edge be found in the right place? Is the edge location reliable? (see Fig. 3.32) Maximum gradient algorithms can help there.

White light as a mixture of different wavelengths does not produce diffraction. The mix prevents the occurrence of diffraction. The results are blurred edges in comparison to the best possible edge in monochromatic light.

Diffraction occurs at each material edge that the light touches. These edges can be

- the test object (distorts gray value transitions in the image)
- dust spots on the lens, sensor ... (enlarges the dust spots)
- f-stops, lens mounting (reduces the resolution of the objective)

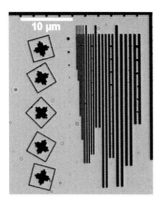

Fig. 3.32 Diffraction on dust spots and geometrical structures (chrome on glass) at a test chart for photolithography. The pixel resolution of 0.688 µm/pixel is on the resolution limit for visible light. Diffraction limits the detectability.

3.4.1.7 **Refraction**

If the test object is transparent and a backlight application is used, refraction can occur. Basically it is connected to the refraction law (see section 3.3.1). The refraction indices are listed in the relevant optical literature.

If light passes a boundary between two transparent materials with different refraction indices (propagation velocities), the light propagation direction is deviated. What is useful for the function of all lenses can cause problems and surprises for illumination and imaging the test object (see Fig. 3.33).

Tip

> If you are working with transparent test objects it can be helpful to approximately do ray tracing from the light source through the

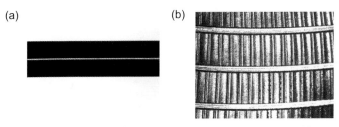

Fig. 3.33 Change of light propagation in transparent test objects in telecentric backlight, (a) glass rod, (b) curved transparent plastic body. Curved transparent objects act like optical elements and refracted light. A parallel glass plate would not influence the brightness, it would appear homogenously bright.

test object, through the objective to the camera sensor. It shows the change in the propagation of light depending on the geometry of the included test object.

3.4.2
Light Color and Part Color

3.4.2.1 **Visible Light (VIS) – Monochromatic Light**
Why does light occur in a special color? Because it consists of a limited range or of a special spectrum of wavelengths.

Why does the test object occur in their specific color if it is illuminated with white light? Because it reflects only the range of light wave of their own color and absorbs or transmits the rest. A blue body, for example, reflects only the light waves of blue light and absorbs the others.

Accordingly it is most effective (brightest!) to illuminate a blue part with blue light and a red part with red light and so forth.

On the other hand there are colors – those which are best possible extinct– the so-called complementary colors. Illuminating a part with their complementary color it will appear dark (see Fig. 3.34).

Monochrome cameras have an enormous effect on the gray value produced in the imaging sensor. Knowing these connections it is possible to emphasize or suppress colors for a contrastful imaging of the test object. This is valid if the test object has a constant color (see Fig. 3.34). On the other hand the color perception of the imaging sensor influences the result (see Section 3.3.1).

Color cameras will produce a mixed color between illumination light color and part color. In this case things get difficult to predict the resulting hue, because this depends on many material specific factors.

In some Machine Vision applications the part color is permanently changing. For these parts white light is the best alternative, because it does not suppress or emphasize any colors. In that case it will be necessary for the set-

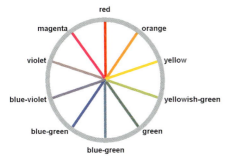

Fig. 3.34 Colors and complementary colors. On the opposite side of the circle can be read the complementary color.

tings of the software to find those tolerant parameters that make it possible to detect parts of all colors (see Fig. 3.35 and 3.36).

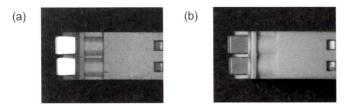

Fig. 3.35 Yellow-greenish connection block with orange terminals, (a) illuminated with red light, (b) illuminated with green light.

The color of incident light can be mixed (additive mix) using the three basic colors red, green, blue (RGB). In the illumination plane a new mixed light color is made, brighter than the initial brightnesses (see Table. 3.9).

Tab. 3.9 Additive mixed illumination colors.

Intensities Blue (%)	Green (%])	Red (%)	Additive mix color
100	100	100	White
100	100	0	Cyan
100	0	100	Magenta
0	100	100	Yellow
100	50	50	Unsaturated blue
0	50	100	Orange
50	100	0	Blue green
100	50	0	Green blue
100	0	50	Violet
50	50	0	Dark blue green

Some new imaging objectives for extremely large fields are adapted to the use of monochromatic light. Due to the built-in lenses and the color correction

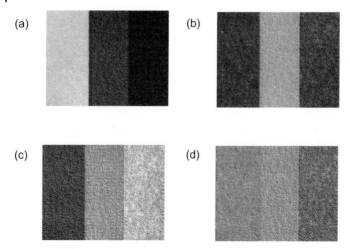

Fig. 3.36 Different color perceptions caused by different illumination colors. The color bars are red, green, blue (from left to right), (a) red illumination, (b) green illumination, (c) blue illumination, (d) white illumination.

it is necessary to use these objectives with monochromatic light to achieve best imaging results. White light would decrease the imaging power of these objectives because of chromatic aberrations.

Usually the imaging objectives of Machine Vision work with monochromatic and white light as well. Nevertheless, it should be considered that a change of the illumination wavelength can also change working distances. Particularly for large image scales (magnifications larger than 1) the illumination wavelength notably determines the working distance. This is based on the variable refraction index of the lenses inside the imaging objective that depends on the wavelength (dispersion). Shorter wavelength (UV, blue) are more refracted than longer (red, IR). This means shorter working distances for UV/blue light than for red/IR light. The consequence is to adapt focus when changing the illumination wavelength.

3.4.2.2 Visible Light (VIS) – White Light

The most natural light is white light as a melange of (all) visible light wavelengths (colors) (see Fig. 3.10). Some classical light sources produce it directly by their principle of origin.

Above all white light is used

- either for the inspection of colored parts with monochrome cameras
- or for color image processing.

Frequently white light is the only alternative to get a grip on the processing of incidental but different colored test objects. This happens typically in industrial production processes, if the part color does not play a role for the function of the part, but parts are delivered from different suppliers and the hue of the part changes caused by the compound of the raw materials.

Related to monochrome Machine Vision solutions colored light of small wavelength ranges will emphasize or tone down single colors (see Section 3.4.2.1 and Fig. 3.36). This produces high contrasts between parts of different colors.

Illumination with white light produces the as moderate as possible contrasts. To work with white light is the compromise to work with different colored test objects. This leads to find parameters for a reliable working software. If the fitting parameters could not be found it is wise to change the monochrome camera into a color camera.

In color Machine Vision applications white light is necessary to reproduce a realistic color perception for the vision system. To get reliable results in these applications, it is important to acquaint oneself with the theory of colors, a special field of optics.

Today (2005) approximately only 5–10 % of all Machine Vision applications are color applications. For the future the number of color application will be become proportionally more.

3.4.2.3 Infrared Light (IR)

Traditional Machine Vision uses IR light in the wavelength range from 780 to 1,000 nm due to the sensitivity and common use of CCD and CMOS sensors. Longer wavelengths are processed with special thermal imagers for temperature measurements.

To work with IR illumination means to use all the considerations for radiometric values and not for photometric values (see Section 3.5.1).

However, it is known that IR is not so effective to the sensor from their spectral response (see Fig. 3.10). Nevertheless, IR light is favorably used in some applications. It is not visible to the human eye and does not disturb the worker (see below). Furthermore, IR light penetrates the silicon substrate of the imager and produces blurred images due to induced charges. The consequence are images with lower contrast.

Together with a daylight suppression filter or bandpass filter IR illumination can avoid from extraneous (visible) light (see Sections 3.6.5.2 and 3.6.5.7). On the other hand is IR lighting based on LED particularly effective in fast running applications where short (down to 1 µs duration) and high energy flash light is needed.

Some details of test objects are viewable only with IR light. It is possible to see the invisible detail. Particularly, the color reproduction is not reproducible.

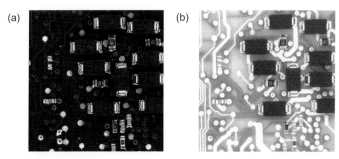

Fig. 3.37 Circuit board with dark green solder resist, illuminated in red (a) and IR, 880 nm (b). The IR light passes the solder resist almost without loss and is reflected on the surface of the supporting material of the circuit board.

Objects that are colored in white light could be transparent to IR light (see Fig.s 3.37 and 3.38). Objects that are dark in white light can reflect IR light and be bright. Rules to determine cannot be designated. Nothing is predictable. To check the color reproduction under IR illumination there is only one way and that is the experiment.

From object to object, from pigment to pigment the reflecting and absorbing properties of parts illuminated with IR light can change for some of them such as

- painted surfaces
- ink on printed materials
- color pigments in plastic materials

Natural materials interact with infrared light too. The *Wood effect* is responsible for the bright appearance of leafs, because the chlorophyll is transparent for IR light, and the infrared light is reflected at air bubbles on the underside of leafs.

Note the influence of infrared light on the used imaging objective. Not all lenses (glass sorts) are highly transparent for the IR light. Some of them reduce the passing IR light. To check this, evaluate the spectral transmission of the objective.

Most imaging objectives for Machine Vision use transmit IR light without a remarkable loss of intensity up to wavelength of 900 nm.

The interaction of the test object and the light causes diffraction at the test object edges. This appears particularly at large magnifications of the objective. The size of diffraction effects is proportional to the used wavelength. The longer the wavelength of the IR light the larger the diffraction effects (see Section 3.4.1.6).

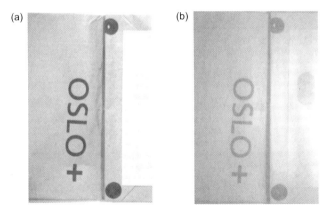

Fig. 3.38 To see the *invisible* detail: panel opening in a vacuum cleaner bag, (a) illuminated with white light, (b) illuminated with the IR light. It is to recognize how the IR light make the contrast of a green printing worse.

Some information for eye protection. Due to the enormous radiation power of LED, nowadays there are some rules to be considered when using the IR light. Preliminary remark: IR light and much more intense than LEDs can produce, is part of the spectrum of the sunlight. It can be found only in combination with the visible light.

But LED lighting are often used in relatively dark factory halls which means that the iris of the human eye is open. Incident isolated IR light from a lighting does not close it like visible light it does. So the full radiation can pass and reach the retina. This can cause eye damages. For the safety of all persons working in the environment of IR lighting it is best to switch on and off IR lighting only for the time it is needed. So nobody is exposed to IR light durably.

The standard EN 171 gives further information about eye protection.

3.4.2.4 Ultraviolet Light (UV)

Ultraviolet light on the opposite side of the spectrum uses wavelength below 380 nm. Machine Vision applications with UV light were made relatively seldom in the past. The light sources were unwieldy, expensive and inflexible and the cameras are expensive (specialized sensors, because the conventionally used silicon is not sensitive to UV). On the other hand the contact with UV light implies some danger for the operator (human eye and skin). Current developments in the LED technology make it possible to produce small, flexible and easy controllable light sources in the range of UV-A light (typical 370 nm with a narrow half wave bandwidth of 10–12 nm). This will inspire the Machine Vision applications for the future.

Two general classes of application of UV illumination are given:

- direct UV illumination in combination with an UV sensitive camera
- fluorescence caused by illumination with UV light

UV illumination for observation with an UV sensitive camera needs special sensors. These sensors use a fluorescence layer at the front glass of the imager to convert the UV light into the visible light (for example, a Lumogen conversion layer converts wavelength from 190 to 380 nm into visible light). Disadvantages of these sensors (layers) are aging and high costs.

Some kind of test objects produce themselves fluorescence under an UV light illumination. This effect converts the incoming UV light on the object (short wavelength < 380 nm) into an emission of visible light from the part (longer wavelength > 380 nm). This effect is known from fluorescence paint illuminated with *black light* or whitener in washing powder or paper illuminated with daylight (with quantity of UV light).

For the color rendering similar relations are valid as known from IR light. Only experiments can give information about the suitability of UV light for the concrete test object.

Note the influence of the used imaging objectives in interaction with ultraviolet light. Not all lenses (glass sorts) are transparent for UV light. Some of them block the UV light. To check this, evaluate the spectral transmission of the objective.

However, consider the protection measures for the human eye and skin using UV too.

3.4.2.5 Polarized Light

The effects of polarization can be explained only by the wave character of light. Almost all light sources emit unpolarized (natural) light (LEDs, incandescent lamps, fluorescent lamps). This means that the light contains light waves that oscillate in all directions round about the normal of the propagation direction (see Fig. 3.39). Only special lasers with Brewster windows can directly emit polarized light.

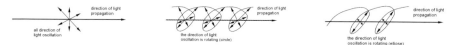

Fig. 3.39 Unpolarized, linear, circular, and elliptical polarized light..

Polarization occurs only on smooth surfaces. On these surfaces, polarization can be produced. The degree of polarization depends on the material and on the incident angle of light for reflection. On the other hand change transmitting materials the polarization (test objects, light filters).

Different types of polarization can occur: linear, circular, elliptical (see Fig. 3.39). Machine Vision typically uses two types of application of polarization:

1. polarized illumination to take advantage of the polarizing properties of the test object in combination with a polarizing light filter in front of the imaging objective (see Fig 3.40)

2. suppression of reflexes from unpolarized illumination reflected from the object using only one polarizing light filter in front of the imaging objective

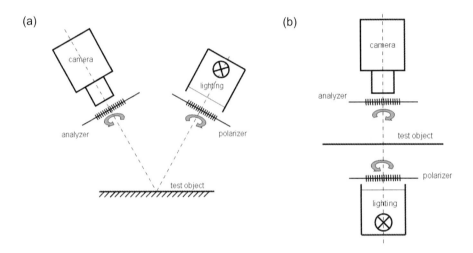

Fig. 3.40 Principle of polarizer and analyzer: (a) top light application and (b) backlight application.

Polarized illumination can be achieved using a polarization filter (linear, circular or elliptical) in front of the light source (see Fig. 3.40). These light filters act as the *polarizer* and produce polarized light (possible in white, monochromatic, band of wavelengths). Not all wavelengths can be polarized by commercial polarizers. For the IR light it is a problem.

After leaving the lighting interacts the polarized light with the test object (transmission or reflection) that changes the polarizing properties in relation to the incoming light. To grasp this in front of the camera objective is mounted another polarizing filter, the *analyzer*. Depending on the rotation of this filter the incoming polarized light from the test object is analyzed (see Fig. 3.40 and 3.41).

The removal of surface reflexes from some materials for example glass, metal, plastic, etc. can be made using unpolarized light with only one polar-

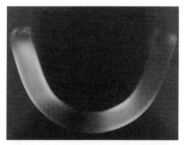

Fig. 3.41 Example for polarization with transmission: the Glass handle of a cup. The transparent glass part between two crossed linear polarizing filters change the polarization direction as a result of stretched and pressed particles inside the glass. Mechanical tension becomes viewable with the principle of tension optics.

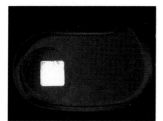 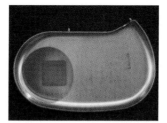

Fig. 3.42 Polarized incident light in combination with polarizing filter in front of the camera. Left: the transparent plastic label inside a stainless steel housing of a pacemaker is viewable because it turns the polarization different from the steel surface. Right: in comparison the same part without polarized light.

ization filter in front of the imaging objective. The surface of the test object polarizes depending on the refraction index of the irradiated material. It is possible to analyze it with a polarization filter (see Fig. 3.40(a) and Fig. 3.42). Under some conditions (material, incident/reflecting angle of light, wave length), it is possible to prevent from disturbing reflections (see Fig. 3.43).

Smaller polarization effects occur on surfaces of electrically conductive materials such as shiny metal parts. Among other things, the differences of polarization effects between conductive and nonconductive materials are usable to suppress the influence of oil films (corrosion protection on metal parts) for the nontactile measurement with Machine Vision.

It goes beyond the scope of this chapter to explain more about polarization. It is a much more complex effect and can be basically looked up from the optical literature.

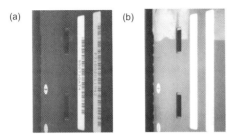

Fig. 3.43 Especially electrically nonconducting materials tend to larger polarization effects: barcode reading of lacquered batteries, (a) with polarization filter, (b) without polarization filter.

3.5
Basic Rules and Laws of Light Distribution

To work effectively with light means to know basically a few rules and laws. This makes possible to selectively control the light and maximizes the effects of the used illumination. This chapter gives a short overview of some influencing factors.

3.5.1
Basic Physical Quantities of Light

To describe the properties of light and light distribution one can generally use the radiometric quantities. This covers the whole range of the light from 15 nm to 800 μm wavelength. If the light is evaluated by a receiver (sensor, eye) with a characteristic spectral response (for example $V(\lambda)$ – curve with sensitivity from 380 nm to 780 nm (see Section 3.3.2) – one can use the photometric quantities. This paper pays attention only to some of the photometric quantities that represent the visible range of light. The UV and IR light are not considered.

Origin of all photometrics related to the International System of Units (SI) is the luminous intensity with the unit candela (cd) (see Fig. 3.44). Today one candela is defined as the luminous intensity that a monochromatic light source with $f = 540$ THz (555 nm (green light)) emits (into all directions) with a radiant intensity of $1/683$ W sr^{-1} (Watts per steradian) (see Fig. 3.45). To illustrate a practical comparison, one candela approximately corresponds to the luminous intensity of one candle light.

Characteristics mostly used for the power of lighting components in Machine Vision are

- the illuminance E for the incident light, ring lights, dark field ring lights

3 Lighting in Machine Vision

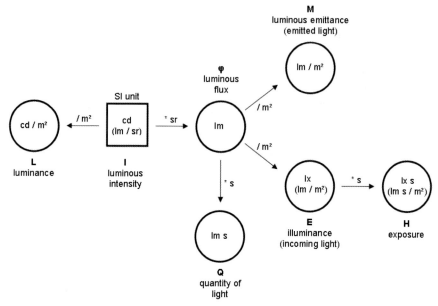

Fig. 3.44 Schematic connection between SI unit luminous intensity and other photometric quantities of lighting engineering. More about lighting units and basics can be found in [10, 11].

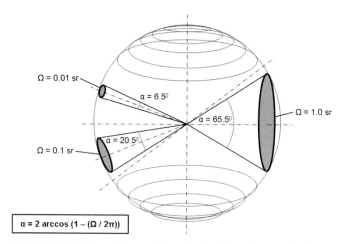

Fig. 3.45 Definition of the solid angle Ω with unit steradian (sr). A light source that emits light into all directions (full sphere) covers 4π sr, a half sphere $2\pi 2$ sr. One steradian (sr) solid angle includes a cone with 65.5° plane angle.

- the luminance L for backlights such as diffuse backlights or luminous objects.

3.5.2
The Photometric Inverse Square Law

Everyone who is working with top light has to do with the photometric inverse square law. The statement of this law is that the illuminance E on the surface of the test object decreases with the reciprocal of the square of the distance d (see Fig. 3.46).

$$E \sim 1/d^2$$

with

E being illuminance at the object and
d the distance lighting – object

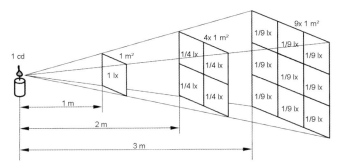

Fig. 3.46 Course of the illuminance at the object depends on the distance of the lighting due to the photometric inverse square law.

In a practical usable approximation is $E = I/d^2$ [3] with I-luminous intensity of the light source. Note that the distance between the test object and the observing camera with objective does not influence the brightness at the image sensor of the camera! This distance only influences optical parameters of the image.

The connection between illuminance and distance is exactly valid only for a point-shaped light source. In approximation it can also be used for two-dimensional light sources (see Table 3.10).

An approximation example: an incident area lighting illuminates the surface of a test object with an illuminance of 10,000 lx at a distance of 100 mm. The distance of the lighting component is doubled to 200 mm. The consequence is that the illuminance at the test objects surface decreases down to 2,500 lx. That is only caused by the change of distance.

But this is only the half truth. Not all the incoming light on the part surface reaches the sensor too. The reflectance, absorbance and transmittance at the

Tab. 3.10 Relative error Δ for a circular Lambert radiator (d – distance lighting – object; r – radius of the light source (radiator)).

r/d	Δ in %
3	10
5	3.8
10	1
15	0.44
30	0.1

object (see Section 3.4) reduce the available light. Furthermore they reduce the light path through the imaging objective (see below) and the sensitivity of the image sensor the available light for the image sensor.

Tip

> To minimize the impact of the photometric inverse square law, try to mount the lighting components for top light applications as near as possible to the test object. Note the effects caused by the brightness distribution of the lighting.

The practical consequences of this effect are far-reaching. Caused by the mechanical construction of the machine or environmental limitations it is not always possible to mount the lighting as near as possible to the test object. Consequences for the practical application can be found in Section 3.8.3.1.

Dynamical effects happen if the distance between the test object and lighting changes from part to part. This occurs due to handling and feed equipment irregularities (for example part on a conveyor belt) and leads to changing brightness in the image. These effects happen for the incident light. Robust algorithms in the image processing software have to cope with such changes in brightness from part to part.

3.5.3
The Constancy of Luminance

The incoming light from a light source generates an illuminance E_{ob} on the surface of the test object. Multiplied with the reflectance ρ of the object results in a luminance of the object L_{ob}. The object shines with the luminance L_{ob} (see Fig. 3.47):

$$L_{ob} = E_{ob} \cdot \rho$$

where

L_{ob} is the resulting luminance of the object,
E_{ob} is the incoming illuminance at the object and
ρ is the reflectance of the object.

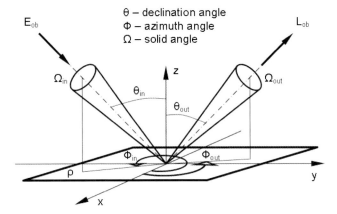

Fig. 3.47 Conversion of illuminance into luminance [12].

The characteristic value of a backlight is the luminance L too. It shines itself with a luminance. The consequence of the above equation means this is independent of the distance.

In contrast to the illuminance is the luminance not dependent on a distance. The luminance is constant in the optical pass of the vision system.

Practically this means

- for top light applications with fixed distance lighting – test object:
  No changes of brightness at the test object, neither from short or long camera distances, nor from short or long focal distances of the imaging objective.

- for backlight applications (see Fig. 3.48):
  No change of brightness of the lighting independent of the distance from the object to the imaging objective.

3.5.4
What Light Arrives at the Sensor – Light Transmission Through the Lens

The following viewings are made without considering the natural vignetting (*cosine to the power of 4-law*) of the imaging objective. A homogenous light distribution at the image sensor plane is assumed. To simplify the relations, a perpendicular view to the object and a relative long working distance in relation to the field of view is supposed.

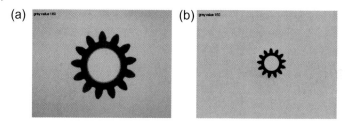

Fig. 3.48 Identical luminance (gray value) on the part surface, (a) Imaged with a short working distance of the objective, (b) imaged with a long working distance.

With some approximations [3] is the illuminance E_{sensor}:

$$E_{sensor} = E_{ob} \cdot \rho \cdot \frac{\tau}{4k^2}$$

with
- E_{ob} the incoming illuminance at the object,
- ρ the reflectance of the object,
- τ the transmittance of light through the objective, (medium value $\tau \sim 0,9$ for an average imaging objective with anti-reflective coating) and
- k the f-stop number at the imaging objective.

The illuminance E_{sensor} arriving at the image sensor is not dependent on any distances and can be directly controlled by the f-stop of the imaging objective and the illuminance of the used lighting. Imaging objectives with anti-reflective coatings can help to increase the throughput of light through the lenses (transmission τ). The following example shows how a camera can be checked for the usability with regard to the sensor sensitivity.

3.3 Example An incident lighting has an illuminance of 10,000 lx at a distance of 100 mm according to the values written in the data sheet. The lighting is mounted in a distance of 200 mm. This means on the surface of the test object come in 2,500 lx (see Section 3.5.2). Let us say that the test object is made of aluminum. The reflectance (see Table 3.8) is approximately $\rho = 0.6$. According to an estimated transmittance of the imaging objective of $\tau = 0.9$ (anti-reflex coated objective) and a f-stop setting of $k = 8$ (read from the imaging objective) all values for the approximate calculation are given.

Using the above equation the illuminance that arrives at the sensor is 5.27 lx. The intended fictitious camera has a minimal sensitivity of 0.3 lx (value from the data sheet of the camera). This means that there is about 17 times more light than minimum necessary. Under these conditions should be the lighting right dimensioned.

Tip

> The lighting should produce minimum a 10 times higher illuminance at the sensor than the documented minimal sensitivity from the data sheet of the imaging sensor/camera. This ensures a reliable function even under changing industrial conditions.

In some data sheets the information about the minimal sensitivity is mentioned using the value of the *exposure H*. To get the needed illuminance it is necessary to divide this value by the exposure time (integration time, shutter time) to get the minimal illuminance at the imaging sensor.

The viewings in this paragraph are made only approximately. For an exact view following should be considered:

illuminance from lighting, f-stop number, wavelength of lighting, incident angle of light, type of lighting (diffuse, direct, telecentric, structured), reflectance of the test object, transmittance of the imaging objective, perspective characteristics of the used objective (entocentric, telecentric, hypercentric), inserted light filters (in the camera too) and so forth.

3.5.5
Light Distribution of Lighting Components

The distribution of illuminance and luminance in the object plane plays a basic role for the analysis with image processing algorithms.

The conventional aim is to achieve a homogenous, temporal and spatial constant lighting or at least a predictable and constant distribution of brightness (structured lighting). Such conditions make the software easier, and the application more reliable and stable.

Different types of lighting, different models and different lighting techniques produce different characteristic light distribution and brightness profiles. But even for one lighting component can the light distribution change with the distance, incident angle and direction. That is why data sheets can only tell a part of the whole light characteristics.

The reasons for inhomogenous lighting can be caused by

- the principle of the lighting technique
- the quality of the lighting component (low-cost components tend to poor light distribution)
- wrong application of the lighting
- strongly changing properties of the test object

The graphical expression of luminance distribution depending on the viewing angle is the luminance indicatrix. It shows the connection between the luminance and observation direction. For a Lambert radiator looks the luminance indicatrix like a half sphere [2]. Other lighting components possess directed radiation characteristics in some other shapes.

Backlights are often characterized only with a perpendicular view to the light emitting surface. The distribution of the luminance gives information about the homogeneity.

Incident light is usually characterized in a similar way. It shines perpendicular to a surface where the illuminance is measured (see Fig. 3.49).

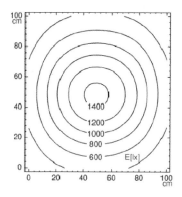

Fig. 3.49 Distribution of the illuminance in 0.5 m distance of a commercial fluorescence light source with two compact radiators of each 11 W (source: www.vision-control.com).

Other light sources with special shapes produce characteristic illumination profiles independent of the distance to the test object. Some typical brightness profiles are shown in Fig. 3.50:

Avoid to prove this directly with the vision system. Many influencing factors are added to this case. Especially the natural vignetting of the imaging objective will falsify the result. These measurements can be made only with specialized instruments.

It is to be seen that the light distribution of light sources can be totally different depending on the selected light source and the desired purpose. The art is to find the light source that emphasizes on the necessary details best as possible. In addition, the light distribution can differ for various designs. One model of a ring light with a constant diameter and working distance lights up totally different with fresnel lens, with a diffuser or without an additional element. All these factors influence the lighting result.

The drop of brightness caused by the lighting (see Fig. 3.50) can cause software problems. A decrease of 40% means a drop from a gray value of 230 to a gray value of 138. Is the software able to tolerate so large differences? Some (a)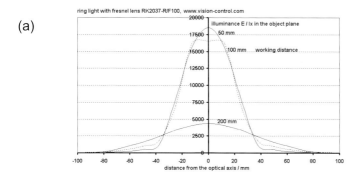

(b)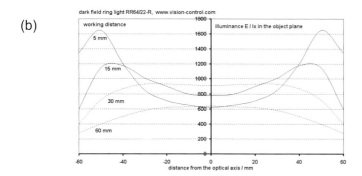

(c)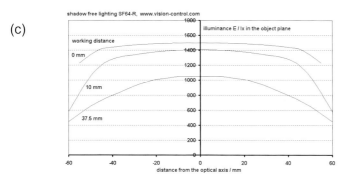

Fig. 3.50 Exemplary distributions of illuminances, (a) ring light with fresnel lens, (b) dark field ring light, (c) shadow-free lighting (source: www.vision-control.com).

inhomogeneities of lighting can be compensated using the shading correction function of the image processing software.

Sometimes a deliberated inhomogenous light distribution is used to compensate other effects like the natural vignetting from the optics. This decrease of illuminance toward the edges of the image can be compensated in a plane, for example, with a dark field ring light that is brighter in the outer illuminated areas (see Figs. 3.50 and 3.51).

Fig. 3.51 Brightness profile of a lighting to compensate natural vignetting of an imaging objective.

Some new possibilities in technology (small discrete light sources, progress in information technology) permit us to control the well-defined distribution of light. These so-called *adaptive lighting* allow us to control each single light source (LED) in intensity and time, even with changing distribution in real time. This makes possible to compensate inhomogeneities in a very sensitive way (see Section 3.8.2.5).

3.5.6
Contrast

All Machine Vision applications are living from the contrast. A good contrast at the object and the image sensor respectively is the base that the algorithms of the image processing software can work. No contrast means no usable information. The contrast is the direct consequence of lighting.

Contrast in the classical definition stands in the context of the human perception and is related to absolute values of luminances:

$$K = (L_{max} - L_{min})/(L_{max} + L_{min})$$

with
- K being the contrast for human perception/interpretation,
- L_{max} the maximal luminance at the object (top light), or at the lighting (backlight) and
- L_{min} the minimal luminance at the object.

Some simple calculations show that the contrast is higher if the luminances found are larger. In comparison, the contrast in images for digital interpretation K_{dig} is defined by the absolute gray value difference within the image (area of interest/relevant image area):

$$K_{dig} = G_{max} - G_{min}$$

with
K_{dig} being the contrast in digital images/gray value difference,
G_{max} the maximal gray value and
G_{min} the minimal gray value.

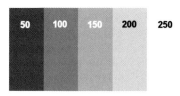

Fig. 3.52 Gray bars with constant gray value differences of 50. That means a constant contrast K_{dig} of 50 from bar to bar. Note the seemingly decreasing contrast from left to right at assessment with the human eye.

Table 3.11 shows the comparison of contrasts K and K_{dig} (values from gray values taken as luminance values) from Fig. 3.52.

Tab. 3.11 Contrast comparison of K and K_{dig}. The values for K correspond with their decreasing course to the human perception.

Transition from gray value to gray value		K	K_{dig}
50	100	0.33	50
100	150	0.2	50
150	200	0.14	50
200	250	0.11	50

The better the contrast, the more reliable are the results of the software (see Fig. 3.53). There is no rule to predict which contrast is necessary, because it depends on the function and the built-in algorithms. Some algorithms (gradient algorithm for edge detection) are known that can still work with contrasts of 3 (from 255) gray values. Some others need gray value differences of more than 30.

Tip

Image processing algorithms work more precisely at contrastful images. Also the processing time decreases for a few algorithms if they can work on contrastful images. And not at least is the processing of contrastful images more reliable.

All efforts to maximize the contrast should be considered to avoid from saturation. The maximal brightness that is found in the image (if necessary

Fig. 3.53 Images of one part: (a) imaged with poor contrast and (b) imaged with strong contrast.

only in a relevant area, see Fig. 3.54) should be not more than 90 % of the maximum value. For a camera with 8 bit gray value resolution means that a maximal gray value of 230. At gray values of 255 (for 8 bit), the image information is lost and no result can be calculated. Depending on the oversize of light and sensor principle even the image information in the neighbourhood can be destroyed by flooding charges (for a CCD).

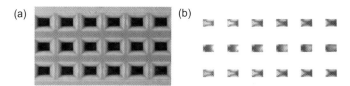

Fig. 3.54 Connector housing with inside contact springs, (a) without saturation. There is nothing to recognize in the holes, (b) local overexposure of the housing makes the connector springs viewable.

3.5.7
Exposure

The exposure forms the connection between lighting, time and camera hardware. The almost general goal will be to generate images with sufficient contrast (light and dark), not over-exposed and not under-exposed. This can be achieved in different manners. The *exposure* opens different possibilities of getting the same brightness in the image. Exposure means in simplification

$$H = E \cdot t$$

with
- H being the exposure,
- E the illumination at the test object and
- t the exposure time/shutter time.

A constant exposure can be achieved with different boundary conditions. Some of them are

3.5 Basic Rules and Laws of Light Distribution

- double (half) exposure time and half (double) illuminance (power) of lighting
- open (close) one f-stop number of the objective and half (double) illuminance (power) of lighting
- insert (remove) a 2x neutral filter and double (half) illuminance (power) of lighting

All these parameters can also be used in combinations. Further allow these variations to use brighter or darker lighting and compensate this. To play with the parameters of exposure means to know more about the application. Note that changes of the exposure time have an effect on the motional blurring of moving parts. Changes of the f-stop have effects on depth of focus, aberrations and more (see Section 4 also).

Tip

Consider the boundary conditions for every application before choosing a lighting. What is possible?

- long or short exposure time (still standing/moving test object)
- small/large f-stop number (small/large depth of focus needed)
- or both

These facts will influence the needed illuminance/luminance, which means lighting power! Table 3.12 shows the example of constant exposure for all combinations of f-stop and exposure time.

Tab. 3.12 Example for constant exposure with time and f-stop variations. f-stop number and exposure time determine the possible applications. If the combination of a large f-stop number and a short exposure time is needed the only way is to increase the lighting power. Each step (one f-stop number more or halfen the exposure time) means to double the light emission.

f-stop	2	2.8	4	5.6	8	11	16
Exposure time (s)	1/3200	1/1600	1/800	1/400	1/200	1/100	1/50
Depth of focus	Small						Large
Sensitivity to movement/vibration	Small						Large

Note that the combination of f-stop number 16 together with a 1/3200 s exposure time needs a 64 times brighter lighting. This is usually limited by the limits of lighting power.

3.6
Light Filters

Light filters are optical elements that influence the light path and remove unwanted or unused parts of light. With their effect they are a bridge between optics and lighting.

3.6.1
Characteristic Values of Light Filters

Besides geometrical values such as socket thread, diameter and height of light filters are characterized by few values [19]. Here only the general values are introduced (see Fig. 3.55):

Transmission (T). Relation from passing light energy and incoming light energy (see Section 3.4.1.4). Sometimes the average transmission is indicated for a limited wavelength ranges.

Cutoff/cuton wavelength (λ_c). The wavelength specifies the location of the transition from high (low) transmission to low (high) transmission. This term is often used for specifying wavelength location of a short-wavelength/long-wavelength pass filter. The criterion of 5% absolute transmission describes the wavelength.

Half-power points (HPP). The wavelength at which a filter transmits one-half of its peak transmission. (For a bandpass with peak transmission of 70%, the HPP are at 35% transmission).

Center wavelength (λ_0). For bandpass filters can be applied $2\lambda_1\lambda_2/(\lambda_1 + \lambda_2)$:, where λ_1 and λ_2 are the bandpass half-power points (HPP). This is not the simple arithmetical average.

Bandwidth (BW). The width of a bandpass filter between specific absolute transmission points. Usually for this the half-power points (HPP) are used. Then it is called half bandwidth (HBW).

The transmission properties of light filters can be found in different descriptions (see Table 3.13):

- filter diagrams for filters with wavelength-dependent transmission
- filter prolongation factors for edge filters
- values for optical density for edge and neutral filters

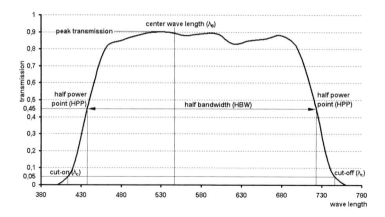

Fig. 3.55 Major characteristics of light filters.

Filter diagrams usually show in logarithmic dependence the transmission of light relative to the wavelength. They are used for wavelength sensitive filters. It has to be read from the diagram how much light the filter blocks for a defined wavelength range.

Filters without/low spectral response are described by their filter factor. This is an average value how much the exposure has to be prolonged to achieve the same brightness in the image as without the light filter. If there are several filters in use as a stack the densities have to be multiplied to get the total density.

A third description of the loss of intensity of light filters is the optical density D. It is the logarithmic reciprocal of the transmission T. The definition is $D = \log 1/T$.

Tab. 3.13 Light transmission through filters: connection between filter factor, optical density and f-stop-difference of the lens.

Light transmission T	Filter factor for exposure time (without change of f-stop no.)	Optical density D	No. of f-stops to open to compensate intensity difference (without change of exposure time)
50% or 0.5	2	0.3	1
25% or 0.25	4	0.6	2
12.5% or 0.125	8	0.9	3
6.25% or 0.0625	16	1.2	4
...	...	...	...
0.1% or 0.00962	1000(1024)	3.0	10

Filters can be used in stacks. If more than one filter is used the filter factors are multiplied or the optical densities are added.

3.4 Example Stack of two light filters with 50% and 25% transmittance.
Calculation with filter factors 2 and 4 → $2 \times 4 = 8$
Calculation with optical density 0.3 and 0.6 → $0.3 + 0.6 = 0.9$

3.6.2
Influences of Light Filters on the Optical Path

All light filters are active optical elements and act like a plain parallel glass plate and lengthen the optical path length. The consequences are longer working distances of the imaging objective. This effect can be particularly strong depending on the place where the light filter is inserted/removed.

Rule of the thumb
As an approximation for an average optical glass (refraction index $n = 1.5$) applies:

> *Filter in front of lens:* increase of working distance = Glass thickness/3
> *Filter in front of sensor:* increase of working distance
> $\qquad\qquad\qquad\qquad$ = Glass thickness/$3\beta^2$

with β being the image scale of the used imaging objective [6].

Tip

> Consider the change of the working distance if a light filter is removed (IR cutting filter from a camera for an application with IR lighting) or inserted (daylight cutting filter to suppress ambient light). This can cause strong effects to the construction/distances of the machine where the camera is built-in.

All inserted light filters not only change the optical path, but also cause artificial astigmatism (aberration). A light filter that is tilted in relation to the optical axis additionally causes a parallel offset of the optical axis because of the refraction on the glass plate surfaces (see Fig. 3.56). For large image scales this can be significant. The connection is for small angles (the paraxial area) [6]:

$$v = \frac{n-1}{n} \cdot d \cdot \varepsilon$$

with
$\quad v \quad$ being the parallel offset,
$\quad n \quad$ the refractive index of the filter,
$\quad d \quad$ the thickness of the filter and
$\quad \varepsilon \quad$ the twisting angle of the filter

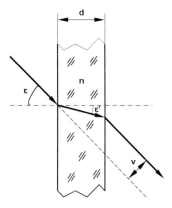

Fig. 3.56 Parallel offset of the optical axis caused by a tilted light filter.

3.6.3
Types of Light Filters

Light filters can be divided into three main classes by their principle of extinction:

Absorbing filters [14]:

- extinct wavelength ranges by absorbing light

- can be made of colored glass, with a gelatine layer between two transparent glass plates or plastic plates (limited optical quality! – streaks, different thicknesses, etc.)

- provide wide bands of transmission (not so exact)

- are available in many hues and made from standard optical glasses

- can be made of fluorescent dyes too (conversion: *new colors for old*)

- are cost effective, mounted in sockets with standardized threads and made from many manufacturers

Interference filters [20]:

- extinct wavelength ranges by interference of light
- use thin metal or dielectric films ($\lambda/4$ thickness) on transparent glass plates
- provide narrow bands of transmission (very accurate)
- are working very effectively. But, the smaller the bandwidth the larger the loss of transmission even in the pass range: 5 nm bandwidth means 70% loss of transmission, 10 nm: 50%, 20 nm: 10%.
- available for many center wavelengths in small intervals
- are expensive and usually not easy to handle for Machine Vision (square glass plates without socket)

Polarization filters

- extinct light with defined directions of oscillation. Light with defined oscillation directions can pass.
- are made of crystal or synthetic materials that are stretched (to align the molecules) to get the directional properties
- available as thin films, glass filters. Plastic plates are more suitable for large areas (only to use in the lighting light path)
- work relatively independent of wavelength within a wide bandwidth
- mounted in sockets with standardized threads
- are turnable to adjust the polarization

Absorbing and interference light filters can be divided into four main classes by their range of extinction of wavelengths [19]:

Short-wavelength pass filters (SWP). Filters that transmit light at wavelengths shorter than the cut-off. Light at wavelengths longer than the cut-off wavelength is attenuated.

Long-wavelength pass filters (LWP). Filters that transmit light at wavelengths longer than the cut-on. Light at wavelengths shorter than the cut-on is attenuated.

Bandpass filters (BP). Filters that transmit light energy only within a selected band of wavelengths. Two classes are given: narrow or wide band-pass filters. A detailed description of function is written in [17].

Neutral density filters. Filters that are relatively little selective to a wavelength range and serve to suppress a defined amount of light. They are characterized by their optical density (see Table 3.13).

Note that some light filters have a limited wavelength range where they work.

3.6.4
Anti-Reflective Coatings (AR)

Every transition of light between two media with different refractive indices such as air and glass causes a loss of light energy. The incident light is reflected from each uncoated average glass surface with an approximate reflectance of 4% (exactly that depends on wavelength, incident angle of light, etc.) (see Fig. 3.57).

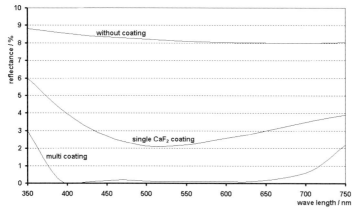

Fig. 3.57 Typical course of reflectances of an uncoated, single coated (CaF_2 coating) and multi coated glass plate (average optical glass, transitions air–glass–air) in dependence of the wavelength [14].

For a filter glass plate this means loss of intensity too. Simultaneously this is connected with a worsening of the image quality (loss of contrast, milky images, ghost images) caused by multiple reflections inside the glass (see Fig. 3.24, backside reflections).

This fact is significant for single lenses, complete objectives and for light filters too. That is why qualitatively better light filters are coated with anti-reflective coatings made from dielectric thin films of $\lambda/2$ thickness. Single or multi layer coatings (metal oxides, CaF_2) are applied to improve the image quality.

Caused by the necessary thickness of $\lambda/2$ for the anti-reflective coating, it is necessary to use different anti-reflective layers for a high qualitative dereflection. Each of them is applied for one narrow wavelength band. The so-called multi coated lenses and filters are expensive. A good compromise for Machine Vision is the use of single layer coatings made from calcium fluoride (CaF_2) (see Table 3.14).

Concerning the employment of light filters as the front element and interface to the industrial environment in Machine Vision applications there are needs for the durability. Besides their optical function, the coatings have to

- protect the glass surface of the filter/lens
- be nonabrasive
- be noncorroding.

3.6.5
Light Filters for Machine Vision

3.6.5.1 UV Blocking Filter

This long-wavelength pass filter blocks the UV part of the light spectrum. Since the most image sensors of Machine Vision (CCD or CMOS) are not sensitive in the UV range of light they almost does not influence the image brightness. The visible light can pass more than 90% means there is no noticeable prolonging factor. The cut-on wavelength depends on the filter model and is typically about 380 nm.

UV blocking filters (see Fig. 3.58) are used as

- protective screens against dirt
- protection of the front lens against cleaning with polluted clothes
- as mechanical barriers against unintentional strokes

Tip

> Use generally an UV blocking filter as a seal of the imaging objective. It is const effectively and protects the costly lens from the rough environmental conditions of the industrial floor.

3.6.5.2 Daylight Suppression Filter

This long-wavelength pass filter blocks the visible light of the spectrum or their parts. Near-infrared light can pass. Since most of the image sensors of Machine Vision (CCD or CMOS) are sensitive in the range of near-infrared light, only this part of light can be used when applying a daylight suppression filter. The cut-on wavelength is model specific. Typical for Machine Vision are cut-on wavelength from about 780 nm.

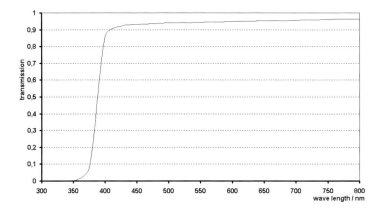

Fig. 3.58 Typical transmission curve of an UV blocking filter [14].

Daylight suppression filters (see Fig. 3.59) are used

- for IR lighting applications
- for independence from extraneous light (lighting of the factory hall) in combination with an IR lighting (needs coordination of cut-on wavelength and wavelength of the lighting)

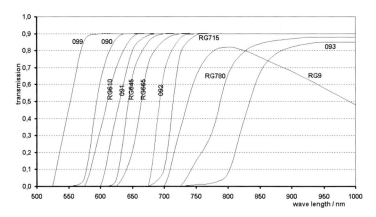

Fig. 3.59 Typical transmission curves of different filter glass sorts of daylight suppression filters [13].

Note: remove the IR suppression filter from the camera before using a daylight suppression filter! Otherwise no light can pass to the image sensor, because both filter characteristics overlap.

If sunlight is part of the ambient light the effect of daylight suppression filters can be strongly limited, because the incoming sunlight contains quantities of the IR light too. These quantities can be partially larger than the IR-lighting itself (typically LED-lighting with their narrow bandwidth of light emission). In these cases there will be no effect of daylight suppression filters. The use of the filter is practical in factory halls with artificial light.

For a better suppression of sunlight it is recommended to use a filter combination (see Section 3.6.5.7).

3.6.5.3 IR Suppression Filter

This wide band pass filter let pass the visible light and blocks the UV light and IR light alike. Many Machine Vision cameras contain these filters to improve the image quality for the visible light, because IR light penetrates deep into the silicon substrate of the image sensors and causes blurred and contrastless images.

Usually the IR suppression filters are separate filters and removable from the camera (reddish green coated glass in front of the sensor chip) for applications with IR light. In some cases the cover glass of the sensor chip is directly coated with the IR suppression layer.

IR suppression filters are in use not only for monochrome cameras but also for color cameras. They ensure that the IR light does not have influence in a different manner on the brightness of the single color pixels (RGB). Thus, the brightness offset for the whole color triples is the same.

IR suppression filters are used

- to achieve contrastful and sharp images (in visible light)
- to suppress uncontrolled influence of the IR light
- to remove brightness offset for color pixels in color cameras.

Note: for IR applications the IR suppression filter must be removed! An application with IR lighting is impossible, if the cover glass of the sensor chip is coated with the IR suppression filter layer (see Fig. 3.60). Furthermore the removal of the IR suppression filter causes changes in the working distance of the imaging objective (see Section 3.6.2).

3.6.5.4 Neutral Filter/Neutral Density Filter/Gray Filter

These unselective light filters reduce the luminous flux (relatively) independent of the wavelengths used. The degree of reduction is specified by the density or the prolongation factor. They are available in different densities (usually with prolongation factors from 2 to 1,000) (see above). Neutral filters are used

- if there is too much light (too bright light source)

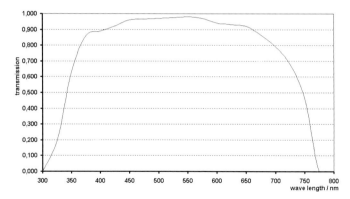

Fig. 3.60 Typical transmission curves of an IR suppression filter glass used for Machine Vision [14].

- to reduce the light intensity without change of the f-stop number (for objectives with fix f-stop respectively with no change of depth of focus!)
- to reduce the light intensity without change of exposure time (no change of motional blurring)

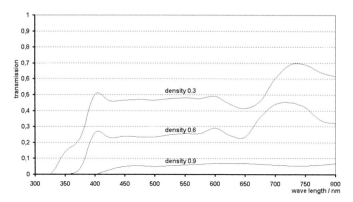

Fig. 3.61 Typical transmission curves of neutral filters with different densities [14].

3.6.5.5 Polarization Filter

Polarization filters are color neutral. They extract light components with special polarization properties: linear, circular or elliptical polarized light (see Section 3.4.2.5). To achieve the right polarization it is necessary to turn (and fix!) the filter in the socket. Typical filter prolongation factors are between 2.3 and 2.8 depending on the angle of rotation. Caused by the used polarizing materials many polarization filters do not work in IR light – see the data sheet! Polarization filters are used

- to suppress reflections
- for special analysis techniques (tension optics)
- for polarizing lighting components
- to adjust infinitely variable brightness (two stacked linear polarizing filters) without any change of f-stop or exposure time

3.6.5.6 Color Filters

Color filters are realized as long-wavelength pass filters or as bandpass filters. In the color that they appear to the human eye, most light energy can pass.

The use of color filters is limited to monochrome image sensors/cameras, because they distort the color information. The course of transmission of these filters is individual, dependent on the wavelength and the optical glass used.

Color filters in combination with white light result in similar effects from contrast as the use of colored light (see Section 3.4.2 cont.). The disadvantage of this construction is that the color filter reduces the radiation of the light source with the filter. It is better to use a light source that already emits effectively in the desired wavelength band.

A light filter with the same color as the part produces a bright part, and a light filter with complementary color produces a dark part (see Section 3.4.2). Color filters (see Fig. 3.59 and Fig. 3.62) are used

- inside color image sensor chips (mosaic filter) to achieve a defined and narrow spectral response of the RGB pixels
- to emphasize or suppress colored image information

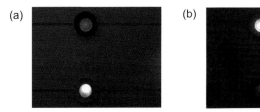

Fig. 3.62 Plausibility check of orange (top position) and green (bottom position) LEDs. Left: with green bandpass color filter. Right: with orange color glass filter.

3.6.5.7 Filter Combinations

Light filters can be stacked or combined, one on the lighting and the other in front of the imaging objective and so forth. The overall effect is to combine different filter characteristics.

As an example hereinafter the combination of a short-wavelength pass filter and a long-wavelength pass filter is explained. The combination simulates a bandpass filter.

An effective suppression of daylight for Machine Vision applications – better than with an IR suppression filter – can be achieved using the camera built-in IR suppression filter – this removes disturbing IR-light. The combination with a red color long-wavelength pass filter 090 (see Fig. 3.59) in front of the imaging objective removes the visible light with a shorter wavelength. The result is a filter characteristic of a band pass filter (see Fig. 3.63) but much cheaper.

If the selected lighting emits within the center wavelength of this band pass, an almost perfect suppression of ambient light is realized. The used lighting will have a much higher light output in this narrow band than the ambient light (usually white light as a mixture of different wavelengths). For the example described in Fig. 3.63, a red light with a 660 nm center wavelength will fit the requirements.

In addition this is particularly effective because the imaging sensors are more sensitive in this wavelength range than in the IR.

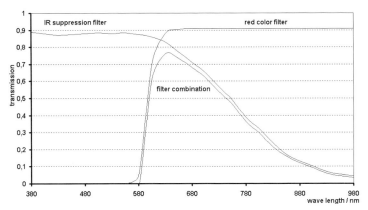

Fig. 3.63 Transmission characteristics of an IR suppression filter and a red color filter and the resulting characteristics of their combination.

3.7
Lighting Techniques and Their Use

3.7.1
How to Find a Suitable Lighting?

This will be the most important question of an engineer who has to select a right lighting set-up for the Machine Vision application. Probably he remembers some clever Machine Vision proverbs such as "better to light than write (software)", "avoid garbage in (bad lighting) that causes *garbage* out (bad result)", "create the BEST image first" and so forth. The general aim will be

- to maximize the contrast of features of interest
- to minimize the contrast of all others
- to minimize external influences.

To get an approach for the selection of lighting a general comparison of human and machine perception of visual information is made.

Human beings own a learning capable biological vision system combined with a skilled handling system with all degrees of freedom. The look of a defect is known from their term and from recollection. The part and the viewer move around until the required lighting is obtained and the feature can be seen, even under the most inadequate lighting conditions. And each part more that was inspected increases the ability to recognize (cognitive vision) – the man learns from experience.

A Machine Vision system usually owns static performance data: predetermined position for image acquisition and predetermined position of lighting, limited range of motion, predetermined functionality of the software. Under these limitations the result shall be found reliably, precisely and stably. This is the chance for lighting to compensate all those shortcomings and result in the demand to find the best possible lighting.

A good starting point for all viewings to select a lighting is the lighting perception with the human eye. Although the biological vision is not comparable with a Machine Vision system it can give first clues for lighting, even if it is known that a man cannot see some details that the machine can see and vice versa. Some exceptions are given where the first check from the eye will fail:

- too small details (not resolvable by the eye)
- lighting in a wavelength range where the eye is not sensitive (UV and IR)
- lighting with flash light or lighting of fast running parts.

Tip

> Observe the object with the human eye using different lighting techniques. What cannot be illuminated viewable for the human eye that is (usually) not viewable for the machine too.

A strategy to find a right lighting can be to think backward. Think of these questions before:

"What I am looking for?", "How the lighting is to move to produce light or shadow at the part in the right place?". "What lights or shadows are disturbing?", "Where do they come from?", "How I have to modify the lighting to emphasize or to suppress these effects?"

To get the first ideas for lighting imagine where I have to put the light that produces the reflexes in the places where I expect it. How large must be the lighting to obtain this? Keep to the basic rules of light propagation: reflection and scattering. The surface angles determine very roughly, where the light is reflected and must be positioned in order to a properly illuminated object. So you get the first approximation for the set-up.

Tip

> Not always an overall lighting (with only one or even more lighting components) is possible, because light works according to the superposition principle. Different lighting overlay. The consequence is low contrast.

That is why divide the complex lighting task into single jobs with single components. Possibly a few images have to grabbed sequentially, each with another lighting method and component. Try to optimize, what lighting are combinable without mutual influence.

Nevertheless, find a right lighting that depends on a multitude of factors. A few standard solutions are possible. However, the lighting still depends on many factors. It is a mixture of systematics, experience, and trial and error.

3.7.2
Planning the Lighting Solution – Influence Factors

Frequently the practically possible lighting set-up differs from the solution found from experiments or from the theory. It depends on many factors besides the inherent lighting characteristics of the component itself. Not each lighting technique found can be put into practice. The environment of the vision system, the machine and the interfaces are restricting and interacting. Those factors can be

The lighting component itself:

- light source properties (see Sections 3.3, 3.4, 3.5)
- power and heat generation (refrigeration necessary)
- available size of illuminated area

The properties of the test object (see Section 3.4).

Lighting environment:

- ambient/extraneous light (constant, changing)
- preferred wavelengths (for example pharmaceutical applications: IR light)
- interactions with the background (constant, changing)

Machine Vision hardware:

- possible operation mode of lighting: static, pulse or flash
- possible synchronization of lighting (trigger of flash, switch on/off, brightness control)
- possible operation mode of the camera: short time/long time shutter
- necessary intensity on the image sensor of the camera (power of lighting)
- spectral sensitivity of the image sensor of the camera (see Section 3.3)
- need of stabilized power supply for lighting

Machine Vision software:

- supported synchronization of lighting (trigger of flash, switch on/off, brightness control)
- necessary minimum gray value contrast
- necessary/tolerable edge widths
- necessary homogeneity of light
- useable/necessary algorithms

Imaging optics:

- spectral transmission of the imaging objective
- working distance of the imaging objective (determines power of lighting needed)
- f-number of the imaging objective (determines the position, where the lighting is mounted)
- use of light filters

Machine environment:

- limitations of the space available
- necessary distances, angles, directions of lighting
- vibrations that forbid the use of some light sources (for example, halogen lamps)
- cable specifications (bending-change strength)

Customer requirements:

- compliance of standards
- company internal forbidden/unwanted components like lasers
- prescribed wavelengths from special branches
- prescribed maintenance intervals/lifetime demands
- costs

All these exemplary factors demonstrate that good initial considerations and good contacts with the design engineer of the whole machine helps us to avoid nasty surprises. The design engineer has to put your lighting set-up into the action at the machine. For an early intervention, it is important to co-ordinate all actions that could have an influence on the lighting. Remember: The design engineer is the master of space and arrangement at the machine!

3.7.3
Lighting Systematics

Above all, the following prior criteria are helpful to systematize lighting techniques for the practical use (see Table 3.14 and Fig. 3.64):

- directional properties of lighting
- direction of the arrangement of lighting
- characteristics of the illuminated field

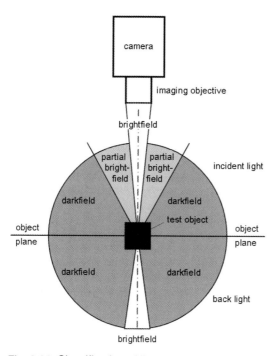

Fig. 3.64 Classification of lighting techniques: spatial arrangement of the lighting.

Tab. 3.14 Classification of lighting techniques: systematic overview. The name of the technique is to read from bottom to top.

Light direction field characteristics	Incident lighting		Transmitted lighting	
	Bright field	Dark field	Bright field	Dark field
Directional properties	Diffuse (1)	Diffuse (5)	Diffuse (7)	Diffuse (10)
	Directed (2)	Directed (6)	Directed (8)	Directed (11)
	Telecentric (3)	–	Telecentric (9)	–
	Structured (4)	–	–	–

3.7.3.1 Directional Properties of the Light

The directional properties of light give the base for the interaction of the lighting component with reflective, transmitting and scattering properties of the test object. They are divided into diffuse, directed, telecentric and structured properties.

Diffuse lighting does not have the preferred direction of light emission. The light leaves the emitting surface in each direction. Frequently the light emission obeys the rules of a Lambert radiator. This means that the luminance indicatrix of a plane light source forms a half sphere (Fig. 3.65) and the luminance is independent of the viewing direction.

Fig. 3.65 Luminance indicatrix of a Lambert radiator. Most diffuse area lighting react like a lambert radiator.

These lighting need no special precautions or preferential directions with installation. For the function it is important to achieve local homogeneity on the part. The illuminated area is directly defined by the size of the luminous field of the diffuse lighting.

The general use of diffuse lighting is to obtain even lighting conditions.

Directed lighting have radiation characteristics that can vary widely (Fig. 3.66). Already the directive characteristics of the single light sources can vary as also the characteristics of clusters of light sources. Generally produce lighting components with overlapping of multiple light sources produce a better light intensity and homogeneity.

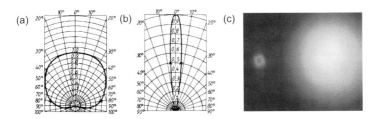

Fig. 3.66 Brightness distribution of LEDs (luminance indicatrix) with different directive properties, (a) cardioid characteristics, (b) beam shaping characteristics, (c) brightness distribution on the surface of an object. Left: from a single LED. Right: from a LED cluster.

Note for the installation that shifting and tilting of the illumination or the object has a strong influence on the brightness and contrast in the image due to the directionality.

The general use of directed lighting is to show edges and surface structures by reflection or shadowing. Even strong contrasts that are desired for lighting pre-processing can be achieved.

Telecentric lighting are special forms of directed lighting with extremely strong directional characteristics by means of an optical system in front of the light source. Due to the arrangement of the light source (typically LED with a mounted pin-hole aperture) in the focal plane of the optics they produce parallel chief rays. Telecentric lighting is not parallel lighting, as the light source is not infinitely small. Parallel, convergent and divergent light rays contribute to illumination (Fig. 3.67).

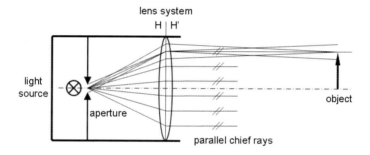

Fig. 3.67 Principle function of a telecentric lighting.

Telecentric lighting are not so sensitive against vibration and adjustment like parallel lighting. And even if they are not fully aligned with a telecentric objective the principle of telecentry still works, but the brightness distribution becomes inhomogenous. That is why it is ensured that telecentric components are stable and defined mounted.

Telecentric lighting work only in combination with telecentric objectives. If this is not taken into account, the view from an entocentric objective (objective with prespective properties) into a telecentric lighting shows only a spot. For the objective seems to be the lighting in infinity – as the name *telecentric* already says.

Though the telecentric lighting uses only one single LED it is much more brighter than a diffuse transmitted lighting that is using many LEDs. This happens because the light source of a telecentric lighting emits the light only in the direction where it is needed.

Telecentric lighting on the basis of LED produce incoherent light. This means the avoidance of speckles (intensity differences from lasers made by interferences of waves from same wavelength and same phase).

Telecentric lighting are mostly in use for transmitted light applications. Different wavelengths (light color) are

- red light for a maximum of brightness (most imagers have a maximum of sensitivity of red light)
- blue light for a maximum of accuracy (size of diffraction effects is proportional to the wavelength)
- IR light for lighting with reduced extraneous light
- IR flash for very short and bright flashes in fast processes

A specific feature besides the directional properties of light is structured lighting. Superimposed to the direction the light can carry various geometrical bright-dark structures some of which are

- single points
- grids of points (point arrays)
- single lines
- groups of parallel lines
- grids of squared lines
- single circles
- concentric rings
- single squares
- concentric squares

The methods of production of these geometrical structures are manifold. From slides, templates, masks, through LCD projectors and laser diodes with diffraction or interference gratings or intelligent adaptive lighting with LED everything is possible.

The general use of the structured light is to project the structure onto the test object. The knowledge of the light pattern and the comparison with the distorted pattern gives detailed information about the 3D structure and the topography of the part.

3.7.3.2 Arrangement of the Lighting

Incident lighting affects the test object from the same side like the imaging objective. Reference line is the optical axis of the imaging objective. From there with +/-90 degrees the range of incident light encloses. The dividing plane is the object plane (see Fig. 3.64). The function needs incident light reflections or at least scattering of the test objective for imaging. Most Machine Vision application must work with the incident light, because frequently the backlight construction is impossible by the mechanical structure of the machine.

Transmitted lighting (backlighting) is positioned at the opposite side of the object plane contrary to the imaging system (see Fig. 3.64).

Transmitted lighting creates sharp and contrastful contours for opaque or low transparent parts. Transparent parts can appear different depending on the lighting technique.

Tip

> Only a few Machine Vision applications can work with transmitted lighting. If so, do not position the part directly on the light emitting surface of the lighting. In time the surface will be scratched and scraped. Take care for a position of the lighting away from the object. This ensures that dust and dirt on the lighting surface is not imaged (out of focus) and cannot disturb the evaluation of the software.

3.7.3.3 Properties of the Illuminated Field

The *brightfield illumination* is named after the brightness appearance of the filed of view (see Fig. 3.64). The field of view is directly illuminated by the lighting – it is bright. It is bright from the incident and the reflected light of a perfect test object (see Fig. 3.68) or directly from the transmitted lighting. Disturbing structures such as defects, scratches or flaws (incident light) or the test object itself (transmitted light) appear dark. Because of the bright illuminated field are applications with bright field illumination usually

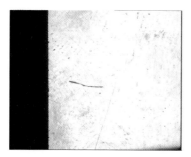

Fig. 3.68 Bright field reflections from a glass surface. The flawless surface appears bright; engraved scratches are dark.

unsensitive against extraneous light.

Partial bright field illuminations are called those illuminations that are located at the transition between bright field and dark field illumination (see Fig. 3.64). There is no clear dividing line/limiting angles between both. This determines the surface roughness of the test object. A typical clue for a partial bright field illumination is that the image appears as a bright field image and the local arrangement means a dark field lighting. The limits are given at the transition from reflecting light (bright field) to scattering light (dark field) on the surface of the test object (see Fig. 3.69). Frequently used components for partial bright field illuminations are ring lights.

Fig. 3.69 Partial brightfield illumination on a brushed part.

Darkfield illumination describes the field brightness of a flawless field of view that is not directly illuminated by the lighting. It remains dark (see Fig. 3.70). No light from the lighting is directly reflected into the imaging objective. The only light that can pass the lens is the scattered light from the surface of the test object. As a consequence homogenous objects appear dark and defects, scratches textures, dust and so forth appear bright.

Fig. 3.70 Darkfield illumination with test object, mirror bar. Only the dust corns and the grinded edges of the mirror bar scatter light.

3.7.4
The Lighting Techniques in Detail

3.7.4.1 Diffuse Bright Field Incident Light (No. 1, Tab. 3.14)
This lighting technique is used for

- homogenous lighting of weakly structured and slightly uneven little formed parts
- specular parts
- opaque/transparent parts (with well reflecting surface)
- creation of low glare
- suppression of small surface structures e.g. machining and tool marks, diffuse surfaces
- low contrast at edges
- smaller contrasts on surfaces of translucent materials

Arrangement

The lighting technique can be achieved with

- coaxial diffuse light, the so-called diffuse on axis light
- tilted camera and tilted diffuse area lighting component

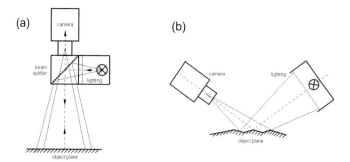

Fig. 3.71 (a) Coaxial diffuse light, (b) tilted camera and tilted diffuse light.

Coaxial light means that the light passes the same optical path like the image information through the imaging objective. It is necessary to use some kind of beam splitter (foil, plate or prism cube). Usually these components divide the transmitted and reflected light into 50/50 parts for lighting purposes.

3.7 Lighting Techniques and Their Use | 161

An inherent characteristic is the appearance of ghost images (double images) from the front and back surface and lower contrast. With some optical efforts this can be compensated.

The light efficiency of coaxial light is very poor. Caused by their principle (2 times passing the 50/50 beam splitter) less than $\frac{1}{4}$ intensity of the light source can reach the camera sensor. In reality it is much less because of the limited reflectance of the test objects. That is why this kind of lighting works fairly well only with extreme powerful light sources and/or good reflecting test objects.

The units for coaxial lighting are mounted between imaging objective and test object. They can be mounted directly in front of the lens or separately. In most cases these components reduce the working distance of the imaging objective (not, if built in the objective). Note that the necessary size of the light source increases with a larger viewing angle of the used imaging objective (see Fig. 3.71(a)). Worst case is that the field of view is not completely illuminated.

For the use of tilted lighting it is noted that the reflection law lies down where the lighting has to be positioned. On the other hand viewing angle of the imaging objective determines the size and distance of the lighting (perspective effect, see Fig. 3.71(b)). The more distant the lighting is, the larger the size of the lighting has to be. And the more powerful – note the photometric inverse square law! (see Fig. 3.72)

Fig. 3.72 The diffuse incident lighting levels differences of brightness from tooling marks on the test object surface: chip sticked on to an aluminum sheet.

Components:

- beamsplitter units with connected diffuse area lighting
- diffuse area lighting

How to select?

1. determine the light color: Note the interaction with the part color.
2. determine the lighting size
   a) the beam splitter lays down the usable lighting size by the field of view of the imaging objective
   b) approximately 5 mm wider than the field of view. For entocentric objectives note the effect of the perspective/viewing angle of the imaging objective. In this case depends the size on the distance of the lighting to the test object.
3. determine the performance of the controlling technique
4. select from catalog

3.7.4.2 Directed Bright Field Incident Light (No. 2, Tab. 3.14)

Directed bright field incident light is similar to the diffuse bright field incident light. But the directional characteristic of light leads to the use for

- even illumination of parts with diffusely reflecting surfaces
- conscious utilization of reflections
- reflectant flat surfaces and deep cavities.

Arrangement

The lighting technique can be achieved with

- coaxial directed light/directed on axis light (see Fig. 3.73(a))
- a tilted direct area lighting component (see Fig. 3.73(b))

To arrange coaxial directed light, it is necessary to ensure that the optical path of the imaging objective is almost perpendicular to the surface details that shall be bright. Deviations of angles of a few degrees are possible depending on the size of the illumination and the surface quality of the test object.

The arrangement of tilted direct area lighting components is strongly connected with the reflecting properties of the test object, light direction, height and angle.

Set-up and selection are similar to diffuse bright field incident light (see Section 3.7.4.1).

Components

- directed area lighting with and without polarizers
- special design as line lights to emphasize line-shaped information
- beamsplitter units with connected directed area lighting

Examples:

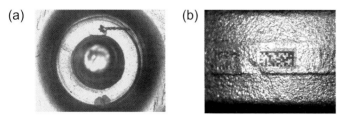

Fig. 3.73 (a) Surface check of an interference fit-pin connection with coaxial incident bright field lighting, (b) data matrix code is recognizable at rough casting parts with directed incident light.

3.7.4.3 Telecentric Bright Field Incident Light (No. 3, Tab. 3.14)

One can consider the telecentric brightfield incident light as a special and extreme case of the directed bright field incident light (see Section 3.7.4.2). It can be used only under some conditions for

- check for even very small surface imperfections and inhomogeneities of specular parts: surface defects show up clearly as dark areas.

This lighting technique only works in combination with a telecentric imaging objective.

Arrangement

The lighting technique can be achieved with

- coaxial telecentric light/telecentric on axis light(see Fig. 3.74)
- a tilted telecentric lighting component

This lighting technique needs mechanically stable conditions. The preferential direction of telecentric light causes the extreme sensitivity of this lighting technique against changes in rotation of the part. Bear in mind that if the part tilts with an angle α, referring to the reflection law the light will be deflected with the double angle 2α.

For coaxial telecentric lighting it is important to hold the reflecting surface of the test object at exactly $90°$ to the optical axis of the imaging objective.

Already a small deviation or a slightly arched test object will result in failure. The size of the tolerated angle depends on the degree of parallelity of the telecentric light and the quality of the telecentric objective. Typical tolerated angles for tilting are 0.5°–1.5°.

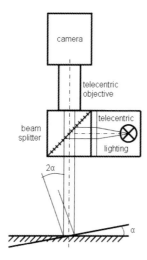

Fig. 3.74 Influence of a tilting object using coaxial telecentric bright field incident light.

Furthermore is telecentric incident light attached to the condition of smooth and shiny surfaces. Roughness destroys the telecentric light by scattering. After this diffusion the amount of reflected light toward the lens is much too small and the image is dark.

Components

- telecentric lighting
- beamsplitter units with connected telecentric lighting (see Fig. 3.75)

How to select?

1. determine light color: Note the interaction with the part color.
2. determine lighting size: The beam splitter lays down the lighting size by the field of view of the imaging objective
3. determine the performance of the controlling technique
4. select from catalog

The example for telecentric bright field incident lighting is shown in Fig. 3.76.

3.7 Lighting Techniques and Their Use | 165

Fig. 3.75 Combination of a telecentric lighting (right bottom) with a beam splitter unit (middle) and a telecentric objective (cylinder left) (source: www.vision-control.com).

Fig. 3.76 Highly specular surface with engraved characters. Only at the perfect reflecting surface parts the image is bright. Disturbing structures destroy the telecentric characteristics of light. So the characters appear dark.

3.7.4.4 Structured Bright Field Incident Light (No. 4, Tab. 3.14)

Structure lighting is used to check 3D surface information with a projected light pattern. The projected light on a nontransparent surface allows us to achieve three-dimensional information using a two-dimensional sensor. By means of light pattern one can

- check the presence of three-dimensional parts
- measure the height of one point
- measure the macroscopic height distribution along a line
- measure the microscopic roughness along a line
- measure the complete topography/flatness of a part.

The interpretation is done comparing mathematically the projected light structure with the deformed light structure at the surface topography of the object.

The discipline of structured lighting is widely split up. Many different principles are known. Common for all is that they are number crunching and need specialized algorithms for processing.

The achievable accuracy of the depth information depends on many factors such as lighting technique, camera resolution and analysing software and others. Generally, the resolution can be increased by using the approach of coded light. The principle can be found in the relevant literature.

Even transparent materials can be checked with structured light. The projection of light patterns on known shaped transparent materials produces predictable (multiple) reflections inside the materials. What is disturbing for the light path through imaging elements (see Section 3.4 – anti-reflective coatings) is useful for the point-shaped 3D – measurement of transparent parts. Derived from this it is possible to determine measures in depth like

- thicknesses of a wall
- parallelity of planes (wedge angle)
- using the effect of multiple and backside reflections

Arrangement

Conditions for the application of structured lighting are that

- the inspected parts have scattering and nonspecular surfaces (for surface measures)
- the inspected parts have reflecting back surface (for transparent materials)
- the inspected parts do not contain viewable undercuts (for inspection of the whole topography). Test objects with undercuts make it impossible to assign the projected lines from the lighting to the interpreted lines on the part.

Three basic light structures and principles are used:

0-dimensional light structure (point-shaped)

The projection of a light point leads to the principle of triangulation. The local drift of the position of the projected light point depending on the height of the test object gives a point-shaped information (see Fig. 3.77).

A set-up made with a telecentric objective takes care of simple mathematical relations using only angular functions.

For Machine Vision the triangulation plays only a subordinated role as an additional information to a complex two-dimensional measuring job. The set-up with a Machine Vision system only for a point-shaped height information is too costly.

For that purpose complete triangulation sensors are manufactured as small automation devices (triangulation head) comprising a point-shaped light source (usually laser), an imaging optics, a light sensitive receiver (position sensitive device (PSD) or CCD/CMOS sensor) and an evaluating electronics.

3.7 Lighting Techniques and Their Use | 167

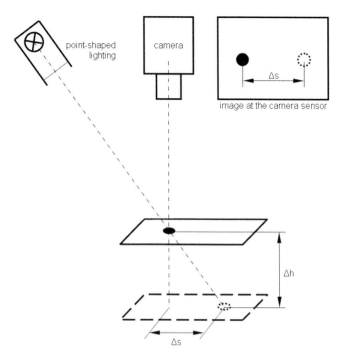

Fig. 3.77 Principle of triangulation.

The output signal can be digital or analog. The working distances and measurement ranges cover the range from millimeters up to meters. Depending on the measurement ranges and working distances accuracies down to 0.01 mm are possible.

A good height resolution of this method can be achieved with a

- short working distance
- wide basic distance/angle (distance between lighting and receiver)
- small measuring range
- highly resolving imaging sensor
- highly resolving evaluating software
- constant reflective/scattering optical properties of the test object

1-dimensional light structure (line).
The projection of a light line leads to the principle of the so-called *light-slit method*. The deformation of a straight line projected at an angle on a surface

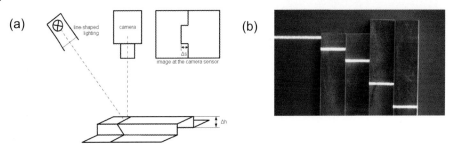

Fig. 3.78 (a) Principle of the light-slit method, (b) application of the light-slit method for the height measurement of stacked blocks.

gives information about the height along a *section* on the surface of the test object (see Fig. 3.78).

The principle was introduced from Schmaltz many decades ago for the visual measurement the microscopical surface roughnesses of parts. Today it can be used for the check of *sections* of larger parts too.

2-dimensional light structures (line-grid-shaped).
The projection of multiple parallel lines or other two-dimensional light structures allows the evaluation of complete surfaces (see Fig. 3.79). Many different procedures are known from the literature. For the calculation of the height properties of the complete surfaces (topography) complex mathematical algorithms are in use.

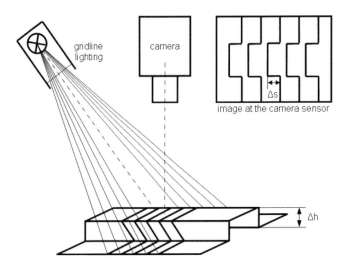

Fig. 3.79 Principle of multiple parallel lines.

Components

- diode lasers with beam shaping optics of various shapes provide almost distance-independent measures of light geometry with structure widths down to 0.05 mm.
- LCD projectors with adaptable light and intelligent sequences of light structures

3.7.4.5 Diffuse Directed Partial Bright Field Incident Light (No. 1 and 2, Tab. 3.14)

This group of lighting can be considered as a sub category of bright field incident light, because the result at the imaging sensor looks the same like this. The effects are made by the scattering surface of the part.

From the exact definition (see Section 3.7.3.3) it is an actual dark field illumination. Partial bright field lighting is used for

- homogenous illumination of three-dimensional parts
- illumination of larger parts with wavy or crushed surface
- emphasizing cracks at angled convex and concave surfaces
- selected lighting to produce shadows from edges and steps
- reduction of shadows and softening of textures on even surfaces
- minimization of the influence of fissures, dust and faults

Arrangements and components

The result of this lighting technique strongly depends on materials, positions and angles (see Fig. 3.80). With reference to the reflection law one can adapt the brightness and brightness distribution on the illuminated surface in combination of

- moving the lighting along the optical axis (changing distance)
- choosing light directions of the component (directed light, directed light with fresnel lens, diffuse light)
- choosing the light emitting diameter of the component
- changing the focal length of the imaging objective

Ring lights

The use of ring lights combined with diffuse reflecting/scattering surfaces are a common and universal lighting technique for partial bright field illumination. Depending on the parts geometry shadows can occur. To control the light distribution, models are available with direct light emission, with included fresnel lenses, with diffusors and with polarizing filters. (see Fig. 3.81).

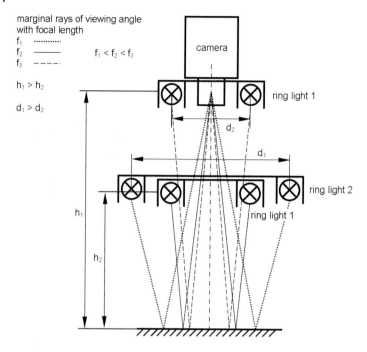

Fig. 3.80 Relationships for distances, angles and focal lengths of partial bright field components.

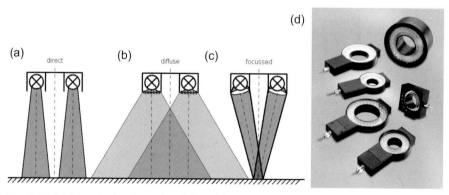

Fig. 3.81 Light emission from ring lights, (a) direct emission, (b) diffused emission, (c) focused emission, (d) different models of LED ring lights (source: www.vision-control.com).

Lighting with through-camera view
This kind of lighting is the plane extension of conventional ring lights but the emitting light surface is much larger. This ensures that even slightly crushed surfaces are illuminated homogenously. For scattering surfaces directed light

emitting models can be used and for greater uneven and moderately formed parts the diffuse light emitting models.

Typical application of lighting with through-camera view are in robotics and the packing industry (see Fig. 3.82).

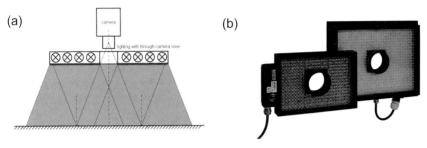

Fig. 3.82 (a) Principle of a lighting with through-camera view. (b) Lighting component with through-camera view (source: www.vision-control.com).

Superdiffuse ring lights and shadow-free lighting

The most extreme case of a diffuse incident lighting is a shadow-free lighting. These lighting components are not only extended in the plane but also in depth. This makes sure that structures in depth are illuminated too. The light emitting area is represented from almost the whole half-space of incident light (see Fig. 3.64). This can be achieved by an emitting cylinder with a special reflecting covering area or with a half-sphere construction, the so-called cloudy day illumination or a dome lighting. This provides light from all angles of the top half sphere (see Fig. 3.83).

Furthermore these lighting are characterized by a hole in the top for the camera to look through. Especially shiny parts will answer this construction with a dark spot in the center of the image (the camera observes itself). To avoid this one can combine a shadow-free lighting with a brightness adjusted coaxial incident bright field lighting (see Section 3.7.4.1) that compensate the dark spot from the camera side.

Shadow-free lighting is used for homogenous lighting of larger uneven and heavily formed and deformed parts. There is no hard influence of reflection or scattering: the overlay of the macrostructure (the shape of the object) and microstructure (surface imperfections) superimposes and the homogeneity increases more with every diffuse reflection. Almost each object can be illuminated homogenously. This eliminates shades and reflections, smoothes the textures and diminishes the influence of dust, reliefs and curvatures.

This kind of lighting technique needs to be very close to the object in order to the function based on their principle.

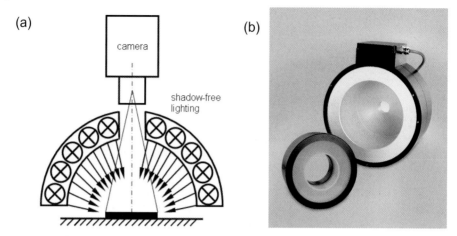

Fig. 3.83 (a) Principle of a shadow-free lighting, (b) superdiffuse ring light and shadow-free lighting component (source: www.vision-control.com).

How to select?

1. preliminary selection of ring light, lighting with through-camera view, superdiffuse or shadow-free lighting

2. determine light color: Note the interaction with the part color.

3. determine lighting size

    a) ring lights, lighting with through-camera view, superdiffuse lighting: Note the reflections (micro and macro). Where must be the lighting positioned that the reflected light meets the viewing angle of the imaging objective? Take the whole light emitting area into account (see Fig. 3.80).

    b) shadow-free lighting: Ensure that a very short working distance is possible. Choose the diameter of lighting no larger than the half of the cylinder or half sphere. Prefer longer focal length for better homogeneity on the edges of the field of view.

4. determine the performance of the controlling technique

5. select from catalog

The examples are shown in Figs. 3.84 and 3.85.

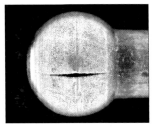

Fig. 3.84 Detection of cracks in a forged shiny ball joint for steering systems. The use of a shadow-free lighting is the only way to check these safety relevant parts with a step-by-step rotation of only three steps of 120°.

(a)  (b)

Fig. 3.85 Reading characters on knitted surfaces, (a) conventional diffuse lighting with low and strongly changing contrasts, (b) shadow-free lighting ensures a contrastful and homogenous image.

3.7.4.6 Diffuse/Directed Dark Field Incident Light (No. 5 and 6, Tab. 3.14)

Dark field incident light is created if the lighting is positioned in a place that the imaging objective cannot see. A perfect even surface will appear dark and only elevations and deepenings appear bright due to scattering. (see Fig. 3.70).

Scattering and diffuse reflecting surfaces produce lower contrasts than a shiny or specular one. Also, the dark field lighting technique is sensitive against dust, dirt and fibers. These disturbances appear in the same way like structures. Dark field incident light is used for

- emphasizing contours, shapes, structures, textures, edges
- emphasizing structural imperfections (scratches, corrugations, cracks)
- emphasize single details
- surface defect detection
- make viewable embossed or engraved contours like characters of OCR/OCV
- lighting with high contrast at opaque objects; especially flat ones gives the silhouette/contour of the object

- translucent objects. They become transparent and the upper and lower surfaces cannot be separated.

- creating contrast with embossed or engraved surfaces and for distinguishing surfaces of differing texture (laser-etched symbology).

Arrangement

The arrangement of dark field lighting components is strongly connected with the reflecting properties of the test object, the light direction, height and angle. The farther the object is, the smaller the dark field effect appears (see Fig. 3.86). That is the reason why dark field illuminations work always with short distances to the part (down to a few millimeters).

Fig. 3.86 The effect of dark field illumination in dependence of different distances to a test object with needled data matrix codes. (a) Lighting is laying on the part surface. (b) 15 mm distance. (c) 30 mm distance.

The width of the dark field effect (bright lines and areas) depends on the quality of the edges and the degree of direction of light. A sharp physical edge produces a thin reflected light line in cooperation with a directed dark field illumination. A well-rounded edge will produce a broad one (see Fig. 3.87). That is the span. On the other hand determines the width of the light emitting ring the width of the bright line. Narrow rings produce thin lines and wide rings broad lines.

Conventionally the dark field illumination components are realized as ring lights for a rotation-independent lighting. If the rotation of the part is known a directed lighting component can be used for the technique of dark field lighting too. The so-called streaking light works as a dark field lighting too (see Fig. 3.88).

A possibility of adapting the structure of a dark field ring light to the necessities of the shape of the test object is to use an adaptive lighting (see Section 3.8.2.5). Sectors can be switched on or off or the illuminating sector can walk around.

Tip

> Directed dark field incident light is often used for surface inspections. The dark field effect of scattering light will be better when

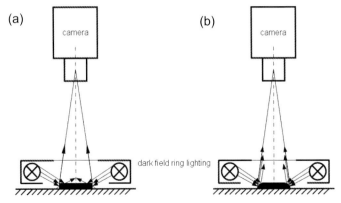

Fig. 3.87 Principle and path of rays of a dark field incident ring light, (a) sharp edged part, (b) well-rounded edged part.

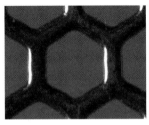

Fig. 3.88 streaking light from left. The directed area lighting component at large angle emphasizes left vertical edges using the dark field effect.

light with shorter wavelength is used. That is why a blue lighting will emphasize scratches better than a red one.

Components

- specialized dark field ring lights
- directed area lighting (for streaking light)
- spot lights (for streaking light)

How to select?

1. Ensure that a short/very short working distance is possible.
2. determine light color: Note the interaction with the part color.
3. Determine the illuminated area
4. determine the performance of the controlling technique

5. select from catalog

The examples are shown in Fig. 3.89.

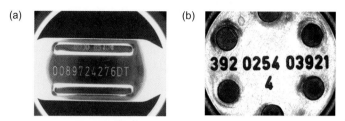

Fig. 3.89 (a) Engraved numbers in a specular metal plate, (b) embossed characters on a plastic part.

3.7.4.7 The Limits of the Incident Lighting

The application of incident light is connected with a few limitations. This should be known to avoid from bad surprises. All incident lighting live from the reflecting and scattering properties and the shape and outer contour of the part. Note that all checked parts are different and – imperfect! The consequences are no perfect reflections, no perfect scattering as estimated and no realistic shape or contour as expected!

Some of the influencing factors of edges found in different places can be (see Section 3.4.1):

- changed manufacturing procedure
- different position/rotation of parts (varying reflections or scattering)
- wear out (cutting) tools (see Fig. 3.27)
- instable porous surfaces
- torn edges
- rounded edges
- contamination

Example stamped parts: parts that are cut with a new and a wear out cutting tool look totally different (edges). The mechanical function of the part for the end user may be right, but the light reflecting properties with the incident light are drastically different.

All this and much more influences on the place where transitions of brightness are found in the image. Note this fact and take it into consideration whether measurement with the incident light is required (see Fig. 3.90).

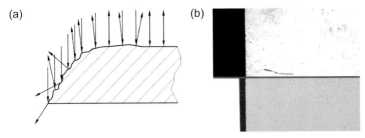

Fig. 3.90 (a) Course of parallel incident light at a typically shaped edge (micro structure: broken edge) not all light can return to the objective on the top, (b) seemingly differences at one part (glass plate with chamfer) illuminated with incident light (top) and transmitted light (bottom). With incident light the complete part is not viewable.

Tip

> Avoid using the incident light for measuring jobs with Machine Vision in any case. The accuracy of measurement obtainable within the range of several to a great many pixels is totally dependent on the micro geometry of the part. (see [31, Bl. 6.1, p. 10])

3.7.4.8 Diffuse Bright Field Transmitted Lighting (No. 7, Tab. 3.14)

This lighting technique provides good contrastful images. It is used for

- opaque parts
- flat to very flat parts gives almost perfect transition from bright to dark (foils, seals, stampings)

Transparent parts give lower contrasts. The more transparent the parts are the smaller the contrasts become. Three-dimensional parts can cause problems due to reflections from the side wall areas. This results in edge transitions with shapes that are difficult to interpret from the software.

Arrangement

The mounting is directly on the opposite side of the imaging objective (see Fig. 3.91). No special requirements for the orientation of the lighting are needed. The distance between part and lighting should be larger than the depth of focus (no imaging of dirt and dust).

Components: diffuse area lighting (see Section 3.7.4.1)

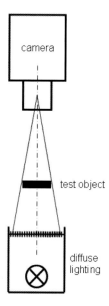

Fig. 3.91 Principle of diffuse bright field transmitted lighting.

How to select?

1. determine light color. Monochrome camera: color almost does not matter – Note the spectral response for the brightest appearing color. Color camera: white.

2. determine lighting size: Approximately 5 mm wider than the field of view. For entocentric objectives note the effect of perspective/viewing angle of the imaging objective. In this case the size depends on the distance of the lighting to the test object.

3. determine the performance of the controlling technique

4. select from catalog

The examples are shown in Fig. 3.92.

3.7.4.9 Directed Bright Field Transmitted Lighting (No. 8, Tab. 3.14)

This relatively seldom used lighting technique needs translucent parts to homogenize the light. Otherwise, the structure of the light source(s) overlays the test object structure. It makes use of different material distributions inside the part (see Fig. 3.93). It can be used for

- check the completeness of translucent parts

(a) 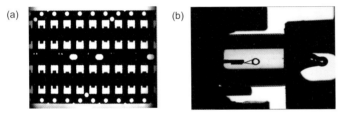 (b)

Fig. 3.92 (a) Lead frame silhouette, (b) inspection of a filament inside the glass bulb.

- targeted lighting of deep holes, small slits and clefts where diffuse or telecentric lighting fails.

Arrangement
Ensure that the light sources inside the lighting component are not viewable for the camera. Avoid overexposure!

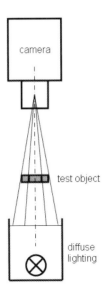

Fig. 3.93 Principle of directed bright field transmitted lighting.

Components: directed area lighting (see Section 3.7.4.2)

How to select?

1. determine light color: For translucent parts use the same color as the part. These wavelength can pass best. For holes and clefts use the most effective wavelength to the sensor.

2. determine lighting size: Approximately 5 mm wider than the field of view. For entocentric objectives note the effect of perspective/viewing angle of the imaging objective. In this case the size depends on the distance of the lighting to the test object.

3. determine the performance of the controlling technique

4. select from catalog

The example is shown in Fig. 3.94.

Fig. 3.94 Green diffuse transparent molded packaging part illuminates from back with a green directed bright field transmitted lighting to check the completeness of molding. This lighting technique gives a much more contrastful image than the incident light.

3.7.4.10 Telecentric Bright Field Transmitted Lighting (No. 9, Tab. 3.14)

This introduction of this lighting technique in the beginning of the 1990s revolutionized the accuracy of Machine Vision. The imaging of extremely well-shaped silhouettes and contours became possible. Together with telecentric objectives, the telecentric bright field transmitted lighting can be used for

- contrastful illumination of transparent to opaque 3D objects
- providing sharp edges even from shiny spherical and cylindrical parts
- applications where accuracy and reliability of results is needed
- very good suppression of extraneous light
- very homogenous and high contrasted illumination of transparent and opaque parts

Arrangement

Telecentric lighting can work only in combination with a telecentric lens (see Section 3.7.3.1). A parallel assembly of both (objective and lighting) is needed. Deviations of approximately $< \pm 1°$ are typically allowed. Connected with this demand is to ensure that the components are mounted vibration free.

The distance between telecentric objective and telecentric lighting does not matter. The distance between the test object and telecentric lighting should be adjusted so that no dust at the lens surface of the lighting is viewable (distance approximately 3× larger than the depth of focus of the objective).

Figures 3.95 and 3.96 demonstrate the effect of conventional lighting components in comparison with telecentric lighting. The right co-ordination of the lighting and imaging system apertures determines the position of the edge. Telecentric lighting ensures independent edge location.

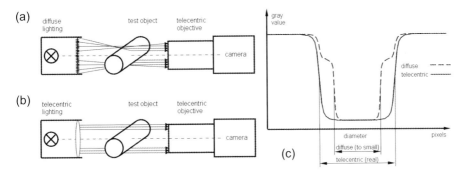

Fig. 3.95 Course of light on the surface of a shiny cylinder: (a) with diffuse transmitted light, (b) with telecentric transmitted light, and (c) course of gray values at the image sensor.

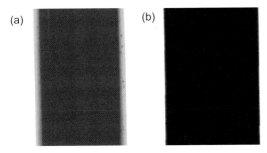

Fig. 3.96 Shiny cylindrical metal part imaged with a telecentric objective. (a) with diffuse transmitted lighting: the large lighting aperture causes undefinable brightness transitions depending on surface quality, lighting size and lighting distance. (b) with telecentric lighting: the small lighting aperture guarantees sharp and well-shaped edges for precise and reliable edge detection.

Occasional problems crop up with lighting for parallel walled parts. If the wall is shiny and the part axis is not parallelly arranged to the axis of the imaging objective disturbing reflections can occur. In some cases help the change of the lighting method and the use of diffuse lighting.

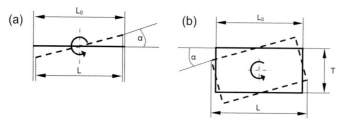

Fig. 3.97 Tilting and projection leads to changed results in the projection. (a) flat part (2D). (b) deep part (3D).

If the part is tilted, even the use of telecentric components does not help to avoid measuring errors. Only the projection of the part is viewable (see Fig. 3.97). The deviation depends on the depth of the part and the tilting angle.

Flat parts:
following Abbe's comparator principle errors of the second order will occur. Seemingly the parts become smaller if they are tilted (see Fig. 3.97(a)):

$$L = L_0 - [L_0(1 - \cos \alpha)]$$

with
- L_0 being the original length of the part,
- L the apparent length and
- α the tilting angle.

For Machine Vision applications can be neglected angles $< |3°|$.

Deep parts:
in this case errors of the first order occur. Seemingly the parts become larger if they are tilted (see Fig. 3.97(b)):

$$L = L_0 - [L_0(1 - \cos \alpha)] + T \sin \alpha$$

with
- L_0 being the original length of the part,
- L the apparent length and
- T the depth of the part
- α the tilting angle.

This effect is also present for small angles and becomes larger with an increasing depth of the object. The consequence for the brightness transition at the sensor are unsymmetrical edges (note the processing by the software).

Components: telecentric lighting (see Fig. 3.98)

Fig. 3.98 Telecentric lighting components of differenet sizes (source: vision-control.com).

How to select?

1. determine light color
   (see Section 3.7.3.1). Color cameras should use white light.

2. determine lighting size
   corresponding to the field of view of the telecentric objective. Note that the size has to be larger than the outside measures of the test object (parallel chief rays).

3. determine the performance of the controlling technique

4. select from catalog

Examples are shown in Fig. 3.99.

Fig. 3.99 (a) quality check for a shiny milled part. (b) diameter measurement of glass rods. (c) check of completeness of sinter parts.

3.7.4.11 Diffuse/Directed Transmitted Dark Field Lighting (No. 10 and 11, Tab. 3.14)

This lighting technique works only for transparent or strongly translucent parts. To see the bright reflexes at the edges of the backside it is necessary that the lighting components are mounted behind the part and the light can pass the part. These are the conditions for the function (see Fig. 3.100). Otherwise this method does not work.

It is applicable to similar kind of parts like for diffuse or directed incident dark field lighting (Section 3.7.4.6). Only the structures at the backside of the part become visible.

Arrangement, components, how to select?

Same statements as diffuse/directed incident dark field lighting (Section 3.7.4.6).

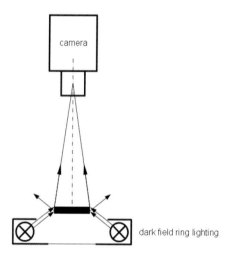

Fig. 3.100 Principle of diffuse / directed transmitted dark field lighting.

3.7.5
Combined Lighting Techniques

The lighting techniques mentioned above are systematically derived from all feasible arrangements of the practice. Each lighting technique was regarded individually.

Realistic parts from industrial processes are often complex. The necessary areas to illuminate covers. However, some lighting techniques are mutually exclusive, or they include different features that cannot be illuminated with only one lighting or one lighting technique.

In those cases it is impossible to apply a pure lighting technique. Compromises and combinations must be made.

(a) *Individual lighting for each feature.*
The simplest but neat variation is to use individual and separate lighting for each feature. It is costly and takes time, because all features are processed one after the other. This is only possible if there are no time limitations. Image acquisition with many cameras or moving the camera with a handling system is necessary for optimal optical and lighting conditions.

(b) *Classify lighting situations.*
More effective is it to use one lighting technique for all similar lighting situations. Image acquisition in combination with a handling system is needed. Repetition must be made for the other lighting techniques in the same manner. It is as time consuming as the variation (a).

(c) *Different lighting techniques in sequence.*
Use a long exposure time and switch through sequentially all necessary lighting within the exposure time of one image. This method is applicable only if all used lighting techniques are feasible in one view.

(d) *Change the lighting technique within the lighting component*
Use of an adaptive lighting (see Section 3.8.2.5) that can change the light structure and direction sequentially from image acquisition to image acquisition or within one image acquisition.

To select the best variation from (a) to (d) note the mutual influence of lighting. To use all lighting techniques together is impossible almost in every case.

And the rest of lighting jobs? Some lighting problems are not solvable. They cannot be solved, particularly if the expected lighting technique is refused under the predetermined mechanical and arrangement conditions. For example, it is impossible to apply a dark field illumination if the minimal possible distance from the object to the lighting is, let us say, 200 mm.

Among the experts are well known the so-called *challenge cups*. These are jobs that circulate in the market and come up from time to time, because nobody could solve the problem. Unexperienced personal will try many times and put in much energy, time and money to find a way. Frustration will be the result. But to recognize the limits is not to learn – it is only to experience.

However, keep in mind the high performance of the human eye. And again and again the lighting engineer will ask the question: *Why does the machine not see what I can see?*

3.8
Lighting Control

Deduced from the demands on Machine Vision lighting (see Section 3.2) and the outstanding properties of LEDs for an easy lighting control (see Section 3.3.3.5) the following comments are made for LED lighting components. In some points they can be transferred to other kind of lighting.

3.8.1
Reasons for Light Control – the Environmental Industrial Conditions

Besides emphasizing on desired details, the aim of a Machine Vision lighting is to be as much as possible independent of disturbing and influencing factors (see Section 3.7.2). Many of them can be lined out by controlling the lighting such as

- voltage variations (instable light)
- machines running with asynchronous cycle times (triggering lighting)
- fast running machines (avoid motional blurring with flash light)
- inspection of different parts/changing the light distribution (adaption of light to the part)
- coordination of different lighting (switchable lighting)
- switching of dangerous lighting
- dynamic response (no delay or afterglow allowed)
- adaption to space limitations (using deflection, long distances)
- disturbing light (reflexes, ambient light)

3.8.2
Electrical Control

3.8.2.1 Stable Operation
Only in a few cases single LEDs work in lighting components such as in telecentric lighting or coaxial lighting for microscope. But even these LED need stable operation.

The required high luminances and illuminances require interconnection of many LEDs. Note that the large area lighting need a few hundred up to a few thousands of LEDs.

If LEDs are combined in series connection typical voltages in industry from 12 to 24 V are rapidly exceeded. This limits the length of the chains, because all LEDs operating with the same current brightness differences occur due to different forward currents of each LED. One single defect LED puts the whole LED chain out of action. A positive feature of this are small operating currents.

If LEDs are working in parallel connection rapidly high currents occur. The nonlinear characteristics of the LEDs cause additional problems for a homogenous brightness. Small operating voltages are advantageous.

It is the art to find a matching combination of series (chains) and parallel connection of them. Different possibilities are given (see Fig. 3.101):

1. serial connection

2. parallel connection of single LEDs without series resistor

    - used for low-cost lighting
    - strong inhomogeneity of brightness

3. parallel connection of chains with resistor

    - compensation of different currents possible

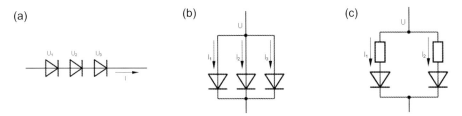

Fig. 3.101 Possible LED connection: (a) series connection, (b) parallel connection, (c) parallel connection with series resistors.

Different LED substrates (LED colors) have different current–voltage characteristics. This makes it impossible to work with those constructions in a color mixed mode (see Fig. 3.102).

Additionally the manufacturing data of the LEDs are not stable so that they even differ in their forward voltage for one model see Fig. 3.103):

To know this structure gives the best base for a defined control. LED lighting with defined control use classified and sorted LEDs (according to the characteristics). This is expensive but precise. Furthermore, an individual series

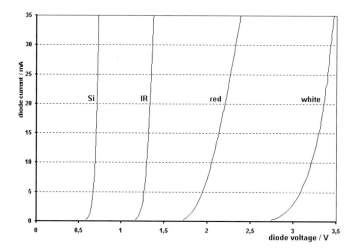

Fig. 3.102 Current–voltage characteristics of different LED substrates (colors).

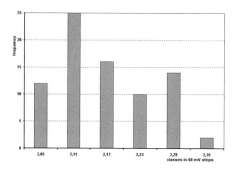

Fig. 3.103 Classes of forward voltages for red LEDs [15].

resistor per chain is integrated. This limits small differences of the forward voltage but produces additional dissipation power and therefore heat. A wide voltage range input 10 to 30 V DC rounds up the stabilization.

Knowing the condition that LEDs are working most effectively if they are cold and that overheating reduces the lifetime and increases aging and brightness drift (see Section 3.3.5.2 and Fig. 3.22), it is desirable that lighting operates cold. This can be achieved with a built-in current control circuitry. This compensates the loss of brightness caused by increasing temperature. So the current control works as a brightness control and is the basic equipment for all controllable lighting.

Tip

>Details about the quality of the temperature compensation and thus for a stable function gives the temperature coefficient of the illuminance (incident light) or the luminance (transmitted light). A good lighting for industrial use should have a coefficient of only a few tenth percent per Kelvin ($-0.4\ldots0.4\%/K$).

Many simple casted LED arrays without housing do not use a controlling circuitry, a thermal compensation or overvoltage protection. The consequence is that each small change of the supplying voltage is passed through directly in a change of brightness. These lighting need a good stabilized and costly power supply.

The application is to recommend to refrigerate these lighting and to use it only in a switch mode. The continuous operation with full capacity slowly destroys the lighting.

3.8.2.2 Brightness Control

To adapt the brightness on the test object, it is necessary to adjust the brightness of the lighting. This does not mean a change of the supplying voltage. For industrial use this must be done on a separate input. Basically two possibilities are in use:

- brightness adjustment with potentiometer. This can be done manually.

- brightness adjustment with typical voltages between 0 and 10 V DC.

This needs an additional controlling voltage. 0 V mean dark and 10 V mean maximum brightness. So the lighting can be controlled (brighter, darker) and more. They can be switched on/off from the far. For steady maximal light output the controlling input channel can be connected with the supplying voltage level. Most lighting permits this.

Typical delay times from connecting the controlling voltage to the light effect are a few milliseconds. Using the power supply input for switching on/off this takes approximately 10–20 times more.

3.8.2.3 Temporal Control: Static-Pulse-Flash

Due to their electrical properties LEDs can be modulated up to the MHz range. Independent of the adjustment of brightness they can operate in three different operation modes:

- *static mode:* the LED current is intended changed slowly or not at all

- *pulse mode*: switching the LED with nominal current in time periods $>$ 100μs

- *flash mode*: overload the LED for very short time periods < 100μs with up to 10 fold nominal current (see Fig. 3.17). Some models allow flash times of smaller than 0.5 μs. The controlling unit ensures the right pulse duty factor to protect the LED chip from thermal destruction.

The lighting component for *static lighting* is simply made with or without a current/voltage stabilization. If they are misused to switch that can cause delay time from switching to full/no brightness of a few up to a few hundred milliseconds. The concrete value depends on the model.

Pulsed lighting or controlled lighting are equipped with a separate pulse input. The included controlling electronics ensures a fast reaction from pulse to light. Typical delay times are in the range of microseconds. Minimal light pulse widths of 10 μs are possible. The triggering pulse can use TTL or PLC level.

Flash lighting electronics are more complex. A few of flash lighting components use a constant power integral circuitry. This means a constant exposure independent of the chosen flash time. The shorter the flash, the powerful it must be. The adjustment of the flash time can be fixed, adjustable with potentiometer, with DIL switches or with a bus interface connection. Flash times shorter than 5 μs make great demands of the energy storage in the lighting and the driving electronics and make those components very expensive apart from the ambitious synchronization that is needed to coordinate all optical hardware and software processes and EMC problems.

Typical delay times from flash trigger impulse to light are 10 μs. Trigger impulses use TTL or PLC level. The controlling unit includes a circuitry to keep the requirements for the pulse-duty factor. This ensures stable lighting and protect the LEDs. Typical are 100 Hz for full load. The temporal variation from one flash to the next is less than a microsecond.

The construction of flash lighting components can be different. Some of them have built-in flash controller, and some of them use external control boxes. The disadvantages of external controllers are the influence of interferences and additional delays.

Some demands that one should make on flash lighting:

- short delay, raise, fall times inside the flash
- flash recurrence rate > frame rate of the used camera
- EMC capable
- long lifetime
- low loss of brightness beyond the lifetime (no compensation possible!)

- changeable flash time at constant light energy

- changeable flash brightness without change of flash time

Modern lighting components allow all operation modes in one lighting component. The operation modes (static, pulse, flash) are electrically adjustable even during operation. The brightness control can be made with a bus interface, digital, analog or manual. Some components save the lighting settings even at a power failure. An integrated temperature / current measurement at the LED module provides the automatic tracking of optical power parameters.

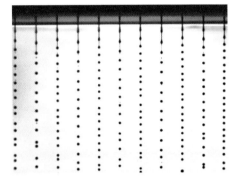

Fig. 3.104 Edge triggering with very low delay times allow us to achieve flashes from 0.5 to 100 μs that are able to image blur free object speeds up to 30 ms^{-1}. The image shows ink drops that are injected under high pressure.

3.8.2.4 Some Considerations for the Use of Flash Light

The short light pulse width of flash lighting makes them an excellent choice for applications where stop motion is required, such as for high-speed inspections. A flash lighting can be used to *freeze* a part on a production line as it passes under a camera for inspection. The consistency at which the flash lighting can be triggered is also a major benefit in these applications (see Fig. 3.104).

The use of flash light enables shorter exposure times (down to 1/200,000 s) than the preset from the imaging sensor (typically 1/30,000 to 1/60,000 s) is given. Furthermore, it is possible to flash into the standard exposure time for sensors that do not have a shutter. Sensors with rolling shutter technology are not suitable for the combination with flash lights – progressive scan are required for that.

Last but not least are flash lighting (and short exposure times) suitable to suppress the influence of ambient light. This needs a few times higher light energy of the flash than the ambient light.

From the theory flash applications are possible for both, for incident and transmitted light. Note the reflectance of the test object (see Section 3.4.1.2). The limited reflectance of many real parts makes it impossible to use the incident flash light. The amount of the reflected light is simply too small. Flash light applications with the transmitted light work generally reliably.

Motional blurring is the effect that the part moves during the exposure or flash time. It can be caused by

- movement of the part during the exposure time
- movement of the camera during the exposure time (vibrations)
- too long exposure time
- too long illumination time (flash time)

Tip

In general a motional blurring of one pixel in the image is accepted for sharp imaging (exceptions possible).

The size of motional blurring can be determined by

$$\text{MB} = s_{\text{exp}}/\text{PR} = s_{\text{exp}} \cdot \frac{\text{no. of pixels}}{\text{FOV}}$$

with

MB	being the motional blurring in the image,
s_{exp}	the movement of the part during the exposure time/flash time,
PR	the pixel resolution,
no. of pixels	the used number of pixels of the sensor in the direction of the movement of the part and
FOV	the length of the field of view of the camera in the direction of the movement of the part.

The above formula illustrates that a better pixel resolution makes an application more sensitive against motional blurring. To avoid this is to shorten the exposure time and/or the flash time.

A remark on brightnesses. Based on the demand for a constant exposure $H = E \cdot t$ (see Section 3.5.7) it is necessary for reaching the same brightness in the image to increase drastically the lighting power (illuminance E) if a flash lighting is used. To generate the same brightness in the image (constant f-stop number), the following exemplary relations are valid:

	Exposure time/flash time	necessary illuminance (for the time of exposure)
Static light	1/50 s = 20 ms	2,000 lx
Flash light	1/20,000 s = 50 µs	800,000 lx

The example shows that the need for such high illuminances can be satisfied only for short times and only from flash light sources.

The demands for enough flash light energy are often connected with the demands for a large depth of focus (see Chapter 4). The connections shows the Table in Section 3.5.7. A stronger closed f-stop for a larger depth of focus will again increase the demand for more light. One larger f-stop number means the need for the double amount of light.

The synchronization of flash lighting is a complex field connected with the cooperation of

- trigger sensors and PLC of the machine
- vision system
- flash lighting
- camera
- software

These connections do not make it easy to investigate problems. Components are often made by different manufacturers and the necessary information of the internal function of the components is not available. That is why only a few influencing factors can be considered at this point. Condition is a precise enough positioning of the test part. The worst case conditions for the whole machine should be considered.

Three synchronization possibilities are given as follows:
(a) flash time > shutter time → safe, but loss of light energy,
(b) flash time = shutter time → not stable working by varying time components
(c) flash time < shutter time → most effective, but needs perfect synchronization

Case (c) is the most interference free (and demanding). The light pulse has to flash matching into the short time window of the light sensitive phase of the imaging sensor (shutter time).

Figure 3.105 shows a typical succession of processes that are to synchronize. A trigger pulse is given from a sensor in the machine. This information is transferred through the PLC (delay!) to the trigger input of the vision system. The vision system processes this information with their software (delay!) and sends a trigger pulse to the flash light (or to the PLC that triggers the flash light (delay!)). Considering the delays up to the start of the light sensitive phase of the camera the image acquisition (including short shutter time) is started. After finishing the light emission of the flash light the camera shutter is closed.

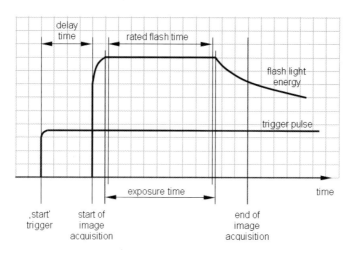

Fig. 3.105 Time diagram of the connected processes during flash light synchronization.

To achieve stable time conditions for synchronization it is necessary to know the time response of the whole system. Especially unregular processes causes problems of synchronization. This includes the mechanical synchronization of the machine too. Reasons for time problems can be

- variable/unknown moments of the trigger impulses
- undefined time response of the controlling unit (PLC, etc.)
- variable/unknown delay times (vision system, flash lighting, camera, PLC)
- variable/unknown flash times (flash duration, constancy)
- variable/unknown shutter times (camera)
- start of image acquisition to early or to late (vision system, camera)
- variable/unknown signal runtimes in the machine.

To get down to bedrock the problems of synchronization it needs special equipment (oscilloscope, optical detectors (for short light pulses), background information to the components and ... perseverance.

How to select a flash lighting? – A rough estimation

1. find a matching lighting technique (see Section 3.7):
   lighting check with static light: possible or impossible?

2. determine light color:
   Note the interaction with the part color.

3. determine lighting size

4. lighting component available as pulsed or flash model?

5. time and brightness considerations (strongly depends on design):
   incident light with flash light: shorter flash times than 1/10,000 s are mostly problematic
   incident light with pulsed light: shorter flash times than 1/2,000 s are mostly problematic
   transmitted light with flash light: possible up to 1/200,000 s
   transmitted light with pulsed light: approximately possible up to 1/20,000 s

6. flash repetition rate attainable (strongly depends on design)

7. select from catalog.

3.8.2.5 Temporal and Local Control: Adaptive Lighting

Traditional lighting components suffer from the fact that they are selected for one special purpose. Once selected only little possibilities modify the use.

Adaptive lighting extends these possibilities. Temporal, local and spectral are all participating light sources (above all LEDs) separately controllable per software. The individual adjustment of local illumination parameters (duration and moment of emission, brightness, wavelength) provides application of specific lighting. This leads to the following advantages:

- different lighting modes in one lighting (static, pulsed, flash)
- very fast automatic lighting control
- state enquiry possible (e.g. failure check for single LEDs)
- simple adaptation of lighting to changing products
- treating the lighting component as an automation device

- no sensitive electromechanical adjustment elements

Some examples for the wide use of adaptive lighting are

- compensation of the natural vignetting of objectives ($\cos^4$ brightness decrease, see Fig.3.108(a))
- avoidance of reflections from the test object (see Fig.3.108(c))
- homogenous lighting of three-dimensional structured objects (see Fig.3.108(b))
- light structures for transmitted lighting
- adaptation of the illuminating wavelength/color mixture to the color of the test object.

Adaptive lighting are based on complex drives (see Fig. 3.106). They are equipped with wide range voltage input, they are switchable and flashable. Flash times and brightnesses can be saved resident in a memory. Basic structures are adaptive base modules (ABM) with on-board-electronics including interface for control per software or keystroke. The base modules can be combined just as you like. This opens up total new possibilities to combine them in (almost) every shape and realize all lighting techniques.

The temporal and local and brightness control of each single LED is done with pulse width modulation (PWM).

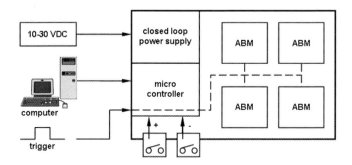

Fig. 3.106 Block diagram of adaptive lighting.

Above all there are two procedures for the input of the lighting parameters (light patterns and times) for adaptive lighting:

- *manually by keystroke*: using the keys at the housing of the adaptive lighting components
- *by programming* (see Fig. 3.107): this can be done from a user interface or directly from a vision system.

During operation the light pattern information is sent from the vision system to the adaptive lighting through an USB or Ethernet interface.

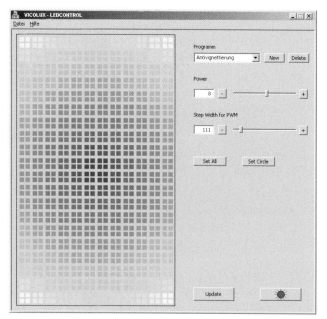

Fig. 3.107 Windows user interface for programming an adaptive lighting. The brightness and flash duration of each single LED can be selected and adjusted by a mouse click. The demonstrated light pattern can be used for vignetting compensation (source: www.vision-control.com).

Fig. 3.108 Different programming examples of a diffuse adaptive area lighting: (a) correction of natural vignetting of an objective, (b) and (c) compensation of reflexes at shiny parts (source: vision-control.com).

All kind of lighting components can be designed as adaptive lighting. Tendencies for the future are adaptive lighting that adapts automatically to the lighting situation.

3.8.3
Geometrical Control

3.8.3.1 Lighting from Large Distances

Except the fact that some lighting techniques need special distances and strongly determine the image of the test object (see Section 3.7), two general lighting situations differ the lighting from large distances:

(a) incident light

or

(b) backlight

For the use of the incident light shows the existence of the photometric inverse square law (see Section 3.5.2) that it is a difficult assignment to illuminate a part from large distances. In addition some of the lighting techniques cannot work from a distance by their principle.

If they can work by their principle, the effect of darkening for larger distances can be compensated by

- Use of a more powerful lighting.
  Note: to compensate the reciprocal square reduction it is necessary to increase the power exponential! For the replacement consider the limits of different light sources. Not all substitute light sources allow all kinds of lighting techniques!

- Opening the f-stop from the imaging objective.
  To open the f-stop one step means that the double luminous flux can pass the objective. Two f-stop steps mean four times more luminous flux and so forth.
  Use this option only if the application allows the subsequent change (reduction) of the depth of focus.

- Extending the exposure time of the camera.
  To double the exposure time means that the double quantity of light can be collected from the image sensor.
  Use this option only if the motional blurring of the test object at the image sensor is smaller

- Removing light absorbing filters if possible (think of the optical consequences)

- Using another wavelength of light that is more effective for the image sensor (see Section 3.3.2).
  Note the effects that happen in interaction with the test object (see Section 3.4)

- Using a objective with a higher light intensity.
  This option is effective only if the application can work with a open f-stop at the objective. Note that the depth of focus is reduced in this case as much as possible.

- Using a higher sensitive camera

- Increasing the gain of the camera (notably noise, nonlinearities, etc.)

- Using highly robust software algorithms that can process a wide range of gray values

- using another lighting technique if all mentioned above measures do not help.

For backlight applications a longer distance between lighting and test object does not have an influence on the image brightness because of the constancy of luminance. Prolonging the distance of the lighting can have an effect on

- The complete illumination of the field of view of the camera objective when using a lens with perspective properties. Compensate it with a larger lighting component.

- Changed reflections at the test object because of three three-dimensional structure of the part (interaction part – light).

3.8.3.2 Light Deflection

Not all applications and machine environment possess a simple structure. They are adapted to complex industrial manufacturing processes. Machine Vision is usually only an add-on device that controls. The consequence for mounting the components is often a limited space.

It is known from the lighting techniques (see Section 3.7) that some of them need space. Other lighting components are bulky itself caused by their optical principle (for example telecentric lighting).

To mediate between both, demands from lighting and machine too one can use deflection units for some lighting. They operate with mirrors or prisms and work follow the optical principles of reflection.

Deflection units are front end elements of lighting and thus the interface to the rough environment of the industrial floor. That is why they should be easy-care and robust. Prisms meet these demands more than mirrors.

Prisms:

The glass surface of a prism is an ideal front seal for the lighting. There are no problems to clean them. They are easy to install if they are built in prism deflection units. The compact size make them match even under limited space conditions. Referring to the reflection law depends their size on the illuminated field and light emitting angle of the lighting (note that the total reflection of a prism is limited by a critical angle). Mostly they are used for a 90° beam deflection.

Prisms are more costly and more heavy than mirrors.

Mirrors :

The cheap and easy availability of mirrors makes them seemingly applicable for Machine Vision. However, one should evade the application of mirrors in the industrial floor wherever possible.

1. To avoid ghost images and low contrast only surface mirrors should be used. But these mirrors have an extremely sensitive surfaces against mechanical and chemical stress.

2. To remove dust is a problem. Frequently mirrors are built in constructions with cavities that attract dust. On the other hand dust should be only blown away to save the mirror surface (see Section 3.1.).

3. Because mirrors are no standard components (they are cut on request) there are no standardized holders and adjustable frames. Each mounting is a costly handicraft.

4. Vibrations and not precise positioning have a multiple higher influence on the beam deflection than at prisms.

3.8.4
Suppression of Ambient and Extraneous Light – Measures for a Stable Lighting

The industrial floor of Machine Vision provides all other than ideal lighting environment like in the lab. The tested lighting from the lab does not work automatically in the factory floor too. This plastically demonstrates the difference between theory and practice. The biggest and imponderable influence comes from ambient and extraneous light. It can be static (illumination from factory hall) or dynamic (incoming sunlight through windows).

The first condition to minimize the influence of ambient light is to choose as powerful as possible lighting referred to as a right selected lighting technique. The aim is that the calculated lighting power (in the spectral band considered) is mightier than the ambient light.

If this is guaranteed, it can be tried to minimize the influence of the ambient light by

- choosing a shorter exposure time
- choosing a larger f-stop number

- choosing a flash lighting

- choosing all these measures together.

Tip

> For a first approximation check the ambient and extraneous lighting conditions at the place where the vision system is later mounted. Check it by measuring with a vision system too, not only by your subjective eye. Get a feeling for the amount of the ambient light. Take this into consideration when choosing a lighting technique, lighting components and software algorithms. Some options for that are automatically denied.

Suppress the influence of ambient light using monochromatic or infrared light in combination with light filters (see Section 3.6).

A simple but very effective measure is to enclose the camera – optics – lighting unit with an opaque housing. This is a tried and tested measure that is used from prestigious companies to achieve the as best possible ambient/extraneous light suppression that works under every condition. If it is not clear where the machine later is installed this measure will be the most secure.

First of all the ambient light is the most important source of lighting disturbances. But to ensure stable lighting conditions a few other measures have to take into consideration too:

- an electrically and optically stabilized lighting component (see Section 3.8.2)

- operate the lighting in the middle of the characteristics. This provides power reserves to compensate other influences

- no (low) aging and drift (see Section 3.3.5.2) that ensure no change of materials (change of hue of the light source, diffusors, etc.)

- mechanical stable mounting of the lighting to avoid vibrations and change of adjustment.

3.9
Lighting Perspectives for the Future

The fast development of small, discrete and powerful semiconductor light sources drastically sped up the present developments of Machine Vision lighting. This process will continue. It is assumed that within the next 3–5 years

LEDs are so powerful that they can completely substitute all usual incandescent lamps for indoor and car applications.

This will have a big effect on Machine Vision lighting. All known lighting components from today will be equipped with those new light sources. "Too dark light" will be an episode of the past. Even the light critical applications of line scan cameras will work with LEDs.

New technologies and materials such as OLEDs (organic light emitting diodes) and electro-luminescent foils will influence the design of lighting components. The technical problems of limited lifetime and low efficiency will be soon overcome. A cold lighting that is thin like film, dimmable and cuttable with a scissors will provide a perfect adaptation to the lighting job.

Today reliable LEDs with UV emission too are available. Built in all known lighting components opens a few new applications where UV light is needed.

The fact that Machine Vision conquers more and more branches will drive the development of industrial compliant lighting with robust construction, assembly areas and wiring. This will lead to standards for Machine Vision lighting components.

State-of-the-art lighting components are needed for shaping light optical components too. Machine Vision lighting technology will indirectly benefit from some new developments of micro optics. Only to mention is the use of micro prism foils.

The interaction between Machine Vision hardware, image processing software and lighting hardware will make possible to use adaptive lighting that adapt their light emission, pattern and sequences themselves. Smarter software algorithms will provide an automatic closed-loop control of lighting to achieve the optimal image. This will lead to standardized protocols for light control. Especially the robotic technology will benefit from this new approach.

Lighting as a long time neglected discipline of Machine Vision became a driving force. In consequence of this the Smart Cameras will get siblings: Smart Lighting will change the future applications of Machine Vision.

References

1 AHLERS, R.-J. (HRSG.), 3D-Bildaufnahme mit programmierbaren optischen Lichtgittern, *Handbuch der Bildverarbeitung*, expert-Verlag, Renningen-Malmsheim, 2000

2 JAHR, I., *Lexikon der industriellen Bildverarbeitung*, Baunach, Spurbuchverlag, 2003

3 JAHR, I., *Opto-elektronisches Messen und Erkennen zur Qualitätssicherung*, course documentation, Essen, Haus der Technik, 1991

4 JAHR, I., *Adaptive Beleuchtungen*, Vortrag, 23. Heidelberger Bildverarbeitungsforum, 2003

5 JACOBSEN, D., KATZMAN, P., *Benefits of Xenon Technology for Machine Vision Illumination*, www.machinevision-online.org,

6 NAUMANN, H., *Bauelemente der Optik, Hanser*, Hanser, 1987

7 SCHUSTER, N., HOTOP, M., *Spektrale Wirksamkeit von Lichtemitterdioden fuer CCD-Matrizen*, msr-Magazin, 10/2000
8 PHILLIPOW, E., *Taschenbuch Elektrotechnik*, Bd. 6, Verlag Technik, Berlin, 1982
9 BAER, R.; ECKERT, M. RIEMANN, A., *VEM-Handbuch Beleuchtungstechnik*, Verlag Technik, Berlin, 1975
10 BAER, R., *Beleuchtungstechnik – Grundlagen*, Verlag Technik, Berlin, 1996
11 HENTSCHEL, H.-J., *Licht und Beleuchtung*, Hüthig Verlag, Heidelberg, 1994
12 NEHSE, U., Beleuchtungs- und Fokusregelungen für die objektivierte optische Präzisionsantastung in der Koordinatenmesstechnik, *Dissertation* TU Ilmenau, Fakultät für Maschinenau, 2001
13 Filter catalogue, B+W Filter, Bad Kreuznach
14 Filter catalogue, Heliopan Lichtfilter-Technik, Graefeling
15 PISKE, C., *Statische und dynamische Ansteuerung von LED's für ortsaufgelöste, adaptive Flächenbeleuchtungen in der industriellen Bildverarbeitung*, Diplomarbeit, FH Lübeck, 2003
16 *The Photonics Handbook*, Laurin Publishing, Pittsfield., 2004, p. 272.
17 MUEHLEMANN, M.: *Tungsten Halogen: when economy, reliability count*, The Photonics Handbook, Laurin Publishing, Pittsfield, 2004.
18 Machine Vision lighting: a first-order consideration The Photonics Handbook, Schott North America, 2004
19 Filters: glossary, equations, parameters, The Photonics Handbook 2004, Laurin Publishing, Pittsfield, page H-304
20 JOHNSON, R. L., Bandpass filters: the transistor's optical cousin, The Photonics Handbook 2004, Laurin Publishing, Pittsfield.
21 *Study of color fidelity of white LEDs*, company paper Vision & Control GmbH,
22 *Aging considerations of LEDs*, company paper Vision & Control GmbH,
23 teaching material *Demands on industrial vision components*, Vision Academy, www.vision-academy.org
24 Information and images from the catalogue "Vision Control components system", Vision Control, Suhl, www.vision-control.com
25 Course documentations *Optics and lighting in Machine Vision*, Vision Academy, Weimar, 2003
26 *Training scripts of the Vision Academy*, Weimar, www.vision-academy.org
27 DIN 5031, *Strahlungsphysik im optischen Bereich und Lichttechnik,* Teil 7: Benennung der Wellenlängenbereiche.
28 DIN EN 60825-1:1994 , *Laser protection classes*
29 DIN 5031, *Strahlungsphysik im optischen Bereich und Lichttechnik*, Blatt 1 bis 10
30 European Standard: *Persönlicher Augenschutz – Infrarotschutzfilter – Transmissionsanforderungen und empfohlene Anwendung*
31 VDI Richtlinie 2617, *Koordinatenmessgeräte mit optischer Antastung*, Bl. 6.1, S. 10

4
Optical Systems in Machine Vision
Dr. Karl Lenhardt, Jos. Schneider Optische Werke GmbH

4.1
A Look on the Foundations of Geometrical Optics

4.1.1
From Electrodynamics to the Light Rays

The optical system transmits information such as intensity distribution, colour distribution, forms and structures into the image space. This information is finally stored in a single plane, the image plane.

The usual and most simple treatment of this information transmission – or *imaging* – is accomplished with the help of *light rays*. With their introduction, and with the knowledge of the laws of reflection and refraction one arrives at the field of geometrical optics, in which the object–image relations follow from purely geometrical considerations.

Now, the concept of light rays is a pure hypothetical assumption which is well suited to work in a first approximation. According to classical physics light consists of electromagnetic waves of a certain wavelength range

$$\text{light = electromagnetic waves with wavelengths } \lambda = 380 - 780 \text{ nm } (1 \text{ nm} = 10^{-9} \text{ m})$$

In order to give a general survey on the underlying principles, we will give a very short outline on the approximations which result in the idea of light rays.

The propagation of electromagnetic waves is described by electrodynamics and in particular by Maxwell's equations. These are a set of vectorial, linear and partial differential equations. They play the role of an axiomatic system, similar to Newton's axioms in mechanics. From this set of axioms one may – at least in principle – calculate all macroscopic phenomena. In the absence of charges and currents in space they may be simplified to the wave equation:

Handbook of Machine Vision. Alexander Hornberg (Ed.)
Copyright © 2006 WILEY-VCH Verlag GmbH & Co. KGaA, Weinheim
ISBN: 3-527-40584-7

Wave equation in homogeneous media:

$$\Delta \vec{E} - \frac{\epsilon \cdot \mu}{c_0^2} \cdot \frac{\partial^2 \vec{E}}{\partial t^2} = \vec{0} \qquad (4.1)$$

With $\vec{E}$ = electric field vector, Δ = Laplace operator, ε = dielectric constant, μ = magnetic permeability, c_0 = velocity of light in vacuum $\approx 3 \times 10^8 \, \mathrm{m\,s^{-1}}$

$$c = \frac{c_0}{\sqrt{\varepsilon \cdot \mu}} = \text{velocity of light in matter}$$

An analogue equation holds for the magnetic field vector $\vec{H}$.

Now the wavelength in the visible region of the electromagnetic spectrum is very small with regard to other optical dimensions. Therefore, one may neglect the wave nature in many cases and arrives at a first approximation for the propagation laws if one performs the limiting case

$$\lambda \longrightarrow 0$$

in the wave equation. Mathematically, this results in the Eikonal equation which is a partial differential equation of first order.

This limiting case ($\lambda \to 0$) is known as *geometrical optics* because in this approximation the optical laws may be formulated by purely geometrical considerations.

From the Eikonal equation one may derive an integral equation, the so-called *Fermat principle*:

> Among all possible ray paths between points P_1 and P_2 the light rays always choose that which makes the product of refractive index and geometrical path a minimum.

$$\text{Wave equation} \to \text{Eikonal equation} \to \text{Fermat's principle}$$

The Fermat principle is, by the way, in a complete analogy with the Hamiltonian principle in classical mechanics and in the same way the Eikonal equation corresponds to the Hamilton–Jacobi differential equation in mechanics. From a formal point of view the light rays may thus be interpreted as particle rays; in fact this has been the interpretation by Isaac Newton!

The conclusions from the limiting case $\lambda \to 0$ for homogeneous media are twofold:

1. Four basic laws of geometrical optics

2. The light rays may be defined as the orthogonal trajectories of the wavefronts and correspond to the direction of energy flow (Pointing vector)

The four basic laws of geometrical optics will be dealt in the next section; here we will give some simple interpretations for wavefronts and light rays.

4.1 Example point-like light source – spherical wavefronts (Fig. 4.1)

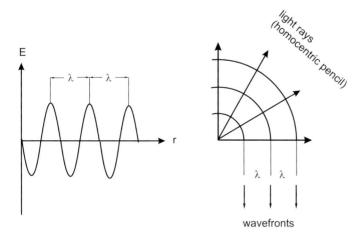

Fig. 4.1 Spherical wavefronts and light rays.

The light rays – being orthogonal to the spherical wavefront – form a homocentric pencil which means that they all intersect at one single point.

4.2 Example plane wavefronts – light source at infinity (Fig. 4.2)

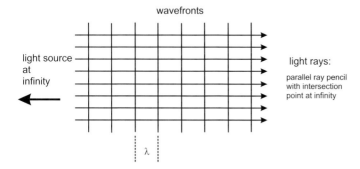

Fig. 4.2 Plane wavefronts and parallel light-ray bundle.

The light rays – orthogonal to the plane wavefronts – now form a parallel bundle which is also homocentric, since there is one single intersection point at infinity.

4.1.2
The Basic Laws of Geometrical Optics

The four basic laws of geometrical optics may be described as follows:

1. rectilinear propagation of light rays in homogeneous media
2. undisturbed superposition of light rays
3. the law of reflection
4. the law of refraction.

The laws of reflection and refraction will be dealt in Section 4.2.

The rectilinear propagation and undisturbed superposition of light rays are very easily demonstrated by a pinhole camera (camera obscura, Fig. 4.3).

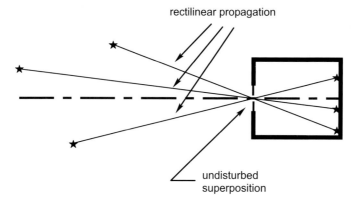

Fig. 4.3 The pinhole camera.

Light rays from different object points propagate rectilinear through a very small hole until they reach the image plane. They intersect at the pinhole with undisturbed superposition.

The image formation is done by *central projection* (Fig. 4.4):

The pinhole acts as the centre of perspective, the projection plane is the image plane and the projecting rays are the light rays. The image y' will become smaller and smaller, when the objects y are more distant from the projection centre. In the limiting case of vanishingly small pinhole this corresponds to the linear camera model in digital image processing. This model may be described mathematically by a projective transformation with rotational symmetry. Points in the object space are associated with the corresponding points in the image plane by a linear transformation. Geometrically, one may construct this image formation in central projection just as with the pinhole camera:

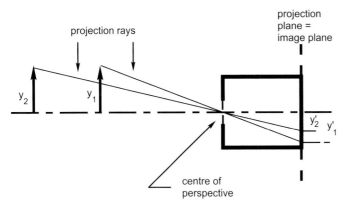

Fig. 4.4 Central projection.

lines joining the object point, the projection centre P and intersecting the projection plane (image plane, sensor plane). The intersection point represents the image point to the corresponding object point (Fig. 4.5).

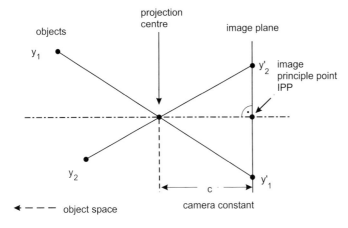

Fig. 4.5 The linear camera model.

This model is hence represented by the projection centre and the position of the image plane relative to it. This position is given by the distance c to the projection centre (*camera constant*) and by the intersection point of the normal to the image plane through the projection centre, sometimes called *image principal point*.

As may be seen, this is a linear image formation: straight lines in the object space are mapped to straight lines in the image plane. However, this imagery is not uniquely reversible. To each object point there exists a unique image point but to one single image point there correspond an infinite number of object points, namely those lying on the projection ray.

The question arises how and to what approximation this linear camera model may by realized by systems with the non-vanishing pinhole diameter, since the energy transport of electromagnetic waves requires a finite entrance opening of the optical system. We have to postpone this question to Section 4.2.12.4 and will deal first with the second set of the basic laws of geometrical optics, namely the laws of reflection and refraction.

4.2
Gaussian Optics

4.2.1
Reflection and Refraction at the Boundary between two Media

We will deal with the boundary between two transparent (non-absorbing) media. The propagation speed of light in matter (c) is always smaller than that in vacuum (c_0). The frequency ν will be the same, so the wavelength λ will become smaller in matter. The refractive index n in a certain medium is defined as the ratio of the light velocity in vacuum (c_0) to that in the medium (c) and hence is always larger than 1.

$$n = \frac{c_0}{c} \tag{4.2}$$

With the refractive indices or the light velocities in different media one may deduce the law of refraction. When a light ray meets a plane boundary between two media it will be split into a reflected ray and a transmitted (refracted) one. Incident, refracted and reflected rays lie within one plane which contains also the normal to the plane boundary between the two media (Fig. 4.6).

$$\alpha_i = -\alpha_r \quad \text{law of reflection} \tag{4.3}$$
$$n_1 \sin \alpha_i = n_2 \sin \alpha_t \quad \text{law of refraction} \tag{4.4}$$

The propagation velocity c of light depends on the wavelength λ and therefore the refractive index n is a function of the wavelength.

$$n = n(\lambda) \tag{4.5}$$

This fact is known as the *dispersion of light*. According to the law of refraction the change of direction of a light ray at the boundary will be different for different wavelengths. White light, which is a mixture of different wavelengths will split up by refraction in different wavelengths or colours (Fig. 4.7).

As Eq.(4.4) shows, the law of refraction is non-linear. Thus we may not expect that we get a linear transformation (especially a central projection) when we image with light rays through lenses.

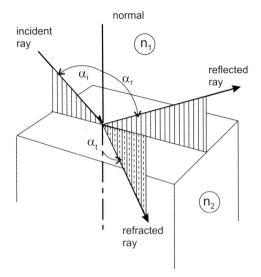

Fig. 4.6 The laws of reflection and refraction.

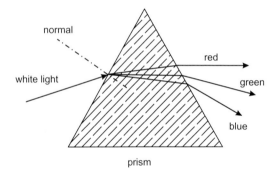

Fig. 4.7 Dispersion of white light as a prism.

It may be shown that with arbitrarily formed refracting surfaces no imaging of the points in the object space to the corresponding points in an image space is possible in general.

This means that a homocentric pencil of rays originating from an object point will not be transformed into a homocentric pencil of rays converging to a single point in the image space (Fig. 4.8).

There is no unambiguous correspondence between object and image points. The next section shows how one can solve this pitfall.

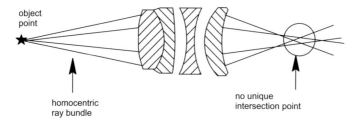

object point

homocentric ray bundle

no unique intersection point

Fig. 4.8 Ray pencils in image space.

4.2.2
Linearizing the Law of Refraction – the Paraxial Approximation

The region where the light rays intersect in Fig. 4.8 will become smaller as the incident angle between the rays and the normal to the surface becomes smaller. This will lead to a further approximation in geometrical optics.

If one develops the trigonometric sine function of Eq. (4.4) into a power series,

$$\sin \alpha = \alpha - \frac{\alpha^3}{3!} + \frac{\alpha^5}{5!} - \cdots \qquad (4.6)$$

and if we choose the incidence angle α so small that we may restrict ourselves to the first member of the series within a good approximation, then we may linearize the law of refraction. From

$$n_1 \sin \alpha_1 = n_2 \sin \alpha_2$$

we get

$$n_1 \alpha_1 = n_2 \alpha_2 \qquad (4.7)$$

This is the case of *Gaussian optics*. It may be shown then, that with a system of centred spherical surfaces (for instance two spherical surfaces which form a lens) one has a unique and reversible correspondence between object points and the related image points. The axis of symmetry is the *optical axis* as the line which joins all centres (vertices) of the spherical surfaces (Fig. 4.9).

The validation of this approximation is restricted to a small region around the optical axis, the *paraxial region*. For angles $\alpha \leq 7°$ the difference between the sine function and the arc of the angle α is about 1%.

Gaussian optics is thus the *ideal for all optical imaging* and serves as a reference for the real physical imaging. All deviations thereof are termed *aberrations* (deviations from Gaussian optics). It should be pointed out that these aberrations are not due to manufacturing tolerances, but the reason is the physical law of refraction.

A large amount of effort of optical designers goes to enlarge the paraxial region so that it may be useful for practical applications. In this case, a single

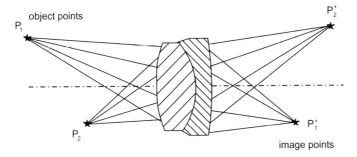

Fig. 4.9 Imaging with the linearized law of refraction.

lens will no longer be sufficient and the expense will rise with the requirements on image quality (Section 4.5).

Before we deal with the laws of imaging in the paraxial region we have to introduce some basic optical conventions.

4.2.3
Basic Optical Conventions

(1) Definitions for image orientations

The images through the optical system have, in general, a different orientation with regard to the object. This image orientation may be defined with the help of an asymmetrical object, for instance with the character of the number 1, Fig. 4.10.

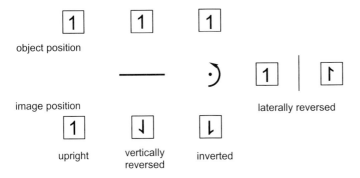

Fig. 4.10 The definitions for image orientations.

Important: for the definition of image orientations one has to look against the direction of light.

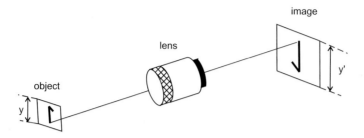

Fig. 4.11 The definition of the magnification ratio β.

(2) Definition of the magnification ratio β

The magnification ratio is defined as the ratio of image size to object size (Fig. 4.11).

$$\beta = \frac{\text{image size}}{\text{object size}} = \frac{y'}{y} \qquad (4.8)$$

Hence this is a linear ratio (not by areas).

(3) Real and virtual objects and images

A *real object or image* really exists; the image may be captured for instance by a screen. A real image point is given by the intersection point of a converging homocentric ray pencil.

A *virtual image* exists only seemingly; the eye or a camera locates the image at this position, but it may not be captured with a screen.

4.3 Example imaging by a mirror (Fig. 4.12). The eye *does not know* anything about reflection; it extends the rays backwards until the common intersection point of the pencil. Virtual image points are the result of backward extended ray pencils. There is no real intersection point; the point is virtual.

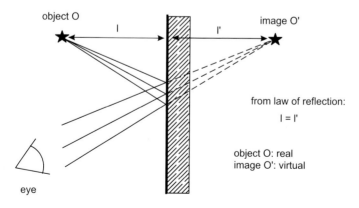

Fig. 4.12 Virtual image by reflection.

Now to *virtual objects*. They may occur when imaging through optical systems with several components in the sense that the image of the first component serves as an object for the following optical component. If the intermediate image is virtual this will be a *virtual object* for the imaging at the following component.

4.4 Example Taking an image with a camera over a mirror (Fig. 4.13). The real object O gives a virtual image by the mirror. This virtual image O' is an object for the imaging with the lens of the camera, so it is a virtual object. The final image is again real and located in the sensor plane.

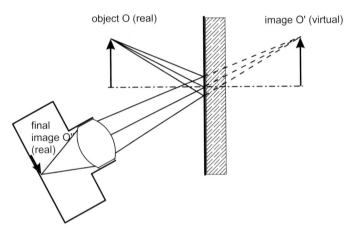

Fig. 4.13 Image by a mirror followed by a camera.

4.5 Example Imaging with a lens followed by a mirror (Fig. 4.14). The image by the lens is an object for the following imaging at the mirror. But this mirror is located in front of the normally real image by the lens! So the intermediate image does not really exist. It *would* be there if the mirror would not exist.

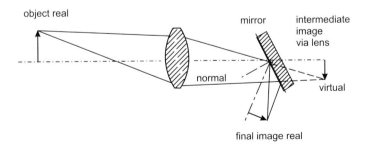

Fig. 4.14 A virtual object.

Hence this intermediate image is a virtual object for the imaging at the mirror. The final image is again real.

(4) The *tilt rule* for the evaluation of image orientations by reflection.

From the law of reflection one gets the following simple procedure for the evaluation of image orientations (Fig. 4.15).

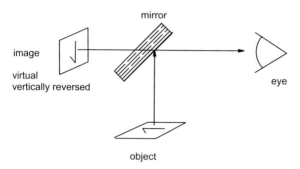

Fig. 4.15 The *tilt rule*.

We take an asymmetrical object; for instance, a character 1 push it along the light ray and tilt it to the surface of the mirror. Then we turn it in the shortest way perpendicular to the light ray. If we look against the direction of light we will see a vertically reversed image.

This rule is very useful if one has to consider many reflections. As an asymmetrical object one may take for instance a ball pen.

4.6 Example for two reflections (Fig. 4.16).
    Even number of reflections: the image is upright or inverted
    Odd number of reflections: the image is vertically reversed or laterally reversed

4.2.4
The Cardinal Elements of a Lens in Gaussian Optics

A parallel pencil of rays, which may be realized by a light source at infinity, passing through a converging lens will intersect at a single point, the image side focal point F' of the lens. A typical and well-known example is the *burning glass* which collects the rays coming from the sun at its focal point. The sun is practically an object at infinity (Fig. 4.17).

Next we will denote image side entities always with a prime. In just the same way one may define an object side focal point F (Fig. 4.18).

The location on the optical axis where a point-like light source gives, after passing through the lens, a parallel bundle of rays will be called the object side focal point F. For practical reasons, we enlarge the paraxial region which

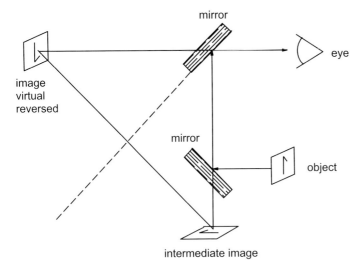

Fig. 4.16 Image position with two reflections.

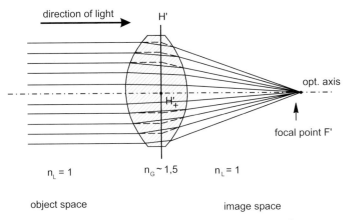

Fig. 4.17 The definition of the image side focal point F' and the principal plane H'.

means that the linear law of refraction is valid in the whole space, which is the idealized assumption of Gaussian optics.

If we extend in Figs. 4.16 and 4.17 the incident rays forward and the refracted rays backward (against the direction of light), then all the intersection points will lie in two planes the object and image side principal planes H and H' respectively. The intersection of these two planes with the optical axis is the object and image side principal point H_+ and H'_+. All the refractive power seems thus to be unified in the principal planes of the lens. In this way they may be a replacement for the refracting spherical surfaces, if we do not con-

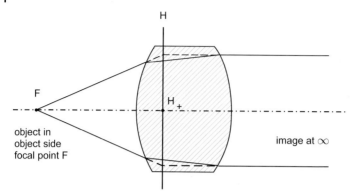

Fig. 4.18 The definition of the object side focal point F and the object side principal plane H.

sider the real ray path within the lens. If one knows the position of H, H' and F, F' then the imaging through the lens is completely determined. In order to see this we have to consider some further important properties of the principal planes.

> The principal planes are conjugate planes with the magnification ratio of $+1$. Conjugate planes or points are those which may be imaged to each other.

If one could place an object on the object side principal plane (for instance a virtual object!), then the image would lie in the image side principal plane with the same magnitude ($\beta = +1$).

4.1 Note *The focal points F and F' are not conjugate to each other. Conjugate points are for instance $-\infty$ and F'. This is an important exception from the general convention that conjugate entities are denoted by the same symbol, without the prime in the object space and with the prime in the image space.*

For a lens which is embedded in the same optical medium on both sides we further have the following rule:

> a ray passing through the point H_+ will leave the lens parallel to it and originating from H'_+.

In the following we always consider optical systems in air.

These rules will be made evident in Section 4.2.9, where we deal with the imaging equations and their related coordinate systems.

With these properties of the principal planes and those of the focal points (the ray parallel to the optical axis will pass through F' and the ray through F will leave the lens parallel to the optical axis) we may determine the image of an object without knowing the real form of the lens. In order to fix the position and magnitude of an image we select three distinct rays from the homocentric bundle of an object point (Fig. 4.19). Considering the property $\beta = +1$ for all points on the object side principal plane, all the rays between H and H' must be parallel to the optical axis.

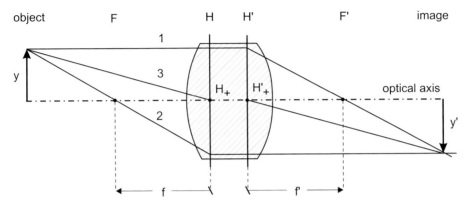

Fig. 4.19 Graphical construction of the image position.

Ray 1 parallel to the optical axis, must pass through F'

Ray 2 passing through the point F, must be parallel to the optical axis

Ray 3 passing through H_+ must leave the lens parallel to it and passing through H'_+ (additional).

Since the points H_+, H'_+ and F, F' completely determine the imaging situation in Gaussian optics, these points are called the *cardinal elements*.

Focal lengths f and f':

The directed distances from H_+ to F and from H'_+ to F' are called the object side focal length f and the image side focal length f', respectively. As a general convention, we assume that the light rays pass from left to right in our drawings. If the directed distance $\overrightarrow{H_+F}$ is against the direction of light, then f will be negative and if the directed distance $\overrightarrow{H'_+F'}$ will be in the direction of light, then f' will be positive and vice versa.

The positions of the cardinal elements with respect to the geometrical dimensions of the lens depend on their thickness d, refractive index n and the radii r_1 and r_2 of the spherical refractive surfaces 1 and 2. Here we will restrict ourselves to the pure presentation of the formula, without proof. Once we know the position of the cardinal elements

with respect to the lens geometry we do not need the lens geometry itself in Gaussian optics.

First, we have to describe the geometry of the lens and the corresponding conventions (Fig. 4.20).

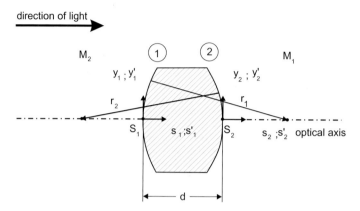

Fig. 4.20 The geometry of the thick lens.

Convention:

1. direction of light from left to right (positive z-axis).
2. S_1, S_2 vertices of the refracting surfaces 1 and 2.
3. Surface numbering is in the direction of light.
4. M_1, M_2 centres of the refracting spherical surfaces.
5. Radii of curvature. These are directed distances from S to M; $\vec{r} = \overrightarrow{SM}$, positive in the direction of light, negative against it. In Fig. 4.20 we have $r_1 > 0, r_2 < 0$.
6. d is the thickness of the lens, $d = \overline{S_1 S_2}$ and is always positive, because the refracting surfaces are numbered in the direction of light.

The position of the cardinal elements is described by the distances from the vertices S_1, S_2; they are taken with reference to the coordinate systems y_1, s_1, y_2', s_2'. Looking in the direction of light, objects or images on the left of the optical axis are positive ($+y_1, y_2'$ directions) while on the right-hand side of the optical axis they are negative (Fig. 4.21).

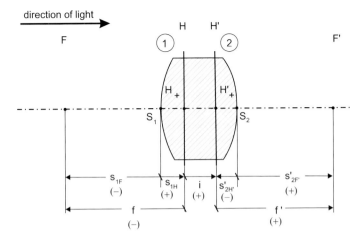

Fig. 4.21 The position of the cardinal elements.

$$f' = \frac{r_1 \cdot r_2 \cdot n}{(n-1) \cdot R} \quad , \quad f = -f' \tag{4.9}$$

$$R = n \cdot (r_2 - r_1) + d \cdot (n-1) \tag{4.10}$$

$$s'_{2H'} = -\frac{r_2 \cdot d}{R} \quad s_{1H} = -\frac{r_1 \cdot d}{R} \tag{4.11}$$

$$s'_{2F'} = f'\left(1 - \frac{(n-1) \cdot d}{n \cdot r_1}\right) \quad , \quad s_{1F} = -f'\left(1 + \frac{(n-1) \cdot d}{n \cdot r_2}\right) \tag{4.12}$$

$$i = d\left(1 - \frac{r_2 - r_1}{R}\right). \tag{4.13}$$

4.2.5
The Thin Lens Approximation

If the thickness of the lens is small compared to other dimensions of the imaging situation (radii, focal length, object distance) we may set $d = 0$ (thin lens approximation). In this case we obtain from Eqs. (4.9–4.13)

$$R = n(r_2 - r_1) \tag{4.14}$$

$$f' = \frac{r_1 r_2}{(n-1) \cdot (r_2 - r_1)} \tag{4.15}$$

$$s_{1H} = s'_{2H'} = i = 0. \tag{4.16}$$

The principal planes coincide and the lens is replaced by a single refracting surface. The ray through the principal point $H_+ = H'_+$ passes straight through (Fig. 4.22).

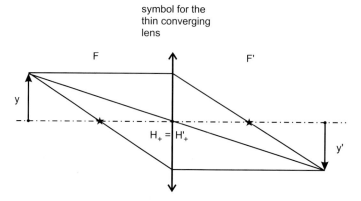

symbol for the thin converging lens

Fig. 4.22 Image construction with the thin lens.

4.2.6
Beam Converging and Beam Diverging Lenses

If in the formula for the focal length we have $r_1 > 0$ and $r_2 < 0$, we get $f' > 0$ and $f < 0$. The image side focal point F' then is on the right-hand side of the lens and F on the left. The lens is beam converging and the image construction is as shown in Fig. 4.19.

Another situation arises if $r_1 < 0$ and $r_2 > 0$ (Fig. 4.23).

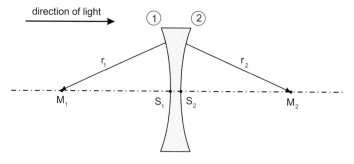

Fig. 4.23 Beam diverging lens.

Here one has $f' < 0$ and $f > 0$. The image side focal point F' is now in front of the lens (taken in the direction of light); the lens is beam diverging (Fig. 4.24).

The direction of rays incident parallel to the optical axis after refraction is such that the backward extension passes through F'! The image side focal length is negative. The imaging through a thin beam diverging lens is as in Fig. 4.25.

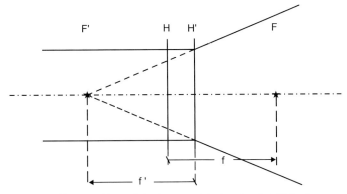

Fig. 4.24 The position of the image side focal point F' for beam diverging lenses.

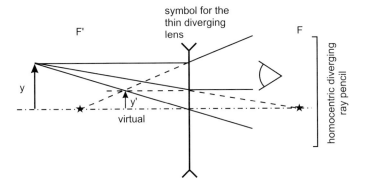

Fig. 4.25 Image construction with the thin beam diverging lens.

4.2.7
Graphical Image Constructions

4.2.7.1 Beam Converging Lenses

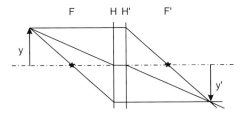

Fig. 4.26 Real object, real image.

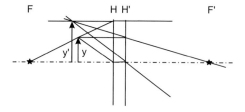

Fig. 4.27 Real object, virtual image.

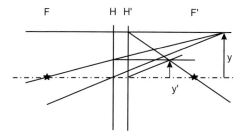

Fig. 4.28 Virtual object, real image.

4.2.7.2 Beam Diverging Lenses

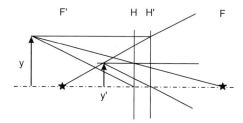

Fig. 4.29 Real object, virtual image.

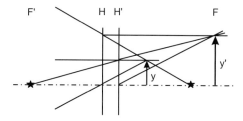

Fig. 4.30 Virtual object, real image.

4.2.8
Imaging Equations and Their Related Coordinate Systems

There are different possibilities for the formulation of imaging equations and correspondingly different sets of coordinate systems related to them.

4.2.8.1 Reciprocity Equation

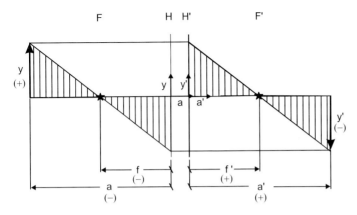

Fig. 4.31 Derivation of the reciprocity equation.

Here the object and image side coordinate systems (Fig. 4.31) are located in the principal planes.

From geometrical considerations and with the sign conventions one has

$$\frac{-y'}{y} = \frac{-f}{-(a-f)} \quad \text{on the object side and} \tag{4.17}$$

$$\frac{-y'}{y} = \frac{a'-f'}{f'} \quad \text{on the image side.} \tag{4.18}$$

Equating the right-hand sides of (4.17) and (4.18) gives

$$\frac{f'}{a'} + \frac{f}{a} = 1 \tag{4.19}$$

and with $f = -f'$

$$\frac{1}{a'} - \frac{1}{a} = \frac{1}{f'} \tag{4.20}$$

With $\beta = y'/y$ we get from Equation (4.17)

$$a = f\left(1 - \frac{1}{\beta}\right) \tag{4.21}$$

and from Equation (4.18):

$$a' = f'(1-\beta) \tag{4.22}$$

and finally from (4.21) and (4.22)

$$\beta = \frac{a'}{a} \tag{4.23}$$

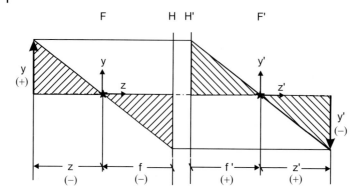

Fig. 4.32 Derivation of Newton's equations.

4.2.8.2 Newton's Equations

Here the origins of the object and image side coordinate systems are located at the object and image side focal points (Fig. 4.32).

From geometrical considerations we get

$$-\frac{y'}{y} = \frac{-f}{-z} \quad \text{on the object side and} \tag{4.24}$$

$$-\frac{y'}{y} = \frac{z'}{f'} \quad \text{on the image side.} \tag{4.25}$$

This gives

$$\beta = -\frac{f}{z} \quad \text{on the object side,}$$

$$\beta = -\frac{z'}{f'} \quad \text{on the image side, finally} \tag{4.26}$$

$$z \cdot z' = f \cdot f'$$

and from Eq. (4.26)

$$\beta = \sqrt{-\frac{z'}{z}} \tag{4.27}$$

where z and z' are the Newton coordinates of the conjugate points O and O', respectively. The advantage of the Newton equations lies in the fact that there are no reciprocal quantities and one may perform algebraic calculations in a somewhat easier way.

The position of the positive principal points H_+, H'_+:

For the conjugated points H_+ and H'_+ we have by definition $\beta = +1$. With Newton's equations we thus get for the position of these points:

$$\begin{aligned} z_{H_+} &= -f \\ z'_{H'_+} &= -f' \end{aligned} \tag{4.28}$$

Position of the negative principal points H_-, H'_-:
The conjugate points for which the magnification ratio β is equal to -1 are called negative principal points H_-, H'_-.

The image has thus the same size as the object, but is inverted. From Newton's equations with $\beta = -1$,

$$\begin{aligned} z_{H_-} &= +f \\ z'_{H'_-} &= +f' \end{aligned} \qquad (4.29)$$

Object and image are located in twice the focal length distances $(2f)$ and $(2f')$ respectively, counted from the corresponding principal planes H and H'. The image is real and inverted.

4.2.8.3 General Imaging Equation

A more general applicable imaging equation is obtained when the origins of the object and image side coordinate systems lie in two conjugate but elsewhere arbitrarily selected points P and P' in object and image space. In Section 4.2.12, we will use these equations with P and P' lying in the entrance and exit pupil of the optical system. Here is now the general derivation (Fig. 4.33).

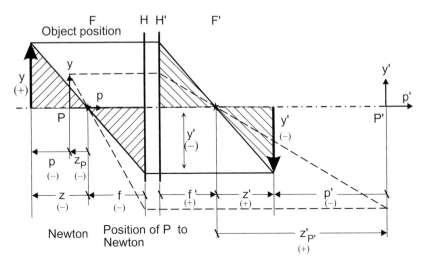

Fig. 4.33 General imaging equation.

From Fig. 4.33:

$$z = p + z_P \qquad p = z - z_P \qquad z_P = (\overrightarrow{FP})_z \qquad (4.30)$$

$$z' = p' + z'_{P'} \qquad p' = z' - z'_{P'} \qquad z'_{P'} = \left(\overrightarrow{F'P'}\right)_z \qquad (4.31)$$

With Newton's equations:

$$z \cdot z' = f \cdot f' = (p + z_P) \cdot (p' + z'_{P'}) \quad (4.32)$$
$$z_P \cdot z'_{P'} = f \cdot f' \quad (P \text{ and } P' \text{ are conjugate}) \quad (4.33)$$

$$\beta_z = -\frac{z'}{f'} = -\frac{f}{z} \qquad z = -\frac{f}{\beta_z} \qquad z' = -\beta_z \cdot f' \quad (4.34)$$

$$\beta_p = -\frac{z'_{p'}}{f'} = -\frac{f}{z_P} \qquad z_P = -\frac{f}{\beta_p} \qquad z'_{p'} = -\beta_p \cdot f' \quad (4.35)$$

Equations (4.30) and (4.31) being substituted into Eqs. (4.34) and (4.35) give:

$$p = f'\left(\frac{1}{\beta_z} - \frac{1}{\beta_p}\right) \quad (4.36)$$

$$p' = f'(\beta_p - \beta_z) \quad (4.37)$$

$$\frac{p'}{p} = \beta_p \cdot \beta_z \quad (4.38)$$

From Eqs. (4.32) and (4.33) we have

$$(p + z_P) \cdot (p' + z'_{p'}) = z_P \cdot z'_{p'} \quad (4.39)$$

and from this

$$\frac{z_P}{p} + \frac{z'_{p'}}{p'} + 1 = 0 \quad (4.40)$$

With Eqs. (4.35) and (4.34), namely $z_P = f/\beta_p$ and $z'_{p'} = -\beta_p \cdot f'$ we get

$$-\frac{\beta_p \cdot f'}{p'} = -\frac{f'}{\beta_p \cdot p} - 1 \quad (4.41)$$

$$\frac{\beta_p}{p'} = \frac{1}{\beta_p \cdot p} + \frac{1}{f'} \quad (4.42)$$

This is the analogous version of the reciprocal equation (4.20) where the origins of the coordinate systems are now generalized to two conjugate, but otherwise arbitrary points P and P'.

If these points coincide with the positive principal points H_+ and H'_+ with $\beta_p = +1$, then we come back to the reciprocal equation:

$$\frac{\beta_p}{p'} = \frac{1}{\beta_p \beta} + \frac{1}{f'} \xrightarrow{\beta_p = +1} \frac{1}{p'} = \frac{1}{p} + \frac{1}{f'} \quad (4.43)$$

With the notation $p' = a', p = a$, Eq. (4.43) changing to Eq. (4.20). Later (Section 4.2.13) we will use these imaging equations under the assumption

P = entrance pupil, P' = exit pupil. Then β_p will be the *pupil magnification ratio*.

4.2.8.4 The Axial Magnification Ratio
The axial magnification ratio is defined as the ratio of the axial displacement of an image, if the object is displaced by a small distance.

$$\text{Axial magnification ratio:}\ \alpha = \frac{dz'}{dz} \approx \frac{\Delta z'}{\Delta z} \tag{4.44}$$

From Newton's equation (4.26) we have

$$z' = \frac{f \cdot f'}{z} \quad \Rightarrow \quad \frac{dz'}{dz} = -\frac{f \cdot f'}{z^2} \tag{4.45}$$

and with $\beta = -f/z$ and $f' = -f$ this results in

$$\alpha = \beta^2 \tag{4.46}$$

4.2.9
The Overlapping of Object and Image Space

From the results of Section 4.2.8 we may conclude the following:

> The imaging with a lens is unidirectional. A displacement of the object along the optical axis results in a displacement of the image in the same direction. With the object and image distances a and a' according to the reciprocal equation, we get the following situations:

$$\begin{array}{llll} a < 0 & \text{real object} & a' < 0 & \text{virtual image} \\ a > 0 & \text{vitual object} & a' > 0 & \text{real image} \end{array}$$

As shown in Fig. 4.34 object and image space *penetrate each other completely*. In the axial direction we have certain stretchings and compressions.

The object and image space are thus not defined geometrically, but by the fact that the object space is defined *before the imaging* and the *image space after the imaging* at the optical system.

4.2.10
Focal Length, Lateral Magnification and the Field of View

From Newton's equations, namely Eq. (4.26), we may see that with a constant object distance the magnification ratio $|\beta|$ becomes larger, with a longer focal

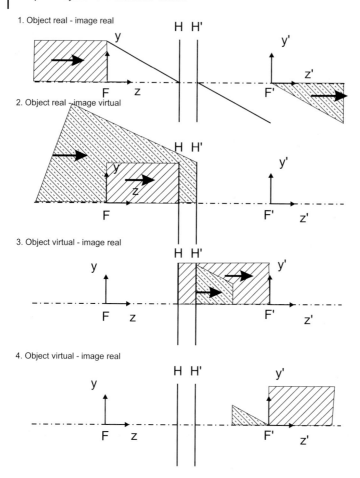

Fig. 4.34 The complete overlapping of object and image space.

Tab. 4.1 Normal focal lengths for different image formats.

Image format	Normal focal length (mm)
16 mm film	20
24×36 mm^2	45 – 50
60×60 mm^2	75 – 80
60×70 mm^2	90
60×90 mm^2	105
90×120 mm^2	135 – 150

length of the lens. As a convention a *normal focal length* is that which is equal to the diagonal of the image format (Table 3.1).

Longer focal length lenses will give a larger magnification ratio (*larger images*). Correspondingly, the object extension which belongs to a certain image format conjugate to it is smaller. This means that the object side field angle becomes smaller (Fig. 4.35).

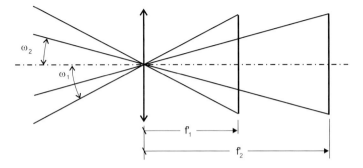

Fig. 4.35 Object side field angle ω and focal length.

According to DIN 19040 one has the following classification of object field angles and focal lengths (Table 4.2).

Tab. 4.2 Classification of optical systems according to the object side field angle ω.

Tele lens, extreme focal length lens	$2\omega < 20°$
Long focal length lens	$20° < 2\omega < 40°$
Normal focal length lens	$40° < 2\omega < 55°$
Wide angle lens	$2\omega > 55°$

4.2 Remark *The object side field angle is taken here with reference to the object side principal point H_+. In Section 4.2.13, we will see that it is more meaningful to refer this angle to the centre of the entrance pupil (EP). In that case this angle will be called the* object side pupil field angle ω_p. *Only in the special case of infinite object distance will the principal point H_+ related field angle ω and the entrance pupil related field angle ω_p be the same.*

4.2.11
Systems of Lenses

In Section 4.2.5 we replaced a lens, which consists of two spherical refracting surfaces, by the principal planes H and H' and the corresponding focal points F and F'. In this way the imaging properties of the lens were completely fixed, as long as we do not consider the real ray path within the lens.

In the same manner we may represent a *system of lenses* by their overall principal planes H_t, H'_t and their corresponding overall focal points F_t and F'_t (the index t stands for the total system).

We will demonstrate this by a graphical construction of these overall cardinal elements in the example of a system of two thin lenses. The extension to thick lenses of arbitrary numbers is straightforward but no principally new insight would be gained by doing so. The basic idea behind this is that most of the renowned optical manufacturers present the Gaussian data of their optical systems in their data sheets. So we may limit ourselves to the overall cardinal elements of a certain optical system when dealing with object-to-image relationships, as long as we are not interested in the real ray paths within that optical system.

Figure 4.36 shows a system of two thin lenses where we want to determine the overall cardinal elements graphically.

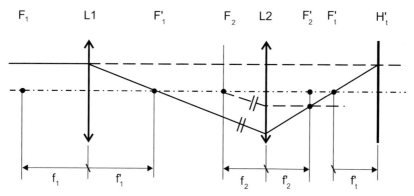

Fig. 4.36 Graphical construction of the overall cardinal elements.

Given are the principal points and focal points of the two single thin lenses. The direction of a light ray incident parallel to the optical axis will be constructed after passing each single lens up to the intersection point with the optical axis. This will give the overall image side focal point F'_t. In the same way, a light ray parallel to the optical axis but with opposite direction will give the overall object side focal point F_t. The intersection points of the backward extended, finally broken rays with the forward extended, incident rays will give the overall principal planes H_t and H'_t respectively.

With this procedure the question arises how we may find out the direction of the ray refracted at the second lens. This is shown for the image side cardinal elements of the system in Fig. 4.37.

The ray refracted at $L1$ (not shown in the figure) will give an intersection point with the object side focal plane of $L2$. All rays which would originate from this point would leave the lens $L2$ parallel to each other because they originate from a point on the focal plane. Within this ray pencil there is a ray passing through $H_{2+} = H'_{2+}$ and consequently leaves the thin lens $L2$ undeviated. This gives the direction of the real beam after refraction at $L2$.

The Gaussian data of an optical system are given with reference to two coordinate systems, the first in the vertex of the first lens surface. This describes

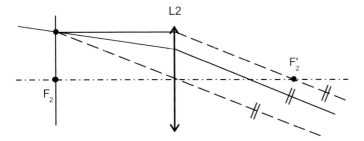

Fig. 4.37 Construction of the ray direction refracted at L2.

the object side entities. The second coordinate system has its origin in the last surface of the last lens of the system and describes the image side Gaussian entities. We will describe this with reference to the German standard DIN 1335 since no international standard with this subject exists (Fig. 4.38).

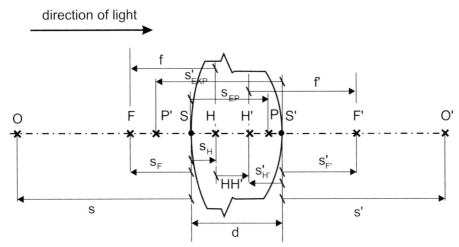

Fig. 4.38 Conventions for describing the Gaussian data of an optical system: e.g. s_F and $s'_{F'}$ are the z-coordinates (in the light direction) of the points F and , F' related to the origins S and S', respectively.

As an example we take a real optical system the Xenoplan 1,4/17 mm from Schneider Kreuznach to show the cardinal elements (Table 4.3):

4.3 Note *In the same way as we described the overall Gaussian data of an optical system* one may apply the above considerations to some *components of the system. It turns out that this is very useful for systems with two principal components in order to understand the particular constellation and properties of such systems. Among those are for example telecentric systems, retrofocus systems, tele objective systems and others. We will come back to this viewpoint in Section 4.2.14.*

Tab. 4.3 The Gaussian data of a real lens system.

Focal length	$f' = 17.57$ mm
Back focal distance	$s'_{F'} = 13.16$ mm
Front focal distance	$s_F = 6.1$ mm
Image side principle point distance	$s'_{H'} = -4.41$ mm
Principal plane distance	$i = HH' = -3.16$ mm
Entrance pupil distance	$s_{EP} = 12.04$ mm
Exit pupil distance	$s'_{EXP} = -38.91$ mm
Pupil magnification ratio	$\beta_p = 2.96$
f/number	$f/nr = 1.4$
Distance of the first lens surface to the last lens surface	$d = 24.93$ mm

4.7 Example We calculate the missing lens data f, s_H and $i_p = PP'$ using the lens data of Table 4.3.

$$f = -f' = -17.57 \text{ mm}$$
$$s'_{H'} = s'_{F'} - f' = 13.16 \text{ mm} - 17.57 \text{ mm} = -4.41 \text{ mm}$$
$$s_H = s_F - f = 6.1 \text{ mm} + 17.57 \text{ mm} = 23.67 \text{ mm}$$
$$s_H = d - i + s'_{H'} = 24.93 \text{ mm} + 3.16 \text{ mm} - 4.41 \text{ mm} = 23.68 \text{ mm}$$
$$i_p = d - s_{EP} + s'_{EXP} = 24.93 \text{ mm} - 12.04 \text{ mm} - 38.91 \text{ mm} = -26.02 \text{ mm}$$

The difference of the results of s_H is a rounding effect. We recognize that the data of $s'_{H'}$ and i in Table 4.3 are obsolete if we know s_F and d.

4.2.12
Consequences of the Finite Extension of Ray Pencils

4.2.12.1 Effects of Limitations of the Ray Pencils

Up to now we used three distinct rays to construct the images from the homocentric ray pencils originating from the object points:

1. ray parallel to the optical axis
2. ray through the object side focal point F
3. ray passing through the principal points H_+, H'_+

We could do this because in Gaussian optics there is a one-to-one relationship between object and image space: a homocentric diverging ray pencil originating from an object point will intersect in a single point, the image point. In reality, however the ray pencils will be limited by some finite openings, for instance the border of lenses or some mechanical components. It thus may be that the selected rays for image construction do not belong to this limited ray pencil. But because these rays *would intersect at the same image point* we may use them nevertheless for the image construction (Fig. 4.39).

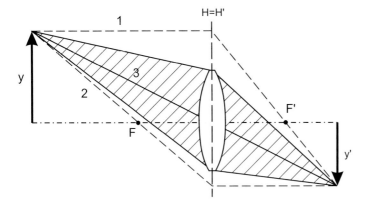

Fig. 4.39 The limitation of ray pencils.

However, these limitations of the ray pencils have other effects which are of utmost importance for the imaging. In order to show this, we use a very simple example of a thin lens with finite extension and in a certain distance an image plane also with finite extension, Fig. 4.40.

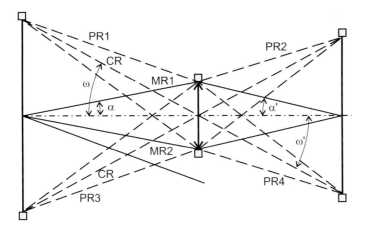

Fig. 4.40 The limitation of ray pencils in a simplified camera.

Only a limited amount of the ray pencils originating from the object points will contribute to the imaging, namely those rays passing through the aperture of the lens. Since this limited ray pencil corresponds to a certain fraction of the spherical wave emitted by the object point, only a fraction of the energy flow will be transmitted. This means that the finite extension of the lens limits the brightness of the image. We therefore call it an *aperture stop* because it limits the angle of the ray pencil which is responsible for the image brightness.

But in our imaging device there is yet another limiting opening namely the one in the image plane. Only that region in the object space may be imaged (sharply) onto the finite image plane, which is *conjugate* to this limited field. In this way the limited image plane restricts the extension of the image and object field. It is therefore called a *field stop*. However this field stop must not necessarily be located in the image plane. With a slide projector this opening is realized by the mask of the slide and hence lies in the object plane.

We thus have identified two important effects of stops:

Aperture stops have influences on the image brightness

Field stops limit the extensions of the object and image field.

Under the concept of stops we always will understand real (physically limiting) openings as for example openings in mechanical parts, the edges of lenses, the edges of mirrors a.s.o.

There are still other effects of stops on the imaging which will be partly discussed in the following sections. If for instance the stops are very small, then the approximation of geometrical optics (*light rays!*) will be invalid, and one has to consider the wave nature of light. This will have consequences for the image sharpness and resolution (Section 4.3).

Table 4.4 shows other important influences of stops on the imaging.

Tab. 4.4 Influences of stops in the ray path.

Aperture stops	Image brightness, depth of field
Very small aperture stops	Diffraction (image sharpness, resolution)
Field stops	Limitation of the field of view
Other (undesired) stops	Vignetting
Position and size of aperture stops	Influence on aberrations (e.g. distortion)

Finally in this section, we have to give some definitions for those rays which limit the ray pencils.

The rays which limit the homocentric pencil for object points on the optical axis will be called *marginal rays* (MR). The angle α of the half-cone with respect to the optical axis is a measure for this homocentric ray pencil.

Rays which limit the field of view are called *chief rays* (CR) (or *principal ray*). These rays proceed from the edge of the object field via the centre of the aperture stop up to the edge of the image field. They are the central rays for the ray pencil emerging from this object point. The corresponding angle ω with reference to the optical axis represents the field of view:

$$\omega = \text{object side field of view}$$
$$\omega' = \text{image side field of view}$$

Rays from object points at the edge of the object field which limit these ray bundles are called *pharoid rays* (PR). They are in analogy with the marginal rays (for object points on the optical axis) and thus determine the image brightness for points at the edge of the field of view.

4.4 Note *By our definition, the pharoid rays are conjugate rays, that means corresponding to the same ray pencil.*

Sometimes rays 1, 2, 3, 4 of Fig. 4.40, which characterize the totality of all rays passing through the optical system are called pharoid rays. In that case these rays are of course not conjugate rays.

4.1 Summary

Marginal ray (MR). It passes from the centre of the object to the edges of the aperture stop and to the centre of the image.

Chief ray (CR). It passes from the edges of the object to the centre of the aperture stop to the edges of the image.

Pharoid ray (PR). It passes from the edges of the object to the edges of the aperture stop up to the edges of the image.

4.2.12.2 Several Limiting Openings

The very simple model of ray pencil limitations introduced in the last section must now be extended to more realistic situations. In real optical systems there are several lenses with finite extension, the lens barrel, the iris diaphragm as limiting openings. It seems to be difficult to decide which of these stops is responsible for the image brightness and which one determines the field of view. In order to decide on this we introduce a more realistic example: the optical system may consist of two thin lenses and an iris diaphragm in between. The image plane IP shall again be limited according to the image format (Fig. 4.41)

In Fig. 4.41 we also introduced the cardinal elements H, H', F, F' of the system. These may be constructed by the rules introduced in Section 4.2.11. From these cardinal elements the position and size of the object plane OP (which is conjugate to the image plane) were found.

In order to decide which of the stops really limits the ray pencils, we use a simple trick: if a limiting stop is touched in the object space by a ray, then also the image of this stop in the image space will be touched by the corresponding ray, since object and image are conjugate to each other. The same is also valid for the opposite direction of light. Hence, we will image all the limiting stops into the object space (from right to left). From the centre of the object plane we look at all these stops (or images of stops in the object space) and may

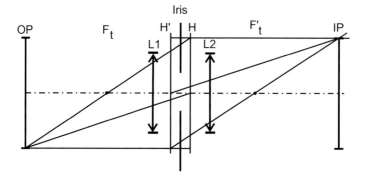

Fig. 4.41 An example for ray pencil limitations.

now decide which one is the smallest. This stop will limit the ray pencil for an object point on the optical axis in the object space and acts as an *aperture stop*. However, since this must not only necessarily be a physical stop but may also be an image of such a stop in the object space, we call it the *entrance pupil* (EP). This is shown in Fig. 4.42.

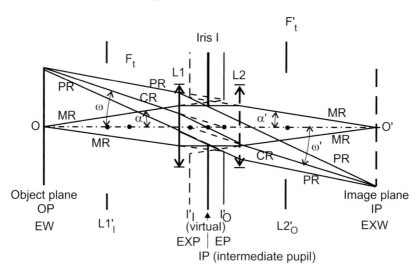

Fig. 4.42 The concept of pupils and windows.

All the limiting stops or the images of these which lie in the object space are drawn with thick lines (object plane OP, L1, L2$'_O$, I$'_O$). From these stops the image of the iris in object space (I$'_O$) is the smallest opening viewed from the point O. This image is virtual *and lies in the object space*, since it was imaged via L1 to the left. *Object and image space intersect each other here!* This is the *entrance pupil* (EP) of the system. The image of the entrance pupil EP in the image space (or what turns out to be the same: the image of the iris I via L2) is then the *exit pupil* (EXP) and limits the ray pencil in the image space. EXP is also virtual. The iris I itself is hence the aperture stop and at the same time *intermediate pupil* IP in an intermediate image space and is of course a real (physical) opening.

All the limiting stops in the image space or the *images* of stops in the image space are drawn with dashed lines. If one has already fixed the entrance pupil EP we must not only construct all these images, but only the image of EP in the image space, which is then the exit pupil EXP.

The same considerations are valid for stops which limit the field of view. In order to find these, one looks from the centre of the entrance pupil EP into the object space. The smallest opening with respect to the centre of EP is then the *object field stop* . But since this must not necessarily be a physical stop we generalize it to the concept of *windows*. Hence this opening is called *entrance window* (EW). The image of the entrance window in the image space is the *exit window* (EXW).

The concept of aperture and field stops from the simple constellation of Fig. 4.40 has thus been generalized to the concept of pupils and windows. Of course we still may speak from the aperture stop if this physical opening or an image of it in the object space acts as entrance pupil. The same considerations apply for the field stops.

For the limiting rays of the pencils the same rules apply as in Section 4.2.12.1, with the generalization that the concept of aperture stops and field stops will be replaced by pupils and windows. If there exist intermediate pupils and windows, they have to be incorporated in the order as the light rays touch them. Virtual pupils and windows are not really touched by the limiting rays but only by their (virtual) extensions. With the example of Fig. 4.42 we have the following definitions.

Marginal ray (MR). Centre of EW to the edges of EP (real up to L1, then virtual, since EP is virtual). From L1 to the edges of the intermediate pupil IP up to L2 (real since IP is real). From L2 to the centre of EXW. The backward extensions of these rays touch virtually the edges of EXP.

Chief rays (CR). Edges of EW to L1 and to the centre of EP (real up to L1, then virtual, since EP is virtual). From L1 to the centre of IP up to L2 (real since IP is real). From L2 to the edges of EXW. The backward extensions of these rays touch virtually the centre of EXP.

Pharoid rays (PR). Edges of EW to L1 and to the edges of EP (real up to L1, then virtual, since EP is virtual). From L1 to the edges of IP up to L2 (real since IP is real). From L2 to the edges of EXW. The backward extensions of these rays touch virtually the edges of EXP. The virtual pharoid rays are drawn only partly since they lie nearly to the real rays. Furthermore, the intermediate window (image of EXW at L2) is not taken into consideration.

Finally, we will show the limitations for the ray pencils of a real optical system. In Table 4.3 we gave all the relevant Gaussian data for this system, Fig. 4.43.

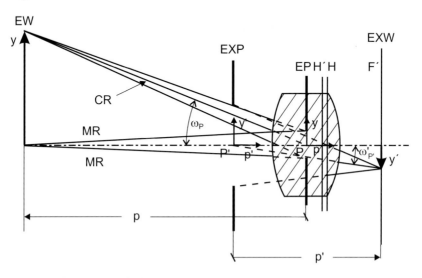

Fig. 4.43 The ray pencils for a real optical system.

Object and image space intersect each other! The exit pupil EXP for instance is geometrically in front of the optical system but belongs to the image space since it is the image of the iris imaged through the optical system into the image space (virtual image!).

4.2.12.3 Characterizing the Limits of Ray Pencils

For systems with a fixed object distance the extension of the ray pencil which corresponds to the marginal rays may be characterized by the angle α of Fig. 4.43, more distinctly by the *numerical aperture A*,

$$A = n \sin \alpha \quad (n = 1 \text{ for systems in air}) \qquad (4.47)$$

In many cases the object distance is variable or very large; then it makes no sense to characterize the ray pencil by the angle α. In this case one uses the *f-number*

$$f/nr = \frac{f'}{\text{Diameter EP}} \qquad (4.48)$$

Relation between f/nr and numerical aperture A:

$$f/nr = \frac{1}{2A} \qquad (4.49)$$

The extension of the object field is characterized by the object side field angle ω_P. It is given by the angle of the *chief rays* with respect to the optical axis. These chief rays pass from the edges of the windows to the centres of the pupils. On the object side, this angle is denoted by ω_P and on the image side by $\omega'_{P'}$. Since the corresponding ray pencils are limited by the pupils, the chief ray will be the central ray of the pencil. Because of the finite extension of the pencils only one distinct object plane will be imaged sharply onto the image plane. All object points in front or behind this *focussing plane* FOP will be imaged as a circle of confusion, the centre of which is given by the intersection point of the chief ray with the image plane, Fig. 4.44.

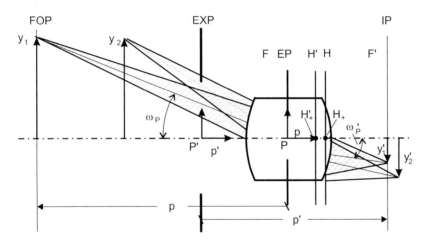

Fig. 4.44 Chief rays as centres of the circle of confusion.

Thus, it is more appropriate to use the chief rays for the characterization of the field of view and *not* the principal *point* rays (Section 4.2.10). The angle ω for the principal *point* rays is, in general, different from ω_p for the object side chief ray. Only for the case of infinitely distant objects these two angles are equal in magnitude (Fig. 4.45).

The extension of the image field is mostly characterized by the image circle diameter d. The image circle is, however, not sharply limited but the brightness decreases towards the border and in the peripheral zone the image is unsharp because of aberrations (deviations from Gaussian optics).

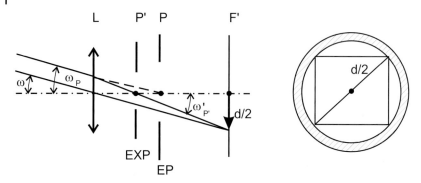

Fig. 4.45 Field angles ω and $\omega_P, \omega'_{P'}$ for infinite object distance.

Imaging equation with the coordinate systems in the centre of the pupils

In the general imaging equations of Section 4.2.8 we now choose as the two conjugate points for the coordinate systems the centre P of the entrance pupil on the object side and the centre P' of the exit pupil on the image side. The magnification ratio between these conjugate points is then a *pupil magnification ratio*

$$\beta_p = \frac{\text{diameter EXP}}{\text{diameter EP}} = \frac{\varnothing_{EXP}}{\varnothing_{EP}} \tag{4.50}$$

The imaging equations related to the centres of the pupils are then given by

$$p' = f'(\beta_p - \beta) \qquad \text{distance EXP-image} \tag{4.51}$$

$$p = f'\left(\frac{1}{\beta} - \frac{1}{\beta_p}\right) \qquad \text{distance EP-object} \tag{4.52}$$

$$\frac{p'}{p} = \beta_p \cdot \beta \qquad \text{magnification ratios} \tag{4.53}$$

$$\frac{\beta_p}{p'} = \frac{1}{\beta_p \cdot p} + \frac{1}{f'} \qquad \text{imaging equation} \tag{4.54}$$

Relation between object and image side pupil field angles

From Fig. 4.43 we have

$$\tan \omega_P = \frac{y}{p} \qquad \text{on the object side and} \tag{4.55}$$

$$\tan \omega'_{P'} = \frac{y'}{p'} \qquad \text{on the image side.} \tag{4.56}$$

Thus

$$\frac{\tan \omega_P}{\tan \omega'_{P'}} = \frac{y \cdot p'}{p \cdot y'} \tag{4.57}$$

with Eq. (4.53) and

$$\frac{y}{y'} = \frac{1}{\beta} \tag{4.58}$$

One finally has
$$\frac{\tan \omega_P}{\tan \omega'_{P'}} = \beta_p \qquad (4.59)$$

If $\beta_p \neq +1$, then ω_P and $\omega'_{P'}$ will be different, as in Fig. 4.42 where $\beta_p \approx 3$!

4.2.12.4 The Relation to the Linear Camera Model

We now may answer the question of Section 4.1.2 on how the linear camera model is related to Gaussian optics. From the results of the last section we have the following statements:

1. The chief rays (CR) are the projection rays.

2. There exist *two* projection centres, one in the centre of the entrance pupil (P) for the object side, and one in the centre of the exit pupil (P') on the image side.

In order to reconstruct the objects from the image coordinates one has to reconstruct the object side field angles. But this may not be done with P' as the projection centre because the chief ray angles ω_p and $\omega'_{p'}$ are in general different.

One rather has to choose the projection centre on the image side such that the angles to the image points are the same as those from the centre of the entrance pupil P to the object points. We denote this image side projection centre by P^* with a distance c to the image plane. Then we require

$$\omega_P = \omega'_{P*} \quad \text{or} \quad \frac{y}{p} = \tan(\omega_P) = \tan(\omega'_{P*}) = \frac{y'}{c} \qquad (4.60)$$

This gives
$$c = \frac{y'}{y} \cdot p = \beta \cdot p \qquad (4.61)$$

This equation describes nothing but a *scaling of the projection distance on the image side*. Since the image is changed by a factor β with respect to the object we have to change the distance of the projection centre as well by the same factor in order to arrive at the same projection angle. This gives the connection to the linear camera model.

The common projection centre is $P = P^*$ (P = centre of EP) and the distance to the image plane is the camera constant c.

$$c = \beta \cdot p = f' \left(1 - \frac{\beta}{\beta_p}\right) \qquad (4.62)$$

In photography this distance is called the *correct perspective viewing distance*. Figure 4.46 illustrates this fact.

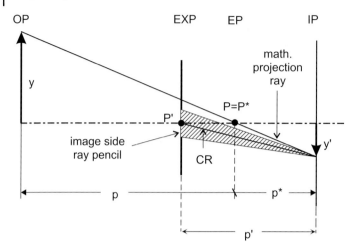

Fig. 4.46 Projection model and Gaussian optics.

4.5 Note *In order to avoid misunderstandings we point out that when dealing with error consideration with variations of the imaging parameters (e.g. variation of image position) one always has to take into consideration the real optical ray pencils. The projection rays of Fig. 4.46 are purely fictive entities on the image side.*

Finally, we make the following comment concerning the limitations of ray pencils in Gaussian approximation:

4.6 Remark *Against this model of Gaussian ray pencil limitation one may argue that the pupil aberrations (deviation of real rays from Gaussian optics) are in general large and thus the Gaussian model may not be valid in reality. But indeed we used only the centres of the pupils in order to derive the projection model. If one introduces canonical coordinates according to Hopkins [1] one may treat the ray pencil limitation and thus the projection model even for aberrated systems in a strictly analogous way. The pupils are then reference spheres and the wavefront aberrations of real systems are calculated relative to them. The reduced pupil coordinates will then describe the extensions of the ray pencils and are – independent of the chief ray angle – unit circles. For a detailed treatment we refer to [1].*

4.2.13
Geometrical Depth of Field and Depth of Focus

As a consequence of the finite extension of the ray pencils only a particular plane of the object space (the focussing plane FP) will be imaged sharply onto the image plane. All other object points in front and behind the focussing plane will be imaged more or less unsharply. The diameters of these circles of confusion depend on the extension of the ray pencils, Fig. 4.47

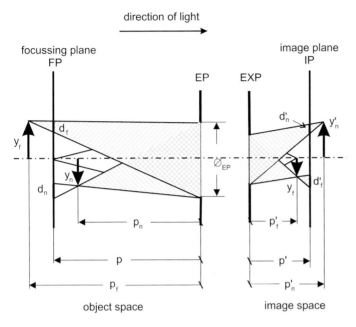

Fig. 4.47 Depth of a field.

d is the diameter of the circle of confusion in the object space. Index f means far point and index n is for **n**ear point. First we consider the situation in the object space. By geometric considerations we have

$$\frac{\varnothing_{EP}}{-p_f} = \frac{d_f}{-(p_f - p)} \tag{4.63}$$

or

$$p_f = \frac{p \cdot \varnothing_{EP}}{\varnothing_{EP} - d_f} \tag{4.64}$$

Multiplying numerator and denominator with β we get with

$$d'_f = -\beta \cdot d_f \tag{4.65}$$

$$p_f = \frac{\beta \cdot p \cdot \varnothing_{EP}}{\beta \cdot \varnothing_{EP} + d'_f} \tag{4.66}$$

and in the same way for the near point

$$p_n = \frac{\beta \cdot p \cdot \varnothing_{EP}}{\beta \cdot \varnothing_{EP} - d'_n} \tag{4.67}$$

In the following, we set $d' = d'_f = d'_n$ with d' as the permissible circle of confusion in the image plane. From these relations we may now calculate the depth of field as a function of the object distance or, alternatively, as a function of the magnification ratio β.

4.2.13.1 Depth of Field as a Function of the Object Distance p

With the imaging equation (4.52) solved for p and with Eq. (4.48) for f/nr we get from Eqs (4.66) and (4.67):

$$p_{f,n} = \frac{f'^2 \cdot p}{f'^2 \pm d' \cdot \left(p + \frac{f'}{\beta_p}\right) \cdot (f/nr)} \tag{4.68}$$

$$T = p_f - p_n = \frac{2p \cdot f'^2 \cdot d' \cdot (f/nr) \cdot \left(p + \frac{f'}{\beta_p}\right)}{f'^4 - d'^2 \cdot (f/nr)^2 \cdot \left(p + \frac{f'}{\beta_p}\right)^2}. \tag{4.69}$$

These equations are exact within the scope of Gaussian optics.
Approximation 1:
Let

$$p \geq 10 \cdot \frac{f'}{\beta_p} \tag{4.70}$$

Then we have from (4.68) and (4.69)

$$p_{f,n} \approx \frac{f'^2}{f'_2 \pm p \cdot d' \cdot (f/nr)} \tag{4.71}$$

$$T \approx \frac{2 \cdot f'^2 \cdot f/nr \cdot d' \cdot p^2}{f'^4 - (f/nr)^2 \cdot d'^2 \cdot p^2} \tag{4.72}$$

Approximation 2:
In addition, we require

$$f'^4 \geq 10 \cdot p^2 \cdot (f/nr)^2 \cdot d'^2 \tag{4.73}$$

Then

$$T \approx \frac{2 \cdot (f/nr) \cdot d' \cdot p^2}{f'^2} \tag{4.74}$$

Within the validity of this approximation the depth of field T is inversely proportional to the square of the focal length (*with a constant object distance p*). Reducing the focal length by a factor of 2 will enlarge the depth of field by a factor of 4.

4.8 Example for the validity of Eq. (4.74):
With $d' = 33$ µm and $f/nr = 8$ we get for

$$f' = 100\,\text{mm} \qquad 1\,\text{m} \leq p \leq 12\,\text{m}$$
$$f' = 50\,\text{mm} \qquad 0.5\,\text{m} \leq p \leq 3\,\text{m}$$
$$f' = 35\,\text{mm} \qquad 0.35\,\text{m} \leq p \leq 1.5\,\text{m}.$$

4.2.13.2 Depth of Field as a Function of β

We replace p in Eq. (4.68) with Eq. (4.52) and introduce f/nr. Then

$$p_{f,n} = \frac{f'^2 \left(1 - \frac{\beta}{\beta_p}\right)}{f' \cdot \beta \pm (f/nr) \cdot d'} \tag{4.75}$$

and

$$T = \frac{2 \cdot \left(1 - \frac{\beta}{\beta_p}\right) \cdot (f/nr) \cdot d'}{\beta^2 + \frac{(f/nr)^2 \cdot d'^2}{f'^2}} \tag{4.76}$$

These equations are exact within the scope of Gaussian optics.

Approximation. We require

$$\beta^2 \geq 10 \cdot \frac{(f/nr)^2 \cdot d'^2}{f'^2} \tag{4.77}$$

and introduce the *effective f-number*

$$(f/nr)_e = (f/nr) \cdot \left(1 - \frac{\beta}{\beta_p}\right)$$

Then

$$T \approx \frac{2 \cdot (f/nr)_e \cdot d'}{\beta^2} \tag{4.78}$$

Within the validity of this approximation the depth of field T is independent of the focal length (with equal β). For the *depth of focus T* we have

$$T' \approx \beta^2 \cdot T \tag{4.79}$$
$$T' = 2 \cdot (f/nr)_e \cdot d' \tag{4.80}$$

4.9 Example for the range of validity

$$d' = 33\,\text{µm (35 mm format)} \qquad f' = 50\,\text{mm}, \ f/nr = 16, \ |\beta| \geq \frac{1}{30}$$
$$d' = 100\,\text{µm (9} \times 12\,\text{cm}^2) \qquad f' = 150\,\text{mm}, \ f/nr = 16, \ |\beta| \geq \frac{1}{30}$$

4.2.13.3 The Hyperfocal Distance

In some cases one has to image objects at (nearly) infinity and in the same case some other objects which are at a finite distance in the foreground. Then it will not be useful to adjust the object distance at infinity. One rather has to adjust the distance such that the far distance p_f of the depth of field is at infinity. This is called the *hyperfocal distance* p_∞. From Eq. (4.75) we see that for $p_f = \infty$ the denominator must vanish, which means

$$\beta_\infty = -\frac{(f/nr) \cdot d'}{f'} \tag{4.81}$$

and from Newton's equations we have

$$\beta_\infty = \frac{z'_\infty}{f'} \tag{4.82}$$

Thus

$$z'_\infty = (f/nr) \cdot d' \tag{4.83}$$

z'_∞ is the Newton image side coordinate, counting from F'. For the object plane conjugate to this we again have with Newton's equations

$$z_\infty = -\frac{(f')^2}{(f/nr) \cdot d'} \tag{4.84}$$

For the pupil coordinates p_∞, p'_∞ (with reference to the pupils) we see from Eq. (4.68) that the denominator must vanish and we get

$$p_\infty = -\frac{(f')^2}{(f/nr) \cdot d'} - \frac{f'}{\beta_p} = z_\infty - \frac{f'}{\beta_p} \tag{4.85}$$

and with Eq. (4.52)

$$p'_\infty = (f/nr) \cdot d' + f' \cdot \beta_p = z'_\infty + f' \cdot \beta_p \tag{4.86}$$

Fig. 4.48 gives an illustration of Newton's- and pupil coordinates.

Finally, we are interested at the *near limit* of the hyperfocal depth of field. With equ (4.75) and the minus sign in the denominator and Eq. (4.81) we get

$$p_n = -\frac{(f')^2}{2 \cdot (f/nr) \cdot d'} - \frac{f'}{2\beta_p} \tag{4.87}$$

comparing it with (4.85)

$$p_n = \frac{p_\infty}{2} \tag{4.88}$$

The near limit of the hyperfocal depth of field is thus half of the hyperfocal distance p_∞!

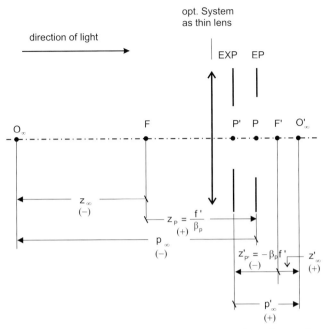

Fig. 4.48 The relations between Newton's and pupil coordinates: $z_P = FP = f'/\beta_p$, $z'_{P'} = F'P' = -\beta_p f'$.

For the near limit of the depth of focus we start with Eq. (4.81) and with

$$p = z - \frac{f'}{\beta_p}, \quad z \cdot z' = -f'^2 \qquad (4.89)$$

One has

$$z'_n = 2 \cdot z'_\infty \left[\frac{1}{1 + \frac{\beta_\infty}{\beta_p}} \right] \qquad (4.90)$$

Usually

$$\beta_\infty = -\frac{(f/nr) \cdot d'}{f'} \ll \beta_p \qquad (4.91)$$

which leads to the approximation

$$z'_n \approx 2 \cdot z'_\infty \qquad (4.92)$$

4.2.13.4 The Permissible Size for the Circle of Confusion d'

The permissible size for the circle of confusion depends on the resolution capability of the detector. For visual applications this is the eye with a resolving power of approx. 1.5′. Together with the correct perspective viewing distance this would give the size d' for the circle of confusion.

But usually one does not look at an image with this distance for instance taken with a telephoto lens, but much closer. The reason is that the eye may detect some details which could not be seen in the object only if the viewing angle is larger than that to the object. With the correct perspective viewing distance we have however the same angle.

It thus makes sense to connect the viewing distance with the final image format. Because a larger taking (format) requires only a smaller magnification for the final print then a smaller taking format, the permissible size for the circle of confusion may thus be larger for the larger sensor (format).

The starting point for the evaluation is an image format of $9 \times 12 \text{ cm}^2$, which is observed at the standardized viewing distance of 25 cm without magnification. For the permissible circle of confusion we then have

$$d' \approx \frac{\text{sensor diagonal}}{1500} = \frac{150}{1500} = 0.1 \text{ mm} \tag{4.93}$$

From the standardized viewing distance (5/3 of image diagonal) this corresponds to an angle of approx. 1.5′ which is the resolving power of the eye. The values of d' for other sensor formats follow from Eq. (3.93) with the corresponding image diagonal. For small diagonals, d' will become smaller but the viewing angle will be approx. the same, Table 4.5.

Tab. 4.5 Standard diameters for the permissible circle of confusion for different image formats.

Format	90x120	60x90	60x60	45x60	24x36	18x24	7.5x10.5	3.6x4.8
$\dfrac{\text{diagonal}}{\text{mm}}$	150	108	85	75	43	30	13	6
$d' = \dfrac{\text{diag}}{1500}$	0.1 mm	72 μm	57 μm	50 μm	29 μm	20 μm	9 μm	4 μm
d'_N stand.	0.1 mm	75 μm	60 μm	50 μm	33 μm	25 μm	15 μm	10 μm
Δ in %	0	-4	-5	0	-12	-20	-40	-60
β for enlarging to diag. 150 mm	1	1.39	1.76	2	3.49	5	11.5	25
viewing angle for $d'_N \cdot \beta$ from 250 mm	1′23″	1′26″	1′27″	1′23″	1′35″	1′43″	2′23″	3′26″

moving pictures less critical

Tab. 4.6 Summary of the formula for the depth of field and depth of focus.

Depth of field as function of

Distance p	Magnification ratio β

geometrically exakt:

$$T = \frac{2p \cdot f'^2 \cdot d' \cdot (f/nr) \cdot \left(p + \dfrac{f'}{\beta_p}\right)}{f'^4 - d'^2 \cdot (f/nr)^2 \cdot \left(\beta + \dfrac{f'}{\beta_p}\right)^2}$$

$$T = \frac{2 \cdot \left(1 - \dfrac{\beta}{\beta_p}\right) \cdot (f/nr) \cdot d'}{\beta^2 + \dfrac{(f/nr)^2 \cdot d'^2}{f'^2}}$$

$$p_{f,n} = \frac{p \cdot f'^2}{f'^2 \pm d' \cdot (f/nr)) \cdot \left(p + \dfrac{f'}{\beta_p}\right)}$$

$$p_{f,n} = \frac{f'^2 \left(1 - \dfrac{\beta}{\beta_p}\right)}{f' \cdot \beta \pm (f/nr) \cdot d'}$$

Approximation 1: $p < 10 f'/\beta_p$

$$T \approx \frac{2 f'^2 \cdot (f/nr) \cdot d' \cdot p^2}{f'^4 - (f/nr)^2 \cdot d'^2 \cdot p^2}$$

$$p_{f,n} \approx \frac{f'^2}{f'^2 \pm p \cdot d' \cdot (f/nr)}$$

Approximation: $\beta^2 \geq 10 \cdot \dfrac{(f/nr)^2 \cdot d'^2}{f'^2}$

$$T \approx \frac{2 \cdot (f/nr)_e \cdot d'}{\beta^2}$$

$$T' \approx \beta^2 \cdot T = 2 \cdot (f/nr)_e \cdot d'$$

Approximation 2:
$10 f'/\beta_p < p < f'/3(f/nr)d'$

$$T \approx \frac{2(f/nr) \cdot d' \cdot p^2}{f'^2}$$

4.2.14
The Laws of Central Projection–Telecentric System

4.2.14.1 An Introduction to the Laws of Perspective

It is a well-known fact that extreme tele and wide angle photos show very different impressions of spatial depth. When taking pictures with extreme telephoto lenses (long focal length lenses) the impression of spatial depth in the image seems to be very flat, whereas with extreme wide angle lenses (short focal lengths) the spatial depth seems to be exaggerated. As an example we look at the two photos of a checker board (Fig. 4.49).

Both photos have been taken with a small format camera (24 × 36 mm), the left one with a moderate wide angle lens (focal length 35 mm) and the right one with a tele-lens of 200 mm focal length. In both pictures the foreground (e.g. the front edge of the checker board) has about the same size. In order to realize this, one has to approach the object very closely with the wide angle

Tab. 4.7 Summary of the formula for the hyperfocal distance.

Hyperfocal distance	
NEWTON coordinates	**Pupil coordinates**
exakt $z_\infty = \dfrac{f'^2}{(f/nr) \cdot d'}$	$p_\infty = z_\infty - \dfrac{f'}{\beta_p}$
$z'_\infty = (f/nr) \cdot d'$	$p'_\infty = z'_\infty + \beta_p f'$
$z_n = \dfrac{1}{2}\left(z_\infty + \dfrac{f'}{\beta_p}\right)$	$p_n = \dfrac{p_\infty}{2}$
$z'_n = 2z'_\infty \dfrac{1}{1 - \dfrac{(f/nr) \cdot d'}{f' \cdot \beta_p}}$	$p'_n = z'_n + \beta_p \cdot f'$
Approximation: $(f/nr) \cdot d' \leq \dfrac{1}{10} f' \cdot \beta_p$	$p'_n = 2z_\infty + \beta_p \cdot f'$
$z'_n = 2 \cdot z'_\infty$	

Fig. 4.49 Wide angle and tele perspective.

lens (approx. 35 cm). With the tele-lens however the distance has to be very large (approx 2 m).

One may clearly observe that with the wide angle lens the chess figures, which were arranged in equally spaced rows, seem to become rapidly smaller, thus giving the impression of large spatial depth. In contrast, with the tele-lens picture the checker figures seem to reduce very little, thus leading to the impression that the spatial depth is much lower. What are the reasons for these different impressions of the same object?

In order to give a clear understanding for these reasons, we look away from all laws of optical imaging by lenses and look at the particularly simple case of a pinhole camera (*camera obscura*).

We call the distance between the pinhole and the image plane the *image width p'*. With this pinhole camera we want to image different regularly spaced objects of the same size into the image plane.

Figure 4.50 shows a side view of this arrangement.

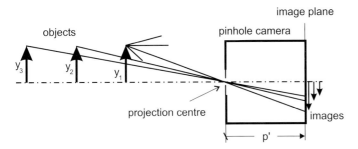

Fig. 4.50 Imaging with a pinhole camera.

Each object point, e.g. also the top of the objects – characterised by the arrows – will emit a bundle of rays. But only one single ray of this bundle will pass through the pinhole onto the image plane. Thus this indicates the top of the object in the image and hence its size. All other rays of this object point are of no interest for this image.

In this manner if we construct the images of all objects we will see that the images will become smaller and smaller the more distant the object is from the pinhole. The pinhole is the centre, where all imaging rays cross each other. Therefore we will call it the *projection centre*.

> To summarize, *the farther away the object located from the projection centre the smaller the image. This kind of imaging is called* **entocentric perspective**.

The characteristic feature of this arrangement is that viewed in the direction where the light travels we have first the object, then the projection centre and finally the image plane. The objects are situated at a finite distance *in front* of the projection centre.

But now what are the reasons for the different impressions of spatial depth? With the wide angle lens, the distance to the first object y is small, whereas with the tele-lens this distance is considerably larger by a factor of 6. We shall simulate this situation now with the help of our pinhole camera (Fig. 4.51).

The picture on the top (case a) shows the same situation as in Fig. 4.50. In the picture at the bottom (case b) the distance to the first object is however 3.5 times larger. In order to image the first object y_1 with the same size as in case (a) we have to expand the image distance by the same factor 3.5 correspondingly. We now see that the three objects y_1, y_2, y_3 have nearly of the same size as with the telephoto lens of Fig. 4.49. Whereas in case (a) the image y'_3 is

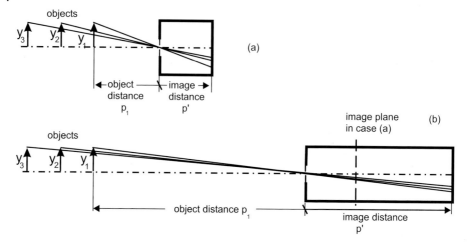

Fig. 4.51 Imaging with different object distances.

roughly half the size of the image y'_1 in case (b) it is three quarter of that size. There would have been no change in the *ratio* of the image sizes if the image distance p' in case (b) were the same as in case (a); only the total image size would become smaller and could be enlarged to the same size.

4.7 Conclusion The only thing that is important for the *ratio* of the image sizes is the distance of the objects from the projection centre.

To gain a clearer picture of the resulting spatial depth impression, we have to arrange the object scene in space, as shown in Fig. 4.52

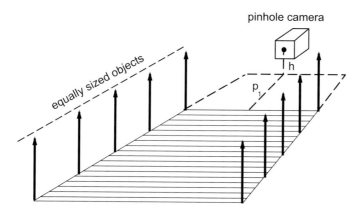

Fig. 4.52 Positional arrangement of the objects in space.

Here we have two rows of equally sized and equally spaced objects (arrows). The objects we may imagine to be the checker figures of Fig. 4.49. The pinhole camera is located centrally and horizontally in between the two object rows at a distance p_1 and height h relative to the objects. For the construction of the images taken with different object distance p_1 we choose to view this scene from above. This is shown in Fig. 4.53.

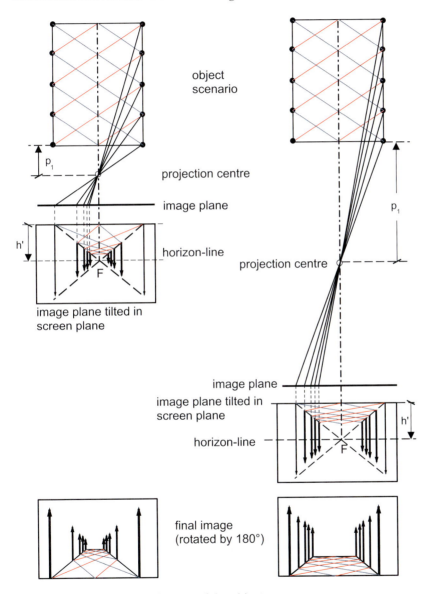

Fig. 4.53 Constructing the images of the object scene.

Here the objects are represented by circles (view from above) and are situated on the dashed background plane. The pinhole camera is shown by its projection centre and its image plane. The image plane is perpendicular to the paper plane and is therefore only seen as a straight line. The projection rays (view from above) are shown only for one row of the objects for the sake of clarity and not to overload the drawings.

The two parallel straight lines, on which the objects are positioned, intersect each other at infinity. Where is this point located in the image plane? It is given by the ray with the same direction as the two parallel straight lines passing through the projection centre and cutting the image plane.

If we tilt the image plane into the plane of the paper, we may now reconstruct all the images.

The horizon line (skyline) is to be found in an image height h' as the image of the height h of the pinhole camera above the ground. The point of intersection of the two parallel lines at infinity is located in the image plane on this horizon and is the *far point F for this direction*. We transfer now all the intersection points of the projection rays with the perpendicular image plane onto the tilted image plane. The connecting line between the intersection point of the nearest object projection ray and the far point gives the direction of convergence of the two parallel object lines in the image. In this way we may construct all the images of different objects. If we now finally rotate the image by 180° and remove all auxiliary lines, we then see very clearly the effect of different perspective for the two object distances. This is in rather good agreement with the photo of the introductory example (Fig. 4.49).

Imagine now we want to measure the heights of different objects from their images. This proves to be impossible because equally sized objects have different image sizes depending on the object distance which is unknown in general. The requirement would be that the images should be mapped with a constant size ratio to the objects. Only in this case would equally sized objects result in equally sized images, independent of the object distance.

In order to see what this requirement implies, we will investigate the image size *ratios* of equally sized objects under the influence of the object distance (Fig. 4.54).

From Fig. 4.54

$$\frac{y'_1}{y_1} = \frac{p'}{p_1} \quad \text{and} \quad \frac{y'_2}{y_2} = \frac{p'}{p_2} \tag{4.94}$$

The image size ratio y'_1/y'_2 will be called m. Dividing the first by the second equation and knowing that $y_1 = y_2$ will result in the following:

$$\frac{y'_1}{y'_2} = m = \frac{p_2}{p_1} \tag{4.95}$$

The image sizes of equally sized objects are inversely proportional to the corresponding object distances.

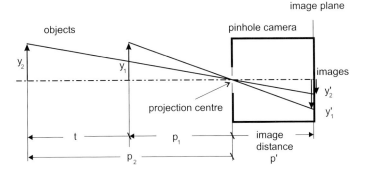

Fig. 4.54 Image size ratio and object distance

In addition to
$$t = p_2 - p_1 = y'_1 \cdot m - y'_1 \tag{4.96}$$
we have
$$p_1 = \frac{t}{m-1}, \quad p_2 = \frac{t \cdot m}{m-1} \tag{4.97}$$

If now p_1 becomes larger and larger, the ratio
$$\frac{p_2}{p_1} = \frac{p_1 + t}{p_1} \tag{4.98}$$

approaches 1, and hence also the image size ratio y'_1/y'_2. In the limit, as p_1 reaches infinity, the image size ratio will be exactly 1.

This is the case of *telecentric perspective*:

> The projection centre is at infinity, and all equally sized objects have the same image size.

Of course we may not realize this in practice with a pinhole camera, since the image distance p' would also go to infinity. But with an optical trick we may transform the projection centre into infinity without reducing the image sizes to zero. In front of the pinhole camera, we position a lens in such a way that the pinhole lies at the (image side) focal point of this lens (see Fig. 4.55).

The image of the pinhole imaged by the lens onto the left side in the object space is now situated at infinity. Only those rays parallel to the axis of symmetry of the pinhole camera will pass through the pinhole, because it lies at the image side focal point of the lens. That is why all equally sized objects will be imaged with equal size onto the image plane.

As a consequence of the imaging of the pinhole, we now have *two projection centres*, one on the object side (which lies at infinity) and other on the side of the images, which lies at a finite distance p' in front of the image plane. Therefore this arrangement is called *object side telecentric perspective*.

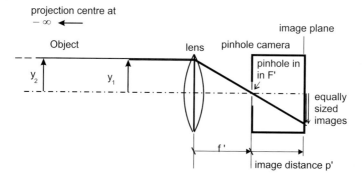

Fig. 4.55 Object side telecentric perspective.

If the pinhole is positioned with regard to the lens in such a way that its image (in object space on the left) is situated at a finite distance in front of the objects, the case of *hypercentric perspective* is realized (see Fig. 4.56).

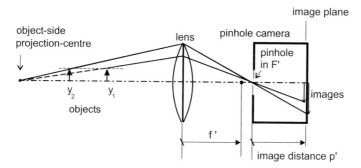

Fig. 4.56 Hypercentric perspective.

The projecting rays (passing through the pinhole) now have such a direction, that their backward prolongations pass through the object side projection centre.

All equally sized objects will now be imaged with larger size the farther they are away from the pinhole camera. Thus we may look *around the corner*!

At the end of this section we will show some examples which are of interest for optical measurement techniques. The object is a plug-in with many equally sized pins. Figure 4.57 shows two pictures taken with a 2/3 inch CCD camera and two different lenses.

The left side picture is taken with a lens with a focal length of 12 mm (this corresponds roughly to a standard lens of 50 mm focal length for the format 36 × 24 mm). Here we clearly see the perspective convergence of parallel lines, which is totally unsuitable for measurement purposes. On the other hand, the

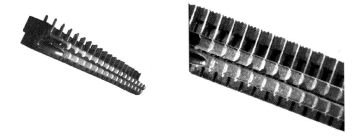

Fig. 4.57 Entocentric and telecentric perspective.

telecentric picture on the right, which is taken from the same position, shows all the pins with equal size and is thus suited for measurement purposes.

The perspective convergence is not only true for horizontal views into the object space, but also for views from the top of the objects. This is shown in the following pictures for the same plug-in (Fig. 4.58)

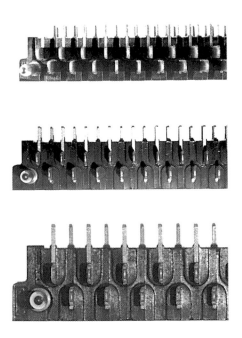

Fig. 4.58 Viewing direction from the top of the object.

The picture with the standard lens ($f' = 12$ mm) shows only one pin exactly from top (it is the fourth counted from left). All others show the left or right sidewalls of the pins. For the picture with a three-times tele lens ($f' = 35$ mm), this effect is already diminished. But only the picture taken with a telecentric lens clearly shows all pins exactly from the top. In this case, the measurement of the distance of all the pins is possible.

4.2.14.2 Central Projection from Infinity – Telecentric Perspective

Object Side Telecentric Systems
Limitation of ray bundles

As has been shown in Section 4.2.12.4 there are two projection centres with Gaussian optics, the centre P of the entrance pupil in object space and the centre P' of the exit pupil in image space. Hence, for object side telecentric systems the object side projection centre P (and therefore the entrance pupil) has to be at infinity (cf. Section 4.2.14.1).

In the simplest case – which we choose for didactic reasons – this may be done by positioning the aperture stop at the image side focal point of the system. The aperture stop is then the exit pupil EXP. The entrance pupil EP is the image of the aperture stop in the object space which is situated at infinity and is infinitely large. For the pupil magnification ratio we therefore have

$$\beta_P = \frac{\varnothing_{\text{EXP}}}{\varnothing_{\text{EP}}} = 0 \tag{4.99}$$

(Fig. 4.59).

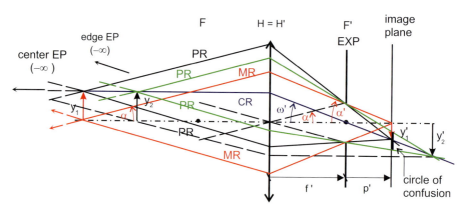

Fig. 4.59 The principle of object side telecentric imaging.

The chief rays (blue) are parallel to the optical axis (virtual from the direction of the centre EP – dashed) passing through the edge of the object to the lens, from there to the centre of EXP, and finally to the edge of the image.

This chief ray is the same for all equally sized objects (y_1, y_2) regardless of their axial position, and it intersects the fixed image plane at one single point.

For the object y_2 it is (in a first approximation) the central ray of the circle of confusion.

> The result is that the image size in a *fixed* image plane is independent of the object distance

The marginal rays MR (red) and the pharoid rays PR (black and green) have an aperture angle α which depends on the diameter of the aperture stop AS (=EXP). Here we may see the characteristic limitation for the imaging with telecentric systems:

> The diameter of the optical system (in the present case, the lens) must at least be as large as the object size plus the size determined by the aperture angle of the beams.

The imaging equations now become

$$p' = -f' \cdot \beta \quad \text{(counted from EXP)} \quad (4.100)$$

$$p = \infty \quad \text{(counted from EP)} \quad (4.101)$$

$$\beta_p = \frac{\tan \omega_p}{\tan \omega'_{p'}} = 0 \Rightarrow \varnothing_{EP} = \infty \quad (4.102)$$

$$c = f' \cdot \left(1 - \frac{\beta}{\beta_p}\right) = \infty \quad \text{(camera constant)} \quad (4.103)$$

The f-number $f/nr = f'/\varnothing_{EP}$ is now formally 0 and $1/\beta_p$ approaches infinity. Therefore the usual expression for the depth of field will be undetermined. We may however transform the equation

$$T = \frac{2d' \cdot (f/nr) \cdot \left(1 - \frac{\beta}{\beta_p}\right)}{\beta^2}$$

$$= \frac{2d' \cdot f' \cdot \left(\frac{1}{\varnothing_{EP}} - \frac{\beta}{\varnothing_{AP}}\right)}{\beta^2} \quad (4.104)$$

$$\lim_{\varnothing_{EP} \to \infty} T = \frac{-2 \cdot f'}{\varnothing_{AP}} \cdot \frac{d'}{\beta} \quad \text{from figure 4.59:} \quad \frac{f'}{\varnothing_{AP}/2} = \sin \alpha \quad (4.105)$$

We have

$$T = -\frac{d'}{\beta \cdot \sin \alpha} \quad (4.106)$$

$A = \sin(\alpha)$ is the object side numerical aperture. With the sine condition

$$\frac{\sin \alpha}{\sin \alpha'} = -\beta \quad (4.107)$$

we get

$$T = \frac{d'}{\beta^2 \cdot \sin \alpha'} \quad (4.108)$$

$A' = \sin(\alpha') =$ image side numerical aperture.

Bilateral Telecentric Systems
Afocal systems

For bilateral (object *and* image side) telecentric systems, the entrance pupil EP and the exit pupil EXP have to be at infinity. This we may achieve only with a two-component system, since the aperture stop has to be imaged in the object space as well as in the image space to infinity. It follows that the image side focal point of the first component must coincide with the object side focal point of the second component ($F_1' = F_2$). According to Section 4.2.11 (system of lenses) we may represent the two components of the system by two thin lenses in a first approximation.

Such systems are called afocal systems since the object and image side focal points (F, F') for the complete system are at infinity now. Figure 4.60 shows the case of two components $L1, L2$ with positive power.

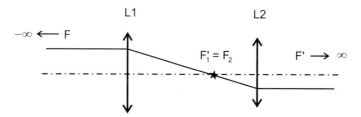

Fig. 4.60 Afocal systems.

A ray parallel to the optical axis leaves the system also parallel to the optical axis. The intersection point lies at infinity; this is the image side focal point F' of the system. The same is true for the object side focal point F of the system (afocal means *without focal points*). A typical example of an afocal system with two components of positive power is the *Keplerian telescope*. Objects which practically lie at infinity will be imaged by this system again to infinity. The images are observed behind the second component with the relaxed eye (looking at infinity). Figure 4.61 shows this for an object point at the edge of the object (e.g. the edge of the moon disc).

The clear aperture of the first component ($L1 =$ telescope lens) now acts as the entrance pupil EP of the system. The image of EP, imaged by the second component ($L2 =$ eye-piece), is the exit pupil EXP in image space (dashed black lines).

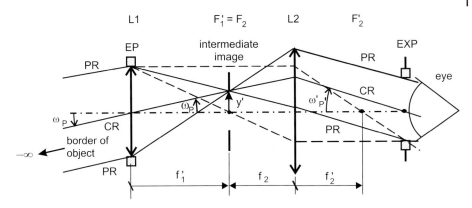

Fig. 4.61 The Kepler telescope.

The object point at infinity sends a parallel pencil of rays at the field angle ω_P into the component L1. L1 produces the intermediate image y' at the image side focal plane F_1'. At the same time this is the object side focal plane for component L2. Hence this intermediate image is again imaged to infinity by the second component L2 with a field angle of $\omega'_{P'}$. The eye is positioned at the exit pupil EXP and observes the image without focussing under the field angle $\omega'_{P'}$. Since the object and the image are situated at infinity (and are infinitely large), the magnification ratio is undetermined. In this case, for the apparent magnification of the image, it depends on how much the tangent of the viewing angle $\omega'_{P'}$ has been changed compared to the observation of the object without telescope from the place of the entrance pupil (tangent of viewing angle ω_P). In this way the telescope magnification is defined:

> The telescope magnification V_F is the ratio of the viewing angle $(\tan \omega'_{P'})$ with instrument to the viewing angle $(\tan \omega_P)$ without instrument.

$$V_F = \frac{\text{tangent of the viewing angle } \omega'_{P'} \text{ with instrument}}{\text{tangent of the viewing angle } \omega_P \text{ without instrument}} \quad (4.109)$$

From Fig. 4.61

$$\tan \omega_P = \frac{y'}{f_1'} \quad (4.110)$$

$$\tan \omega'_{P'} = \frac{y'}{f_2} = -\frac{y'}{f_2'} \quad (4.111)$$

Hence the telescope magnification V_F is given by

$$V_F = -\frac{f_1'}{f_2'} \quad (4.112)$$

With a large focal length f_1' and a short eye-piece focal length f_2' we will get a large magnification of the viewing angles. This is the mean purpose of the Keplerian telescope.

Imaging with afocal systems at finite distances

With afocal systems one may however not only observe objects at infinity, but we may also create real images in a finite position. This is true in the case of a contactless optical measurement technique and has been the main reason for choosing the Keplerian telescope as a starting point.

In order to generate real images with afocal systems we have to consider the following fact: we will get only real final images when the intermediate image of the first component is *not* situated in the region between $F_1' = F_2$ and the second component. Otherwise, the second component would act as a magnifying lupe and the final image would be virtual. This is explained in Fig. 4.62.

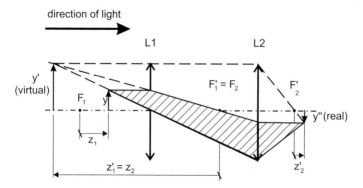

Fig. 4.62 Real final images with afocal systems.

The object is now situated between the object side focal points F_1 and $L1$. Because of this, the intermediate image y' is virtual and is located in front of the object. This virtual intermediate image is then transformed by $L2$ into the real final image y''.

For the mathematical treatment of the imaging procedure, we now apply Newton imaging equations one after the other to components $L1$ and $L2$.

Intermediate image at component $L1$:

$$\beta_1 = -\frac{z_1'}{f_1'} = -\frac{f_1}{z_1} \tag{4.113}$$

This intermediate image acts as an object for the imaging at the second component L2 (in Fig. 4.62). The intermediate image is virtual, hence for the second component a *virtual object*!

With the relation

$$F_1' = F_2 \quad \text{(afocal system)} \tag{4.114}$$

we have

$$z_2 = z_1' \tag{4.115}$$

Imaging at the second component (L2):

$$\beta_2 = -\frac{f_2}{z_2} = -\frac{z_2'}{f_2'} \tag{4.116}$$

Overall magnification ratio:

$$\beta = \beta_1 \cdot \beta_2 \tag{4.117}$$

$$\beta = -\frac{z_1'}{f_1'} \cdot \left(\frac{f_2}{z_2}\right) \tag{4.118}$$

With $z_1' = z_2$ and $f_2 = -f_2'$, this gives:

$$\beta = -\frac{f_2'}{f_1'} \tag{4.119}$$

With the telescope magnification V_F,

$$\beta = \frac{1}{V_F} \tag{4.120}$$

The overall magnification ratio is independent of the object position (characterized by z_1) and constant.

This is valid – in contrast to the object side telecentric imaging (where the image size has been constant only for a *fixed* image plane, cf. Section 4.2.14.2.1) – *for all image planes* conjugate to different object planes. If the image plane is slightly tilted, there is – to a first approximation – no change in the image size, because the image side chief ray is parallel to the optical axis. The image size error produced by the tilt angle only depends on the cosine of this angle.

The *image positions* may be calculated with the help of the second form of the Newton equations:

$$z_1 = \frac{f_1' \cdot f_1}{z_1'} \quad , \quad z_2' = \frac{f_2' \cdot f_2}{z_2} \tag{4.121}$$

Dividing the second by the first equation fields,

$$\frac{z_2'}{z_1} = \frac{f_2' \cdot f_2}{f_1' \cdot f_1} \cdot \frac{z_1'}{z_2} \tag{4.122}$$

With $z_1' = z_2$ we have

$$z_2' = \frac{f_2' \cdot f_2}{f_1' \cdot f_1} \cdot z_1 = (-\beta)^2 \cdot z_1 = \beta^2 \cdot z_1 \tag{4.123}$$

For the *axial* magnification ratio α we have

$$\alpha = \frac{dz'_2}{dz_1} = \beta^2 = \text{const} \qquad (4.124)$$

Limitation of ray bundles

We now position the aperture stop AS at $F'_1 = F_2$. The entrance pupil EP (as the image of the aperture stop in the object space) is now situated at $(-)$ infinity, and the exit pupil EXP (as the image of the aperture stop in the final image space) is situated at $(+)$ infinity. With that we may construct the ray pencil limitations as in Fig. 4.63.

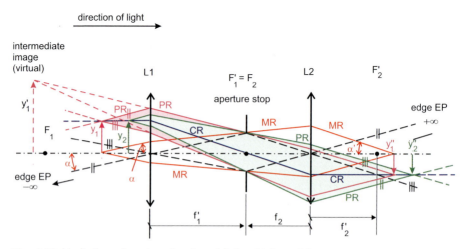

Fig. 4.63 Limitation of ray bundles for a bilateral telecentric system.

The *principal rays* MR (blue) are again the projection rays and represent the central ray of the imaging ray pencils. They start parallel to the optical axis (coming virtually from the direction of the centre of EP at infinity – dashed) passing through the edges of the objects y_1 and y_2 up to component L1, from there through the centre of the aperture stop (which is conjugate to EP, EXP) to component L2 and finally proceeding parallel to the optical axis to the edge of the images and pointing virtually to the centre of EXP at infinity (dashed).

The *marginal rays* MR (red) have an aperture angle α which depends on the diameter of the aperture stop. They start at the axial object point (pointing backwards virtually to edge of EP) up to L1, from there to the edge of the aperture stop up to component L2 and proceeding finally to the image (pointing virtually to the edge of EXP at infinity).

The *pharoid rays* PS (pink and green) also have an aperture angle α. They are coming virtually from the edge of EP (dashed) to the edge of the objects, from there real up to component L1, then through the edge of the aperture stop up to L2 and finally to the edge of the image (pointing virtually to the edge of EXP at infinity-dashed).

Here we may see again very clearly that because of the bilateral telecentricity equally sized objects will give equally sized images, even for different image planes.

Pupil magnification ratio β_p

Since the entrance pupil EP and EXP are at infinity (and are infinitely large), we may not use the term *magnification ratio* but should speak correctly of the pupil magnification β_p. The principle rays of the Keplerian telescope (cf. Fig. 4.61) are now the marginal rays of the bilateral telecentric system, since the positions of pupils and objects/images have been interchanged. Therefore, we may define the pupil magnification β_p in an analogous way to the telescope magnification V_F:

$$\beta_p = \frac{\text{tangent of marginal ray } \alpha' \text{ of EXP}}{\text{tangent of marginal ray } \alpha \text{ of EP}} = \frac{\tan \omega'_F}{\tan \omega_F} = V_F \quad (4.125)$$

$$\beta_p = \frac{\tan \alpha'}{\tan \alpha} = \frac{\frac{1}{2}\varnothing_{AB}}{f_2} \cdot \frac{f'_1}{\frac{1}{2}\varnothing_{AB}} = -\frac{f'_1}{f'_2} = V_F \quad (4.126)$$

Depth of field

Starting point is the formula which expresses the depth of field in dependence of the magnification ratio β (cf. Section 4.2.13.2).

$$T = \frac{2 \cdot d'}{\beta^2} \cdot (f/nr)_e \quad (4.127)$$

we have

$$(f/nr)_e = \frac{f'}{\varnothing_{EP}} \left(1 - \frac{\beta}{\beta_p}\right) \quad (4.128)$$

$$\frac{(f/nr)_e}{\beta} = \frac{f'}{\varnothing_{EP}} \cdot \left(\frac{1}{\beta} - \frac{1}{\beta_p}\right) = \frac{p}{\varnothing_{EP}} \quad (4.129)$$

Hence

$$T = \frac{2 \cdot d'}{\beta} \cdot \frac{(f/nr)_e}{\beta} = \frac{2 \cdot d'}{\beta} \cdot \frac{p}{\varnothing_{EP}} = \frac{d'}{\beta} \cdot \frac{z_p}{\varnothing_{EP}} \quad (4.130)$$

With

$$\lim_{p \to \infty} \frac{\varnothing_{EP}}{z_p} = \sin \alpha \quad (4.131)$$

This gives for the object side *depth of field*:

$$T = \frac{d'}{\beta \cdot \sin \alpha} \quad (4.132)$$

with
$$\frac{\sin \alpha'}{\sin \alpha} = \frac{1}{\beta} \quad \text{and} \quad T' = \beta^2 \cdot T \tag{4.133}$$

we have
$$T' = \frac{d'}{\sin \alpha'} \tag{4.134}$$

4.3
The Wave Nature of Light

4.3.1
Introduction

We come back to the starting point of geometrical optics: because of the very small wavelength of light ($\lambda \approx 0.5$ µm) we introduced in the wave equation (4.1) the limiting case $\lambda \to 0$ as approximation. This resulted in the Eikonal equation and the concept of light rays.

As a further approximation we introduced for the paraxial region the *Gaussian optics*: the homocentric ray pencil from a certain object point is transferred to a homocentric ray pencil converging at a well-defined image *point*.

This is valid only under the assumption that the spatial dimensions are large compared to the wavelength of light. If one looks at regions of the order of the magnitude of the wavelength in the neighbourhood of the image 'point' or at the limits of the geometrical shadow, then the limiting case $\lambda \to 0$ is invalid and geometrical optics is no more meaningful.

Since this wave propagation is in general not rectilinear as required by the light rays, it is termed as *diffraction*.

First investigations to the propagation of waves were undertaken by Christian Huygens (1678!). He postulated that each point on a wavefront is the source of a secondary spherical wave. Then the wavefront at a later time will be given by the envelope of all the secondary wavelets (Fig. 4.64).

With these assumption he was however not able to give a satisfactory explanation of the wave propagation. Much later (1818) Augustin Fresnel introduced Young's principle of interference in diffraction theory. In this way he could calculate the diffraction phenomena with good accuracy. However he had to introduce some strange assumptions on the Huygens wavelets:

1. The phase of the wavelets differs by $\pi/2$ from the original wave front at this point

2. The amplitude is proportional to $1/\lambda$

3. Existence of an obliquity factor

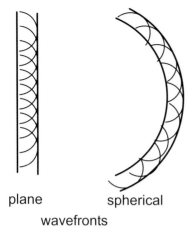

plane spherical
  wavefronts

Fig. 4.64 Reconstruction of the wavefront by secondary wavelets.

These assumptions could be confirmed by Kirchhoff (1882) and Sommerfeld (1894) in the development of a scalar diffraction theory.

4.3.2
The Rayleigh–Sommerfeld Diffraction Integral

As implied by the wave equation (4.1), the electric disturbance is a vectorial quantity. We neglect this vectorial nature by dealing with one component only. This scalar theory will give accurate results only if the following conditions are met.

1. The diffracting aperture must be large compared to the wavelength
2. The diffracted field must not be observed too close to the aperture

In any case all polarization phenomena are neglected, since these essentially depend on the vectorial nature of light. We then may write the field amplitude for a monochromatic wave as

$$u(\vec{r}, t) = U_0(\vec{r}) \cos\left[2\pi\nu t + \phi(\vec{r})\right] \tag{4.135}$$

$U_0(\vec{r})$ and $\phi(\vec{r})$ are the amplitude and phase of the wave at position $\vec{r}$, respectively. This may be expressed in an exponential form as

$$u(\vec{r}, t) = \operatorname{Re}\left[U_0(\vec{r}) \cdot e^{i[2\pi\nu t + \phi(\vec{r})]}\right] = \operatorname{Re}\left[U(\vec{r}) \cdot e^{i2\pi\nu t}\right] \tag{4.136}$$

where Re means the real part, and $U(\vec{r})$ is the complex amplitude and a function of the position $\vec{r}(x, y, z)$.

The real disturbance $u(\vec{r}, t)$ must satisfy the scalar wave equation

$$\Delta u - \frac{1}{c^2}\frac{\partial^2 u}{\partial t^2} = 0 \tag{4.137}$$

(4.136) in (4.137) with $c = \lambda \cdot \nu$ gives

$$\Delta U + \frac{4\pi^2}{\lambda^2} U = 0 \tag{4.138}$$

With

$$\left|\vec{k}\right|^2 = k^2 = \frac{4\pi^2}{\lambda^2}, \quad k = \text{wave number} \tag{4.139}$$

one has

$$\Delta U(\vec{r}) + k^2 U(\vec{r}) = 0 \tag{4.140}$$

This is the time-independent Helmholtz equation. Only are the relative phases of the wave of importance not the absolute phases given by t since the monochromatic wave exhibits absolutely synchronous time behaviour in all points of space. The particular solutions of Eq. (4.140) are:

1. plane wave in the x-direction

$$U(x) = U_0 e^{-ikx} \tag{4.141}$$

2. spherical wave

$$U(|\vec{r}|) = \frac{U_0}{r} \cdot e^{-ikr} \tag{4.142}$$

We shortly describe the conditions which lead to the Rayleigh-Sommerfeld formulation of scalar diffraction:

1. with the Helmholtz equation (4.140)
2. with Greens theorem of vector analysis and
3. with the appropriate boundary conditions (the Greens function introduced by Sommerfeld without discrepancy)

one gets the amplitude of the wave at a point $P(\vec{r}_0)$ behind an aperture of an infinite opaque screen as a superposition of spherical waves which originate from the wavefront in the diffracting aperture, Fig. 4.65.

The diffraction integral has the form

$$U(\vec{r}_0) = -\frac{1}{i\lambda} \iint_S U(\vec{r}_1) \frac{e^{-i\vec{k}\cdot(\vec{r}_0-\vec{r}_1)}}{|\vec{r}_0-\vec{r}_1|} \cos(\vec{n}, \vec{r}_0 - \vec{r}_1) \, dS \tag{4.143}$$

The integration has to be taken over the whole surface S of the aperture.

The first factor under the integral is the (complex) amplitude of the incident wavefront at point $\vec{r}_1$; the second factor is the spherical wave.

The following conditions must hold:

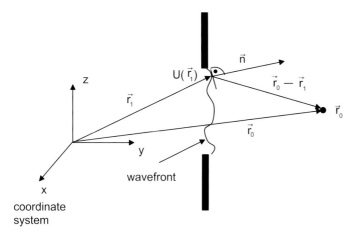

Fig. 4.65 Diffraction by a plane screen.

1. The amplitude of the wave immediately behind the limiting aperture is zero whereas within the aperture it remains undisturbed.

2. The distance of the observation point $\vec{r}_0$ from the screen is very large compared to the wavelength: $|\vec{r}_0 - \vec{r}_1| \gg \lambda$.

Equation (4.143) gives the explanation for the seemingly arbitrary assumptions of Fresnel:

1. phase advance of the secondary wavelets by

$$-\frac{1}{i} = +i = e^{+i\frac{\pi}{2}}$$

2. amplitudes proportional to $1/\lambda$

3. obliquity factor as cosine between the normal $\vec{n}$ of the wavefront at $\vec{r}_1$ and $\vec{r}_0 - \vec{r}_1$ as the direction of point $U(\vec{r}_0)$ viewed from $\vec{r}_1$:

$$\cos(\vec{n}, \vec{r}_0 - \vec{r}_1) \quad .$$

4.8 Note *The diffraction integral may be interpreted as* **superposition integral** *of linear system theory (refer to section 4.4)*

$$U(\vec{r}_0) = \iint_S U(\vec{r}_1) h(\vec{r}_0, \vec{r}_1) \, dS \qquad (4.144)$$

with

$$h(\vec{r}_0, \vec{r}_1) = -\frac{1}{i\lambda} \frac{e^{-i\vec{k}\cdot(\vec{r}_0 - \vec{r})}}{|\vec{r}_0 - \vec{r}_1|} \cos(\vec{n}, \vec{r}_0 - \vec{r}_1) \qquad (4.145)$$

as impulse response of the system of wave propagation. The impulse response corresponds to the Huygens elementary waves or: the response of the system of wave propagation to a point-like excitation is a spherical wave.

4.3.3
Further Approximations to the Huygens–Fresnel Principle

We make the following assumptions:

1. plane diffracting aperture Σ
2. plane observation surface B parallel to it.
3. distance z of these two planes much larger than the dimensions of Σ and B

$$z \gg \text{dimension of } \Sigma$$
$$z \gg \text{dimension of } B$$

(Fig. 4.66).

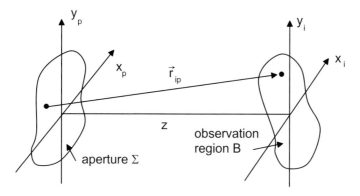

Fig. 4.66 Geometry of the diffraction constellation.

This gives

$$\cos\left(\vec{n}, \vec{r}_{ip}\right) \approx 1 \quad ,$$

with an accuracy of 1% for an angle of 8°. Then the diffraction integral will give

$$U(x_i, y_i) = -\frac{1}{i\lambda} \iint_{\mathbb{R}^2} P(x_p, y_p) \cdot U(x_p, y_p) \cdot \frac{e^{-i\vec{k}\cdot\vec{r}_{ip}}}{r_{ip}} dx_p dy_p \qquad (4.146)$$

The function $P(x_p, y_p)$ takes into consideration the boundary conditions

$$P(x_p, y_p) = \begin{cases} 1 \text{ within the diffracting aperture} \\ 0 \text{ outside the diffracting aperture} \end{cases}$$

With our geometrical assumptions 1 to 3 we may replace $\vec{r}_{ip} = \vec{r}_0 - \vec{r}_1$ in the denominator of the third factor in the integral equ (4.146) by z, *not however* in the exponent of the third factor, since $|\vec{k}|$ is a large number $|\vec{k}| = \frac{2\pi}{\lambda}$ and the phase error would be much larger than 2π.

4.3.3.1 Fresnel's Approximation
The expression in the exponent will be developed into a power series:

$$r_{ip} = \sqrt{z^2 + (x_i - x_p)^2 + (y_i - y_p)^2} = z\sqrt{1 + \left(\frac{x_i - x_p}{z}\right)^2 + \left(\frac{y_i - y_p}{z}\right)^2} \tag{4.147}$$

With

$$\sqrt{(1+b)} = 1 + \frac{1}{2}b - \frac{1}{8}b^2 + \dots \qquad (|b| < 1)$$

we get

$$z\sqrt{1 + \left(\frac{x_i - x_p}{z}\right)^2 + \left(\frac{y_i - y_p}{z}\right)^2} \approx z\left[1 + \frac{1}{2}\left(\frac{x_i - x_p}{z}\right)^2 + \left(\frac{1}{2}\frac{y_i - y_p}{z}\right)^2\right] \tag{4.148}$$

and the diffraction integral will give

$$U(x_i, y_i) = -\frac{e^{ikz}}{i\lambda z} \iint_{\mathbb{R}^2} P(x_p, y_p)\, U(x_p, y_p)\, e^{i\frac{k}{2z}\left[(x_i - x_p)^2 + (y_i - y_p)^2\right]} dx_p dy_p \tag{4.149}$$

If we expand the quadratic terms in the exponent we get

$$U(x_i, y_i) \cdot e^{-i\frac{\pi}{\lambda z}(x_i^2 + y_i^2)} \tag{4.150}$$

$$= c \iint_{\mathbb{R}^2} P(x_p, y_p)\, U(x_p, y_p)\, e^{i\frac{\pi}{\lambda z}(x_p^2 + y_p^2)} e^{-i\frac{2\pi}{\lambda z}(x_i x_p + y_i y_p)} dx_p dy_p$$

with c = complex factor.

Interpretation of Eq. (4.150):
Apart from the corresponding phase factors $-i\frac{k}{2z}(x_i^2 + y_i^2)$ and $+i\frac{k}{2z}(x_p^2 + y_p^2)$ which appear as factors for the diffracted and incident wave, the complex amplitude in the plane z is given by a *Fourier transform* of

the wave in plane $z = 0$ where the spatial frequencies r and s have to be taken as

$$r = \frac{x_i}{\lambda \cdot z}, \quad s = \frac{y_i}{\lambda \cdot z} \tag{4.151}$$

in order to get correct units in the image plane. (For the meaning of spatial frequencies see Section 4.4.)

The mentioned phase factors may be interpreted as optical path difference, so that the corresponding wavefronts will be curved surfaces. The deviation of the wavefront from the x–y plane for each point is equal to the path difference introduced by the phase factors. For the image plane

$$e^{-i\frac{\pi}{\lambda}\left(\frac{x_i^2+y_i^2}{z}\right)} = e^{+i\frac{\pi}{\lambda}c} \quad (c = \text{constant}) \tag{4.152}$$

one has

$$z = -\frac{1}{c}\left(x_i^2 + y_i^2\right) \tag{4.153}$$

This is a rotation paraboloid opened in the $-z$-direction. In the same way for the plane $z = 0$ (index p) one gets a rotation paraboloid opened in the $+z$-direction.

However we know that these paraboloids are quadratic approximations to spherical surfaces!

If thus the complex amplitude of the incident wave is given on a spherical surface, then the complex amplitude of the diffracted wavefront is also given on a spherical surface by simple Fourier transformation.

The Fraunhofer Approximation

An even simpler expression for the diffraction pattern arises if one chooses more stringent approximations than those of the Fresnel case.

If we assume

$$z \gg \frac{k\left(x_p^2 + y_p^2\right)_{max}}{z} \quad \text{and} \quad z \gg \frac{k\left(x_i^2 + y_i^2\right)_{max}}{2} \tag{4.154}$$

then

$$U(x_i, y_i) = c \iint_{\mathbb{R}^2} U(x_p, y_p) e^{-i\frac{2\pi}{\lambda z}(x_p x_i + y_p y_i)} dx_p dy_p \tag{4.155}$$

This is simply the Fourier transform of the aperture distribution evaluated at spatial frequencies $r = x_i/\lambda z$, $s = y_i/\lambda z$.

The requirement of the Fraunhofer approximation is quite stringent; however one may observe such diffraction patterns at a distance closer than required by Eq. (4.154), if the aperture is illuminated by a spherical converging wave or if an (ideal) lens is positioned between the aperture and the observing plane.

4.3.4
The Impulse Response of an Aberration Free Optical System

An ideal lens, which fulfils completely the laws of Gaussian optics will transfer a diverging homocentric ray pencil (with its intersection point at a certain object point) into a converging homocentric pencil with its intersection point at the ideal (Gaussian) image point.

As pointed out in Section 4.1.1, the light rays are the orthogonal trajectories of the wavefronts. Thus for a homocentric pencil of rays, the corresponding wavefronts are spherical wavefronts with their centre at the intersection point of the homocentric ray pencil. Such a system which converts a diverging spherical wave into a converging spherical wave will be called *diffraction limited*.

Speaking in terms of imaging, a diverging spherical wave originating from a certain object point and travelling to the entrance pupil will be transformed by the diffraction limited optical system into a converging spherical wave originating from the exit pupil and travelling towards the ideal image point, Fig. 4.67.

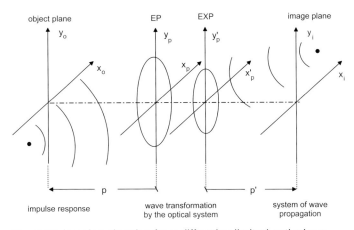

Fig. 4.67 Imaging situation for a diffraction limited optical system.

The problem of finding the impulse response (for the diffraction limited optical system) may thus be dealt within three steps:

1. The impulse response of the system of wave propagation to a point-like disturbance (object point) is a spherical wave, namely a Huygens elementary wave as shown in Section 4.3.3.

2. This wave enters the optical system and will be transformed in an ideal manner. This transformation by the diffraction limited system must be formulated mathematically.

3. The spherical wave originating from and limited by the exit pupil will be dealt within Fresnel approximation in order to end with the amplitude distribution in the image plane.

This point of view, to consider the wave propagation from the finite exit pupil, was introduced by Rayleigh (1896) and is essentially equivalent to the considerations of Abbe (1873), who analysed the relations in the finite entrance pupil.

We will consider the three indicated steps now.

1. **impulse response**
   According to Section 4.3.3 we have for the impulse response in Fresnel approximation and with Fig. 4.67:

$$U_{EP}(x_p, y_p) = -\frac{1}{i\lambda p} e^{-i\frac{k}{2p}\left[(x_p-x_0)^2+(y_p-y_0)^2\right]} \quad (4.156)$$

2. **ideal lens transformation** $t_l(x_p, y_p)$
   This transformation will be given as a result only; detailed considerations may be found in [2]

$$t_l(x_p, y_p) = c\, P(x_p\beta_p, y_p\beta_p) \cdot e^{+i\frac{k\beta_p^2}{2z'_{p'}}(x_p^2+y_p^2)} \quad (4.157)$$

   with

$$P(x_p\beta_p, y_p\beta_p) = \begin{cases} 1 & \text{inside the pupil} \\ 0 & \text{outside the pupil} \end{cases}$$

   and

$$\beta_p = \text{pupil magnification ratio}$$
$$z'_{p'} = \overrightarrow{(F'P')}_z \text{ distance from the focal point } F' \text{ of the lens to the centre of the exit pupil (Equation (4.31))}$$

3. **Amplitude distribution in the image plane**
   The diffraction integral in the Fresnel approximation for our constellation is

$$U(x_i, y_i) = \frac{1}{i\lambda p'} \iint_{\mathbb{R}^2} U(x'_p, y'_p)\, e^{i\frac{k}{2p'}\left[(x_i-\beta_p x_p)^2 + (y_i-\beta_p y'_p)^2\right]} dx'_p dy'_p$$

$$(4.158)$$

with
$$U(x'_p, y'_p) = U_{EP}(x_p, y_p)\, t_l(x_p, y_p)$$

and substituting from Eqs. (4.156) and (4.157) into Eq. (4.158) gives

$$\begin{aligned}U(x_i, y_i) &= \frac{1}{\lambda^2 \cdot p \cdot p'} \exp\left\{i \cdot \frac{k}{2p'}\left(x_i^2 + y_i^2\right)\right\} \exp\left\{-\frac{ik}{2p}\left(x_o^2 + y_o^2\right)\right\} \\ &\cdot \iint_{\mathbb{R}^2} P(x_p\beta_p, y_p\beta_p) \exp\left\{\frac{ik}{2}\left(\frac{\beta_p^2}{z'_{p'}} - \frac{1}{p} + \frac{\beta_p^2}{p'}\right)\left(x_p^2 + y_p^2\right)\right\} \\ &\cdot \exp\left\{ik\left[\left(\frac{x_o}{p} - \frac{\beta_p x_i}{p'}\right)x_p + \left(\frac{y_o}{p} - \frac{\beta_p \cdot y_i}{p'}\right)y_p\right]\right\} dx_p dy_p \end{aligned} \quad (4.159)$$

The first two exponential terms are independent of the pupil coordinates and could therefore be drawn in front of the integral. They may again be interpreted as phase curvatures over the object and image plane in the sense that they will vanish if one takes as image and object surfaces the corresponding spherical surfaces. In the case of incoherent imaging the system is linear with respect to intensities, which means to the square of the modulus of the complex amplitude. In this case these terms may also be omitted for the transmission between object and image *planes*.

The third exponential term vanishes because the position of the image plane has been chosen according to the imaging equations of Gaussian optics:
From Eq. (4.35):

$$z'_{p'} = -\beta_p f' \quad (4.160)$$

and the third exponential term may be written as

$$\frac{ik\beta_p}{2}\left(\frac{\beta_p}{p'} - \frac{1}{\beta_p p} - \frac{1}{f'}\right) = 0 \quad (4.161)$$

The bracket term vanishes because of the imaging equation (4.42).

Thus the impulse response of the diffraction limited optical system is given by

$$\begin{aligned}U(x_i, y_i) = h(x_i, y_i, x_0, y_0) &= \frac{1}{\lambda^2 pp'} \iint_{\mathbb{R}^2} P(x_p\beta_p, y_p\beta_p) \\ &\cdot \exp\left\{ik\left[\left(\frac{x_0}{p} - \frac{\beta_p x_i}{p'}\right)x_p + \left(\frac{y_0}{p} - \frac{\beta_p y_i}{p'}\right)y_p\right]\right\} dx_p dy_p\end{aligned} \quad (4.162)$$

and with Eq. (4.38):

$$\frac{p'}{p} = \beta_p\, \beta$$

we finally have

$$h(x_i, y_i, x_0, y_0) = \frac{1}{\lambda^2 \cdot p \cdot p'} \iint_{\mathbb{R}^2} P(x_p \beta_p, y_p \beta_p) \qquad (4.163)$$

$$\cdot \exp\left\{-i\frac{k \cdot \beta_p}{p'}\left[(x_i - \beta x_0)x_p + (y_i - \beta y_0)y_p\right]\right\} dx_p dy_p$$

The impulse response (point spread function) is given by the Fourier transform of the pupil function with the centre of the image at

$$x_i = \beta x_0 \quad , \quad y_i = \beta y_0$$

4.3.4.1 The Case of Circular Aperture, Object Point on the Optical Axis

For circular apertures (as usual in optics) it is useful to introduce polar coordinates in the pupil- and image plane. We just give the result for the intensity distribution in the image plane of a diffraction-limited optical system; details are given in [3]

$$I = \left[\frac{2J_1(v)}{v}\right]^2 \qquad (4.164)$$

With J_1 = Bessel function of first order,

$$v = \frac{2\pi}{\lambda}\sin\alpha r = \frac{\pi}{\lambda(f/nr)_e} r \qquad (4.165)$$

$$r = \sqrt{x'^2 + y'^2} \qquad (4.166)$$

This is the well-known *Airy disc*, which is rotationally symmetric and depends only on the numerical aperture $\sin\alpha$ (or the effective f/nr) and on wavelength λ.

Figure 4.68 shows a cross section of the Airy disc for the normalized radius v.

The extension of the central disc up to the first minimum is given by

$$r_0 = 1.22 \cdot \lambda \cdot (f/nr)_e \qquad (4.167)$$

This is usually taken as the radius of the diffraction disc. In this area one has 85% of the total radiant flux. The first bright ring has 7% of total flux.

4.3.5

The Intensity Distribution in the Neighbourhood of the Geometrical Focus

The problem of finding the intensity distribution in the region near the geometrical focus was first dealt by Lommel [4] and Struve [5]. In what follows

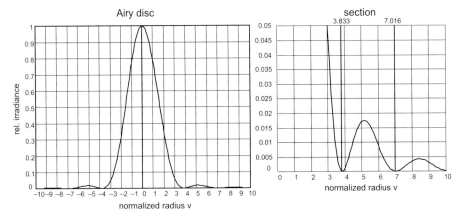

Fig. 4.68 The Airy disc in geometrical focus, $v = \frac{\pi r}{\lambda f/nr}$.

we will give a short outline based on [6]. It will be useful to introduce reduced coordinates. Instead of r and z we use

$$v = k \sin \alpha r \quad \text{(as with the Airy disc)} \tag{4.168}$$
$$u = k \sin^2 \alpha z \tag{4.169}$$

and

$$\sin \alpha = \frac{1}{2 \cdot (f/nr)_e} \tag{4.170}$$

Then the intensity distribution near the Gaussian image plane may be described by two equivalent expressions:

$$I(u,v) = \left(\frac{2}{u}\right)^2 \left[U_1^2(u,v) + U_2^2(u,v)\right] \cdot I_0 \tag{4.171}$$

and

$$I(u,v) = \left(\frac{2}{u}\right)^2 \Big[1 + V_0^2(u,v) + V_1^2(u,v) \\
- 2V_0(u,v) \cos\left[\frac{1}{2}\left(u + \frac{v^2}{u}\right)\right] \\
- 2V_1(u,v) \sin\left[\frac{1}{2}\left(u + \frac{v^2}{u}\right)\right]\Big] I_0 \tag{4.172}$$

U_n, V_n are the so-called *Lommel functions*, an infinite series of Bessel functions, defined as follows:

$$U_n(u,v) = \sum_{s=0}^{\infty} (-1)^s \left(\frac{u}{v}\right)^{n+2s} J_{n+2s}(v) \tag{4.173}$$

$$V_n(u,v) = \sum_{s=0}^{\infty} (-1)^s \left(\frac{v}{u}\right)^{n+2s} J_{n+2s}(v) \tag{4.174}$$

I_0 is the intensity in the geometrical focus $u = v = 0$.

Equ (4.171) converges for points

$$\frac{u}{v} = \sin\alpha \frac{z}{r} < 1$$

This means when the observation point is within the geometrical shadow region, and equ (4.172) converges for points

$$\frac{u}{v} > 1$$

that is for points in the geometrically illuminated region. For points on the shadow limit we have $u = v$.

From these equations and with the properties of the Lommel functions one may derive the following conclusions:

1. The intensity distribution is rotational symmetric around the u-axis (and hence around the z-axis).

2. The intensity distribution near the focus is symmetrical with respect to the geometrical image plane

Figure 4.69 shows the intensity distribution as lines of constant intensity for the upper right half of the light pencils only (because of the mentioned symmetry properties).

Special cases:

From Eqs. (4.171) and (4.172) for $u = 0$ (geometrical focal plane) one has

$$\lim_{u \to 0} \left[\frac{U_1(u,v)}{u}\right] = \frac{J_1(v)}{v} \qquad \lim_{u \to 0} \left[\frac{U_2(u,v)}{u}\right] = 0$$

which gives

$$I(0,v) = \left[\frac{2J_1(v)}{v}\right]^2$$

which is the intensity distribution of the *Airy disc*.

Furthermore from Eq. (4.172) one may derive the *intensity distribution along the optical axis*. For $v = 0$ (i.e. $r = 0$) it is

$$V_0(u,0) = 1 \quad \text{and} \quad V_1(u,0) = 0 \tag{4.175}$$

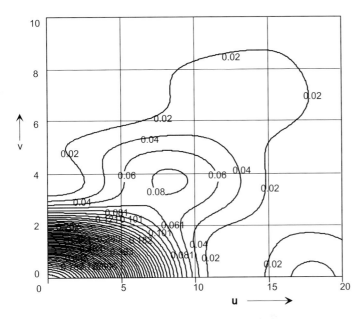

Fig. 4.69 Isophotes of intensity distribution near focus.

and the intensity is

$$I(u,0) = \frac{4}{u^2}\left[2 - 2\cos\frac{u}{2}\right] \tag{4.176}$$

which gives

$$I(u,0) = \left\{\frac{\sin\frac{u}{4}}{\frac{u}{4}}\right\}^2 = \left[\text{sinc}\frac{u}{4}\right]^2 \tag{4.177}$$

The result is that the maximum intensity of the point spread function changes with defocusing (along the optical axis) with a sinc-square function. The first minimum is at

$$u = 4\pi$$

in agreement with the isophotes of Fig. 4.69. Figure 4.70 shows the intensity distribution along the optical axis.

The ratio of the maximum intensity of the defocused point spread function to that of the Gaussian image plane is a special case of the *Strehl ratio*.

It may be shown that with pure defocusing a Strehl ratio of 80% corresponds to a wavefront aberration (deviation from the Gaussian reference sphere) of $\lambda/4$. This value for the Strehl ratio is thus equivalent to the Rayleigh criterion and is considered to be the permissible defocusing tolerance for diffraction limited systems.

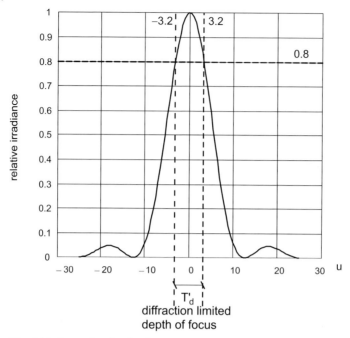

Fig. 4.70 Intensity distribution of a diffraction limited system along the optical axis.

The sinc²-function of Eq. (4.177) decreases by 20% for $u \approx 3.2$. Thus one has for the diffraction limited depth of focus with

$$z = \frac{\lambda}{2\pi} \frac{1}{\sin^2 \alpha} u \quad , \quad \frac{1}{\sin \alpha} = 2(f/nr)_e \qquad (4.178)$$

and with

$$u \approx \pm 3.2 \qquad (4.179)$$

$$\Delta z' = \pm 3.2 \frac{\lambda}{2\pi} \frac{1}{\sin^2 \alpha} \qquad (4.180)$$

or

$$\Delta z' \approx \pm 2\lambda \, (f/nr)_e^2 \qquad (4.181)$$

4.3.6
The Extension of the Point Spread Function in a Defocused Image Plane

Equation (4.181) gives the limits of defocusing at which the point spread function may still be considered for practical purposes as diffraction limited. Then the question arises immediately: what extension has the point spread function at these limits? In order to decide this we need a suitable criterion for that extension. For the Gaussian image plane this criterion was the radius r_0 of the central disc, which includes 85% of the total intensity flux. We generalize this and define:

> The extension of the point spread function in a defocused image plane is given by the radius which includes 85% of the total intensity.

Then the above question may be reformulated as

> What are the limits of the point spread function as a function of defocusing and when will they coincide with the geometrical shadow limits – defined by the homocentric ray pencil?

This problem may be solved by integration of Eqs (4.171) and (4.172) over v. We only show the results in the form of contour line plots of fractions of the total energy, Fig. 4.71.

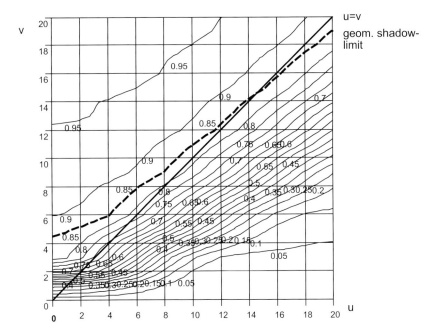

Fig. 4.71 Contour line plots of fractions of total energy.

The extension of the point spread function as a function of defocusing u is given as defined above by the value 0.85 (the thick broken line in Fig. 4.71). The geometrical shadow limit (limit of the Gaussian ray pencil) is given by $u = v$ which is the thick straight line.

In Fig. 4.72 the 0.85 line is approximated by a hyperbolic function. We may draw the following conclusions from this approximation:

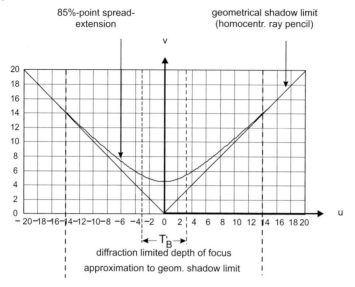

Fig. 4.72 Approximation of the point spread extension and geometrical shadow limit.

1. The extension of the point spread function is always larger than or equal to the geometrical shadow limit.

2. The geometrical shadow limit is reached for $u \approx 14$.

3. For the extension of the point spread functions and the corresponding defocusing values in the focal plane ($u = 0$), at the diffraction limit ($u \approx 3.2$) and at the approximation to the geometrical shadow limit ($u \approx 14$) we get with

$$r = \frac{\lambda \left(f/nr\right)_e}{\pi} v \quad , \quad z' = \frac{2\lambda \left(f/nr\right)_e}{\pi} u \qquad (4.182)$$

Tab. 4.8 Diffraction limited point spread extension in three different image planes.

	u	z'	v	r	r/r_0
Focus	0	0	3.83	$r_0 = 1.22\lambda(f/nr)_e$	1
Diffraction limit	$u_d \approx 3.2$	$z_d = 2\lambda(f/nr)_e^2$	$v_d \approx 5.5$	$r_d \approx 1.9\lambda(f/nr)_e$	~ 1.43
Shadow limit	$u_s \approx 14$	$z_s = 8.9\lambda(f/nr)_e^2$	$v_s \approx 14$	$r_s \approx 4.45\lambda(f/nr)_e$	~ 3.6

4.3.7
Consequences for the Depth of Field Considerations

Diffraction and Permissible Circle of Confusion
The diameter of the diffraction point spread function approximates the geometrical shadow limit at $u = v \approx 14$. This gives

$$d'_s = 9 \cdot \lambda \cdot (f/nr)_e \quad \text{(index s for shadow)} \tag{4.183}$$

Thus there is an upper limit for the f-number in order not to exceed the diameter d'_G of the permissible circle of confusion:

$$(f/nr)_e \leq \frac{d'_G}{4.5} \tag{4.184}$$

for $\lambda = 0.5\,\mu m$ and d'_G in micrometre.

Table 4.9 shows the maximum effective f-number for different circles of confusion, at which one may calculate with geometrical optics.

Tab. 4.9 Maximum permissible f-number for purely geometric calculations.

d'_G (μm)	150	100	75	60	50	33	25
max. $(f/nr)_e$	33	22	16	13	11	7.3	5.6

The Extension of the Point Spread Function at the Limits of the Depth of Focus
The 85% point spread extension as a function of defocusing has been approximated in Section 4.3.5 by a hyperbolic function. This function is

$$v = \sqrt{u^2 + 4.5^2} \tag{4.185}$$

and

$$u = \frac{2\pi}{\lambda}\sin^2\alpha\,\Delta z' = \frac{\pi}{(f/nr)_e^2} \cdot \Delta z' \quad (\lambda = 0.5\,\mu m) \tag{4.186}$$

where $\Delta z'$ is the deviation from the Gaussian image plane. On the other hand, for the limits of the geometrical depth of focus we have by Eq. (4.80) with $\Delta z' = T'/2$:

$$\Delta z' = (f/nr)_e \cdot d'_G \tag{4.187}$$

Thus

$$u = \frac{\pi \cdot d'_G}{(f/nr)_e} \quad \left(d'_G \text{ in } \mu m\right) \tag{4.188}$$

For the coordinate v according to Eq. (4.165):

$$v = \frac{\pi \cdot r_d}{\lambda \cdot (f/nr)_e} \quad r_d = \frac{d'_d}{2} \tag{4.189}$$

$$v = \frac{\pi \cdot d'_d}{(f/nr)_e} \quad \left(\lambda = 0.5\mu m;\ d'_d \text{ in } \mu m\right) \tag{4.190}$$

or
$$d'_d = \frac{(f/nr)_e}{\pi} \cdot v \qquad (4.191)$$

With the hyperbolic equation (4.185):
$$d'_d = \frac{(f/nr)_e}{\pi} \cdot \sqrt{u^2 + 4.5^2} \qquad (4.192)$$

and with u according to Eq. (4.188) finally
$$d'_d \approx \sqrt{d'^2_G + 2(f/nr)^2_e} \qquad (4.193)$$

Figure 4.73 shows d'_d as a function of the effective f-number for $d'_G = 75$ and 100 μm.

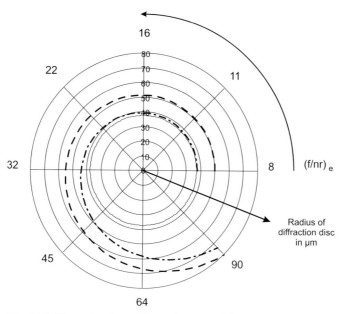

Fig. 4.73 The extension of the point spread function at the limits of the geometrical depth of focus.

Useful Effective f-Number

For rather large magnification ratios β (because of $(f/nr)_e = (f/nr) \cdot \left(1 - \frac{\beta}{\beta_p}\right)$) and with small image formats one exceeds quickly the limit

$$(f/nr)_e \leq \frac{d'_G}{4.5} \quad .$$

Then one should work with the useful effective f-number $(f/nr)_{eu}$, which gives at the *diffraction-limited* depth of focus a point spread diameter equal to the permissible circle of confusion

$$2r_d = 3.8 \cdot \lambda \cdot (f/nr)_{eu} = d'_G \tag{4.194}$$

$$(f/nr)_{eu} = \frac{d'_G}{1.9} \quad (\lambda = 0.5\mu m)$$

Then the depth of focus is equal to the diffraction limited depth of focus:

$$T'_d = \pm 2 \cdot \lambda \cdot (f/nr)^2_{eu} \tag{4.195}$$

4.10 Example

$$d'_G = 33\mu m \quad (f/nr)_{eu} = 18 \quad \beta_p \approx 1 \quad \beta = -1$$

$$(f/nr)_e = (f/nr) \cdot \left(1 - \frac{\beta}{\beta_p}\right) = 2 \cdot (f/nr)$$

$$(f/nr)_u = \frac{1}{2}(f/nr)_{eu} \approx 9$$

$$T'_d \approx \pm 0.32 \text{ mm}!$$

Tab. 4.10 Diffraction and depth of focus.

point spread function in geom. image plane	$d'_d = 2.44\lambda(f/nr)$
diffraction limited depth of focus	$\Delta z' = \pm 2\lambda(f/nr)^2_e$
maximum eff. f/Nr for geometrical considerations	$(f/nr)_e \leq d'_G / 4.5$
85 % intensity diameter for point spread function at the limits of geom. depth of focus	$d'_d \approx \sqrt{d'^2_G + 2(f/nr)^2_e}$
usable effective f-number corresponding diffraction limited depth of focus	$(f/nr)_{eu} = d'_G/1.9$ $T'_d = \pm 2\lambda \cdot (f/nr)^2_{eu}$

4.4
Information Theoretical Treatment of Image Transfer and Storage

Having supplemented the model of geometrical optics (and in particular that of Gaussian optics) by the scalar wave theory, we now proceed with the important topic of information theoretical aspects, not only for the information transfer, but also for the storage of the information with digital imaging devices. In the first step we will introduce the general concept of physical systems as linear invariant filters.

4.4.1
Physical Systems as Linear, Invariant Filters

A physical system in general may be interpreted as the transformation of an input function into an output function. This physical system may act by totally different principles. As examples we note:

- electrical networks: here the entrance and output functions are real functions of voltage or currents of the one-dimensional real variable t (time).

- optical imaging systems, where in the case of incoherent radiation the input and output functions are real positive quantities (intensities) of two independent variables, the space coordinates u and v.

It will come out that there is a close analogy between these two systems. We restrict ourselves to *deterministic* (non-statistical) systems, which means that a certain input function will be transformed into an unambiguous output function. Then the system may be represented by a *black box* and its action may be described by a mathematical operator $S\{\}$, which acts on the input function and thereby produces the output function (Fig. 4.74).

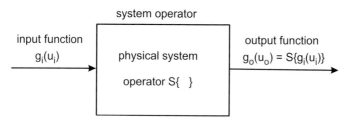

Fig. 4.74 The physical system as a mathematical operator.

An important subclass of the deterministic systems is the *linear systems*. Linear systems exhibit the properties of *additivity* and *homogeneity*, independent of their physical nature. Hereby additivity means that the response of the system to a sum of input functions is given by the sum of the corresponding output functions

$$S\{g_{i1}(u_i) + g_{i2}(u_i) + \cdots\} = S\{g_{i1}(u_i)\} + S\{g_{i2}(u_i)\} + \cdots \quad (4.196)$$

Homogeneity is the property that the response to an input function which is multiplied by a constant is given by the product of the constant c with the response to the input function:

$$S\{c \cdot g_i(u_i)\} = c \cdot S\{g_i(u_i)\} = c \cdot g_o(u_o) \tag{4.197}$$

These two properties may be combined to the *linearity* of the system:

$$S\{c_1 g_{i1}(u_i) + c_2 g_{i2}(u_i) + \cdots\} = c_1 \cdot S\{g_{i1}(u_i)\} + c_2 \cdot S\{g_{i2}(u_i)\} + \cdots \tag{4.198}$$

linearity of the system = additivity + homogeneity

In optics, the linearity of the system is given by the fact that the propagation of waves is described by the wave equation, which is a *linear* partial differential equation. For these a superposition principle is valid, which means that the sum and multiple of a solution is again a solution of the differential equation and thus expresses nothing but the linearity of the system.

With the mentioned properties of linear systems one may decompose a complicated input function into simpler, elementary functions. As an input function we take a sharp (idealized) impulse. With electrical networks where the independent variable is the time t, this corresponds to an impulse-like voltage peak at a certain *time*. In optics, the independent variable is the spatial coordinate u (for the sake of simplicity we consider first a one-dimensional situation). Here the idealized impulse is strictly a point-like object at a certain *position*.

Mathematically, this idealized impulse may be represented by Dirac's δ-function. This function represents an infinitely sharp impulse with the following properties:

$$\delta(u) = \begin{cases} 0 & (u \neq 0) \\ \infty & (u = 0) \end{cases} \tag{4.199}$$

and

$$\int_{-\infty}^{+\infty} \delta(u) = 1 \tag{4.200}$$

(Fig. 4.75).

One more important property of the δ-function is the shifting property:

$$\int_{-\infty}^{+\infty} f(u) \cdot \delta(u - u_1) \, du = f(u_1) \tag{4.201}$$

The integral produces the value of the function $f(u)$ at position u_1 of the shifted δ-function (Fig. 4.76).

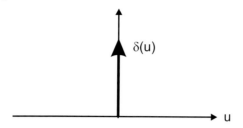

Fig. 4.75 The Dirac impulse.

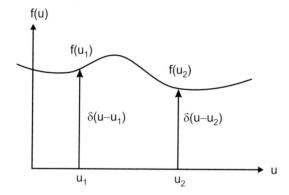

Fig. 4.76 The shifting property of the δ-function.

With this, one may decompose an input signal g_i into an infinitely dense series of δ-functions, weighted with $g_i(\xi)$:

$$g_i(u_i) = \left\{ \int_{-\infty}^{+\infty} g_i(\xi) \cdot \delta(u_i - \xi) \, d\xi \right\} \quad (\xi = \text{auxiliary variable}) \quad (4.202)$$

The output function as the systems response is then

$$g_o(u_o) = S \left\{ \int_{-\infty}^{+\infty} g_i(\xi) \cdot \delta(u_i - \xi) \, d\xi \right\} \quad (4.203)$$

Here $g_i(\xi)$ are nothing but the weighting coefficients for the shifted Dirac impulses. Because of the linearity of the system we may then write

$$g_o(u_o) = \int_{-\infty}^{+\infty} g_i(\xi) \, S \{\delta(u_i - \xi)\} \, d\xi \quad (4.204)$$

The response of the system S at position u_o of the output to a Dirac impulse with coordinate ξ at the input shall be expressed as

$$h(u_o, \xi) = S \{\delta(u_i - \xi)\} \quad (4.205)$$

This is the *impulse response* of the system.

The response of the system to a complicated input function is then given by the following fundamental expression:

$$g_o(u_o) = \int_{-\infty}^{+\infty} g_i(\xi) \cdot h(u_o, \xi) \, d\xi \quad \text{(superposition integral)} \qquad (4.206)$$

The superposition integral describes the important fact that a linear system is completely characterized by its response to a unit impulse (impulse response).

However, in order to describe the output of the system completely, we must know the impulse response for all possible positions (this means: for all possible values of the independent variable) of the input pulse. An important subclass of linear systems are those systems for which the impulse response is independent of the absolute position of the input impulse.

Invariant, Linear Systems

In order to introduce the concept of invariant linear systems, we start from an electrical network (Fig. 4.77).

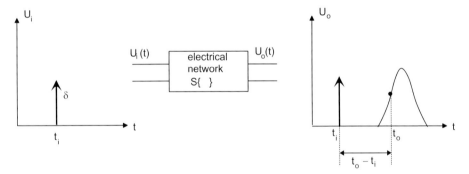

Fig. 4.77 Electrical networks as invariant systems.

Here the impulse response at time t_o to an impulse at time t_i depends only on the *time difference* $(t_o - t_i)$. At a later time t'_o we will get the same response to an impulse at the input at time t'_i if the difference $t'_o - t'_i$ is the same. Here we have a *time invariant* system.

In optics we use, instead of the time coordinate, a (normalized) spatial coordinate u (all considerations are one-dimensional for the moment). Then *space-invariance* means that the intensity distribution in the image of a point-like object only depends on the distance $(u_o - u_i)$ and not on the absolute position of the object point in the object plane (Fig. 4.78).

In optics, space invariance is called isoplanasic condition. The intensity distribution in the image of a point (the so-called point spread function PSF) is the

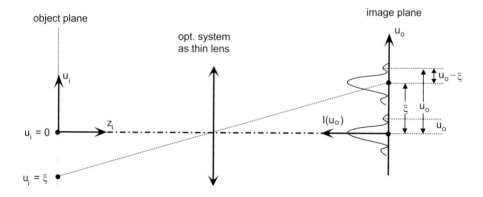

Fig. 4.78 Space invariance with optical systems.

same, independent of its position in the image plane. The impulse response and output function in Eqs. (4.205) and (4.206) may now be written as

$$h(u_o, \xi) = h(u_o - \xi) \qquad (4.207)$$

$$g_o(u_o) = \int_{-\infty}^{+\infty} g_i(\xi) h(u_o - \xi) d\xi \qquad (4.208)$$

Under the condition of space invariance, the superposition integral will be simplified to a *convolution integral*, which is a convolution between the input function (object function) and the impulse response (point spread function PSF). Equation (4.208) may be written symbolically as

$$g_o = g_i * h \qquad (4.209)$$

The condition of space invariance is fulfilled in optics only for limited spatial regions, the *isoplanatic regions*. Optical systems are in general rotationally symmetric and therefore the impulse response (PSF) is constant for a certain distance from the centre of the image, which is the intersection point of the optical axis with the image plane. This means that the impulse response is invariant for an image circle with a given image height $\xi = h'$ (Fig. 4.79).

If we change the radius h' of the image circle gradually, then the point spread function will change continuously as a function of h'. Within a small region $\Delta h'$ we may treat the point spread function as being independent of h'.

During manufacturing there will be tolerances for the constructional parameters of the optical system. These tolerances may destroy the rotational symmetry. Then the isoplanatic regions will be restricted to certain sectors of the image circle area (Fig. 4.80).

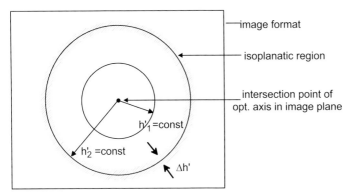

Fig. 4.79 Isoplanatic regions for rotationally symmetric optical systems.

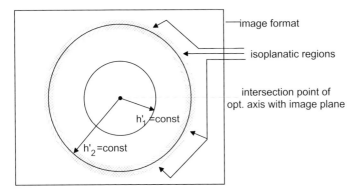

Fig. 4.80 Isoplanatic regions with a decentred optical system.

Then one has to measure the impulse response (PSF) for every isoplanatic region!

4.1 Definition The isoplanatic region of an optical system is the spatial region where the impulse response (the point spread function PSF) is constant within the measurement accuracy.

Within an isoplanatic region the system is completely described by the convolution of the input function with the corresponding impulse response.

For the convolution integral there is a theorem of Fourier analysis:

A convolution of two functions in the spatial domain corresponds – after a Fourier transformation (FT) – to a multiplication of the transformed functions in the frequency domain, Eq. (4.210):

if $g_a(u_a) = g(u) * h(u)$

and $FT^-\{g(u)\} = G(r)$, $FT^-\{h(u)\} = H(r)$ (4.210)

then $FT^-\{g_a(u_a)\} = G_a(r) = G(r) \cdot H(r)$.

For the principles of Fourier theory, see references [7]–[9]. The Fourier transform is defined as

$$F(r) = FT^-\{f(u)\} = \int_{-\infty}^{+\infty} f(u) e^{-i2\pi ur} du \quad (4.211)$$

with the inverse transform

$$f(u) = FT^+\{F(r)\} = \int_{-\infty}^{+\infty} F(r) e^{+i2\pi ur} dr \quad (4.212)$$

The Fourier transform is to be understood as an integral operator, which acts on a function $f(u)$ of the independent variable u (for instance space – or time – coordinate) and transforms it into a function in the frequency domain, with the independent variable r.

With the Fourier transformation we thus make a transition to another *representation* of the same physical system in some other *representation space* (Fig. 4.81).

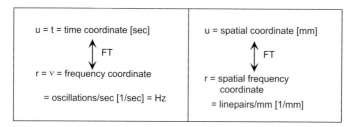

Fig. 4.81 Transition to different representation domains.

As Eq. (4.212) shows, we may interpret the Fourier transform as an alternative decomposition of the function $f(u)$ into other elementary functions.

4.11 Example time function $f(t)$:

$$f(t) = \int_{-\infty}^{+\infty} F(v) e^{i2\pi vt} dv \quad (4.213)$$

A *good behaving* function of time may be decomposed into an infinitely dense sum (integral) of harmonic vibrations. Each harmonic component is weighted with a factor $F(v)$ which is uniquely determined by $f(t)$.

4.12 Example space function $f(u)$

In this case we have from Eq. (4.212)

$$f(u) = \int_{-\infty}^{+\infty} F(r) e^{i 2\pi u r} dr \qquad (4.214)$$

An intensity distribution in the object may be decomposed into an infinitely sum of *harmonic waves*. Each wave is weighted with a certain factor $F(r)$ which is uniquely determined by $f(u)$. The alternative elementary functions are in this case the harmonic waves with the spatial frequencies r. (Harmonic intensity distribution with period $1/r$.)

Note to the Representation of Harmonic Waves

We take two components of Eq. (4.214) with spatial frequency r_n and write them as

$$a_n \cdot e^{i 2\pi r_n u} \quad \text{and} \quad a_{-n} \cdot e^{-i 2\pi r_n u} \qquad (4.215)$$

a_n and a_{-n} are conjugate complex since $f(u)$ is real:

$$a_n = c_n e^{+i\varphi_n}, \quad a_{-n} = c_n e^{-i\varphi_n} \qquad (4.216)$$

with

$$\cos \alpha = \frac{1}{2} \left(e^{+i\alpha} + e^{-i\alpha} \right)$$

we have

$$\frac{1}{2} \left[a_n e^{+i 2\pi r_n u} + a_{-n} e^{-i 2\pi r_n u} \right] = \frac{1}{2} c_n \cdot \left[e^{+i(2\pi r_n u + \varphi_n)} + e^{-i(2\pi r_n u + \varphi_n)} \right]$$
$$= c_n \cdot \cos(2\pi r_n u + \varphi_n) \qquad (4.217)$$

This is a harmonic wave with period $u_n = \frac{1}{r_n}$ and phase shift φ_n, Fig. 4.82. With the convolution theorem we now have

$$g_o(u) = g_i(u) * h(u)$$
$$\text{FT} \downarrow \quad \text{FT} \downarrow \quad \text{FT} \downarrow \qquad (4.218)$$
$$G_o(r) = G_i(r) \cdot H(r)$$

A convolution in the space domain u corresponds to a multiplication in the frequency domain r. The Fourier transform of the impulse response $h(u)$ is the *transfer function* $H(r)$ of the system (of the isoplanatic region, respectively).

$$H(r) = \int_{-\infty}^{+\infty} h(u) e^{-i 2\pi r u} du \qquad (4.219)$$

$$h(r) \overset{+FT-}{\longleftrightarrow} H(r) \qquad (4.220)$$

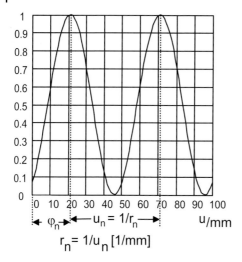

Fig. 4.82 The harmonic wave as a component n of the Fourier integral, Equation (4.214).

$G_i(r)$ is the decomposition of the object function $g_i(u)$ into a sum of harmonic waves (angular spectrum of plane waves).

Figure 4.83 gives an overview on the two representation domains.

representation = spatial domain
independent variable = u (spatial coordinate) elementary function = Dirac-pulse system described by: impulse response
Fourier-transform
independant variable = r (spatial frequency) elementary function = plane waves system described by: transfer function
representation = frequency domain

Fig. 4.83 Relationship between the two representations for optical systems.

4.4.2
The Optical Transfer Function (OTF) and the meaning of spatial frequency

Equation (4.218) means that the convolution in the spatial domain may be transformed – by application of the Fourier transform – into a simple multiplication in the spatial frequency domain. The elementary functions are then the complex exponential functions which may be interpreted as harmonic waves.

The system acts on the components (elementary functions) by multiplying each with a complex number $H(r)$, hence with the transfer function at coordinate r. The action of the system is thus restricted to a change of amplitude and to a phase shift of the waves, Fig. 4.84.

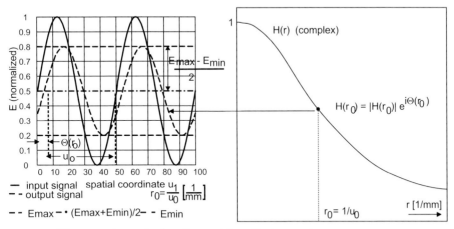

Fig. 4.84 The optical transfer function and the effect on a plane wave component at frequency r_o.

The transfer function is thus a (complex) low pass filter. The amplitude is attenuated *but frequency and mean intensity remain the same.* Additionally the wave will suffer a phase change.

$$\text{mean intensity} = \frac{E_{max} + E_{min}}{2} \quad (4.221)$$

$$\text{amplitude} = \frac{E_{max} - E_{min}}{2} \quad (4.222)$$

$$\text{modulation} = \frac{\text{amplitude}}{\text{mean intensity}} = \frac{E_{max} - E_{min}}{E_{max} + E_{min}} \quad (4.223)$$

$$= |H(r_o)| = \text{MTF}$$

$$\text{phase shift} = \arctan\left[\frac{\text{imaginary part of } H(r_o)}{\text{real part of } H(r_o)}\right] \quad (4.224)$$

$$= \Theta(r_o) = \text{PTF}$$

Therefore we may decompose the complex optical transfer function (OTF) into a modulation transfer function (MTF) as magnitude and into a phase transfer function (PTF) as the phase of the complex transfer function.

Note on the relation between the elementary functions in the two representation domains:
The Fourier transform of the Dirac pulse (which is the elementary function in the space domain) is a constant in spatial frequency domain, Fig. 4.85.

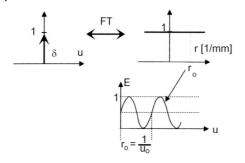

Fig. 4.85 The Fourier transform of the Dirac pulse.

In the frequency domain all the harmonic waves are present with the same magnitude. The totality of all these waves in the frequency domain corresponds to the Dirac pulse in the spatial domain.

The δ-function is thus the sample signal for the optical system and the Fourier transform of the impulse response (point spread function PSF) will determine to which extent the harmonic waves will be attenuated and shifted.

4.4.3
Extension to the Two-Dimensional Case

4.4.3.1 The Interpretation of Spatial Frequency Components (r,s)

The response of the optical system to a Dirac impulse in the two-dimensional object plane is a two-dimensional intensity distribution $I(u,v)$. This is the point spread function depending on the spatial coordinates u,v in the image plane.

The optical transfer function (OTF) is then the two-dimensional Fourier transform of the point spread function depending on the two independent variables, the spatial frequencies r and s with units linepairs/mm.

$$\text{OTF}(r,s) = \iint_{\mathbb{R}^2} I(u,v) e^{-i2\pi(ur+vs)} du\, dv \quad (4.225)$$

$$I(u,v) = \iint_{\mathbb{R}^2} \text{OTF}(r,s) e^{+i2\pi(ur+vs)} dr\, ds \quad (4.226)$$

Symbolically:

$$\text{OTF}(r,s) \xleftrightarrow{-FT+} I(u,v) \quad (4.227)$$

What is the meaning of the two-dimensional spatial frequencies? In order to answer this, we look at one single elementary wave of the form

$$e^{i2\pi(ur_0+vs_0)} \quad (4.228)$$

The real part is
$$I(u,v) = \cos[2\pi(ur_0 + vs_0)] \tag{4.229}$$
The imaginary part introduces a phase shift of the wave, as shown in section 4.4.2 for the one-dimensional case. We will omit it here for simplicity.

The function $I(u,v)$ represents a two-dimensional, cosine-intensity distribution. For each frequency pair (r_0, s_0) (which is represented in the Fourier plane by a point with coordinates (r_0, s_0)) the positions of the constant phase are given by
$$u \cdot r_0 + v \cdot s_0 = \text{const} + n \quad (n = 0,1...) \tag{4.230}$$
or
$$v = -\frac{r_0}{s_0} \cdot u + \frac{\text{const} + n}{s_0} \tag{4.231}$$
There is a periodicity with 2π and the positions of equal phase are straight lines with the slope
$$\tan \alpha = -\frac{r_0}{s_0} \tag{4.232}$$
Thus this elementary function represents a plane wave with the normal direction
$$\tan \Theta = \frac{s_0}{r_0} \tag{4.233}$$
since the normal vector is perpendicular to the lines of constant phase. The length L of a period of the wave is
$$\frac{1}{L} = \sqrt{r_0^2 + s_0^2} \tag{4.234}$$
(Fig. 4.86).

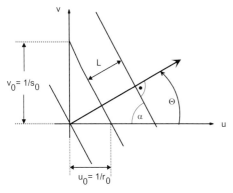

Fig. 4.86 A plane wave with spatial frequency components r_0, s_0.

Figure 4.87 finally gives an overview on the relationships between the two representation domains.

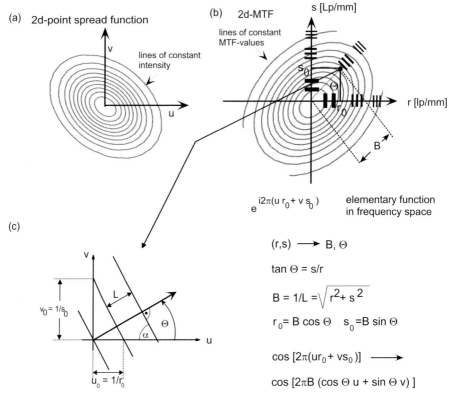

Fig. 4.87 Relationships between the representations in space and spatial frequency domain.

Figure 4.87(a) is a contour line image (lines of constant intensity) of the 2D point spread function and Fig. 4.87(b) the corresponding contour line image (lines of constant MTF) in the spatial frequency domain. Each point r_0, s_0 in the frequency domain corresponds to a plane wave in the spatial domain and vice versa.

4.4.3.2 Reduction to One-Dimensional Representations

A 2D Fourier transform is numerically tedious and in most cases it will be sufficient to characterize the two-dimensional OTF by two cross sections.

A cross section through $MTF(r, s)$ is given by the Fourier transform of an intensity distribution which is obtained by integration of the 2D point spread function perpendicular to the direction of the cross section.

We demonstrate this by the example of a cross section along the r-axis ($s = 0$):

$$\text{OTF}(r,s)|_{s=0} = \text{OTF}(r) = \iint_{\mathbb{R}^2} I(u,v) e^{-i2\pi(ru+sv)} du\, dv \bigg|_{s=0} \quad (4.235)$$

$$\text{OTF}(r) = \int_{-\infty}^{+\infty} \left[\int_{-\infty}^{+\infty} I(u,v) dv \right] e^{-i2\pi ru} du \quad (4.236)$$

$$\text{OTF}(r) = \int_{-\infty}^{+\infty} L(u) e^{-i2\pi ru} du \quad (4.237)$$

with

$$L(u) = \int_{-\infty}^{+\infty} I(u,v) dv \quad (4.238)$$

as integration of the point spread function intensity along the v direction. $L(u)$ is called the *line spread function*.

It may be shown that the same is true for any cross section with the direction α.

A scan of the 2D PSF with a narrow slit and detection of the resulting intensity results in an integration in the direction of the slit.

The slit scans the PSF perpendicular to its extension and thus gives the line spread function $L(u)$. If the slit is oriented tangential to the image circle with radius h' (this means in the direction u) then the scan direction is perpendicular, in direction of v. The integration of the 2D PSF is in direction u and after Fourier transformation of this line spread function one gets spatial frequencies with the lines of constant phase being *tangential* to the image circle and the direction of the cross section of the 2D OTF being perpendicular to the slit, which is in the direction of scan, Fig. 4.88.

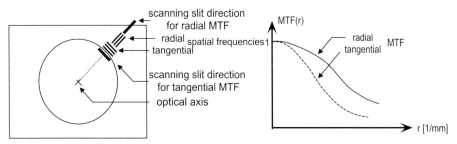

Fig. 4.88 Radial and tangential spatial frequencies and the corresponding cross sections of the transfer function.

An analogue situation arises for the orientation of the slit perpendicular to the image circle, in the v-direction. The scan direction is then the u-direction and after Fourier transformation one has spatial frequencies with lines of constant phase in the radial direction to the image circle.

In order to characterize the transfer function of an optical system one usually takes these two directions of cross sections. Then the MTF for tangential structures is plotted as a function of spatial frequency (for each isoplanatic region) with dashed lines and that for radial structures in full lines, see Fig. 4.88.

4.4.4
Impulse Response and MTF for Semiconductor Imaging Devices

We consider the impulse response of a single pixel. Within the sensitive area of this pixel – which must not coincide with its geometrical extension – the response to a Dirac impulse at position $\tilde{\xi}$ in the form of the electrical signal is given by the pixel sensitivity function $S(\tilde{\xi})$, Fig. 4.89.

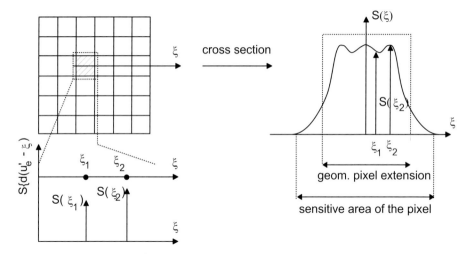

Fig. 4.89 The pixel sensitivity function.

The impulse response within the sensitive area of a pixel is thus space variant and the system is not space invariant in the sub-pixel region (Section 4.4.6). A complete description of the system is only possible if one knows the impulse response for *all* values of $\tilde{\xi}$ within the sensitive area. This is given by the pixel sensitivity function $S(\tilde{\xi})$. In what follows we assume that all pixels of the sensor have the same sensitivity function $S(\tilde{\xi})$, which means that the system is completely described by $S(\tilde{\xi})$ and the pixel distance d. How does this affect the output function of the linear system (semiconductor imaging device)?

Figure 4.90 shows the imaging constellation.

We assume the optical system to be ideal in the sense of Gaussian optics, which means that the image in the sensor plane is geometrically similar and

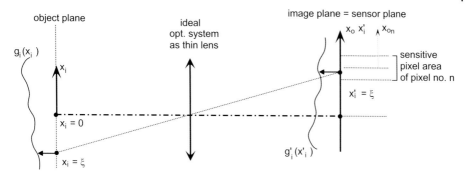

Fig. 4.90 The output function of the linear system.

each object point will be imaged to an exact point. By this assumption we may neglect the influence of the optical transfer function of the lens; we will take this into consideration in Section 4.4.5.

The image is reversed and changed by the magnification ratio $\beta < 0$. In order to describe the input function in the plane of the sensor we have to introduce a coordinate transformation.

$$x'_i = \beta \cdot x_i \tag{4.239}$$

$$g_i(x_i) \rightarrow g'_i(x'_i) \tag{4.240}$$

The weighting of the object function $g'(x'_i)$ with the pixel sensitivity function and integration will give the output function of the nth pixel at position x_{on}.

$$g_o(x_{on}) = \int_{\xi = x_{on}-d/2}^{x_{on}+d/2} g'_i(\xi) \cdot S(\xi) \, d\xi \tag{4.241}$$

This could be, for instance, the response at the output of the horizontal shift register of a CCD for pixel number n in kilo-electrons.

If we shift the pixel continuously along the coordinate x_o then the output function will be

$$g_o(x_o) = \int_{\xi = x_o-d/2}^{x_o+d/2} g'_i(\xi) \cdot S(x_o - \xi) \, d\xi = \int_{-\infty}^{+\infty} g'_i(\xi) \cdot S(x_o - \xi) \, d\xi \tag{4.242}$$

This is a convolution of the object function $g'_i(\xi)$ with the pixel sensitivity function $S(\xi)$, Fig. 4.91.

Although we assumed the same sensitivity function $S(\xi)$ for all pixels, the output function $g_o(x_o)$ is not space invariant in the sub-pixel region ($0 \leq \varphi \leq$

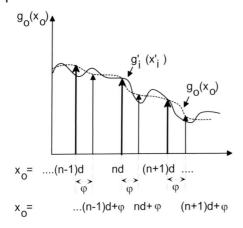

Fig. 4.91 Convolution with the pixel sensitivity function.

d) as shown in Fig. 4.91

$$g_o(x_o = n \cdot d) \neq g_o(x_o = n \cdot d + \varphi) \qquad (4.243)$$

The output signal of the system consists of a Dirac comb, weighted by the convolution function, the amplitudes of this Dirac comb *not* being space invariant in the sub-pixel region. The consequences will be dealt in Section 4.4.6

The transfer function of the pixel system array follows from Eq. (4.242) with the convolution theorem.

$$\begin{array}{c} g_o(x) = g'_i(x) * S(x) \\ \updownarrow FT \quad \updownarrow FT \quad \updownarrow FT \\ G_o(r) = G'_i(r) \cdot \mathrm{MTF}_{\mathrm{sen}}(r) \end{array} \qquad (4.244)$$

Here, the Fourier transform of the pixel sensitivity function is the *transfer function of the pixel*.

4.4.5
The Transmission Chain

With the concept of the pixel transfer function one may formulate now the imaging and optoelectronic conversion of an object scene as *incoherent* transmission chain. Incoherent in this context means that the system is linear with respect to *intensities*.

The total transfer function is equal to the product of the single transfer functions of the m embers of the transmission chain (with incoherent coupling), Fig. 4.92.

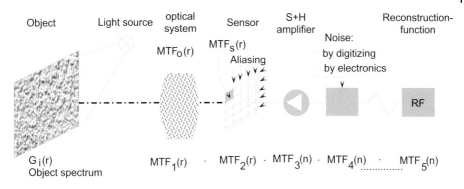

Fig. 4.92 The transmission chain.

The spatial distribution of image information is transferred by the image sensor into a temporal variation of a signal (video signal). There is a one-to-one correspondence between spatial and temporal frequencies. For the standardized video signal for instance the limiting frequency of 5 MHz corresponds with the usage of a 2/3 inch image sensor to a spatial frequency of 30 lp mm^{-1}. The reconstruction function represents further steps of information processing up to the final image. These may contain analogue devices, analogue-to-digital conversion, digital image processing, digital-to-analogue conversion for the display device.

4.4.6
The Aliasing Effect and the Space Variant Nature of Aliasing

We will investigate the requirements and limits for the application of semiconductor imaging devices. A Dirac impulse in the object plane is imaged by the optical system (point spread function) onto the sensor. In order to simplify the considerations, we take a linear sensor array, a line scan sensor. This consists of a linear arrangement of light sensitive surfaces (pixel) with typical dimensions of 2.5–25 µm. The incident photons generate electrical charges which are accumulated during the exposure time and then read out of the device by trigger pulses via a shift register.

The intensity distribution over a single pixel will be spatially integrated by the sensitive area of the pixel. In our one-dimensional consideration this would be the sensitivity function $S(u)$. For simplicity, we assume $S(u)$ to be a rectangular function with width b. The pixel distance is denoted by d.

The spatial integration by the pixels may be modelled by a convolution between the one-dimensional line spread function $h(u)$ with the rectangular sensitivity function [rect(1/b)]. This is shown in Fig. 4.93. Since the pixels are lo-

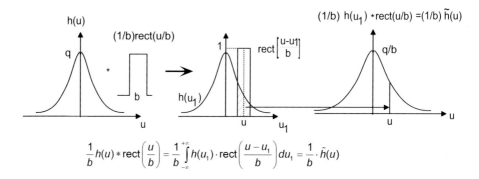

$$\frac{1}{b}h(u) * \text{rect}\left(\frac{u}{b}\right) = \frac{1}{b}\int_{-\infty}^{+\infty} h(u_1) \cdot \text{rect}\left(\frac{u-u_1}{b}\right) du_1 = \frac{1}{b}\cdot \tilde{h}(u)$$

$h(u)$ is area normalized, the factor $1/b$ is introduced because of the normalization of the MTF

Fig. 4.93 Convolution of the impulse response with the rectangular sensitivity function of the pixel.

cated at discrete distances d, one has to multiply the convolved function $\tilde{h}(u)$ with a Dirac comb with distances d (Fig. 4.94).

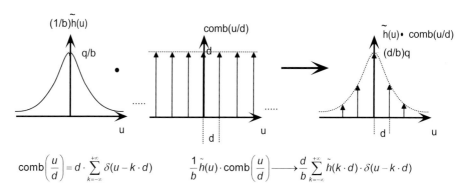

$$\text{comb}\left(\frac{u}{d}\right) = d \cdot \sum_{k=-\infty}^{+\infty}\delta(u - k\cdot d) \qquad \frac{1}{b}\tilde{h}(u)\cdot\text{comb}\left(\frac{u}{d}\right) \longrightarrow \frac{d}{b}\sum_{k=-\infty}^{+\infty}\tilde{h}(k\cdot d)\cdot\delta(u-k\cdot d)$$

Fig. 4.94 Discretizing with pixel distance d.

Figures 4.95 and 4.96 show the same situation in the spatial frequency domain with the independent variable r [1/ mm]. The modulus of the Fourier transform of $h(u)$ is the modulation transfer function MTF_o of the optical system. According to the convolution theorem of the Fourier theory, a convolution in the spatial domain corresponds to a multiplication of the Fourier transforms in the spatial frequency domain. The Fourier transform of the rectangular sensitivity function is a sinc function with the first zero value at $r = 1/b$. This value will become larger for smaller values of b.

More general: The Fourier transform of the pixel sensitivity function $S(u)$ is the optical transfer function of the line-scan array and is a multiplicative factor for the total transfer function $\text{MTF}_t(r)$.

Figure 4.96 shows the influence of the discrete pixel distances d in the spatial frequency domain. Now a multiplication in the spatial domain corresponds to a convolution with the Fourier transformed Dirac comb in the frequency domain. This is again a Dirac comb with distances $1/d$. The effect is a periodic repetition of the total transfer function $\mathrm{MTF}_t(r)$. These functions will overlap each other in certain regions and give rise to incorrect MTF values. This phenomenon is called *aliasing effect* (cross talk) and may only be reduced with smaller pixel distances d ($1/d$ becomes larger).

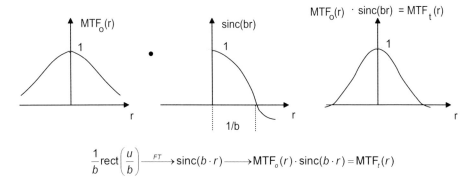

Fig. 4.95 Multiplication of the transfer functions.

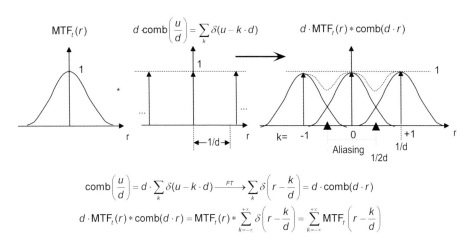

Fig. 4.96 Convolution with Dirac comb in the spatial frequency domain.

For band-limited systems, which means systems where the MTF is zero above a certain highest spatial frequency, the aliasing effect may be avoided completely if one scans with double the rate of the highest spatial frequency (Nyquist sampling theorem).

Now one has to take into account the finite length of the line scan array. In the spatial domain this means that we have to multiply the function $\tilde{h}(u)$ comb (u/d) with a rectangular function of width L, Fig. 4.97.

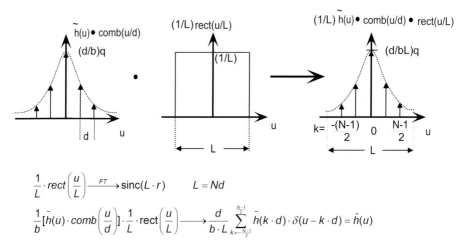

$$\frac{1}{L} \cdot \text{rect}\left(\frac{u}{L}\right) \xrightarrow{FT} \text{sinc}(L \cdot r) \qquad L = Nd$$

$$\frac{1}{b}[\tilde{h}(u) \cdot \text{comb}\left(\frac{u}{d}\right)] \cdot \frac{1}{L} \cdot \text{rect}\left(\frac{u}{L}\right) \longrightarrow \frac{d}{b \cdot L} \sum_{k=-\frac{N-1}{2}}^{\frac{N-1}{2}} \tilde{h}(k \cdot d) \cdot \delta(u - k \cdot d) = \hat{h}(u)$$

Fig. 4.97 Limitation of the impulse response by the length of the array: in spatial domain.

The result in the spatial frequency domain is now a convolution with a small sinc function, first zero at $1/L$, L is large, Fig. 4.98.

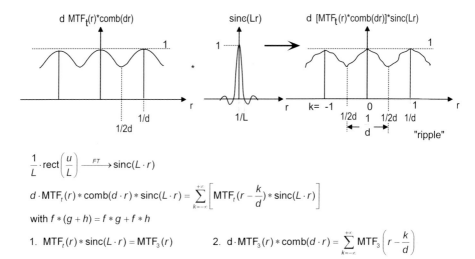

$$\frac{1}{L} \cdot \text{rect}\left(\frac{u}{L}\right) \xrightarrow{FT} \text{sinc}(L \cdot r)$$

$$d \cdot \text{MTF}_t(r) * \text{comb}(d \cdot r) * \text{sinc}(L \cdot r) = \sum_{k=-\infty}^{+\infty} \left[\text{MTF}_t\left(r - \frac{k}{d}\right) * \text{sinc}(L \cdot r) \right]$$

with $f * (g + h) = f * g + f * h$

1. $\text{MTF}_t(r) * \text{sinc}(L \cdot r) = \text{MTF}_3(r)$ 2. $d \cdot \text{MTF}_3(r) * \text{comb}(d \cdot r) = \sum_{k=-\infty}^{+\infty} \text{MTF}_3\left(r - \frac{k}{d}\right)$

Fig. 4.98 Limitation of the impulse response by the length of the array: in the spatial frequency domain.

4.4 Information Theoretical Treatment of Image Transfer and Storage

The last step has nothing to do with the aliasing problem, but will be introduced in order to get a discrete spatial frequency spectrum, which is then a discrete Fourier transform.

This will be done by introducing a periodic repetition of the function in the spatial domain. The basic period is the discrete function array which is cropped with the length L (Fig. 4.97). Convolution with a Dirac comb with impulse distances L gives the desired result. One basic period has thus N Dirac impulses with distance d and $L = N \cdot d$, Fig. 4.99.

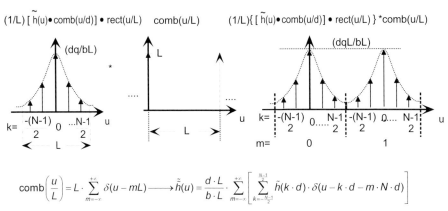

Fig. 4.99 Periodic repetition in the spatial domain u with basic period L.

Transferring to the spatial frequency domain gives a multiplication with a Dirac comb with impulse distance $1/L = 1/Nd$. Since the extension of one basic period in the spatial frequency domain is $1/d$, one has again N Dirac impulses within that period, as in the spatial domain, Fig. 4.100.

This final function represents a discrete Fourier transform of the impulse response function generated by the line scan array.

4.2 Summary Because of the periodic scanning with the pixel distance d there is a periodic repetition of the total transfer function in the spatial frequency domain with distance $1/d$. Since the transfer function in optics is band limited only at very high spatial frequencies there exist cross talking effects of the higher terms into the basic term for spatial frequencies less than the Nyquist frequency $1/(2d)$. Since the object spectrum is not band limited either, these cross talking terms (which may be interpreted as a mirror image of the basic term with respect to the Nyquist frequency) represent parasitic lower frequency terms within the Nyquist band pass . These cross talking effects must therefore be interpreted as *additive noise which depends on spatial frequency*.

In this context we cite from the paper of Park and Rahman [10]:

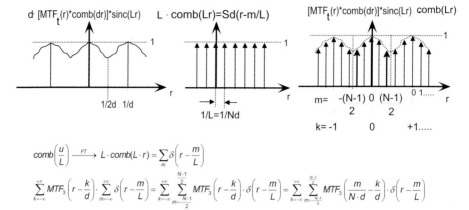

Fig. 4.100 Discretization in the spatial frequency domain (discrete Fourier transform).

The folklore that sampling effects are important only if the scene is *periodic* is false (Note: in this case one will see disturbing moiré-effects). It is certainly true that sampling effects are most *evident* if the scene has small periodic features for example a bar target whose frequency is close to the Nyquist frequency, because in that case familiar, obvious sampling artefacts are produced. It is wrong, however, to conclude that the absence of such obvious visual artefacts proves the absence of sampling effects... However except in special *periodic* cases, aliasing may not be recognized as such because sampling effects are largely indistinguishable from other possible sources of noise.

In this paper, we argue that for the purposes of system design and digital image processing, aliasing should be treated as scene-dependent additive noise.

4.13 Example A real CCD imaging sensor

- Pixel numbers: 2000×3000
- Pixel dimensions (b): $12\ \mu m \times 12\ \mu m$
- Pixel distance (d): $12\ \mu m$

The sensor format corresponds to 35 mm film: $24 \times 36\ mm^2$. To simplify, we assume again that the sensitivity function be rectangular:

$$S(u) = \text{rect}\left(\frac{u}{b}\right) \quad .$$

Then the transfer function of the CCD is a sinc function. The Nyquist frequency is at 41 lp/mm. The first zero for the pixel MTF is at $r \approx 83\,\text{lpmm}^{-1}$ and the modulation at $40\,\text{lpmm}^{-1}$ is approximately 66%. A good optical system may have a modulation of 60% at $40\,\text{lpmm}^{-1}$. Then the total transfer function at $40\,\text{lpmm}^{-1}$ is approximately 40%.

If one enlarges the sensor format to a final print of DIN A4 and looks at this print with a distance of 250 mm, then the eye may not resolve more than $6\,\text{lpmm}^{-1}$. If we transfer this with the corresponding magnification ratio to the image side, this gives a frequency of $40\,\text{lpmm}^{-1}$ which corresponds roughly to the Nyquist frequency of the sensor. Here we have to account for considerable aliasing and the sensor is not well suited for the required image quality.

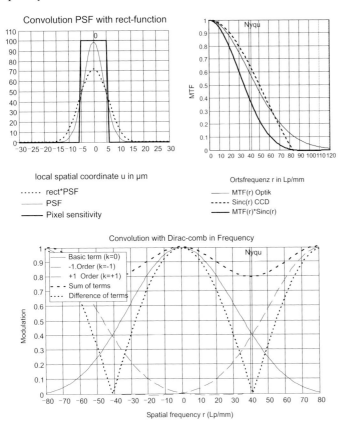

Fig. 4.101 PSF, MTF and aliasing, example 1.

4.14 Example hypothetical sensor (future generation?)

- Pixel numbers: 4000×6000

- Pixel dimensions (b): 6μm × 6 μm
- Pixel distance (d): 6 μm
- image format: 24 × 36 mm^2

Sensor transfer function

- First zero position of MTF at: 167 lp/mm
- MTF at 40 lpmm^{-1}: 91% ($\sim$ 80% with realistic $S(u)$)
- MTF at the Nyquist frequency (83 lpmm^{-1}): 66%

Optical system transfer function (modelled as the Gaussian function)

- Specified as in example 1: $r = 40\,\text{lpmm}^{-1}$, MTF = 60%
- This gives (according to Gauss function): $r = 83\,\text{lpmm}^{-1}$, MTF = 11%

Total transfer function

- MTF at 40 lpmm^{-1}: $\sim$ 50%
- MTF at 83 lpmm^{-1}: $\sim$ 6%

Conclusion: This sensor would fulfil the quality requirements completely, even the enlargement for the final print could be higher (up to 30 × 40 cm for $r_{max} = 70\,\text{lpmm}^{-1}$), Fig. 4.102

4.4.6.1 The Space Variant Nature of Aliasing

Up to now we made some simplifying assumptions when we dealt with the effects of discrete sampling by the pixel structure: on the one hand we approximated the pixel sensitivity by a rectangular function, and on the other hand we did not take into account how a small local shift of the input function (or what is the same: of the sensor) affects the Fourier transform.

The latter will be investigated in detail now and we assume again the one-dimensional case for simplicity.

The impulse response function of the optical system is again $h(u)$. This will be convolved with the finite pixel sensitivity function as before (giving $\tilde{h}(u)$) and then multiplied with a Dirac comb of distance d. But now we introduce a small displacement φ of the Dirac comb with respect to the impulse response $h(u)$:

$$\text{comb}\left(\frac{u}{d}\right) = \sum_{k=-\infty}^{+\infty} \delta(u - k \cdot d - \varphi) \qquad (4.245)$$

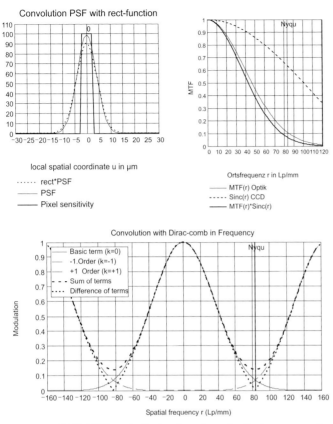

Fig. 4.102 PSF, MTF and aliasing for a hypothetical sensor optic combination (example 2).

Because of the periodic nature of the scanning we may restrict ourselves to the values of φ

$$0 \leq \varphi \leq d \tag{4.246}$$

This gives

$$\tilde{h}(u) \operatorname{comb}\left(\frac{u}{d}\right) = [h(u) * S(u)] \cdot \left[\sum_{k=-\infty}^{+\infty} \delta(u - k \cdot d - \varphi)\right] \tag{4.247}$$

According to the shift theorem of Fourier analysis, the Fourier transform is given by

$$FT\left\{\tilde{h}(u) \operatorname{comb}\left(\frac{u}{d}\right)\right\} = FT\left\{\tilde{h}(u)\right\} * \sum_{k} \delta\left(r - \frac{k}{d}\right) e^{i2\pi \varphi r} \tag{4.248}$$

The Fourier transform of $\tilde{h}(u)$ is again

$$FT\left\{\tilde{h}(u)\right\} = \operatorname{MTF}_o(r) \cdot \operatorname{sinc}(b \cdot r) = \operatorname{MTF}_t(r)$$

From (4.248)

$$\text{MTF}_t(r) * \sum_{k=-\infty}^{+\infty} \delta\left(r - \frac{k}{d}\right) e^{i2\pi\varphi k/d} = \text{MTF}_t\left(r - \frac{k}{d}\right) e^{i2\pi\varphi k/d} \quad (4.249)$$

To the relative shift in the spatial domain there corresponds now a phase term in the spatial frequency domain r. Figure 4.103 shows this situation for the special case of a *symmetric* point spread function which gives, by Fourier transformation, a *real* OTF ($\text{MTF}_t(r)$).

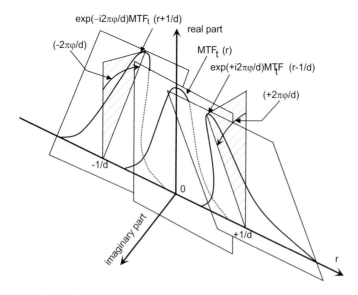

Fig. 4.103 The space variant nature of aliasing.

With the introduction of a small shift φ in the spatial domain the replica functions $(\pm k)$ in the spatial frequency domain rotate about the r-axis anticlockwise for $(+k)$ and clockwise for $(-k)$, with larger angles as k becomes larger. In the overlapping region the functions will be added by their real and imaginary parts separately. But since the real and imaginary parts depend on the rotation angle and hence on the shift φ, *also the sum in the overlapping region will critically depend on the shift φ in the spatial frequency domain!*

In other words,

> Even for very small shifts ($\varphi < d$) the optical transfer function is *no more space invariant* in the overlapping region and the isoplanatic condition is violated. The reason for this is – as shown – the aliasing effect with the finite pixel distance d.

Thus we have to redefine the isoplanatic condition [11]:

An imaging system is called isoplanatic if the Fourier transform of
the impulse response is independent of the position of the impulse
in the object plane

In the same way we have to redefine the isoplanatic condition in the spatial frequency domain:

The isoplanatic region of an imaging system is the region where
the Fourier transform of the impulse response may be considered
as constant within the measurement accuracy, if the point-like light
source in the image plane is shifted within the range $\varphi \leq d$.

Within the validity of the so-defined isoplanatic region the multiplication rule for the OTF of incoherently coupled systems is valid.

Since an aliased system may not completely be described by the Fourier transform of the impulse response, one has to give additionally a measure for the aliasing effect.

There are several possibilities of defining such a measure. ISO 15529 [13] introduces three different measures

1. Aliasing function
2. Aliasing ratio
3. Aliasing potential

1. Aliasing function $AF(r)$:
The difference between the highest and lowest values of $|MTF_t(r)|$ when a shift $0 \leq \varphi \leq d$ is introduced.

2. Aliasing ratio $AR(r)$:
The ratio of $AF(r)$ to $|MTF_t(r)|_{AV}$ where $|MTF_t(r)|_{AV}$ is the average of the highest and lowest values of $|MTF_t(r)|$ as a shift $(0 \leq \varphi \leq d)$ is introduced. $AR(r)$ can be considered as a measure of the noise-to-signal ratio where $AF(r)$ is a measure of the noise component and $|MTF_t(r)|_{AV}$ is a measure of the signal.

3. Aliasing potential AP
The ratio of the area under $MTF_t(r)$ from $r = r_N$ to $r = 2r_N$, to the area under $MTF_t(r)$ from $r = 0$ to r_N ($r_N = 1/2d =$ Nyquist frequency).

Figure 4.104 shows the aliasing measures $AF(r)$ and $AR(r)$ for the constellation given in Fig. 4.101. The aliasing potential in this case is

$$AP \approx 18,5\%$$

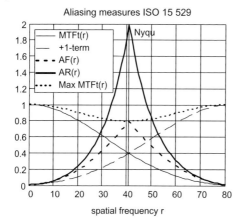

Fig. 4.104 Aliasing measures for the constellation of Figure 4.101.

4.5 Criteria for Image Quality

4.5.1 Gaussian Data

These data are the *backbone* of an optical system; insofar we will count them to the image quality data. In Section 4.2.11 we gave an overview on the most important data, namely in Table 4.3.

4.5.2 Overview on Aberrations of Third Order

In Section 4.2.2 we introduced Gaussian optics by linearizing the law of refraction. In the development of the sine function (4.6)

$$\sin \alpha = \alpha - \frac{\alpha^3}{3!} + \frac{\alpha^5}{5!} - \cdots$$

we restricted ourselves to the first member as linear term.

If one now takes into account the second member additionally, which is of third order, then a detailed analysis shows [12] that certain basic types of deviation from Gaussian optics will occur. Since Gaussian optics is the ideal of optical imaging, these types are called *aberrations* (deviations from Gaussian optics) of third order.

We will shortly mention them only, since for the end user of optical systems the quality criteria given in the following sections are of more importance.

Monochromatic aberrations of third order (Seidel aberrations)

- spherical aberration
- astigmatism
- coma
- field curvature
- distortion

The first three of these have influence on the form and extent of the point spread function (PSF); the last two represent a deviation of the position of the PSF with respect to Gaussian optics.

Chromatic aberrations:

Another group of aberrations is concerned with dispersion, i.e. the dependence of the refractive index with wavelength. This is already true for the Gaussian quantities, so that these aberrations are termed as *primary chromatic aberrations*. These are

- longitudinal chromatic aberration
- lateral chromatic aberrations

The longitudinal chromatic aberrations result in an axial shift of the image point for different wavelengths. The lateral chromatic aberrations are concerned with the chief ray, which is now different for different wavelengths. This results in a lateral shift of the image point for a wavelength λ with respect to the position for a certain reference wavelength λ_0.

In addition all monochromatic aberrations depend on the refractive index and thus depend also on wavelength. The result is that one has a chromatic variation of all monochromatic aberrations (*chromatic aberrations of higher order*).

4.5.3
Image Quality in the Space Domain: PSF, LSF, ESF and Distortion

For the end user, the effects of aberrations on the point spread function (PSF), the line spread function (LSF), the edge spread function (ESF) and on the deviation of the image position with respect to Gaussian optics (distortion) are by far more important than the aberrations themselves. All these quantities depend on the local coordinates u, v in the space domain.

However, it is not very easy to handle two-dimensional quantities such as the PSF. For most applications it is sufficient to consider one-dimensional line spread functions as introduced in Section 4.4.3.2. The LSF is the integration of the two-dimensional PSF in a certain direction, Fig. 4.105

$$LSF(u) = \int_{-\infty}^{+\infty} PSF(u,v)\,dv \qquad (4.250)$$

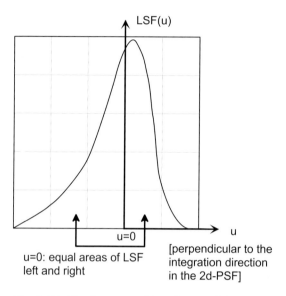

Fig. 4.105 The line spread function LSF(u).

The origin of the LSF is chosen so as to define the radiometric centre. This is true, if the ordinate divides the LSF into two equal areas (Fig. 4.105).

In order to have a more complete characterization of the PSF one takes two LSFs with two integration directions perpendicular to each other, one in the radial direction (for tangential structures) and the other in the tangential direction (for radial structures) to the image circle, Fig. 4.106.

From the viewpoint of application, there is a further important object structure, namely the edge spread function (ESF). This is the image of an ideal edge, i.e. a sudden dark to light transition. The transition in the image of such an ideal edge will be more or less continuous, since the LSF has finite extension.

Indeed the ESF is connected with the LSF as shown in Fig. 4.107.

The area of the LSF up to the coordinate u_1 is A_1 and is equal to the value of the ESF at this coordinate u_1. This is valid for all values of u. Therefore the final value of the ESF represents the total area of the LSF and the value at $u = 0$ is half of the total area according to the definition of the coordinate origin of the LSF. Usually the LSF is normalized to unit area, so that the value

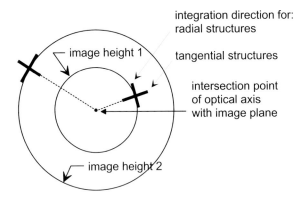

Fig. 4.106 Integration directions for LSFs.

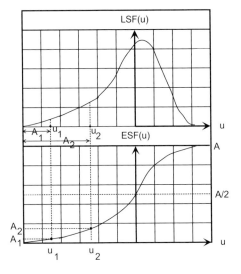

Fig. 4.107 The relation between LSF and ESF.

of ESF at $u = 0$ is $1/2$.

$$\text{ESF}(u) = \int_{-\infty}^{u} \text{LSF}(u)du \qquad (4.251)$$

Sub-pixel edge detection is a standard procedure in digital image processing and is used to locate the position of objects. Therefore, it is important to know the profile of the ESF and even more important to know the variation of the ESF with image height and orientation of the edge. Hereby, it is not so much important to have a steep gradient but rather to have a uniform ESF over image height and orientation. Figure 4.108 shows the ESF of a perfect (diffraction limited) system, as well as those of our sample lens 1.4/17 mm (on axis and for the image height of 5 mm for $f/nr = 8$). The radial ESF for the image height of 5 mm mm is practically identical with the ESF on axis and is omitted therefore.

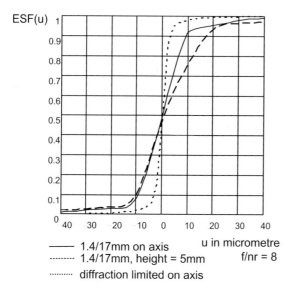

Fig. 4.108 Different edge spread functions (ESFs).

Distortion [14]

Distortion is the geometric deviation of the image point with respect to the ideal position of Gaussian optics (offence against the linear camera model). Since the optical system is nominally rotational symmetric, the theoretical distortion is also the same. Thus it will be sufficient to plot distortion as a function of the image height h'.

$$D(h') = \frac{h'_D - h'_G}{h'_G} \cdot 100 [\%] \qquad (4.252)$$

with

$$h'_D = \text{distorted image position}$$
$$h'_G = \text{Gaussian image position}$$

Figure 4.109 shows the distortion of the lens 1.4/17 for different magnification ratios over the relative image height.

Because of manufacturing tolerances of real optical systems there will be centring errors which destroy the nominal rotational symmetry. In photogrammetry, this real distortion will be calibrated with the help of a power

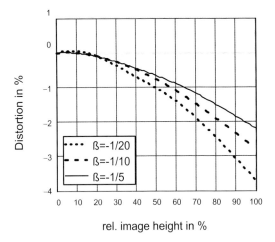

Fig. 4.109 Distortion over the relative image height.

series expansion. For a very high accuracy (in the μm range) a power series with only two coefficients is not sufficient, even not for the reconstruction of nominal distortion! Other forms for the representation of the distortion function may be more efficient (with lower numbers of coefficient than a power series).

4.5.4
Image Quality in Spatial Frequency Domain: MTF

The functions $\text{PSF}(u,v)$ and $\text{LSF}(u)$ in the space domain have their counterparts in the spatial frequency domain, the two-dimensional OTF (as 2D Fourier transform of the PSF) and the one-dimensional OTF (as the Fourier transform of the LSF). The modulus of the OTF is the modulation transfer function MTF, which describes the modulation in the image plane as a function of the line pairs/mm, see Section 4.4.3 and Fig. 4.110.

The modulation becomes lower for finer structures (larger number of lp/mm) and will be zero for a certain number of lp/mm. Hence the optical system is essentially a low pass filter. For the impression of sharpness however, this limit (*resolution limit*) is not a good measure. More important is a high modulation over the entire frequency range, *up to a frequency limit which depends on the application.*

As we have shown in section 3.4.6, because of the regularly spaced pixel structure of the sensor, there exists a highest spatial frequency r_N (*Nyquist frequency*), where for higher frequencies the image information will be falsified

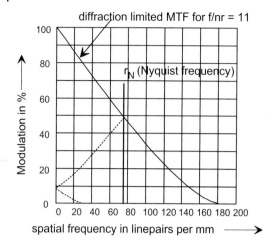

Fig. 4.110 Modulation as a function of spatial frequency.

and leads to a spatial frequency-dependent additive noise.

$$r_N = \frac{1}{2d} \quad \text{example } d = 7\mu m \quad , \quad r_N = 72\,\text{lp/mm}$$

(see Fig. 4.110).

It thus *makes no sense* to require a high modulation at the Nyquist frequency, since this will only enlarge the additive noise!

Parameters Which Influence the Modulation Transfer Function

One MTF is only valid for a certain isoplanatic region and one has to plot quite a lot of MTF curves in order to characterize the optical system. Therefore one uses a different representation in which the modulation is plotted over the image height for some meaningful spatial frequencies as parameter, Fig. 4.111.

This representation is normally used in data sheets and corresponds to the international standards ISO 9334 [15] and ISO 9335 [16]. The dashed lines correspond to the MTF for tangential orientation and the full lines to the radial one. Furthermore the MTF depends on the f-number, the magnification ratio and – very important – on the spectral weighting function within the used wavelength region.

> *Without the information on spectral weighting the MTF data are meaningless!*

For wide angle lenses with large pupil field angles ω_p the MTF for radial orientation decreases with the cosine of the *object side* field angle ω_p and for tangential orientation with the third power of ω_p ($\cos^3 \omega_p$).

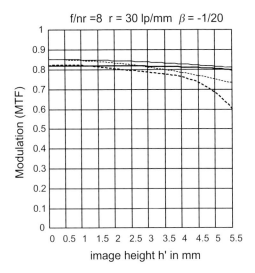

Fig. 4.111 MTF as a function of image height.

The examples in Fig. 4.111 are for a perfect (diffraction limited) lens for 30 lp/mm and $f/nr = 8$ (thin lines) and for our sample lens 1.4/17 mm (thick lines) for radial (full lines) and tangential (dashed lines) orientations.

4.5.5
Other Image Quality Parameters

4.5.5.1 Relative Illumination (Relative Irradiance) [17]

The decrease of illumination with the image height h' depends on several components: the basic term is a decrease with the forth power of the cosine of the *object side* field angle ω_p ($\cos^4$-law, sometimes called *natural vignetting*). This effect may be changed by distortion and by pupil aberrations (area of the entrance pupil changes with the field angle ω_p). On the other hand, the light cones may be reduced by mechanical parts of the lens barrel or by the diameter of the lenses, which gives an additional decrease of illumination (*mechanical vignetting*). This effect will be reduced when stopping down the lens aperture and vanishes in general when stopping down by two to three f/numbers. Figure 4.112(a) shows the $\cos^4 \omega_p$-law and Fig. 4.112(b) our sample lens 1.4/17 mm for three magnification ratios and f-numbers 1.4 and 4.

Spectral Transmittance [18]

The absorption of optical glasses and the residual reflections at the lens surfaces are responsible for wavelength-dependent light losses. The residual reflections may be reduced by anti-reflection coatings (evaporation with thin

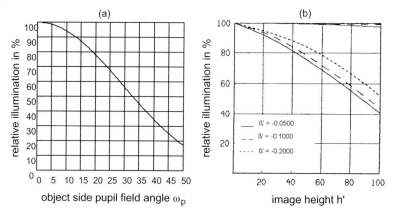

Fig. 4.112 Natural vignetting (a) and relative illumination for the lens 1.4/17 mm (b).

dielectric layers). Figure 4.113 shows the spectral transmittance of our sample lens 1.4/17 mm as a function of wavelength.

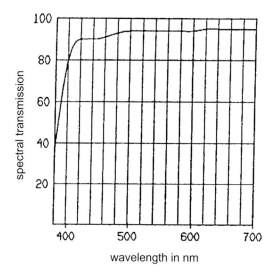

Fig. 4.113 Spectral transmittance.

If one wants to use the whole spectral sensitivity range of semiconductor imaging sensors (∼ 400–1000 nm) one has to use lenses with sufficient spectral transmission over this region and also sufficient image quality. Usual antireflection coatings, as for photographic lenses, are designed for the visible region only and show transmission *reducing* properties in the near infrared. Because of this, the rays will suffer multiple reflection within the optical system which may cause undesired ghost images, secondary iris spots and veiling glare.

4.5.5.2 Deviation from Telecentricity (for Telecentric Lenses only)

Figure 4.114 shows an example for a bilateral telecentric lens. The deviations are given in micrometre over the relative image height for a variation in object position of ±1 mm. The parameters are three different object-to-image distances. The deviations are shown for the object side as well as for the image side. The latter are practically zero here.

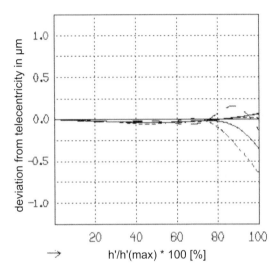

Fig. 4.114 Deviation from telecentricity.

4.5.6
Manufacturing Tolerances and Image Quality

The manufacturing is inevitably connected with tolerances. For the manufacturing of lenses these are typically as follows:

- deviations from the spherical surfaces of lenses (measured in $\lambda/2$ units)
- thickness tolerances for lenses and air gaps (typically 0.01 to 0.1 mm)
- lateral displacement of lenses (typically 5 to 100 μm)
- tilt of lens surfaces (typically 0.5 to 5 arcmin)
- refractive index tolerances of optical glasses.

Here the interaction between mechanics (lens barrel) and optics (lens elements) is most important. Image quality parameters are differently sensitive to these tolerances. Table 4.11 gives a rough classification of the sensitivity of quality parameters on manufacturing tolerances, as well as some typical values one has to account for. Of course these depend on the practical application and the manufacturing efforts. Lenses for amateur photography exhibit much higher quality fluctuations than special optical systems for semiconductor production for instance. Correspondingly, the costs may differ by several orders of magnitude. The table is valid approximately for professional photo quality.

Tab. 4.11 Sensitivity of optical quality parameters to manufacturing tolerances.

parameter	sensitivity				remarks / typical deviations
	++	+	-	--	
1. focal length			x		typ. 1 % max. of nominal value
2. rel. aperture			x		5 % max. according to standard [19]
3. f-number		x			caused by mechanical inaccuracies of the iris Standard [19]: up to 50 % depending on f-number !
4. ESF	x				
5. MTF	x				depending on spatial frequency, f-number fixed focal length: 10 % absolute (typically); zoom lenses: 10-20 % absolute (typically)
6. distortion	x				without special measures: 10-30 μm with special measures: ≤ 5 μm
7. rel. illumination				x	
8. spectral transmission				x	depending on the stability of antireflection manufacturing process

Measurement Errors due to Mechanical Inaccuracies of the Camera System

The corresponding influences may be mentioned here only by some headings. Nevertheless they are as important as the variation of optical quality parameters, e.g. for contactless measurement techniques [20], Table 4.12.

4.6
Practical Aspects: How to Specify Optics According to the Application Requirements?

Not every optical system may be used for whatever application.

The universal optical system does not exist!

Tab. 4.12 Influence of mechanical inaccuracies of the camera system.

	cause	effect
1.	unprecize (non reproducible) image position (e.g. by repetative focussing)	change of image size (heading: bilateral telecentric lenses)
2.	unprecize object position (e.g. parts tolerances)	change of image size (heading: object side telecentric lenses)
3.	image plane tilt (e.g. unprecize sensor position)	introduction of asymmetrical distortion terms
4.	object plane tilt	distortion terms, change of image size
5.	deviation of optical axis from mechanical reference axis ("Bore sight")	as 3. and 4.

Therefore one has to specify the most important parameters in order to find the best suited optics for the particular application – also with regard to coast/performance relation.

The major part of these parameters are fixed by the geometrical imaging situation and by some spatial restrictions. Other parameters are defined by the used image sensor as well as by the light source/filter combination. The following gives a list of the most important parameters.

- object–image distance
- magnification ratio defines the focal length
- required depth of field
- relative aperture (f-number) influences each other
- minimum relative illumination
- wavelength range (spectral transmission)
- maximum permissible distortion
- minimum MTF values at defined spatial frequencies/image heights
- **environmental influences**
    - temperature/humidity
    - mechanical shocks
    - mechanical vibrations
    - dust
    - veiling glare by external light sources
- **mechanical interfaces**

- maximum diameter and length
- maximum weight
- interface to the camera (C-mount, D-mount, etc.)

We will explain the procedure with the help of an example:

Example for the Calculation of an Imaging Constellation
1. **Image sensor data**
   - $\frac{1}{2}$ inch FT-CCD, sensor size 4.4×6.6 mm^2
   - pixel distance: 7μm
   - pixel size 7×7 μm^2

2. **Object data**
   - object distance ≈ 500 mm
   - object size: 90×120 mm^2
   - object depth: 40 mm

3. **Calculation of the magnification ratio and focal length**
   - from object and sensor size:

   $$|\beta| = \frac{\text{sensor size}}{\text{object size}} = \frac{4.4\,\text{mm}}{90\,\text{mm}} \approx 1/20$$
   $$\beta = -1/20 \quad (\text{inverted image})$$

   - from object to image distance and from the imaging equation

   $$e = f'\left(\frac{1}{\beta} - \frac{1}{\beta_p}\right) \quad (\beta_p \approx 1)$$
   $$f' \approx 23\,\text{mm}$$

4. **Calculation of the depth of field**
   - the depth of field is given by the depth of the object and with the depth of field formula:

   $$T = \frac{2(f/nr)_e \cdot d'_G}{\beta^2}$$

- definition of the permissible circle of confusion: for sub-pixel edge detection the range of the ESF shall be 4-7 pixels:

$$d'_G = 28\mu m \quad (4 \text{ pixels})$$

Thus

$$(f/nr)_e \geq 3 \cdot \frac{T \cdot \beta^2}{2 \cdot d'_G} = 1.8 \quad !$$

- control for the validity of the depth of field approximation:

$$|\beta| \geq 3 \cdot \frac{(f/nr) \cdot d'_G}{f'}$$

$$\frac{1}{20} \geq \frac{1}{150} \quad \text{(fulfilled)}$$

- condition for permissible geometrical calculation:

$$(f/nr)_e \leq \frac{2.8}{4.5} \approx 6.2 \quad \text{(fulfilled)}$$

- diffraction limited disc of confusion:
  - at the limits of the depth of field:

$$d'_d = \sqrt{(d'_G)^2 + 2 \cdot (f/nr)_e^2} \approx 28.5\mu m \quad [(f/nr)_e = 4.0]$$

  - in focus

$$d'_d = 1.22 \cdot (f/nr)_e \approx 5\mu m$$

5. **Definition of image quality parameters**

    - spectral weighting: VIS (380-700 nm) (IR-Cut-Off filter required)
    - Nyquist frequency of the sensor:

    $$r_N \approx 70 \, lp/mm$$

    - specification of MTF data:

spatial frequency r (lp/ mm)	20	40	60
MTF [%]	≥ 80	≥ 50	≥ 20

    for image heights $h' \leq 3.5$ mm

6. **Specification of distortion**

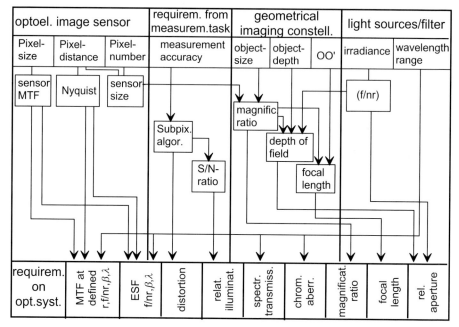

Fig. 4.115 Requirements for the optical system in dependence of the measurement task.

- required measurement accuracy in the object: ±0.05 mm
  → absolute distortion in image

$$|\Delta y'| \leq \frac{0.05}{20}\,\text{mm} = 2.5\,\mu\text{m}$$

→ calibration of distortion required

7. **Comparison with manufacturer data**

   There exist some simple optical design programs which might considerably facilitate such calculations, e.g. [21]

   Figure 4.115 shows schematically the interdependences between the requirements of the optical system and the parameters of the measurement situation.

References

1. HOPKINS H. H., *Canonical and real space coordinates in the theory of image formation*, Applied Optics and Optical Engineering, Vol. IX, Chapter 8, Academic Press, New York (1983)
2. GOODMAN J. W., *Introduction to Fourier Optics*, McGraw-Hill, New York (1968)
3. BORN M., WOLF E., *Principles of Optics*, Chapter 8.5, Pergamon Press, Oxford (1975)
4. LOMMEL E., Abh. Bayer.Akad., 15, Abth. 2, (1995), 233
5. STRUVE H., Mém. de l'Acad. de St. Petersbourgh (7), 34 (1986), 1
6. BORN M., WOLF E., *Principles of Optics*, Chapter 8.8, Pergamon Press, Oxford (1975)
7. BRIGHAM E. O., *The Fast Fourier Transform*, Prentice-Hall, Englewood Cliffs, NJ (1974)
8. BRACEWELL R. N., *The Fourier transform and its applications*, McGraw-Hill Kogakusha, International student edition, 2nd ed. (1978)
9. LIGHTHILL M. J., *Einführung in die Theorie der Fourieranalysis und der verallgemeinerten Funktionen*, Bibliographisches Institut Mannheim (1966)
10. PARK S.K., RAHMEN Z., *Fidelity analysis of sampled imaging systems*, Opt. Eng. 38(5) 786-800 (1999)
11. WITTENSTEIN W. ET AL., *The definition of the OTF and the measurement of aliasing for sampled imaging systems*, Optica Acta, 29(1) 41-50, (1982)
12. WELFORD W.T., *Aberrations of the Symmetrical Optical System*, Academic Press, London (1974)
13. ISO 15529, Optics and Photonics – Optical transfer function – Principles of measurement of modulation transfer function (MTF) of sampled imaging systems
14. ISO 9039:1995: Optics and optical instruments – Quality evaluation of optical systems – Determination of distortion
15. ISO 9334:1995: Optics and optical instruments – Optical transfer function – Definition and mathematical relation ships
16. ISO 9335:1995: Optics and optical instruments – Optical transfer function – Principles and procedures of measurement
17. ISO 13653:1996: Optics and optical instruments – General optical test methods – Measurement of the relative irradiance in the image field
18. ISO 8478:1996: Photography – Camera lenses – Measurement ISO spectral transmittance
19. ISO 517:1996: Photography – Apertures and related properties pertaining to photographic lenses – Designations and measurements
20. www.schneiderkreuznach.com/knowhow/reports (English version), Optical measurement techniques with telecentric lenses
21. www.schneiderkreuznach.com/knowhow/downloads

5
Camera Calibration

R. Godding, AICON 3D Systems GmbH

5.1
Introduction

In recent last years, the use of optical measurement systems in industrial measurement technology has been continuously increased. On the one hand, there are digital cameras (partly with telecentric optics) which are used for the determination of two-dimensional measuring quantities; on the other hand, the 3D measurement technology has greatly come to the fore. This development was favoured by the improvement of the sensor technology of digital cameras so that nowadays cameras with up to 5000×4000 pixels can be used as well as ever faster processors enabling the fast processing of large data quantities.

The mode of operation of the systems, however, is greatly differing.

One method consists in using one-camera systems by means of which almost any objects are taken (partially with several hundred images) which are then evaluated offline. Besides, there are numerous systems in use in which a great number of cameras are simultaneously acquiring an object, the evaluation being made online. Most applications require a high accuracy for the evaluation.

This accuracy can be reached only if corresponding mathematical models form the basis of the cameras used for all calculations, i.e. that the camera parameters are known and are taken in consideration accordingly.

In photogrammetry the subject of which has been the derivation of three-dimensional measuring quantities from images for many years, extensive work has always been carried out in the field of camera modelling. In connection with that the determination of the significant camera parameters is often referred to as camera calibration.

Here, there are basically two approaches. While in some cases the actual object information and the camera parameters can be simultaneously determined with corresponding models (simultaneous calibration), the camera sys-

Handbook of Machine Vision. Alexander Hornberg (Ed.)
Copyright © 2006 WILEY-VCH Verlag GmbH & Co. KGaA, Weinheim
ISBN: 3-527-40584-7

tems first are calibrated separately in other applications. The camera parameters are then assumed to be constant in the subsequent measurement. Approaches combining both methods are used as well.

In addition, the methods of camera calibration can also be used to obtain quality information of a camera system or information about the measuring accuracy that can be obtained with a measuring system.

The following chapters describe methods of calibration and orientation of optical measurement systems, focusing primarily on photogrammetric methods since these permit a homologous and highly accurate determination of the parameters required.

5.2
Terminology

The terminology in the following chapters is based on the photogrammetric definitions. Sometimes this terminology may differ from the terms, which are used in other disciplines, e.g. in machine vision.

5.2.1
Camera, Camera system

All mechanical, optical and electronic components, which are necessary to produce an image used for measurement purposes, can be described as a camera or camera system. The camera mainly consists of a housing, which holds a lens system, sometimes equipped with an additional filter on the one side and a sensor on the other side. Earlier the images were stored on film, which led to additional components (e.g. components for the definition of the film flatness during the exposure time). Today, a lot of different digital sensor types are used for the imaging.

In addition, the electronic components, which have an influence on the geometry of the final digital image (e.g. the A/D converter) have to be added to the definition of the camera. These components can be outside the camera housing, e.g. framegrabbers within the computer systems. Some cameras (especially consumer cameras) perform a kind of image preprocessing within the camera, which can be not influenced by the user. Even these preprocessing algorithms sometimes have an influence on the geometry of the digital image.

5.2.2
Coordinate systems

Three kinds of coordinate systems are mainly used for the description of the complete parameter set for calibration. The first coordinate system is a two-dimensional system defined in the sensor plane and is used for the description of the measured image information. This system is often called sensor or image coordinate system.

The next kind of system is used for the description of the camera parameters, it can be called as interior coordinate system. The last system is a higher-order system, frequently called the world coordinate system or object coordinate system. The last two systems are cartesian, right-handed systems.

5.2.3
Interior Orientation and Calibration

The interior orientation describes all parameters, which are necessary for the description of a camera or a camera system itself.

In photogrammetric parlance calibration refers to the determination of these parameters of interior orientation of individual cameras. The parameters to be found by calibration depend on the type of camera used and on the mathematical camera model, which is used for the description. Various camera models with different numbers of parameters are available. Once the imaging system has been calibrated, measurements can be made after the cameras have been duly oriented.

5.2.4
Exterior and Relative Orientation

The exterior orientation describes the transformation between the coordinate system of the interior orientation (or multiple transformations in case of camera systems with more than one camera) and the world coordinate system.

The relative orientation describes the transformation between different interior coordinate systems in case of a multi-camera configuration.

Both transformations require the determination of three rotational and three translational parameters - i.e. a total of six parameters and can be described with 4×4 homogeneous transformation matrices.

5.2.5
System Calibration

In many applications, fixed setups of various sensors are used for the measurement. Examples are online measurement systems in which, for example, several cameras, laser pointers, pattern projectors, rotary stages, etc., may be used. If the entire system is considered as the actual measurement tool, then the simultaneous calibration and orientation of all the components involved may be defined as the system calibration.

5.3
Physical Effects

5.3.1
Optical System

Practically all lenses have typical radial symmetrical distortion that may vary greatly in magnitude. On the one hand, some lenses used in optical measurement systems are nearly distortion-free [14], on the other hand, wide-angle lenses, above all, frequently exhibit a distortion of several 100 μm at the edges of the field of view. Fisheye lenses belong to a class of their own; they frequently have extreme distortion at the edges. There are special mathematical models to describe those special lens types [25].

Since radial symmetrical distortion is a function of the lens design, it cannot be considered as an aberration. In contrast to that, centering errors often unavoidable in lens making cause aberrations reflected in radial asymmetrical and tangential distortion components [4]. Additional optical elements in the light path, such as the different kind of filters at the lens and the protective filter at the sensor, also leave their mark on the image and have to be considered in the calibration of a system.

5.3.2
Camera and Sensor Stability

The stability of the camera housing has one of the most important impacts on the measurement accuracy. The lens system and sensor should be in a stable mechanical connection during a measurement or between two calibrations of a system. From the mechanical point of view, the best construction may be a direct internal connection between the lens and the sensor which is independent of the outer housing. In addition, this connection can have a provision for mounting the camera on a tripod or carrier. This allows a handling of the camera without too much influence on the stability of the interior orientation.

These ideal conditions are not realized for most of the cameras, which are used for measurement purposes. Normally there is a connection between the lens or sensor and the outer housing so that a rough camera handling can influence the interior orientation. Besides, the sensor is often not fixed well enough to the housing so that a small movement between the sensor and the lens is possible. An influence of thermal effects caused by electronic elements is also possible.

Due to their design, digital sensors usually offer high geometrical accuracy [19]. On the other hand, sensors have become larger in size. Digital sensors with sizes of 24mm × 36mm and larger are commonly used for measurement purposes. Most camera models imply the planarity of the sensor, which normally cannot be guaranteed for larger sensors.

5.3.3
Signal Processing and Transfer

When evaluating an imaging system, its sensor should be assessed in conjunction with all additional electronic devices, e.g., necessary frame grabbers. Geometrical errors of different magnitude may occur during A/D conversion of the video signal, depending on the type of synchronization, above all if a pixel-synchronous signal transfer from camera to image storage is not guaranteed [2]. However, in the case of a pixel-synchronous data readout, the additional transfer of the pixel clock pulse ensures that each sensor element will precisely match a picture element in the image storage. Most of the currently used cameras work with digital data transfer, e.g., with Camera Link, IEEE 1394 or USB 2.0 [15] so that a lot of problems (pixel shift, timing problems) are not relevant. Very high accuracy has been proved for these types of cameras. However, even with this type of transfer the square shape of individual pixels cannot be taken for granted. As with any kind of synchronization, most sensor-storage combinations make it necessary to consider an affinity factor; in other words, the pixels may have different extensions in the direction of lines and columns. Especially some consumer or professional still video cameras are provided with special sensor types (e.g. no square pixels) and an integrated computation system. Even those sensors must be considered in a mathematical model.

5.4
Mathematical Calibration Model

5.4.1
Central Projection

In principle, image mapping by an optical system can be described by the mathematical rules of central perspective. According to these, an object is imaged in a plane so that the object points P_i and the corresponding image points P'_i are located on straight lines through the perspective center O_j (Fig. 5.1). The following holds under idealized conditions for the formation of a point image in the image plane

$$\begin{bmatrix} x_{ij} \\ y_{ij} \end{bmatrix} = \frac{-c}{Z^*_{ij}} \begin{bmatrix} X^*_{ij} \\ Y^*_{ij} \end{bmatrix} \tag{5.1}$$

with

$$\begin{bmatrix} X^*_{ij} \\ Y^*_{ij} \\ Y^*_{ij} \end{bmatrix} = D(\omega, \varphi, \kappa) \begin{bmatrix} X_i - X_{oj} \\ Y_i - Y_{oj} \\ Z_i - Z_{oj} \end{bmatrix} \tag{5.2}$$

where

X_i, Y_i, Z_i are the coordinates of an object point P_i in the object coordinate system K,

X_{oj}, Y_{oj}, Z_{oj} the coordinates of the perspective center O_j in the object coordinate system K,

$X^*_{ij}, Y^*_{ij}, Z^*_{ij}$ the coordinates of the object point P_i in the coordinate system K^*_j,

x_{ij}, y_{ij} the coordinates of the image point in the image coordinate system K_B, and $D(\omega, \varphi, \kappa)_j$ the rotation matrix between K and K^*_j as well as

c the distance between perspective center and image plane, the system K^*_j being parallel to the system K_B with the origin in the perspective center O_j [31].

The above representation splits up the optics in such a manner that in 5.1 it is primarily the image space (interior) parameters and in 5.2 primarily the object space (exterior) parameters, i.e., the parameters of exterior orientation, that take effect.

This ideal concept is not attained in reality where a multitude of influences are encountered due to the different components of the imaging system. These can be modeled as deviations from the strict central perspective. The following section describes various approaches to mathematical camera models.

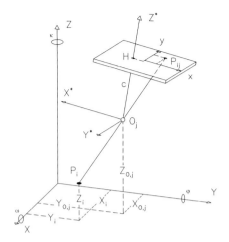

Fig. 5.1 Principle of central perspective [7].

5.4.2
Camera Model

When optical systems are used for measurement, modeling the entire process of image formation is decisive for the accuracy to be attained. Basically, the same ideas apply, for example, to projection systems for which models can be set up similarly to imaging systems.

The basic coordinate system for the complete description is the image coordinate system K_B in the image plane of the camera. In electro-optical cameras, this image plane is defined by the sensor plane. Here it is entirely sufficient to place the origin of the image coordinate system in the center of the digital images in the storage (Fig. 5.2). The centering of the image coordinate system is advantageous to the determination of the distortion parameters, which is described later.

Since the pixel interval in column direction in the storage is equal to the interval of the corresponding sensor elements, the unit *pixel in column direction* may serve as a unit of measure in the image space. All parameters of interior orientation can be directly computed in this unit, without conversion to metric values.

The transformation between a pixel and a metric coordinate system (which is sometimes used to have units independently from the pixel size) is very simple

$$x_{ij} = x_{ij} \cdot Fx + Tx \quad (5.3)$$
$$y_{ij} = Ty - y_{ij} \cdot Fy \quad (5.4)$$

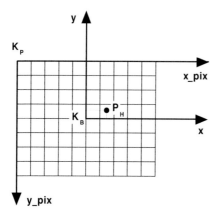

Fig. 5.2 Sensor coordinate system.

where Tx and Ty are half of the sensor resolution in x and y, and Fx and Fy are calculated from the quotient of the sensor resolution and the sensor size in x and y.

5.4.3
Focal Length and Principal Point

The reference axis for the camera model is not the optical axis in its physical sense, but a principal ray which on the object side is perpendicular to the image plane defined above and intersects the latter at the principal point $P_H(x_H, y_H)$. The perspective center O_j is located at the distance c_K (also known as calibrated focal length) perpendicularly in front of the principal point [24].

The original formulation of Eq. 5.1 is thus expanded as follows:

$$\begin{bmatrix} x_{ij} \\ y_{ij} \end{bmatrix} = \frac{-c_K}{Z_{ij}^*} \begin{bmatrix} X_{ij}^* \\ Y_{ij}^* \end{bmatrix} + \begin{bmatrix} x_H \\ y_H \end{bmatrix} \qquad (5.5)$$

5.4.4
Distortion and Affinity

The following additional correction function can be applied to Eq. 5.5 for radial symmetrical, radial asymmetrical and tangential distortion.

$$\begin{bmatrix} x_{ij} \\ y_{ij} \end{bmatrix} = \frac{-c_K}{Z_{ij}^*} \begin{bmatrix} X_{ij}^* \\ Y_{ij}^* \end{bmatrix} + \begin{bmatrix} x_H \\ y_H \end{bmatrix} + \begin{bmatrix} dx(V, A) \\ dy(V, A) \end{bmatrix} \qquad (5.6)$$

We may now define dx and dy differently, depending on the type of camera used, and are made up of the following different components:

$$dx = dx_{sym} + dx_{asy} + dx_{aff} \tag{5.7}$$
$$dy = dy_{sym} + dy_{asy} + dy_{aff} \tag{5.8}$$

5.4.5
Radial symmetrical distortion

The radial symmetrical distortion typical of a lens can generally be expressed with sufficient accuracy by a polynomial of odd powers of the image radius (x_{ij} and y_{ij} are henceforth called x and y for the sake of simplicity):

$$dr_{sym} = A_1(r^3 - r_0^2 r) + A_2(r^5 - r_0^4 r) + A_3(r^7 - r_0^6 r) \tag{5.9}$$

where

dr_{sym} is the radial symmetrical distortion correction,

r the image radius obtained from $r^2 = x^2 + y^2$,

A_1, A_2, A_3 are the polynomial coefficients, and

r_0 is the second zero crossing of the distortion curve,

so that we obtain

$$dx_{sym} = \frac{dr_{sym}}{r} x \tag{5.10}$$

$$dy_{sym} = \frac{dr_{sym}}{r} y \tag{5.11}$$

A polynomial with two coefficients is generally sufficient to describe radial symmetrical distortion. Expanding this distortion model, it is possible to describe even lenses with pronounced departure from perspective projection (e.g. fisheye lenses) with sufficient accuracy; in the case of very pronounced distortion, it is advisable to introduce an additional point of symmetry $P_S(x_S, y_S)$. Fig. 5.3 shows a typical distortion curve.

For numerical stabilization and far-reaching avoidance of correlations between the coefficients of the distortion function and the calibrated focal lengths, a linear component of the distortion curve is split off by specifying a second zero crossing [28].

A different formulation is proposed in [18] for determining radial symmetrical distortion, which includes only one coefficient. We thus obtain the following formula:

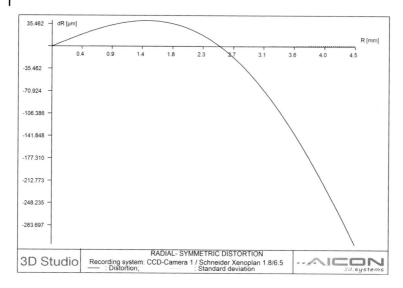

Fig. 5.3 Typical radial symmetrical distortion curve of a lens.

$$dr_{sym} = r\frac{1 - \sqrt{1 - 4Kr^2}}{1 + \sqrt{1 - 4Kr^2}} \quad (5.12)$$

where K is the distortion coefficient to be determined.

5.4.6
Radial Asymmetrical and Tangential Distortion

To cover radial asymmetrical and tangential distortion, a number of formulations are possible. Based on [6], these distortion components may be formulated as follows [4]:

$$dx_{asy} = B_1(r^2 + 2x^2) + 2B_2xy \quad (5.13)$$
$$dy_{asy} = B_2(r^2 + 2y^2) + 2B_1xy \quad (5.14)$$

In other words, these effects are always described with the two additional parameters B_1 and B_2.

This formulation is expanded by [5], who adds parameters to describe overall image deformation or the lack of image plane flatness.

$$dx_{asy} = (D_1(x^2 - y^2) + D_2 x^2 y^2 + D_3(x^4 - y^4))x/c_K$$
$$+ E_1 xy + E_2 y^2 + E_3 x^2 y + E_4 xy^2 + E_5 x^2 y^2 \qquad (5.15)$$
$$dy_{asy} = (D_1(x^2 - y^2) + D_2 x^2 y^2 + D_3(x^4 - y^4))y/c_K$$
$$+ E_6 xy + E_7 x^2 + E_8 x^2 y + E_9 xy^2 + E_{10} x^2 y^2 \qquad (5.16)$$

In view of the large number of coefficients, however, this formulation implies a certain risk of too many parameters. Moreover, since this model was primarily developed for large-format analog imaging systems, some of the parameters cannot be directly interpreted for applications using digital imaging systems. Equations 5.8 and 5.9 are generally sufficient to describe asymmetrical effects. Fig. 5.4 shows typical effects for radial symmetrical and tangential distortion.

5.4.7
Affinity and Nonorthogonality

The differences in length and width of the pixels in the image storage caused by synchronization can be taken into account by an affinity factor. In addition, an affinity direction may be determined, which primarily describes the orthogonality of the axes of the image coordinate system K_B. An example may be a line scanner that does not move perpendicularly to the line direction. These two effects can be considered as follows:

$$dx_{aff} = C_1 x + C_2 y \qquad (5.17)$$
$$dx_{aff} = 0 \qquad (5.18)$$

Fig. 5.5 depicts an example of the effect of affinity

5.4.8
Variant Camera Parameters

In particular for cameras with mechanical instabilities or thermal problems it can be helpful to use special mathematical models, which allow some parameters to be variant during the measurement. That implies that those parameters are different for each image during a measurement process, while other parameters are assumed as stable.

A usual practice in that case is setting the focal length C_k and the coordinates of the principle point x_H and y_H as variant parameters for each image.

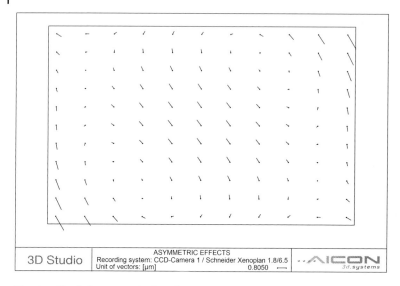

Fig. 5.4 Radial symmetrical and tangential distortion.

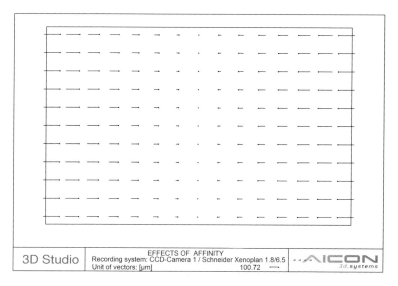

Fig. 5.5 Effects of affinity.

That strategy is only possible for special applications, e. g. photogrammetric measurements with a lot of images. In those cases it is possible to get results with high accuracy, in spite of problems with mechanical instabilities [22].

5.4.9
Sensor Flatness

The mathematical model described above is based on a plane sensor, which cannot be implied in reality, especially for cameras with a large sensor format. In order to compensate for those effects, a finite-elements correction grid based on anchor points can be used, which is described by [17]. The sensor is subdivided by a rectangular grid and for each node of the grid corrections are computed so that each measured coordinate can be improved by the interpolated corrections. Fig. 5.6 shows a typical correction grid for a high-resolution camera (Mamiya DCS 645, sensor size 36mm × 36mm). It could be shown that the use of such models can improve the accuracy of 3D coordinates significantly.

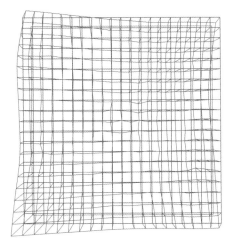

Fig. 5.6 Correction grid for a camera with large sensor.

5.4.10
Other Parameters

The introduction of additional parameters may be of interest and can improve the accuracy for special applications. Formulations described in [11] and [8] also make allowance for distance-related components of distortion within the photographic field. However, these are primarily effective with medium and large image formats and the corresponding lenses, and are of only minor importance for the wide field of digital uses.

A different camera model is used [16] in which two additional parameters have to be determined for the oblique position of the sensor.

5.5
Calibration and Orientation Techniques

5.5.1
In the Laboratory

Distortion parameters can be determined in the laboratory under clearly defined conditions. That possibility is often used for the calibration of high resolution cameras for aerial photogrammetry.

In the goniometer method, a highly precise grid plate is positioned in the image plane of a camera. Then the goniometer is used to measure the grid intersections from the object side and to determine the corresponding angles. Distortion values can then be obtained by comparing the nominal and actual values.

In the collimator technique, test patterns are projected onto the image plane by several collimators set up at defined angles to each other. Here too, the parameters of interior orientation can be obtained by a comparison between nominal and actual values, though only for cameras focused on infinity [24].

Apart from this restriction, there are more reasons speaking against the use of the aforementioned laboratory techniques for calibrating digital imaging systems, including the following:

- The scope of equipment required is high.

- The interior orientation of the cameras normally used is not stable, requiring regular recalibration by the user.

- Interior orientation including distortion varies at different focus and aperture settings, so that calibration under practical conditions appears more appropriate.

- There should be a possibility of evaluating and recalibrating optical measurement systems on the measurement side to avoid system maintenance costs that are too high.

5.5.2
Using Bundle Adjustment to Determine Camera Parameters

All parameters required for calibration and orientation may be obtained by means of photogrammetric bundle adjustment. In bundle adjustment, two so-called observation equations are set up for each point measured in an image, based on Eqs. 5.2 and 5.4. The total of all equations for the image points of all corresponding object points results in a system that makes it possible to determine the unknown parameters [29]. Since this is a nonlinear system of equations, approximate values for all unknown parameters are necessary. The computation is made iteratively by the method of least squares, the unknowns being determined in such a way that the squares of deviations are minimized at the image coordinates observed. Newer approaches work with modern algorithms like balanced parameter estimation [9]. Bundle adjustment thus allows simultaneous determination of the unknown object coordinates, exterior orientation and interior orientation with all relevant system parameters of the imaging system. In addition, standard deviations are computed for all parameters, which give a measure of the quality of the imaging system.

5.5.2.1 Calibration based Exclusively on Image Information

This method is particularly well-suited for calibrating individual imaging systems. It requires a survey of a field of points in a geometrically stable photogrammetric assembly. The points need not include any points with known object coordinates (control points); the coordinates of all points need only be known approximately [29]. It is, however, necessary that the point field is stable for the duration of image acquisition. Likewise, the scale of the point field has no effect on the determination of the desired image space parameters. Fig. 5.7 shows a point field suitable for calibration.

Fig. 5.7 Reference plate for camera calibration.

The accuracy of the system studied can be judged from the residual mismatches of the image coordinates as well as the standard deviation of the unit of weight after adjustment (Fig. 5.8). The effect of synchronization errors, for example, becomes immediately apparent, for instance by larger residual mismatches of different magnitude in line and column direction.

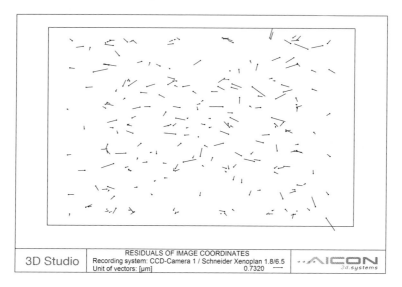

Fig. 5.8 Residual after bundle adjustment.

Figure 5.9 gives a diagrammatic view of the minimum setup for surveying a point array with which the aforementioned system parameters can be determined. The array may be a three-dimensional test field with a sufficient number of properly distributed, circular, retroreflecting targets. This test field is first recorded in three frontal images, with camera and field at an angle

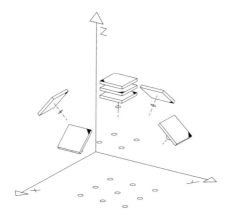

Fig. 5.9 Imaging setup for calibration [14].

of 100° for determining affinity and 200° for determining the location of the principal point. In addition, four convergent images of the test field are used to give the assembly the necessary geometric stability for determinating the object coordinates and to minimize correlations with exterior orientation.

Optimum use of the image format is a precondition for the determination of distortion parameters. However, this requirement does not need to be satisfied for all individual images. It is sufficient if the image points of all images cover the format uniformly and completely.

If this setup is followed, seven images will be obtained roughly as shown in Fig. 5.10, their outer frame standing for the image format, the inner frame for the image of the square test field and the arrow head for the position of the test field. It is generally preferable to rotate the test field with the aid of a suitable suspension in front of the camera instead of moving the camera for image acquisition. The use of retroreflecting targets and a ring light guarantees a proper, high-contrast reproduction of the object points, which is indispensable for precise and reliable measuring. Complete, commercially available software packages offer far-reaching automated processes for the described calibration task, in the most cases using special coded targets for a automatic point number determination.

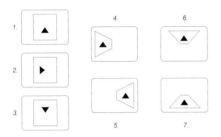

Fig. 5.10 Positions of reference plate during calibration.

5.5.2.2 Calibration and Orientation with Additional Object Information

Once the interior orientation of an imaging system has been calibrated, its orientation can be found by resection in space. The latter may be seen as a special bundle adjustment in which the parameters of interior orientation and the object coordinates are known. This requires a minimum of three control points in space the object coordinates of which in the world coordinate system are known and the image points of which have been measured with the imaging system to be oriented.

In addition to a simple orientation, a compete calibration of an imaging system is also possible with a single image. However, since a single image does not allow the object coordinates to be determined, suitable information within the object has to be available in the form of a three-dimensional control point array [30]. Today such reference fields can be easily constructed be using carbon fiber elements. In any case the control pattern should completely fill the measurement range of the cameras to be calibrated and oriented to ensure good agreement between the calibration and measurement volumes.

The effort is considerably smaller if several images are available. For a two-image assembly and one camera, a spatial array of points that need to be approximately known only, plus several known distances (scales) as additional information distributed in the object space will be sufficient, similar to section 5.5.2.1. In an ideal case, one scale on the camera axis, another one perpendicular to it and two oblique scales in two perpendicular planes parallel to the camera axis are required (Fig. 5.11). This will considerably reduce the effort on the object side, since the creation and checking of scales is much simpler than that of an extensive three-dimensional array of control points.

A similar setup is possible if the double-image assembly is recorded with several cameras instead of just one. This is, in principle, the case with online measurement systems. An additional scale is then required in the foreground of the object space, bringing the total number of scales to five (Fig. 5.12).

If at least one of the two cameras can be rolled, the oblique scales can be dispensed with, provided that the rolled image is used for calibration [30].

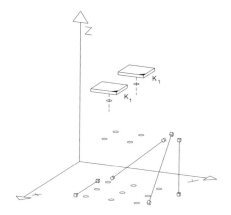

Fig. 5.11 Scale setup for calibrating one camera.

The setups described are applicable to more than two cameras as well. In other words, all the cameras of a measurement system can be calibrated if the above mentioned conditions are created for each of the cameras. At least two cameras have to be calibrated in common, with the scales set up as described. Simultaneous calibration of all cameras is also possible, but then the scale information must be simultaneously available to all the cameras. If all cameras

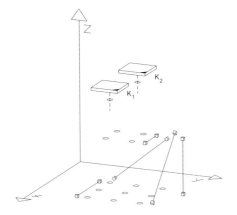

Fig. 5.12 Scale setup for calibrating two cameras.

are also to be calibrated in common, this will have to be done via common points.

With digital, multi-camera-online systems another calibration method is possible, which is based on the above described theory. It is possible to make the calibration only with one scale, which is moved in front of the measurement system. For each position images with all cameras are recorded and the image coordinates are measured. For each camera, all measured image coordinates are subsumed in one image so that we get one virtual image for each camera with the coordinates of all measurements. If raw calibration values are known approximate object coordinates can be computed. Together with the knowledge of the distance of the scale (which is known for each snap of the system) enough information for the calibration is available.

5.5.2.3 Extended System Calibration

As we have seen from the last sections, a joint calibration and orientation of all cameras involved and thus the calibration of the entire system is possible if certain conditions are met. With the aid of bundle adjustment, the two problems can be solved jointly with a suitable array of control points or a spatial point array of unknown coordinates plus additional scales. The cameras then already are in measurement position during calibration. Possible correlations between the exterior and interior orientations required are thus neutralized because the calibration setup is identical to the measurement setup.

Apart from the imaging systems, other components can be calibrated and oriented within the framework of system calibration. A technique described in [13] in which a suitable procedure in an online measurement system allows both the interior and exterior orientation of the cameras involved as well as the orientation of a rotary stage to be determined with the aid of a spatial point array and additional scales. The calibration of a line projector within a measurement system using photogrammetric techniques was, for example, presented by Strutz [26].

The calibration of a recording system with one camera and 4 mirrors is described in Putze [23]. This system generates a optical measurement system with four virtual cameras. The advantage of the setup is that especially in the case of high speed image recording the accuracy of a determination of 3D coordinates is not influenced by synchronization errors from different cameras. This principle allows a much cheaper solution than using four single cameras.

5.5.3
Other Techniques

Based on the fact that straight lines in the object space have to be reproduced as straight lines in the image, the so-called plumb-line method serves to determine distortion. The method is based on the fact that the calibrated focal length and the principal point location are known [12].

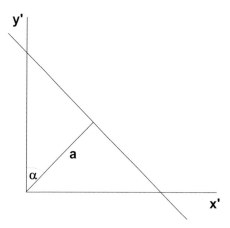

Fig. 5.13 Principle of plumb-line method.

According to Fig. 5.13, each of the straight-line points imaged are governed by the relationship

$$x' \sin \alpha + y' \cos \alpha = a \tag{5.19}$$

where x' and y' can be expressed as follows:

$$x' = x_{ij} + dx_{sym} + dx_{asy} \tag{5.20}$$
$$y' = y_{ij} + dy_{sym} + dy_{asy} \tag{5.21}$$

where $dx_{sym}, dy_{sym}, dx_{asy}, dy_{asy}$ correspond to the formulations in 5.10, 5.11 and 5.13-5.16. It is an advantage of this method that, assuming suitable selection of the straight lines in the object, a large number of observations is available for determining distortion, measurement of the straight lines in the image lending itself to automation. A disadvantage is the fact that simultaneous determination of all relevant parameters of interior orientation is impossible.

A method is presented in [18] in which an imaging system was likewise calibrated and oriented in several steps. The method requires a plane test field with known coordinates, which generally should not be oriented parallel to the image plane. Modeling radial symmetrical distortion with only one coefficient and neglecting asymmetrical effects allows the calibration to be based entirely on linear models. Since these do not need to be resolved interactively, the method is very fast. It is a disadvantage, however, that here too it is impossible to determine all the parameters simultaneously and that, for example, the location of the principal point and the pixel affinity have to be determined externally.

A method decribed in [16] permits cameras to be calibrated and oriented with the aid of parallel straight lines projected onto the image. A cube of known dimensions is required for the purpose as a calibrating medium. Vanishing points and vanishing lines can be computed from the cube edges projected onto the image and are used to determine the unknown parameters.

A frequently used formulation for determining the parameters of exterior and interior orientation is the method of direct linear transformation (DLT) first proposed by Abdel-Aziz and Karara [1]. This establishes a linear relationship between image and object points. The original imaging equation is converted to a transformation with 11 parameters that initially have no physical importance. By introducing additional conditions between these coefficients it is then possible to derive the parameters of interior and exterior orientation, including the introduction of distortion models [3]. Since the linear formulation of DLT can be solved directly, without approximations for the unknowns, the technique is frequently used to determine approximations for bundle adjustment. The method requires a spatial test field with a minimum of six known control points, a sufficient number of additional points being needed to determine distortion. However, if more images are to be used to determine interior orientation or object coordinates, nonlinear models will have to be used here too.

Other direct solutions for the problems of calibration and orientation based on the methods of projective geometry and DLT and attractive for the applications of machine vision are shown by [10].

5.6 Verification of Calibration Results

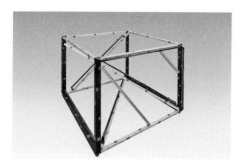

Fig. 5.14 Spatial test object for VDI 2634 test.

After the calibration and orientation has been made the system is ready for measuring. The final acceptance and verification of the measurement system can be done according to the German guideline VDI 2634 by measuring a spatial object with a range of 2m × 2m × 1.5m with at least 7 different measuring lines distributed in a specific manner [20]. The measuring lines are realized, e.g., by highly accurate carbon fiber scales with some external calibrated distances (e.g., by a coordinate measuring machine (CMM)). The accuracy of the system can be verified by a comparison between the measured and the nominal values of the distances so that the quality of the optical measurement system is reflected by the error of the length measurement. Figure 5.14 shows a spatial test object for a verification test, Fig. 5.15 a typical result.

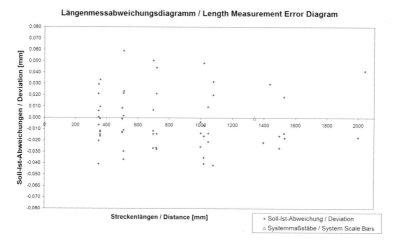

Fig. 5.15 Length measurement error diagram.

5.7 Applications

5.7.1 Applications with Simultaneous Calibration

The imaging setup for many photogrammetric applications allows a simultaneous calibration of cameras. It is an advantage of this solution that no additonal effort is required for an external calibration of the cameras and that current camera data for the instant of exposure can be determined by bundle adjustment. This procedure, however, is possible only if the evaluation software offers the option of simultaneous calibration. As an example, let us look at the measurement of a car for the determination of deformations during a crash test (Figs. 5.16).

Fig. 5.16 Photogrammetric deformation measurement of a crashed car.

A total of 80 photos were taken before and after the crash with a Nikon D2x digital camera (Fig. 5.17) with a resolution of appr. 4000 × 3000 sensor elements. The AICON 3D Studio software was used for the complete evaluation. This allows the fully automatic determination of 3D coordinates, starting with the measurement of image points right up to computation of all unknown parameters. In addition to target sizes and the 3D coordinates of all measured points in the world coordinate system, these include the camera parameters and all camera stations. For the evaluation of the crash test, a complete deformation analysis was made within the software. For this example the coor-

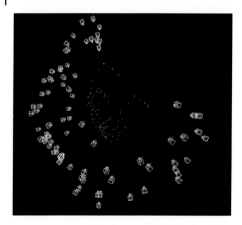

Fig. 5.17 Camera positions for crash car measurement.

dinates have an RMS value of appr. 0.04 mm in each of the three coordinate axes.

Most photogrammetric applications for high-precision 3D industrial metrology work are based on simultaneous calibration. Numerous other uses can be found in the aviation industry (measuring aircraft components and fixtures), in the aeronautical industry (measuring satellites and antennas), in ship building and in civil engineering (measuring of tunnels).

5.7.2
Applications with precalibrated cameras

5.7.2.1 **Tube Measurement within a Measurement Cell**
Some applications are running in an environment with fixed cameras, nevertheless these camera systems have to be calibrated and the calibration has to be checked to ensure reliable measurement results.

One example of such a system is the AICON TubeInspect system, a measurement system which incorporates advanced technology for the high-precision measurement of tubes, the determination of set-up and correction data and quality assurance of different kind of tubes. The system acquires the tube with 16 high-resolution cameras (IEEE 1394), which are firmly mounted on a stable steel frame. All cameras are pre-calibrated. The calibration and orientation is checked with 30 illuminated reference targets with known positions, which cover the whole measuring space (Fig. 5.18). If the system recognizes a significant difference between nominal and measured values a new calibration of some interior orientation or a new camera orientation is made automatically. With that calibration strategy it is possible to obtain accuracies from ±1 mm within a measurement range of 2,500 mm × 1,000 mm × 500 mm.

Fig. 5.18 Tube inspection system with fixed cameras and reference points.

5.7.2.2 Online Measurements in the Field of Car Safety

Another example of a measurement system with pre-calibrated cameras is an online positioning system with four cameras. The AICON TraceCam system is used for positioning tasks especially in car safety applications, e.g., for the positioning of dummies before a crash test is conducted.

Four high resolution cameras (IEEE 1394) are mounted on a stable base plate, made of carbon fiber. The system is mounted on a workstation cart and can be easily moved to different measurement locations. If the system detects deviations during the measurement that are too high a new system orientation may be necessary, which can be done by the user on the test location directly with a carbon fibre reference plate.

Fig. 5.19 Dummy positioning system with 4 cameras.

5.7.2.3 Other Applications

Other photogrammetric applications for the 3D capture of objects can be found, for example, in accident photography and in architecture. In these fields, scale drawings or rectified scale photos (orthophotos) are primarily obtained from the digital images. The cameras used are generally calibrated for different focus settings using the methods described above. Special metric lenses, which guarantee reproducible focus setting by mechanical click stops of the focusing ring, keep the interior orientation constant for prolonged periods. The data of interior orientation are entered in the software and thus used for plotting and all computations. This guarantees high-precision 3D plotting with minimum expense in the phase of image acquisition. Other applications can be found in [21]

References

1 Y. J. ABDEL-AZIZ, H. M. KARARA. *Direct Linear Transformation from Comparator Coordinates into Object Space Coordinates in Close-Range Photogrammetry*. Symposium of the American Society of Photogrammetry on Close-Range Photogrammetry. Falls Church, Virginia 1971.

2 W. BÖSEMANN, R. GODDING., W. RIECHMANN. *Photogrammetric Investigation of CCD Cameras*. ISPRS Symposium, Com. V. Close-Range Photogrammetry meets Machine Vision, Zurich, proc. SPIE 1395, pp. 119-126.1990.

3 H. BOPP, H. KRAUS. *Ein Orientierungs- und Kalibrierungsverfahren für nichttopographische Anwendungen der Photogrammetrie*. Allgemeine Vermessungs-Nachrichten (AVN) 5/87, pp. 182-188, 1978.

4 D. C. BROWN. *Decentering distortion of lenses*. Photogrammetric Engineering 1966, pp. 444-462.

5 D. C. BROWN. *The Bundle Adjustment - Progress and Perspectives*. International Archives of Photogrammetry 21(III). paper 303041, Helsinki 1976

6 A. CONRADY. *Decentered Lens Systems.* Royal Astronomical Society. Monthly Notices. Vol. 79. pp. 384-390.1919.

7 J. DOLD. Photogrammetrie in: Vermessungsverfahren im Maschinen- und Anlagenbau, edited by W. Schwarz, Schriftenreihe des Deutschen Vereins für Vermessungswesen DVW. 1994

8 J. DOLD. *Ein hybrides photogrammetrisches Industriemeßsystem höchster Genauigkeit und seine Überprüfung.* Schriftenrteihe Universität der Bundeswehr, Heft 54 München, 1997

9 M. FELLBAUM. PROMPT - *A New Bundle Adjustment Program using Combined Parameter Estimation.* International Archives of Photogrammetry and Remote Sensing, Vol XXXI, Part B3, pp. 192 - 196, 1996.

10 W. FÖRSTNER. *New Orientation Procedures.* International Archives of Photogrammetry and Remote Sensing. 3A, pp. 297-304, 19th ISPRS Congress. Amsterdam 2000.

11 C. FRASER, M. SHORTIS. *Variation of Distortion within the Photographic Field.* Photogrammetric Engineering and Remote Sensing, Vol. 58, 1992/6, pp. 851-855.1992.

12 J. FRYER. *Camera Calibration in Non-Topographic Photogrammetry*, in: Handbook of Non-Topographic Photogrammetry, American Society of Photogrammetry and Remote Sensing, 2nd edition, pp. 51-69. 1989.

13 R. GODDING, T. LUHMANN. *Calibration and Accuracy Assessment of a Multi-Sensor Online Photogrammetric System.* International Archives of Photogrammetry and Remote Sensing. Com. V. Vol. XXIX, pp. 24-29, 17th ISPRS Congress. Washington 1992.

14 R. GODDING. *Ein photogrammetrisches Verfahren zur Überprüfung und Kalibrierung digitaler Bildaufnahmesysteme.* Zeitschrift für Photogrammetrie und Fernerkundung, 2/93, pp. 82-90. 1993

15 R. GODDING. *Integration aktueller Kamerasensorik in optischen Messsystemen,* Publikationen der Deutschen Gesellschaft für Photogrammetrie, Fernerkundung und Geoinformation, Band 14, pp. 255-262, Potsdam 2005

16 R. GERDES, R. OTTERBACH, R. KAMMÜLLER. *Kalibrierung eines digitalen Bildverarbeitungssystems mit CCD-Kamera.* Technisches Messen 60, 1993/6, pp. 256-261. 1993

17 H. HASTEDT, T. LUHMANN, W. TECKLENBURG. *Image-variant interior orientation and sensor modelling of high-quality digital cameras.* ISPRS Symposium Comm. V, Korfu, 2002.

18 R. LENZ. *Linsenfehlerkorrigierte Eichung von Halbleiterkameras mit Standardobjektiven für hochgenaue 3D-Messungen in Echtzeit.* Informatik Fachberichte 149. 9th DAGM Symposium Brunswick, pp. 212-216. 1987.

19 R. LENZ. *Zur Genauigkeit der Videometrie mit CCD-Sensoren.* Informatik Fachberichte 180. 10th DAGM Symposium Zurich. pp. 179-189. 1988.

20 T. LUHMANN, K. WENDT. *Recommendations for an acceptance and verification test of optical 3D measurement systems*; International Archives of Photogrammetry and Remote Sensing. Vol. 33/5, pp. 493-499, 19th ISPRS Congress. Amsterdam 2000.

21 T. LUHMANN. *Nahbereichsphotogrammetrie in der Praxis – Beispiele und Problemlösungen,* Wichmann Verlag. Heidelberg 2002

22 H.-G. MAAS. *Ein Ansatz zur Selbstkalibrierung von Kameras mit instabiler innerer Orientierung;* Publikationen der DGPF, Band 7, München 1998

23 T. PUTZE. *Geometric modelling and calibration of a virtual four-headed high speed camera-mirror system for 3-d motion analysis applications.* Optical 3D Measurement Techniques VII, Vol 2 , Vienna 2005, pp. 167-174

24 RÜGER, PIETSCHNER, REGENSBURGER. *Photogrammetrie - Verfahren und Geräte.* VEB Verlag für Bauwesen, Berlin 1978.

25 E. SCHWALBE. *Geometric Modelling and Calibration of Fisheye Lens Camera Systems.* (presented at 2nd Panoramic Photogrammetry Workshop) International archives of Photogrammetry, Remote Sensing and Spatial Information Sciences.Vol. XXXVI, Part 5/W8

26 T. STRUTZ. *Ein genaues aktives optisches Triangulationsverfahren zur Oberflächenvermessung.* Thesis, Magdeburg Technical University. 1993.

27 VDI, 2000: Optische 3D-Messsysteme. VDI/VDE-Richtlinie 2634, Blatt 12, Beuth Verlag, Berlin.

28 W. WESTER-EBBINGHAUS. *Photographisch-numerische Bestimmung der geometrischen*

Abbildungseigenschaften eines optischen Systems. Optik 3/1980, pp. 253-259.

29 W. WESTER-EBBINGHAUS. *Bündeltriangulation mit gemeinsamer Ausgleichung photogrammetrischer und geodätischer Beobachtungen.* Zeitschrift für Vermessungswesen 3/1985, pp. 101-111.

30 W. WESTER-EBBINGHAUS. *Verfahren zur Feldkalibrierung von photogrammetrischen Aufnahmekammern im Nahbereich.* DGK-Reihe B, No. 275, pp. 106-114. 1985

31 W. WESTER-EBBINGHAUS. *Mehrbild-Photogrammetrie - Räumliche Triangulation mit Richtungsbündeln.* Symposium Bildverarbeitung '89. Technische Akademie Esslingen, pp. 25.1-25.13. 1989

6
Camera Systems in Machine Vision

Horst Mattfeldt, Allied Vision Technologies GmbH

6.1
Camera Technology

6.1.1
History in Brief

Historically the mechanical scanning of two-dimensional images by rotating disc with spiral holes (Fig. 6.1 by PAUL NIPKOW of Berlin)

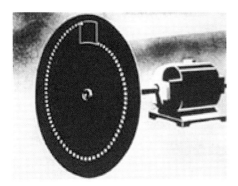

Fig. 6.1 NIPKOW Dics (BR-online).

was replaced in the early 1930s by electronic scanning methods using electron tubes (Fig. 6.2). MANFRED VON ARDENNE and VLADIMIR ZWORYKIN were the first to demonstrate *Television*.

Scanning and amplifying tubes dominated until they were smoothly replaced by transistors in the sixties of the 20th century. First the amplifiers became solid state.

Then the invention of integrated circuits in the seventies led to the development of a silicon scanning technology, the CCDs, to be explained later, still

Handbook of Machine Vision. Alexander Hornberg (Ed.)
Copyright © 2006 WILEY-VCH Verlag GmbH & Co. KGaA, Weinheim
ISBN: 3-527-40584-7

Fig. 6.2 Video camera in the mid-1930s.

dominating the most markets until today. However, there is a second popular technique for image sensors, based on CMOS technology, gaining ground.

6.1.2
Machine Vision versus Closed Circuit Television (CCTV)

First let us try to differentiate between a machine vision camera and a standard (Closed Circuit Television ⇒ CCTV) camera:

- Mechanically a machine vision camera preferably does not have an integrated lens but a standardized adapter with a distance of 17.526 mm (12.5 mm in the case of a CS-Mount) from sensor to thread (called a C or CS-mount) to put different characteristic's lenses on.

- The housing should be robust and equipped with various mounting possibilities so that it can be coupled very flexibly with the housing machine or equipment.

- The connectors should be standardized and lockable.

- Electrically it should accept external synchronization as well as internal operation.

- It should work in continuous modes but it also should accept external trigger/strobing so that the image output can be synchronized to moving objects.

- It should be quick enough to follow the trigger frequency in the case of higher speed application.

- Electrical parameters, such as mode or gain and exposure time (electronic shutter) should be changeable, either by a potentiometer and DIP-switches or, preferably, by software.

- Image output must be stable with no amplitude fading or weakening or image jitter over time and temperature.

Thus, we see that there is not one dominating criterion but a smooth gradation from a machine vision to a CCTV camera. The latter will benefit from the ability to output the best image, no matter what the lighting and ambient conditions (at day/night, sunlight). This imposes an enormous challenge on automatic image control loops in the camera, on the capabilities of the sensor itself, and on the use of special lenses, whereas a machine vision camera works preferably under controlled lighting environment.

This is true even with the most recent advances in adaptive machine vision algorithms able to compensate for lighting changes.

Sometimes based on requirements, machine vision applications can use economical CCTV (board level) cameras; but generally the more demanding the application, the higher the request for a true machine vision camera.

6.2
Sensor Technologies

The essential and most important part of a camera is the sensor, which is used for the generation of the image. Trivial? The sensor is the eye of the camera.

The camera's eye has the task of capturing the image and translating it into information, which can either be preprocessed in the camera, or transmitted to a host, or a monitor, and processed.

This is, of course, easier said than done. The human eye uses

- A spatially ultra-high resolution

- A parallel (nontemporally scanning) image capturing scheme,

which is by far too complicated to get copy with today's technology. Thus, the compromises for "industrial eyes" are following:

- Limited resolution in pixels (picture elements) for both spatial dimensions (preferably with equal spacing ($\Rightarrow$ square pixels))

- Scanning temporally with defined frames per second suitable for following the motion of the object or the changes of the objects.

- Serialization of the sensor's pixel output, be it analog or digital.

6.2.1
Spatial Differentiation: 1D and 2D

Here we first distinguish between one-dimensional and two-dimensional image sensors.

The 1D sensor is a line sensor, whereas a 2D one is an area sensor (consequently, a light barrier should be named a 0-D sensor!).

Although an image is two-dimensional, having a discrete height and width, one can also think of it in terms of an endless image having only a defined width but unknown or unlimited height.

An example for the latter is an image of a wooden board, whose height varies with the height of the tree it is being cut off from.

A line image sensor will be advantageous here because it will scan the width of the image line by line while the wooden board is moved (preferably with constant or known velocity!) under the optics. The wood board's end also limits the height of the image.

Line cameras are not dealt further in this chapter because of their very special nature and lower significance in the market.

6.2.2
CCD Technology

CCDs (charge coupled devices), invented by the American Bell laboratories in the sixties (for analog storage and shift registers!) became the dominant scanning device because silicon is also light sensitive.

In a CCD, the photo effect is used to generate electrons out of photons, this charge collected and held in virtual tiny buckets, forming individual picture elements (pixels)). Using various gating clocks it is possible to move this charge serially towards the output of the device (usually one), where it is converted into a current or voltage.

Currently, Sony of Japan and Kodak of USA are important CCD manufacturers in terms of volume; followed by probably a dozen famous others.

The technology for 2D image sensors, being adopted by these various manufacturers, has emerged and diversified into various technological and architectural substructures.

The differences arise in the way the lines are scanned and the manner in which charge storage is handled and outputed.

Each substructure will be explained shortly with its main advantages and disadvantages, which are summarized from the machine vision point of view as follows:

- Full frame
- Frame transfer sensor
- Interline transfer
    - Interlaced scan interline transfer
    - Progressive scan interline transfer

6.2.3 Full Frame Principle

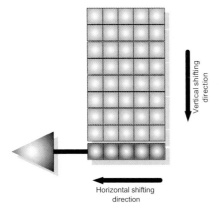

Fig. 6.3 Full frame principle.

Advantages:

- Highest fill factor vs. chip size
- Highest sensitivity
- Suitable for scientific applications

Disadvantages:

- Very susceptible to smear during readout phases (needs mechanical shutter to close sensor while shifting out data)
- No electronic shutter (needs strobe or LCD/mechanical shutter)
- No electronic asynchronous shutter possible (trigger/shutter on demand)

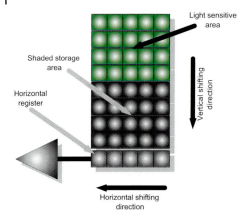

Fig. 6.4 Frame transfer principle.

6.2.4
Frame Transfer Principle

Advantages:

- Fill factor 100%
- Highest sensitivity
- Suitable for scientific applications

Disadvantages:

- Chip size
- Susceptible to smear during readout phases (needs mechanical shutter to close sensor while shifting out data)
- No electronic shutter (needs strobe or LCD/mechanical shutter)
- No electronic asynchronous shutter possible (trigger/shutter on demand)

6.2.5
Interline Transfer

Advantages:

- Electronic (asynchronous) shutter
- Suitable for stationary and moving industrial applications

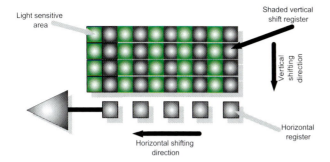

Fig. 6.5 Interline transfer principle.

- No flash needed
- No mechanical/LCD-shutter

Disadvantages:

- Lower fill factor
- Lower sensitivity (micro lenses improve this feature)
- Lower IR sensitivity

Interline technology itself splits into the already mentioned two categories, depending on the way the lines are read out:

- Interlaced scan readout
    - Field integration
    - Frame integration
- Progressive scan readout

6.2.5.1 Interlaced Scan Interline Transfer

Interlaced scan is still the standard video technology in use for more than 70 years:

Interlaced stands for the fact that an image is scanned and displayed consecutively by two half images called fields where one field displays the odd lines, and the other field displays the even lines, simply because at the beginning it was not possible to speed up with all the lines and frame rates. The first field, drawn in solid black, starts with half-a-line, whereas the second field (dashed) starts with a full line (Fig. 6.6). The flyback (of the electron beam of the CRT or tube) is drawn in gray. The vertical fly back is shown only in principle; it actually goes zigzag, because it takes more line times.

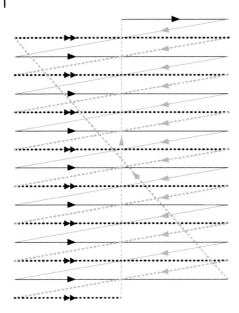

Fig. 6.6 Interlaced scanning scheme.

Usually one-chip color cameras must (and b/w color cameras can) add the content of two adjacent lines together to derive the color information (and to increase sensitivity).

In the first field, for example we add line one and two, while in the second field, we add two and three as shown in Fig. 6.7:

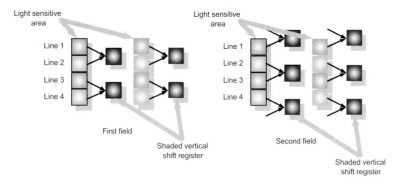

Fig. 6.7 Interlaced sensor with field integration.

With this field integration mode it is also possible to benefit from an electronic shutter.

The drawback is the loss in vertical resolution because adding two lines together is also effectively a low pass filter.

The other method is called frame integration (Fig. 6.8): The advantage is the higher vertical resolution. Because the integration of the two fields overlap, it is possible to use a flash and to freeze an image in full vertical resolution. An electronic shutter is not possible in this mode.

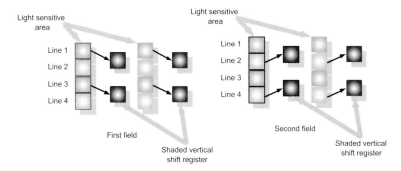

Fig. 6.8 Interlaced sensor with frame integration.

Interlaced sensors may have, depending on their horizontal resolution, rectangular pixels dimensions (e.g., 9.8 μm × 6.3 μm), opting for high sensitivity, but making either rescaling or calibration efforts necessary for dimensional measurement machine vision tasks.

6.2.5.2 Frame Readout

Above 2 Mega pixels, there are currently no progressive scan sensors available from Sony (at least at the time of writing in Q3/2005).

In order to achieve the highest possible sensitivity for 3, 5, and 8 Mega pixel CCD sensors, all sensors are equipped inevitably with micro lenses, and are read out in the so-called *frame readout mode*.

This is a special frame integration interlaced readout field mode, very similar to the interlaced mode of conventional video cameras.

Whereas the 3- and 5-Mega pixel sensor has a two-field readout, the 8-Mega pixel sensor is equipped with a three-field readoutscheme.

With this architecture the vertical transfer register can be more compactly built because it needs to hold only one half or even only a third of all pixels after the charge transfer command.

The conversion from interlaced to progressive takes place in the camera's internal memory.

Special sub-sampling modes allow faster progressive scan readouts, while maintaining the same imaging conditions. These modes can be effectively used for focus and aperture adjustment operations.

The following sample is taken from the data sheets of the 5 Mega pixel sensor Sony ICX-282AQ, detailing the various modes.

Frame readout is done with two fields and has the two primary colors, red and blue, read out in separate fields.

The progressive mode reads out two of four lines, thus achieving a moderate speed increase in progressive scan with mega-pixel resolution.

The readout schemes of these modes are shown in Fig. 6.9:

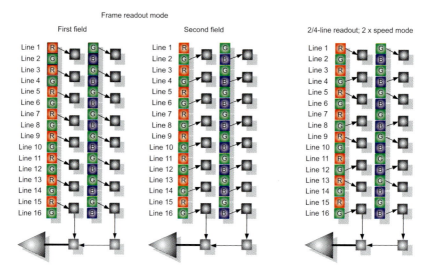

Fig. 6.9 Readout scheme of ICX-282AQ

In the first field, the odd lines with only R-G color filters are read out, in the second field the even lines with G-B color filters follow.

This implies that either the object must be stationary or the following prerequisites are necessary to handle moving image acquisition:

- Strobe light and
- Ambient light reduced by proper light shields or
- Mechanical external shutter
- LCD-optical shutter

Otherwise, striped and/or colored artifacts will occur.

6.2.5.3 Progressive Scan Interline Transfer

Progressive scan is a straightforward method but nevertheless, it is the latest development in CCD: Each light sensitive pixel has its storage companion, which again reduces the fill factor and sensitivity a little bit, but there are sophisticated micro lenses to compensate for this drawback. With the shift pulse the charge of each pixel in each line is transferred, making the time between two shift impulses the frame time and usually also the shutter time. Most progressive scan sensors have square pixels, which enables an easy transformation between pixel count and distance measurement without x–y calibration.

Figure 6.10 illustrates the principle:
In each picture each pixel's storage can be placed (with the charge transfer command) into the respective vertical storage register, which can hold the charge of all lines.

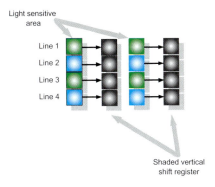

Fig. 6.10 Progressive scan interline transfer.

Very advantageous here is the fact that after a shift command, the sensor can be kept in the reset mode so that integration can start (virtually the shutter can be electronically opened) by a control signal. This enables control over the integration time, so that

- Moving objects do not blur (more light is needed!)
- Moving objects are shuttered precisely from image to image at the same position and are not *jumping*.
- Auto shutter can be built in so that image brightness becomes less dependent on the illumination.

Compare the CCD readout technologies using Table 6.1:

Tab. 6.1 Summary and comparison of modes.

	Progressive scan Noninterlaced	Interlaced Field integration	Frame integration
Vertical resolution	Good	Lower	Good
Temporal resolution	Good	Good	Lower

In conclusion, progressive scan is the best compromise for most of the applications, because it is the most flexible.

Machine vision systems (frame grabbers , displays) favor progressive cameras, because they make the conversion from interlaced to progressive obsolete.

6.3
CCD Image Artifacts

Although this technology was specially optimized over decades for perfect image generation, it may suffer from some general image artifacts.

6.3.1
Blooming

When a CCD gets excessively saturated by extremely bright objects, it may show blooming (Fig. 6.11). This is when the charge cannot be held at an individual pixel's place but floods over the array.

Fig. 6.11 Blooming due to excessive sunlight hitting the CCD sensor.

6.3.2
Smear

Smear can be seen as a vertical line above and under a bright spot in an image (Fig. 6.12). This is due to *cross talk* of electrons moving from the pixel to the vertical shift register, while the image is shifted out.

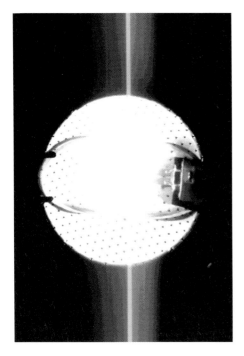

Fig. 6.12 Smear of CCD image sensor due to bright spot.

Smear is affected by the ratio of the readout time to the shutter (integration) time.

Short shutter times and/or long readout times increase smear.

The following two methods are helpful for reducing smear:

Using mechanical shutters can block the light after exposure time, while the image is shifted. Alternatively the use of a flash is recommended so that it dominates over the ambient light.

6.4 CMOS Image Sensor

Often described as the rival technology for solid-state image sensing is the CMOS (Complementary metal oxide semiconductor) image sensor.

It is newer although the underlying technology is not new. Only the use of CMOS silicon for imaging purposes has seen a lot of progress over the last decade.

The basic principle is again the photovoltaic effect. But whereas the CCD holds and moves the charge per pixel to one (or more) output amplifier(s), the

CMOS sensor converts the charge to voltage already in the pixel. By crossbar addressing and switching, the voltage can be read out from the sensing area.

Table 6.2 (by Cypress/FillFactory) explains it.

Tab. 6.2 CCD vs. CMOS comparison by Cypress/FillFactory.

	CCD	CMOS
Photo detection	Buried diode	Photo diode
Technology	Nonstandard	Standard
Charge conversion	At output	In pixel
Read out	Charge transfer	Voltage multiplexing
Supply and Biasing	Multiple supplies needed	Single supply

It should be mentioned that there are integrating CMOS sensors as well as *nonintegrating sensors*.

The first uses the parasitic capacitance of the transistor for storage of the pixel's brightness until it is read out. Thus, the image builds up integrating over the time between two resets (readouts) of the sensor.

A *nonintegrating sensor* can be thought of as an array of photodiodes, which can be arbitrarily x–y addressed. Thus we can read out the momentary brightness of a pixel independent of the others. This can be useful for real-time tracking applications.

Once the target is found in the image, one only needs to read out the neighborhood of the target to track the possible movement, which speeds up the search proportional to the reduction in number of pixels.

Dealing with moving objects introduces the problem of spatial distortion because the pixels can only be read out serially one after the other with pixel clock frequency. This distortion needs to be taken into account.

6.4.1
Advantages of CMOS Sensors

Because CMOS is basically used in digital circuits, it is obvious that analog to digital conversion, addressing, windowing, gain, and offset adjustments can be added easily to the chip, aiming for the ultimate goal of a camera on a chip.

Surveillance cameras are already reaching this goal, but machine vision is a few steps behind. A higher level of integration is one aim, lower power consumption (a third to a tenth) is the other.

By using variable logarithmic photodiodes (Linlog^{TM} Photonfocus) or multiple reset thresholds (knee points), some manufacturers can achieve up to 120 dB (compared to 60 to 80 dB for CCD) in the intra-scene dynamic range.

The graph (by FillFactory) of the relative light intensity shows that one knee point (dark blue line) effectively combines the rapid increase in output signal for long integration time (gray line) which would lead immediately into saturation, with the smooth increase in short integration time (bright line, Fig. 6.13).

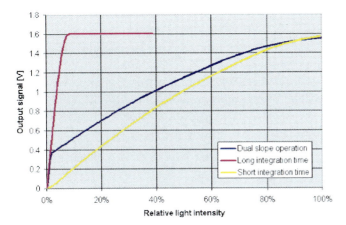

Fig. 6.13 Nonlinear response curve with one knee point (by Cypress/FillFactory).

Figure 6.14 shows an image in linear mode on the left, and another image with one additional knee point to increase the intra-scene dynamic range on the right.

Fig. 6.14 High dynamic range mode with one knee point.

Other advantages:
Using windowing techniques only the pixels of interest need to be read out, so that the chip inherently becomes proportionally faster.
Using multiplexed output channels (up to 32 output channels), it is possible to achieve an extremely high pixel rate of more than 2 Gigapixel/s with frame rates of more than 1000 fps.
This makes CMOS sensors very suitable for high speed imaging systems.

6.4.2
CMOS Sensor Shutter Concepts

Every pixel of a CMOS imager consists of several transistors.

A pixel is called active (APS) when it comes combined with an amplifier. It needs a minimum of three transistors for an APS. In machine vision, it is useful to have an electronic shutter.

Even with the simplest three-transistor pixel it is possible to have an electronic shutter in the so-called rolling (curtain) shutter architecture (Fig. 6.14). The image is reset line by line a short time after reading the respective line. The time difference between any two lines is the integration or the shutter time.

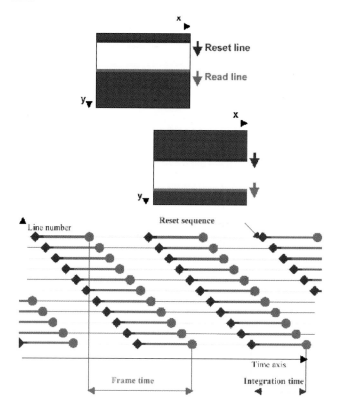

Fig. 6.15 Rolling shutter visualization (by Cypress/FillFactory).

This may be perfect for stationary applications, but it can introduce severe artifacts for moving applications, as can be seen Fig. 6.16.

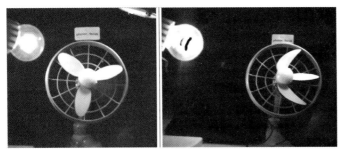

Fig. 6.16 Rolling shutter (right) vs. global shutter (left) (Courtesy Photonfocus).

On the right, the blades of the fan are scanned by a rolling shutter sharply deforming them, while on the left they have been scanned with a global (synchronous for all pixels or snapshots) shutter.

The left image also shows the effect of Photonfocus' LINLOG^{TM} technology, which helps to enhance the dynamic range.

The need for a global shutter makes a fourth transistor necessary for the global resetting of every pixel. With every additional transistor as a side effect, sensitivity or the structural FPN-noise may increase. Figure 6.16 shows the structure of the IBIS5 pixel by FillFactory.

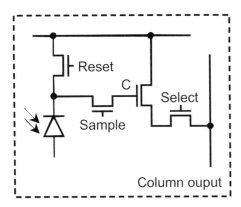

Fig. 6.17 Active pixel structure (by Cypress/FillFactory).

Even with this four transistor cell, the sensor, in global shutter mode, can either integrate or readout at a time. This fact has a clear impact on the achievable frame rate, which will go down when the shutter time goes up. This is explained in Fig. 6.17 which shows in red the constant frame readout time and the variable integration time to be added for the total frame time. The interleaving of integration of a new frame, while reading out the actual one, possible with a CCD, also called the pipelined global shutter, again requires

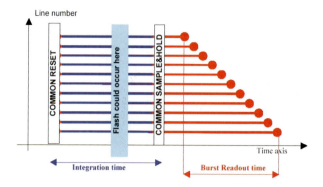

Fig. 6.18 Serialization of integration and read-out (by Cypress/FillFactory).

additional circuitry. At the time of writing this was, for instance, available for certain sensors from MICRON.

Another important fact to discuss is how well the sensor is able to keep ambient light from integrating after the electronic shutter is closed. This is called *shutter efficiency* or *parasitic sensitivity*. A sensor is not suitable for an application, if it presents blurred moving contours or a vertical gradient in background gray levels.

6.4.2.1 Comparison of CMOS versus CCD

Table 6.3 gives, supplied by FillFactory, a summarizing comparison of the two technologies can be given with the following table, as supplied by FillFactory:

Tab. 6.3 Quantitative comparison of CCD vs. CMOS (by Cypress/FillFactory).

CCD	CMOS
++ Very low noise	+ Low noise
+ Optimal image quality	+ Optimal image quality
− Slower read out	+ High speed read out—random addressing
− High power consumption	+ Low power consumption
− Complicated timing and driving	+ Single supply and digital interface
− No co-integration logic	++ Co-integration of logic—smart sensors

Fixed pattern noise and dark signal nonuniformity (DSNU) is higher with CMOS, calling probably for correction.

Blemish pixels are pixels outside of the normal operating range, appearing too dark or too bright, which calls for image makeup.

Some of these corrections can be effectively made in the camera itself, e.g., in the AVT Marlin F-131, which is the model with the FillFactory IBIS 5A-1300 sensor.

With this camera, the DSNU function applies an additive correction to every pixel in order to equalize the dark level of the pixels. This function also enables correction of single- and double-blemished pixels by replacing them with their neighbourhood pixels.

6.4.2.2 Integration Complexity of CDD versus CMOS Camera Technology

One clear advantage of CMOS technology is the higher level of integration.

With CMOS it is possible to integrate analog front end functionality and the timing generator on the sensor, achieving more compact solutions with less power consumption.

Figure 6.19 depicts this integration.

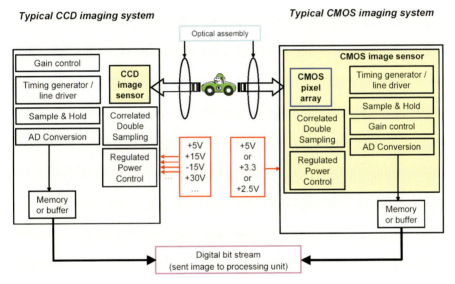

Fig. 6.19 CCD and CMOS camera building blocks (by Cypress/FillFactory).

On the left you see that a CCD camera has most of the building blocks separate, whereas on the right the blocks are part of the CMOS sensor itself.

Sensor overview:

Table 6.4 provides an overview of progressive scan sensors approved by Allied Vision Technologies as being well-suited to machine vision applications:

6.4.2.3 Video Standards

Class follows the nomenclature in the PC industry:

VGA resolution is standard and similar to EIA (RS170) TV video standard for the USA, and Japan.

SVGA resolution is similar to CCIR standard used in Europe and in many other parts of the world.

Tab. 6.4 AVT camera: sensor data.

Class	Techn.	Manu-facturer	Sensor Type	Sensor Size	Microlens	Frame rate	Chip size $h \times v$ (mm^2)	Pixel size $h \times v$ (μm^2)	Eff. pixels
VGA b/w	CCD	SONY	ICX-414AL	1/2"	Yes, HAD	< 75 fps	7.48×6.15	9.9×9.9	659×494
VGA color	CCD	SONY	ICX-414AQ	1/2"	Yes, HAD	< 75 fps	7.48×6.15	9.9×9.9	659×494
SVGA b/w	CCD	SONY	ICX-415AL	1/2"	Yes, HAD	< 50 fps	7.48×6.15	8.3×8.3	782×582
SVGA color	CCD	SONY	ICX-415AQ	1/2"	Yes, HAD	< 50 fps	7.48×6.15	8.3×8.3	782×582
XGA b/w	CCD	SONY	ICX-204AL	1/3"	Yes, HAD	< 30 fps	5.8×4.92	4.65×4.65	1034×779
XGA color	CCD	SONY	ICX-204AK	1/3"	Yes, HAD	< 30 fps	5.8×4.92	4.65×4.65	1034×779
SXGA b/w	CCD	SONY	ICX-205AL	1/2"	Yes, HAD	< 15 fps	7.6×6.2	4.65×4.65	1392×1040
SXGA color	CCD	SONY	ICX-205AK	1/2"	Yes, HAD	< 15 fps	7.6×6.2	4.65×4.65	1392×1040
SXGA b/w	CMOS	FillFactory	IBIS5A	2/3"	No	< 25 fps	8.6×6.9	6.7×6.7	1280×1024
SXGA color	CMOS	FillFactory	IBIS5A	2/3"	No	< 25 fps	8.6×6.9	6.7×6.7	1280×1024
SXGA color	CCD	SONY	ICX-285AL	2/3"	Yes, EXviewHAD	< 15 fps	10.2×8.3	6.45×6.45	1392×1040
SXGA b/w	CCD	SONY	ICX-285AQ	2/3"	Yes, ExviewHAD	< 15 fps	10.2×8.3	6.45×6.45	1392×1040
UXGA color	CCD	SONY	ICX-274AL	1/1.8"	Yes, SuperHAD	< 12 fps	8.5×6.8	4.4×4.4	1628×1236
UXGA b/w	CCD	SONY	ICX-274AQ	1/1.8"	Yes, SuperHAD	< 12 fps	8.5×6.8	4.4×4.4	1628×1236
OF-320C	CCD	SONY	ICX-262AQ	1/1.8"	Yes, SuperHAD	< 5.5 fps	8.10×6.64	3.45×3.45	2088×1550
OF-510C	CCD	SONY	ICX-282AQ	2/3"	Yes, SuperHAD	< 3.5 fps	9.74×7.96	3.4×3.4	2588×1960
OF-810C	CCD	SONY	ICX-456AQ	2/3"	Yes, SuperHAD	< 3.5 fps	9.79×7.93	2.7×2.7	3288×2472

The higher resolution PC standards have no analog in TV standards and can be used moreover with varying frame frequencies.

6.4.2.4 Sensor Sizes and Dimensions

Sensor sizes are given mostly in fractions of inches but also in mm.

At first glance it is confusing that the dimension in mm is smaller than that in inches:

This is (another) relict of the image tube days:

A tube with a diameter of half an inch had a usable image area of 6.4 mm × 6.8 mm, which gives a diameter of 8 mm.

Figure 6.20 shows the dimensions of some other common imager sizes.

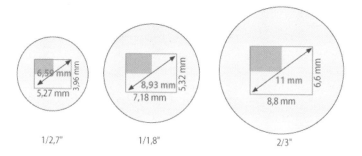

Fig. 6.20 Sensor sizes.

The aspect ratio is usually 4×3 ($H \times V$); other areas of interest can be created by windowing techniques in the system or in the camera, assuming that the sensor supports it.

6.4.2.5 Sony HAD Technology

HAD (hole accumulation Diode) is a Sony terminology for the silicon structure of the CCD.

Figure 6.21 (by Sony) illustrates the concept: Note that the microlenses on top of each pixel focus the light such that it is concentrated on the light sensitive part but not hitting the transfer section, which considerably enhances the sensitivity.

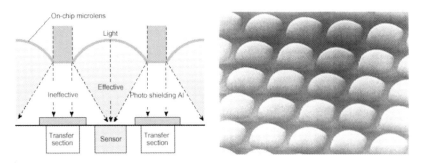

Fig. 6.21 HAD principle of Sony CCD sensors (courtesy Sony).

6.4.2.6 Sony SuperHAD Technology

SuperHAD stands for improvements in the size of the lens, so that a higher sensitivity is achieved.

Figure 6.22 (by Sony) shows the differences:

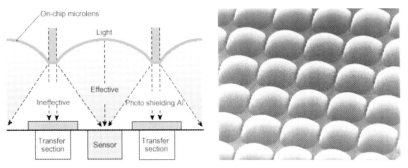

Fig. 6.22 SuperHAD principle of Sony CCD sensors (courtesy Sony).

6.4.2.7 Sony EXView HAD Technology

Finally ExView HAD stands for the latest sensitivity enhancements, due to silicon and a second microlens.

This is illustrated in Fig. 6.23 (by Sony).

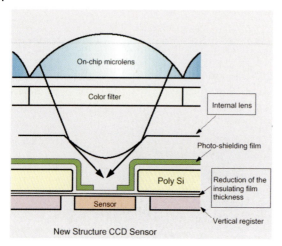

Fig. 6.23 EXView HAD principle of Sony CCD sensors (courtesy Sony).

Why so many different models? How to choose among them?

Sensors differ roughly in

- Resolution
- Chip size
- Sensitivity
- color or b/w
- Speed
- Special features (windowing (area of interest), binning (sub sampling))

Each sensor and associated camera has its best fit:

- A higher resolution shows more details, but can make the camera less sensitive (pixel is smaller) or more costly (chip is more expensive).
- Bigger pixels, on the other hand, have higher sensitivity. Fewer pixels per image can be read out faster so that the frame rate is higher.
- It is generally recommended that the resolution should be only as high as necessary.
- More pixels usually need more processing power/time.
- The effective pixel size goes along with the inherent ability to collect more electrons, a prerequisite, together with low noise electronics, for good dynamic range and high sensitivity.

6.5 Block Diagrams and their Description

There is presumably no single camera for all conceivable applications, but there are (usually confusingly many) choices of different camera types and technologies.

Consequently, the intention of this chapter is to help the reader to choose that camera, which has the ideal performance and price match for the intended application.

Apart from the sensor, various types of cameras differ in the way they process and output the image data from the sensor.

Three groups of cameras were created and a reference candidate for each group was chosen:

- Progressive scan analog image processing with analog output, mostly with black and white cameras

    - Sony XC-HR57/58

- Digital image processing with analog output, mostly with color cameras

    - Sony color camera building blocks

- Digital image processing with digital output (b/w and color)

    - Digital output to be RS 422
    - FireWire^{TM}
        * AVT Marlin series black and white
        * AVT Marlin series color model
    - USB
    - Camera Link
    - GigaBit Ethernet

Bsaed on the data path from the sensor to the output, specific blocks are needed and are described in more detail in the paragraphs to follow.

The block diagrams illustrate the data paths.

6.5.1 Block Diagram of a Progressive Scan Analog Camera

6.5.1.1 CCD Read-Out Clocks

This camera uses a *progressive scan* b/w CCD-sensor, which means that all lines are output consecutively in one image (Fig. 6.24). (This is in contrast to

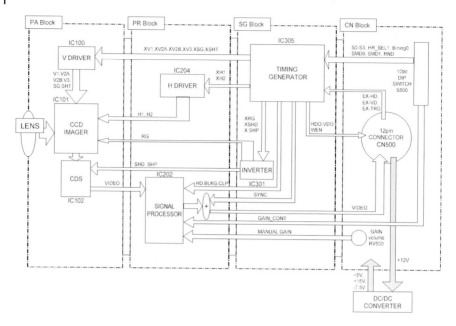

Fig. 6.24 Block diagram of a Sony b/w camera.

the so-called *interlaced* scan sensors where odd and even lines are output in subsequent half images (fields) and interleaved with the help of the idleness of the eye or knitting facilities in the frame grabber.)

Figure 6.25 with information from the Sony sensor's data sheet gives an overview:

There are separate drivers for the horizontal ($H_{\Phi 1}$, $H_{\Phi 2}$, and ΦRG) and the vertical clocks ($V_{\Phi 1}$, $V_{\Phi 2}$, and $V_{\Phi 3}$) needed to drive a CCD sensor.

The charge from all pixels is shifted by the SG-pulse (a special condition of the V-pulse) into the vertical shift register (which terminates the shutter/exposure and thus separates the images).

The SG-pulse occurs at a defined line count within the vertical interval but is a very short signal. After shifting a new image exposure could begin immediately, which makes the pipelining nature a basic feature of a CCD. The actual start is controlled by the ΦSUB-pulse, which acts as a (asynchronous) reset.

With the help of three different phased clocks, the charge is *jostled* to the horizontal shift register, which effectively holds one complete line of pixels.

This line of pixels is then moved with pixel clock frequency to the output amplifier, which is reset after every pixel.

The complete timing is controlled by an ASIC (application specific integrated circuit), adequately called the timing generator.

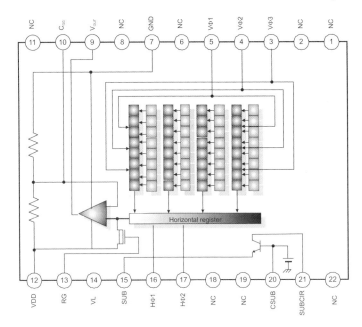

Fig. 6.25 Sony ICX-415AL sensor structural overview.

CCD Binning Mode

Binning is basically a special shifting mode:
   Whereas the normal readout is a sequence of
   V-Shift -> H-Shift ->V-Shift,
   Vertical binning can be achieved by a
   V-Shift -> V-Shift -> H-Shift sequence.
   This adds the charge of two vertically adjacent pixels together, creating a vertical binning effect.
   Vertical binning (Fig. 6.26(a)) increases the light sensitivity of the camera by a factor of two. At the same time this normally improves signal to noise ratio by about 3 dB.

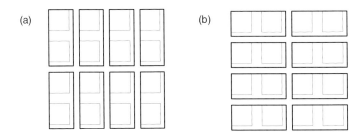

Fig. 6.26 (a) Vertical binning, (b) Horizontal binning.

Vertically shifting out more than one line and finally dumping that charge away can also be used for windowing (area of interest) functionality.

Dumping is also faster, so that the frame frequency goes up as a side effect.

The data sheets of the individual cameras tell, how fast exactly for which model and which mode.

Horizontal binning (Fig. 6.26(b)) can be made in the horizontal shift register by shifting two pixels into the amplifier without resetting.

In horizontal binning adjacent horizontal pixels in a line are combined in pairs.

The light sensitivity of a camera is increased by a factor of two (6 dB). Signal to noise separation improves by approx. 3 dB. Horizontal resolution is lowered, depending on the model.

It is obvious that horizontal binning does not increase the speed.

Also it can be seen that binning and color is not possible or harder to achieve and adds more complexity to the circuitry.

6.5.1.2 Spectral Sensitivity

The spectral sensitivity of this camera is given in Fig. 6.27, taken from the camera's manual. It also illustrates that there is considerable sensitivity in the IR spectrum, as well as in the UV.

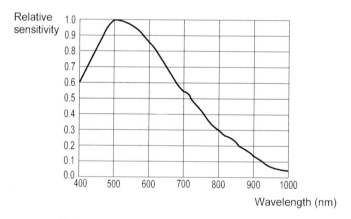

Fig. 6.27 Relative spectral sensitivity of Sony XC-HR57/58, taken from the technical manual of Sony.

In combination with the lens it can be necessary to block the IR spectrum with an IR-Cut filter so that it does not reach the sensor when the application works with the visible spectrum or to block the visible spectrum, when the application works in the IR spectrum.

Although this reduces overall sensitivity it can, depending on the lens, contribute to contrast and sharpness of the image, just because some lenses may need different focal adjustments for the two spectra.

Using the UV spectrum requires special (and rather expensive) UV lenses, made of quartz glass and is nevertheless limited by the cover glass of the sensor, which would have to be removed with extraordinary diligence.

It should be noted that generally the signal of a CCD-imager is an analog signal in terms of levels; it is discrete only in terms of spatial quantization in the form of pixels.

6.5.1.3 Analog Signal Processing

The CDS (correlated double sampling) block is an important block in all CCD-cameras and is used for the elimination of certain noise components. Figure 6.28 (by Analog Devices) gives a detailed explanation.

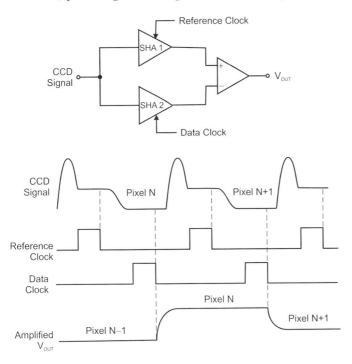

Fig. 6.28 CDS processing (Analog Devices).

By taking two different samples in time of the CCD signal, one at the reset stage and one at the signal stage, and subtracting them, any noise source that is correlated to the two samples will be removed.

A slowly varying noise source that is not correlated will be reduced in magnitude. Noise introduced in the output stage of the CCD consists primarily of kT/C noise from the charge-sensing node, and $1/f$ and white noise from the output amplifier.

The kT/C noise from the reset switch's ON-resistance is sampled on the Sense node, where it remains until the next pixel. It will be present during both the reference and data levels, so it is correlated within one pixel period and will be removed by the CDS.

The CDS will also attenuate $1/f$ noise from the output amplifier, because the frequency response of the CDS falls off with decreasing frequency. Low frequency noise introduced prior to the CDS from power supplies and by temperature drifts will also be attenuated by the CDS. But wideband noise introduced by the CCD will not be reduced by the CDS.

Further analog signal processing comprises of low pass filtering, to eliminate the pixel clock, manual and or automatic gain/shutter and optional gamma enhancement, to accommodate the nonlinear display curve of some displays, all this carried out in one single IC, significantly called the signal processor.

The video signal is then completed by the sync impulses and fed to a 75-Ohm line driver, so that effectively one single wire carries all the information needed for (continuous) display of video.

Fed from an external single 12 V DC, a low noise DC–DC converter is used to generate the three different voltages, which are needed in this camera to drive the CCD and the glue logic.

Depending on the sensor, up to seven different voltages need to be generated in some CCD-cameras, making the power supply quite complex and power consumption an issue.

6.5.1.4 Camera and Frame Grabber

For more advanced modes, like external trigger or multi camera modes, additional signal exchange with the frame grabber comes into play:

Most machine vision cameras offer two different modes for the adaptation to frame grabbers. Either can the camera run as a *master* supplying video and synchronization signals via the standardized 12-pin HiRose connector so that the frame grabber is the slave having to synchronize with its *PLL* (Phase Locked Loop) to the horizontal and vertical timing of the camera.

Alternatively, the frame grabber is the master synchronizing the camera.

The latter is easier when you have more cameras connected which all act as slaves synchronized to the frame grabber.

The camera's timing generator needs to adapt this situation by either inputting or outputting synchronization signals HD and VD (Horizontal and Vertical Drive).

When the camera is the master and runs in, say, the external trigger shutter mode, it signalizes with the so-called WEN impulse when a valid image is output. This is important for the frame grabber, to grab the right image (e.g., corresponding to a flash).

Because of the complexity of certain modes, make sure that the frame grabber supports the camera in the desired mode and look up the technical manual to reassure yourself.

Due to the high component integration, it is possible today to build these cameras in very miniaturized dimensions, e.g., in a square of only 29 × 29 × 32 mm (exclusive C-Mount and lens). Figure 6.29 shows an example in the interlaced camera Sony XC-ES50.

Fig. 6.29 The extremely small Sony XC-Es50 camera.

6.5.2
Block Diagram of Color Camera with Digital Image Processing

The block diagram of a highly integrated solution from Sony, comprising of only three chips for a one chip (interlaced) color camera shows the hybrid nature of most color cameras nowadays: Image processing is done after analog to digital conversion in sophisticated video-DSP, outputting either analog reconstructed video, or digital signals directly for flexible connectivity to displays or computers.

6.5.2.1 Bayer™ Complementary Color Filter Array

To start with the color sensor, it needs to be mentioned that most 1-chip color sensors (be it CCD or CMOS) use the concept proposed by Bryce E. Bayer of KODAK, who got a patent 30 years ago for the idea of placing selectively transmissive filters on top of the pixels of a b/w sensor, such that each filter occurs regularly and repeatedly. This is called the BAYER™ pattern.

Interlaced color sensors mostly use the complementary color filters cyan, yellow, magenta, and green, instead of the primary color filters (red, green, and blue). It is also important that the order of Mg and G is altered for every second line.

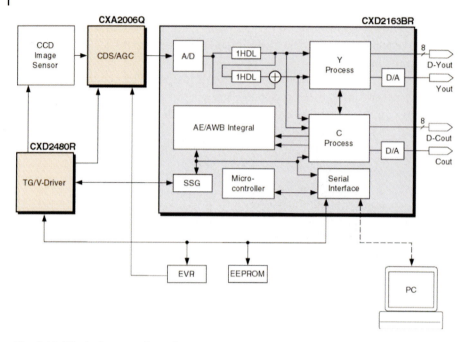

Fig. 6.30 Block diagram of a color camera.

6.5.2.2 Complementary Color Filters Spectral Sensitivity

The main advantage of using complementary color filters (Fig. 6.31) is that this enhances sensitivity in comparison to using primary color (RGB) filters. The screenshot (Fig. 6.32) is of a typical interlaced color sensor (ICX418Q) from Sony.

It shows the transmission curves of the four color filters (yellow, cyan, magenta, and green) relative to the visible spectrum of the human eye, which ranages roughly from 400 nm to 700 nm.

It can be seen that especially yellow and magenta is transmissive over almost half of the visible spectrum, so that lesser energy is filtered out than with the RGB primary color filter, where roughly two-thirds of the energy is filtered out per primary color.

As magenta is not a spectral color (but a mixture of red and blue) the curve shows two maxima.

The fourth color, green, is beneficial for color resolution and gamut.

6.5.2.3 Generation of Color Signals

The output signal Luminance (Y) and the two Chrominance (C) signals ($R - Y$) and ($B - Y$) can be generated relatively easy by vertically averaging the

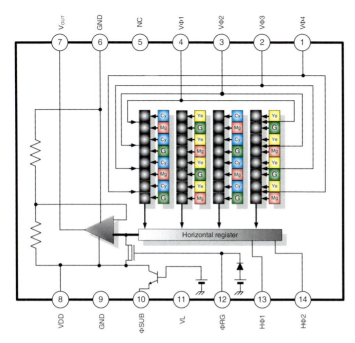

Fig. 6.31 Architecture of Sony CCD complementary color sensor.

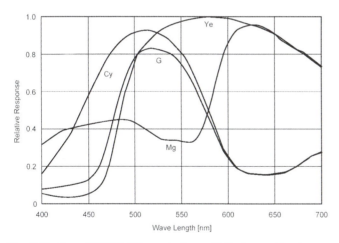

Fig. 6.32 Sony ICX254AK spectral sensitivity (according to datasheet Sony semiconductor corporation) for CyYeMgG complementary color filter array.

charges of two adjacent lines in the analog domain (by *field readout* of the CCD).

It is now important that due to the changed ordering of Mg and G in every second line, the vertical averaging of the first two adjacent lines gives:

$$(Cy + Mg) \quad \text{and} \quad (Ye + G) \; ,$$

and the second two lines give:

$$(Cy + G) \quad \text{and} \quad (Ye + Mg) \; .$$

As an approximation by Sony, the Y signal is created by adding horizontally adjacent pixels, and the chroma signal is generated by subtracting these adjacent pixel signals.

$$Y = \frac{1}{2}(2B + 3G + 2R) = \frac{1}{2}((G + Cy) + (Mg + Ye))$$

with $(R + G) = Ye, (R + B) = Mg, (G + B) = Cy$

$$R - Y = (2R - G) = ((Mg + Ye) - (G + Cy))$$

is used for the second chroma (color difference) signal.

For the first line pair, the Y signal is formed from these signals as follows:

$$Y = \frac{1}{2}((G + Ye) + (Mg + Cy)) = \frac{1}{2}(2B + 3G + 2R)$$

This is balanced since it is formed in the same way as for the first line pair.

Similarly, the second chroma (color difference) signal is approximated as follows:

$$-(B - Y) = -(2B - G) = (G + Ye) - (Mg + Cy)$$

In other words, the two chroma signals can be alternatingly retrieved from the sequence of lines from $R - Y$ and $-(B - Y)$.

This is also true for the second field.

Complementary filtering is thus a way of achieving higher sensitivity at a slight expense of color resolution.

Using a fourth color green is advantageous when one aims to display as many colors as the human eye. Nevertheless, complementary color filter array is somewhat inferior in terms of resolvable colors.

Keeping luminance (Y) and color information separate, as this example shows, is beneficial for the video quality due to the fact that cross color and cross luminance effects ($\Rightarrow$ colored stripes in some patterned jacket of some news anchorman) are greatly reduced.

This is of even greater importance, because the chipset can do extensive horizontal and vertical aperture correction for detail enhancement, as seen in the block diagram of the integrated two video delay lines.

The image is output, either analog or digital, again demonstrating the hybrid nature of this architecture.

With the capability of internal and external synchronization, this architecture is well-suited to many applications not only in surveillance but also in machine vision.

External asynchronous image triggering is though not possible, indicating the need for a more flexible camera architecture for the challenging machine vision applications.

6.6 Digital Cameras

In this section the AVT Marlin series of digital cameras is used to explain the properties of digital cameras.

The block diagrams of the AVT Marlins are examples of digital cameras with IEEE1394 interface with the following design goals:

- Modular concept regarding the sensors, both CCD and CMOS, so that a family of products can be built
- Flexible hardware concept by the use of a combination of powerful microprocessors and FPGA (field programmable gate array)
- Smart real-time image preprocessing functions
- Buying out functionality of the frame grabber, which becomes obsolete due to the IEEE1394 interface
- Best possible image quality
- Best value for money

Going through the building blocks, we learn how the design goals were finally met.

We start with the sensor as the most important part, floating down with the data towards the output.

6.6.1 Black and White Digital Cameras

Figure 6.33 shows a typical block diagram of a digital black and white camera.

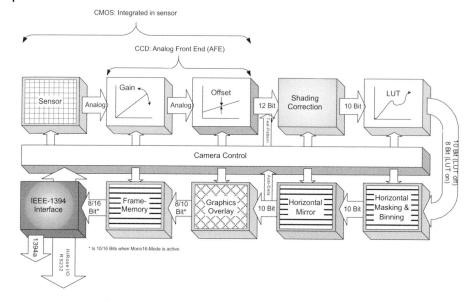

Fig. 6.33 Block diagram b/w camera.

6.6.1.1 B/W Sensor and Processing

B/W cameras are equipped with monochrome sensors. A typical spectral sensitivity is shown in Fig. 6.34:

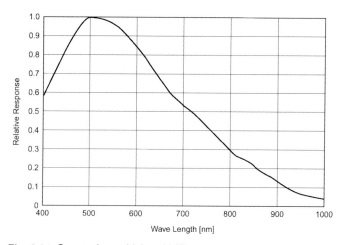

Fig. 6.34 Spectral sensitivity of MF-046B without cut filter and optics.

An example of a real signal from a sensor, clocked at 30 MHz, is shown in Fig. 6.35:

6.6 Digital Cameras

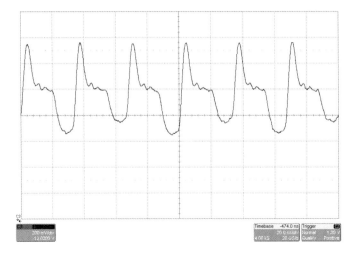

Fig. 6.35 CCD signal entering CDS stage.

In addition, both b/w as well as color cameras are equipped with an IR cut filter in order to keep unwanted IR energy from hitting the sensor.

The spectrum of such a filter is given in Fig. 6.36.

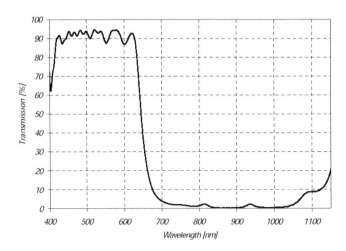

Fig. 6.36 Spectral sensitivity of Jenofilt 217.

Specific applications may require another spectral characteristic, e.g., with less steeper curves, lower transmission loss in the passing region and/or shifted to a longer wavelength.

All color related circuitry is bypassed for b/w. Detailed description as per color camera.

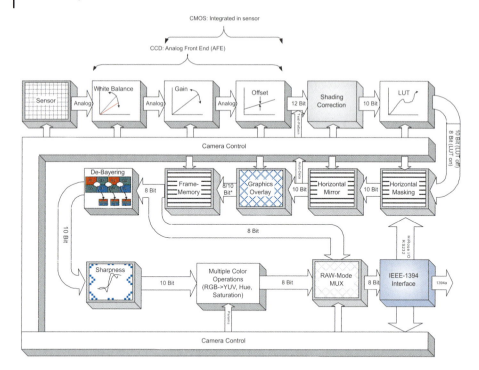

Fig. 6.37 Block diagram of color camera

6.6.2
Color Digital Cameras

6.6.2.1 Analog Processing
The signal from the CCD sensor first must be processed in the analog domain, before it gets converted to digital numbers. Processing is slightly different for b/w and 1 chip-color sensors.

6.6.2.2 One-Chip Color Processing
To start with color sensor, it needs to be mentioned again that most 1-chip color sensors (be it CCD and CMOS) use the concept proposed by BRYCE E. BAYER of KODAK, who got a patent 30 years ago for his idea of placing selectively transmissive filters on top of the pixels of a b/w sensor, such that each filter occurs repeatedly with the luminance (greenish) filter dominating. This is called the BAYER pattern. Here we use the variant, which uses the so-called *primary colors*, red, green and blue, according to the *Tristimulus Theory of color Perception* (Fig. 6.38). We notice that the structure follows insights into the physiology of the eye that it is advantageous for good image reconstruction to have higher luminance than chrominance resolution. So the green filter dom-

inates because it dominates in luminance (Y) as stated in the known formula for Y:

$$Y = 0.3 \cdot R + 0.59 \cdot G + 0.11 \cdot B$$
$$U = -0.169 \cdot R - 0.33 \cdot G + 0.498 \cdot B + 128$$
$$V = 0.498 \cdot R - 0.420 \cdot G - 0.082 \cdot B + 128$$

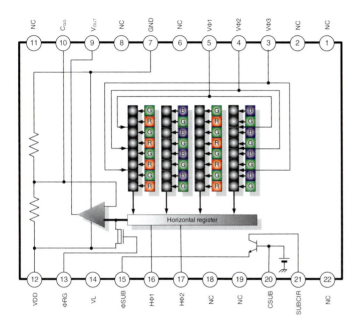

Fig. 6.38 Architecture of the Sony CCD primary color sensor.

Figure 6.39 shows the spectral sensitivity of a sensor with primary color filter array, such as used in the AVT Marlin F-046C.

It can be seen that each of the three primary color filters remove almost two-third of the visible spectrum per color, resulting in the lower sensitivity of a color camera compared to that of a b/w camera for the same light exposure.

It should be noted here that it is not possible with whatever combination of three primaries to display or resolve all the colors the human eye can resolve.

This is due to the fact that certain colors would require *negative lobes* in the spectrum to be generated.

The fact that probably not all colors are (correctly) displayed calls for additional color correction matrices, which is one of the smart functionality of the AVT color family.

Four pixels of two sensor lines are needed as minimum for the reconstruction of all colors.

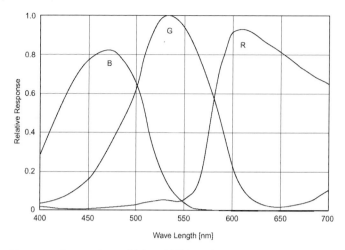

Fig. 6.39 Spectral sensitivity of ICX415Q without cut filter and optics.

Certain models apply a more complex 3 x 3 matrix for an even more accurate debayering.

6.6.2.3 Analog Front End (AFE)

CCD sensors output analog signals pixel per pixel. The analog front end subsumes the analog circuitry with optional color preprocessing and high qualitative digitizing, including analog/digital black level clamping in one chip.

Next to the sensor the AFE is crucial for the performance of a digital camera.

Analog Devices manufactures a broad range of devices, ideally suited for this task.

CMOS sensors usually have a higher integration level and have this functionality built into the sensor itself, making an external AFE obsolete.

The diagram in Fig. 6.40 is taken from the datasheet of such an AFE and illustrates the following details:

The analog color signal, coming in pulse amplitude modulation from the sensor is in the form of the BAYER^{TM} color pattern sequence. It is initially clamped to restore a DC level to fit for the analog circuits, processed in the already discussed CDS (correlated double sampler) then entering the PxGA^{TM} before further amplification and digitization.

The PxGA^{TM} can switch the gain on a pixel-per-pixel basis, thereby compensating for the different transmission losses of the color filter array.

The gain of the two channels with lower outputs can thus be amplified relative to the channel with the highest output. At the output stage all signals are equal in the case of a noncolored template.

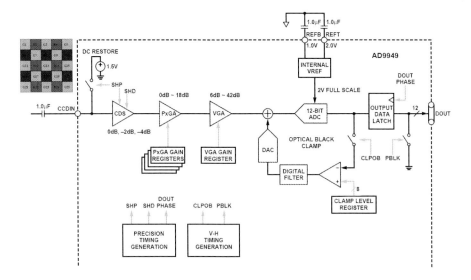

Fig. 6.40 Block diagram of AFE (Source: Analog Devices).

From the users' point of view, the white balance settings are made in a specific register normed by the IIDC.

The values in the *U/B_Value* field produce changes from green to blue; the *V/R_Value* field from green to red as illustrated in Fig. 6.41. This means it requires only two sliders to change whatever is the gain of the three channels.

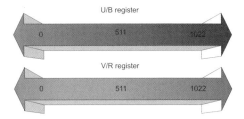

Fig. 6.41 U/V slider range

6.6.2.4 One Push White Balance

Color cameras have not only manual but also one push white balance.

For white balance, a total of six frames are processed, and a grid of at least 300 samples is equally spread over the work area. This area can be the field of view or a subset of it. The RGB component values of the samples are added and are used as actual values for both the One Push and the automatic white balance.

This feature uses the assumption that the RGB component sums of the samples shall be equal, i.e., it assumes that the average of the sampled grid pixels is to be monochrome (*gray world assumption*). This algorithm holds well for most scenery, and fails only when there is only one color in the scene, because this color will desaturate.

6.6.2.5 Automatic White Balance

There is also an auto white balance feature realized, which continuously optimizes the color characteristics of the image.

As a reference, it also uses a grid of at least 300 samples equally spread over the area of interest or a fraction of it.

6.6.2.6 Manual Gain

As shown in Fig. 6.40, all cameras are equipped with a gain setting, allowing the gain to be "manually" adjusted on the fly by means of a simple command register write.

The gain operates in the analog domain, but has the advantage that it is precisely digitally adjustable within a range of 0–24 dB.

The increment is ~0.0354 dB per step so that auto gain functionality with fine granularity is possible.

Because of the fact that there is an analog/digital black clamping circuitry following the gain amplifier, changing gain does not change the black level and an effective black level adjustment (called brightness) is realized.

6.6.2.7 Auto Shutter/Gain

In combination with auto white balance, all CCD-models are equipped with auto shutter/gain feature.

When enabled, auto shutter/gain adjusts the shutter and/or gain within the default shutter/gain limits or within the limits set in the advanced register, in order to reach the brightness set in the auto exposure register as reference. Increasing the auto exposure value increases the average brightness in the image, and vice versa.

The applied algorithm uses a proportional plus integral controller (PI controller) to achieve minimum delay with zero overshot.

It also has priority control, which means that it first uses the shutter and then the gain to make an image brighter, or first reduces the gain and then the shutter to make an image darker.

6.6.2.8 A/D Conversion

As mentioned earlier, A/D conversion is another crucial part of the AFE. The selection criteria are:

- Bit depth: 8, 10, 12 or more bit

- Dynamic and accurate, no missing codes
- Low noise
- Good linearity (differential (DNL) and incremental (INL))
- Pixel-clock frequency, suitable for the sensor's speed
- Power consumption
- Physical size

Clearly, the AFE must not limit the performance of the sensor. This forces the pixel clock to match at least that of the sensor. This calls for up to 36 MHz for certain cameras.

We then have to guarantee that the potential dynamic of the sensor is not sacrificed by insufficient bit depth, which in turn calls quickly for 12 bits, knowing that some digital calculations (e.g., shading correction) will have to multiply the digital signal by a factor of two, thus leading to missing codes, when there is no reserve in bit depth.

Finally the AFE must not add noise to the sensor's signal. A SNR (signal to noise ratio) of at least 72 dB is required for this.

Low power consumption and a small physical size is required to build a small footprint camera can be built, which stays cool so that it has a relatively long lifetime.

It is a generally accepted fact in electronics that the average lifetime drops by 50% for a temperature rise of 10 degrees.

6.6.2.9 Lookup Table (LUT) and Gamma Function

The AVT cameras provide the functionality of a user-defined lookup table (LUT). The use of these Lookup tables (LUTs) allows any function (in the form Output = F(Input) to be stored in the camera's RAM and to apply it on the individual pixels of an image at run-time.

The address lines of the RAM are connected to the incoming digital data, these in turn point to the values of functions which are calculated offline, e.g., with a spreadsheet program. This function needs to be loaded into the camera's RAM before use.

One example of using a LUT is the Gamma LUT: Output = $(Input)^{0.5}$. This is the default gamma LUT, as used in all CCD models. It is known as compensation for the nonlinear brightness response of many displays, e.g., CRT monitors. The lookup table converts the 10 most significant bits (MSBs) from the digitizer to 8 bits.

Due to the fact that different displays may have different gamma curves, a freely programmable LUT is an important advantage.

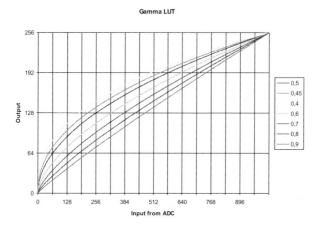

Fig. 6.42 Programmable LUT: Gamma LUTs.

The example was created in MS Excel™, converted to a csv (comma separated value) file and downloaded into the camera, using the AVT camera viewer. There is a universal buffer in the camera, which is used for any data exchange, so that this functionality is also available under user application with only a little programming.

Because of a high speed of the 1394 interface, it takes only a few hundred microseconds for the data to download. Downloading is *transparent*, so it can take place even when the camera is working, because it uses only asynchronous bandwidth, which is not used for image transfer.

But even more specific contrast variations, such as

- threshold (binarization): $Y = 0 (x < A); 255$,
- windowing: $Y = 255 (A < X < B); 0$,
- inversion: $Y = 1023 - X$,
- negative offset: $Y = X - Offset$,
- digital gain: $Y = X(X < 256); 255$,

are very simple to generate.

6.6.2.10 Shading Correction

Shading correction is used to compensate for nonhomogeneities caused by lighting or optical characteristics within specified ranges.

To correct an image, a multiplier between 1 and 2 is calculated for each pixel in 1/256 steps – this allows for shading to be compensated by up to 50%.

All this processing can be done in real time in the FPGA of the camera and is thus another example of a *smart feature*.

The camera allows correction data to be generated automatically in the camera itself.

Upon generation of the shading image in the camera, it can be uploaded to the host computer for nonvolatile storage purposes.

Figure 6.43 describes the process of automatic generation of correction data. The surface plots were created using the software ImageJ.

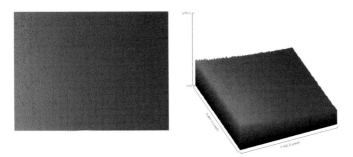

Fig. 6.43 Shading correction: Source image with nonuniform illumination, right: Surface plot made with ImageJ.

On the left you see the source image of a background with nonuniform illumination. The plot on the right clearly shows the brightness level falling off to the right.

By defocusing the lens, high-frequency image data is to be removed from the source image; not to be included in the shading image.

After automatic generation starts, the camera pulls in the number of frames. An arithmetic mean value is calculated from them to reduce noise.

This is followed by a search for the brightest pixel in the mean value frame. A factor is then calculated for each pixel to be multiplied by, giving it the gray value of the brightest pixel.

All these multipliers are saved in a *shading reference image*.

After the lens has been focused again, the image on the left will be seen, but now with a uniform gradient. This is also made apparent in the graph on the right in Fig. 6.44.

6.6.2.11 Horizontal Mirror Function

Many AVT cameras are equipped with an electronic mirror function, which mirrors pixels from the left side of the image to the right side and vice versa. The mirror is centered to the actual FOV center.

This function is especially useful when the camera is looking at objects with the help of a mirror or in certain microscopy applications.

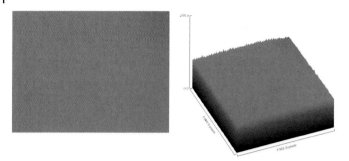

Fig. 6.44 Example of shaded image, right: Surface plot, made with ImageJ.

It is an example for the use of FPGA RAM-cells to build a line memory for the reverse reading out of a CCD-line.

6.6.2.12 Frame Memory and Deferred Image Transport

Normally an image is captured and transported in consecutively pipelined steps. The image is taken, read out from the sensor, digitized and sent over the 1394 bus.

As many AVT cameras are equipped with built-in image memory, this order of events can be paused or delayed by using the so-called *deferred image transport* feature.

The memory is arranged in a FIFO (First in First out) manner. This makes addressing for individual images unnecessary.

Deferred image transport is especially useful for multi-camera applications where a multitude of cameras grab a certain number of images without having to take into account the available bus bandwidth, DMA- and camera (ISO)-channels. Image transfer is controlled from the host computer by addressing individual cameras and reading out the desired number of images. Functionality is controlled by a special register.

6.6.2.13 Color Interpolation

As already mentioned, the color sensors capture color information via the so-called primary color (RGB) filters placed over the individual pixels in a *BAYER mosaic* layout. An effective Bayer-RGB color interpolation already takes place in all AVT color cameras, before they are converted to the YUV format.

Color processing can be bypassed by using the so-called RAW image transfer.

The RAW-mode is primarily used to

- save bandwidths on the IEEE-1394 bus
- achieve higher frame rates

- use different BAYER demosaicing algorithms on the PC

The purpose of color interpolation is the generation of one red, green and blue value for each pixel. This is possible with only two lines:

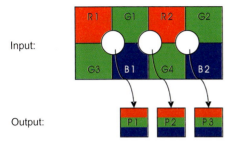

Fig. 6.45 Bayer demosaicing (interpolation).

$$P1_R = R1 \qquad P2_R = R2 \qquad P3_R = R2$$
$$P1_G = \frac{G1 + G3}{2} \qquad P2_G = \frac{G1 + G4}{2} \qquad P3_G = \frac{G2 + G4}{2}$$
$$P1_B = B1 \qquad P2_B = B1 \qquad P3_B = B2$$

Bayer demosaicing inherently introduces color alias effects at the horizontal and vertical borders, which cannot be completely suppressed. There are many algorithms known with different complexities, some clearly overriding FPGA resources.

The photos in Fig. 6.46 show alias occurring at a two pixel wide contour.

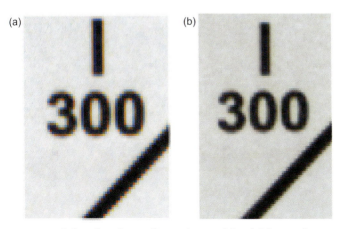

Fig. 6.46 Color alias due to Bayer demosaicing (a) large aliasing, (b) little aliasing.

With adaptive (edge interpretating) techniques in the host PC, it is possible though to achieve almost perfect contours while using a RAW image from the camera.

6.6.2.14 Color Correction

Color correction is performed on all color CCD models before YUV conversion and mapped via a matrix as follows:

$$\begin{bmatrix} R^* \\ G^* \\ B^* \end{bmatrix} = \begin{bmatrix} C_{RR} & C_{GR} & C_{BR} \\ C_{GR} & C_{GG} & C_{BG} \\ C_{RB} & C_{GB} & C_{BB} \end{bmatrix} \begin{bmatrix} R \\ G \\ B \end{bmatrix}$$

Sensor specific coefficients C_{xy} are scientifically generated to ensure that GretagMacbeth™ Color Checker reference colors are displayed with highest color fidelity and color balance.

6.6.2.15 RGB to YUV Conversion

The conversion from RGB to YUV is made using the following formulae:

$$\begin{bmatrix} Y \\ U \\ V \end{bmatrix} = \begin{bmatrix} 0.3 & 0.59 & 0.11 \\ -0.169 & -0.33 & 0.498 \\ 0.498 & -0.420 & -0.082 \end{bmatrix} \begin{bmatrix} R \\ G \\ B \end{bmatrix} + \begin{bmatrix} 0 \\ 128 \\ 128 \end{bmatrix}$$

6.6.2.16 Binning versus Area of Interest (AOI)

Besides the mentioned *binning* in the two directions, which reduces the amount of pixels but not the displayed image angle or field of view, there is another way to reduce the amount of pixel data when only a certain section (area of interest) of the entire image is of interest.

While the size of the image read out for most other video formats and modes is fixed by the IIDC specification, thereby determining the highest possible frame rate, in Format_7 mode the user can set the *upper left corner* and *width and height* of the section (Area of Interest) he is interested in to determine the size and thus the highest possible frame rate.

At a lower vertical resolution the sensor can be read out faster and thus the frame rate is increased.

Setting the AOI is done in defined registers, which are part of the IIDC (formerly called DCAM) specification.

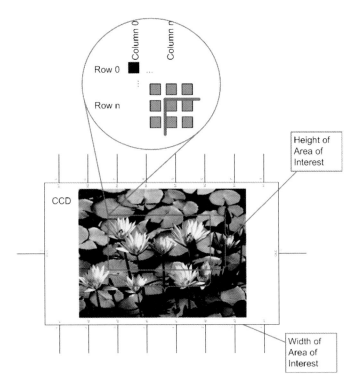

Fig. 6.47 Area of Interest.

6.7
Controlling Image Capture

The cameras support the SHUTTER_MODES specified in IIDC V1.3. For all models this shutter is a global shutter; all pixels are exposed to the light at the same moment and for the same time span.

In continuous modes the shutter is opened shortly before the vertical reset happens, thus acting in a frame-synchronous way.

Combined with an external trigger, it becomes asynchronous in the sense that it occurs whenever the external trigger occurs. Individual images are recorded when an external trigger impulse is present. This ensures that even fast moving objects can be grabbed with no image lag and with minimal image blur.

The external trigger is fed as a TTL signal through Pin 4 of the HiRose connector.

6.7.1
Hardware Trigger Modes

The cameras support IIDC conforming Trigger_Mode_0 and Trigger_Mode_1 and special Trigger_Mode_15.

Trigger_Mode_0 sets the shutter time according to the value set in the shutter (or the extended shutter) register (Fig. 6.48). Trigger_Mode_1 sets the shutter time according to the active low time of the pulse applied (or the active high time in the case of an inverting input).

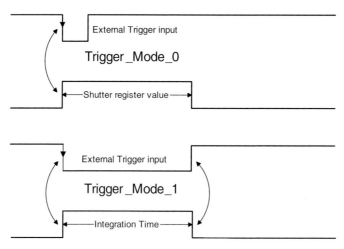

Fig. 6.48 Trigger_mode_0_1, low_active setting register F0F00830h.

Trigger_Mode_15 (Fig. 6.49) is a bulk trigger, combining one external trigger event with a continuous or one shot or multi-shot internal trigger. One external trigger event can be used to trigger a multitude of internal image intakes. This is especially useful for:

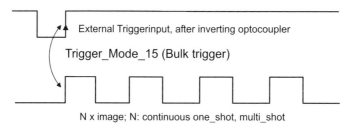

Fig. 6.49 Trigger_Mode_15.

- Exactly grabbing one image based on the first external trigger.

- Filling the camera's internal image buffer with one external trigger without overriding images.

- Grab an unlimited amount of images after one external trigger (Surveillance)

6.7.1.1 Latency (Jitter) Aspects

The following sections describe the time response of the camera using a single frame (One Shot) command. As set out in the IIDC specification, this is a software command that causes the camera to record and transmit a single frame. In the case of a hardware trigger, the One Shot command decoding is skipped so that the camera reacts almost immediately to the trigger edge:

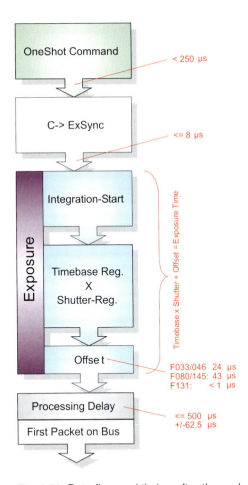

Fig. 6.50 Data flow and timing after the end of exposure.

For example, when the camera is idle, i.e., not transmitting an image, it resets its complete timing within one pixel clock; when the camera is transmitting, the trigger is synchronized with the horizontal clock of the CCD, introducing an extremely low trigger jitter.

After the exposure, the sensor is read out; some data is written into the FRAME_BUFFER before being transmitted to the bus.

The time from the end of exposure to the start of transport on the bus is 500 ± 62.5 μs

This time *jitters* with the cycle time of the bus (125 μs).

It can be seen that the cameras react very quickly to external hardware or software events. There is no extra conceptual 1394-bus latency or delay because of the frame memory.

In all normal modes, the cameras allocate as much bus bandwidth as needed to transport all data read from the CCD with no bigger temporal storage than one cycle buffer (4K).

6.7.2
Pixel Data

Pixel data are transmitted as isochronous data packets in accordance with the 1394 interface described in IIDC v. 1.3 (Table 6.5). The first packet of a frame is identified by the "1" in the sync bit (sy) of the packet header. Communication is secured by a 32-bit cyclic redundancy checksum (CRC) for both the header and the data itself.

Tab. 6.5 Isochronous data block packet format: Source: IIDC v. 1.3 specification.

0–7	8–15	16–23		24–31	
data_length	tg	channel	tCode	sy
header_CRC					
Video data payload					
data_CRC					

The video data for each pixel are outputted in either 8 or 16-bit format.

According to the allocated bandwidth for certain formats, the host knows how much data in how many packets will come until the image is complete. This calculation and interpretation is done in the receiving DMA (direct memory access) machine and thus does not need CPU resources. An interrupt (event) is generated per image to indicate to the software that there is a new image waiting in a buffer within the memory of the host for processing.

Tables 6.6, 6.7, and 6.8 provide a description of the video data format for the different modes. (Source: IIDC v. 1.3 specification)

Tab. 6.6 YUV 4:2:2 and YUV 4:1:1 format: Source: IIDC v. 1.3 specification.

<YUV(4:2:2)>

U-(K+0)	Y-(K+0)	V-(K+0)	Y-(K+1)
U-(K+2)	Y-(K+2)	V-(K+2)	Y-(K+3)
U-(K+4)	Y-(K+4)	V-(K+4)	Y-(K+5)
U-(K+Pn-6)	Y-(K+Pn-6)	V-(K+Pn-6)	Y-(K+Pn-5)
U-(K+Pn-4)	Y-(K+Pn-4)	V-(K+Pn-4)	Y-(K+Pn-3)
U-(K+Pn-2)	Y-(K+Pn-2)	V-(K+Pn-2)	Y-(K+Pn-1)

< YUV(4:1:1) >

U-(K+0)	Y-(K+0)	Y-(K+1)	V-(K+0)
Y-(K+2)	Y-(K+3)	U-(K+4)	Y-(K+4)
Y-(K+5)	V-(K+4)	Y-(K+6)	Y-(K+7)
U-(K+Pn-8)	Y-(K+Pn-8)	Y-(K+Pn-7)	V-(K+Pn-8)
Y-(K+Pn-6)	Y-(K+Pn-5)	U-(K+Pn-4)	Y-(K+Pn-4)
Y-(K+Pn-3)	V-(K+Pn-4)	Y-(K+Pn-2)	Y-(K+Pn-1)

Tab. 6.7 Y8 and Y16 format: Source: IIDC v. 1.3 specification.

<Y(Mono) format>

U-(K+0)	Y-(K+1)	V-(K+2)	Y-(K+3)
U-(K+4)	Y-(K+5)	V-(K+6)	Y-(K+7)
U-(K+Pn-8)	Y-(K+Pn-7)	V-(K+Pn-6)	Y-(K+Pn-5)
U-(K+Pn-4)	Y-(K+Pn-3)	V-(K+Pn-2)	Y-(K+Pn-1)

<Y(Mono16) format>

High byte	Low byte	High byte	Low byte
U-(K+0)		Y-(K+1)	
U-(K+2)		Y-(K+3)	
U-(K+Pn-4)		Y-(K+Pn-3)	
U-(K+Pn-2)		Y-(K+Pn-1)	

6.7.3
Data Transmission

Transmission over the 1394a bus is structured as shown in Fig. 6.51. The green packets are the broadcasted cycle start packets, which are needed for all participants with isochronous (ISO) communication. The example shows three cameras, transmitting a portion of their image, whenever the bus is free, indicated by a short ISO-gap. Asynchronous transfer is only allowed after a longer gap, so that image transmission is already prioritized in the hardware scheme.

Tab. 6.8 Data structure: Source: IIDC v. 1.3 specification.

<Y,R,G,B>
Each component has 8-bit data. The data type is Unsigned Char.

	Signal level (Decimal)	Data (Hexdecimal)
Highest	255	0×FF
	254	0xFE
	⋮	⋮
	1	0×01
Lowest	0	0×00

<U,V>
Each component has 8-bit data. The data type is Straight Binary.

	Signal level (Decimal)	Data (Hexdecimal)
Highest (+)	127	0×FF
	126	0×FE
	⋮	⋮
	1	0×81
Lowest	0	0×80
	−1	0×7F
	⋮	⋮
	−127	0×01
Highest (−)	−128	0×00

<Y(Mono16)>
Each component has 16-bit data. The data type is Straight Short.

Y	Signal level (Decimal)	Data (Hexdecimal)
Highest	65535	0×FF
	65534	0×FE
	⋮	⋮
	1	0×01
Lowest	0	0×00

Iso data is broadcasted over the bus, so that more than one receiver (viewer) can listen, but thus not acknowledged, whereas asynchronous communication is acknowledged by a bi-directional communication between two sources and destinations.

It should be noted here that 1394b uses a different coding technology for transmission, so that the gaps will vanish, making even more efficient use of the doubled bandwidth.

6.7.4
IEEE-1394 Port Pin Assignment

The IEEE-1394 plug is designed for industrial use and has the following pin assignment as per specification:

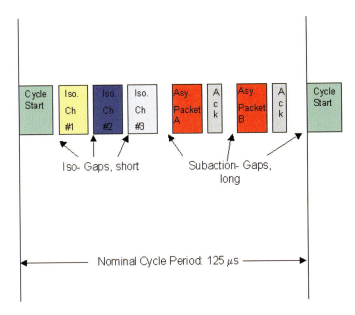

Fig. 6.51 IEEE-1394a data transmission.

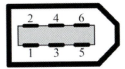

Fig. 6.52 IEEE-1394 connector pin assignment.

Copper cables are available with lengths of 4.5 m or 10 m. Repeaters can be used to add more cable segments (up to 70 m). Plastic or glass optical fiber (POF/GOF) converters can be used to build completely isolated camera connections up to 20m (POF) or to extend the distance up to several 100m at full speed (GOF).

Tab. 6.9 IEEE-1394 pin assignment.

Pin	Signal	Pin	Signal
1	Cable power	4	TPB+
2	Cable GND	5	TPA−
3	TPB−	6	TPA+

6.7.5
Operating the Camera

Power for the camera is supplied either via the FireWire™ bus or the HiRose connector.

The bus is able to source up to 1.5 A of current, making a separate power supply obsolete in many cases. The input voltage must be within the following range: Vcc min.: +8 V Vcc max.: +36 V

6.7.6
HiRose Jack Pin Assignment

The HiRose plug is also designed for industrial use and in addition to providing access to the digital inputs and outputs on the camera, it also provides an RS-232 serial interface, for example, the firmware update or free programmable serial communication (Table 6.10).

Figure 6.53 shows the pinning as viewed from the pin direction.

Fig. 6.53 HIRose connector pin assignment.

Tab. 6.10 HiRose pinning.

Pin	Signal	Use	Pin	Signal	Use
1	External GND	GND for RS232 only	7	GPInput GND GND	Common GND for inputs
2	Power IN (CCD-models only)		8	RS232 RxD	
3			9	RS232 TxD	
4	GP Input 1 (default trigger)	TTL, Edge, progr.	10	OutVCC	Common VCC for outputs
5			11	GPInput 2	TTL
6	GP Output 1 (default IntEna)	Open emitter	12	GPOutput 2	Open emitter

6.7.7
Frame Rates and Bandwidth

An IEEE-1394 camera requires bandwidth to transport images.

The IEEE-1394a bus has a very large bandwidth of at least 32 MB/s for transferring (isochronously) image data. Up to 4096 bytes (or around 1000 quadlets = 4 bytes) can thus be transmitted per cycle.

Depending on the video format settings and the configured frame rate, the camera requires a certain percentage of maximum available bandwidth. Clearly the bigger the image and the higher the frame rate, more is the data to be transmitted.

The following tables indicate the volume of data in various formats and modes to be sent within one cycle (125 μs) at 400 MB/s of bandwidth .

Tables 6.11, 6.12, and 6.13 are divided into three formats: F_0 up to VGA, F_1 up to XGA, and F_2 up to UXGA.

Tab. 6.11 Format_0.

Format	Mode	Resolution	60 fps	30 fps	15 fps	7.5 fps	3.75 fps
0	0	160 x 120 YUV(4:4:4) 24 bit/pixel		1/2H 80p 60q	1/4H 40p 30q	1/8H 20p 15q	
	1	320 × 240 YUV(4:2:2) 16 bit/pixel		1H 320p 160q	1/2H 160p 80q	1/4H 80p 40q	1/8H 40p 20q
	2	640 × 480 YUV(4:1:1) 12 bit/pixel		2H 1280p 480q	1H 640p 240q	1/2H 320p 120q	1/4H 160p 60q
	3	640 × 480 YUV(4:2:2) 16 bit/pixel		2H 1280p 640q	1H 640p 320q	1/2H 320p 160q	1/4H 160p 80q
	4	640 × 480 RGB 24 bit/pixel		2H 1280p 960q	1H 640p 480q	1/2H 320p 240q	1/4H 160p 120q
	5	640 × 480 Y(MONO8) 8 bit/pixel	4H 2560p 640q	2H 1280p 320q	1H 640p 160q	1/2H 320p 80q	1/4H 160p 40q
	6	640 × 480 Y(MONO16) 16 Bit/pixel		2H 1280p 640q	1H 640p 320q	1/2H 320p 160q	1/4H 160p 80q
	7	Y(MONO16) Reserved					

They enable you to calculate the required bandwidth and to ascertain the number of cameras that can be operated independently on a bus and in which mode.

Tab. 6.12 Format_1.

Format	Mode	Resolution	60 fps	30 fps	15 fps	7.5 fps	3.75 fps	1.875 fps
1	0	800 × 600 YUV(4:2:2) 16 bit/pixel		5/2H 2000p 1000q	5/4H 1000p 500q	5/8H 500p 250q	6/16H 250p 125q	
	1	800 × 600 RGB 24 Bit/pixel			5/4H 1000p 750q	5/8H 500p 375q		
	2	800 × 600 Y(MONO8) 8 bit/pixel	5H 4000p 1000q	5/2H 2000p 500q	5/4H 1000p 250q	5/8H 500p 125q		
	3	1024 × 768 YUV(4:2:2) 16 bit/pixel			3/2H 1536p 768q	3/4H 768p 384q	3/8H 384p 192q	3/16H 192p 96q
	4	1024 × 768 RGB 24 bit/pixel				3/4H 768p 576q	3/8H 384p 288q	3/16H 192p 144q
	5	1024 × 768 Y(MONO) 8 bit/pixel		3H 3072 768q	3/2H 1536p 384q	3/4H 768p 192q	3/8H 384p 96q	3/16H 192p 48q
	6	800 × 600 Y(MONO16) 16 bit/pixel		5/2H 2000p 1000q	5/4H 1000p 500q	5/8H 500p 250q	5/16H 250p 125q	
	7	1024 × 768 Y(MONO16) 16 bit/pixel			3/2H 1536p 768q	3/4H 768p 384q	3/8H 384p 192q	3/16H 192p 96q

Tab. 6.13 Format_2.

Format	Mode	Resolution	60 fps	30 fps	15 fps	7.5 fps	3.75 fps	1.875 fps
2	0	1280 × 960 YUV (4:2:2) 16 bit/pixel				1H 1280p 640q	1/2H 640p 320q	1/4H 320p 160q
	1	1280 × 960 RGB 24 bit/pixel				1H 1280p 960q	1/2H 640p 480q	1/4H 320p 240q
	2	1280 × 960 Y(MONO8) 8 bit/pixel			2H 2560p 640q	1H 1280p 320q	1/2H 640p 160q	1/4H 320p 80q
	3	1600 × 1200 YUV(4:2:2) 16 bit/pixel				5/4H 2000p 1000q	5/8H 1000p 500q	5/16H 500p 250q
	4	1600 × 1200 RGB 24 bit/pixel				5/8H	1000p 750q	5/16H 500p 375q
	5	1600 × 1200 Y(MONO) 8 bit/pixel			5/2H 4000p 1000q	5/4H 2000p 500q	5/8H 1000p 250q	5/16H 500p 125q
	6	1280 × 960 Y(MONO16) 16 bit/pixel				1H 1280p 640q	1/2H 640p 320q	1/4H 320p 160q
	7	1600 × 1200 Y(MONO16) 16 bit/pixel				5/4H 2000p 1000q	5/8H 1000p 500q	5/16H 500p 250q

As an example, VGA MONO8 @ 60 fps requires four lines (640 × 4 = 2560 pixels/byte) to transmit every 125 μs: this is a consequence of the sensor's line time of about 30 μs, so that no data need to be stored temporarily. It takes 120 cycles (120 × 125 μs = 15 ms) to transmit one frame, which arrives every 16.6 ms from the camera. Again there is no need for data to be stored temporarily.

Thus, around 64% of the available bandwidth is used.

The third table shows that an MF-145B2 @ 7.5 fps has to send 1280 pixels or 1 line of video per cycle. The camera thus uses 32% of the available bandwidth. This allows up to three cameras with these settings to be operated independently on the same bus.

In video Format_7, frame rates are no longer fixed but can be varied dynamically by the parameters described below.

The following formula (6.1) is used to calculate the highest frame rate in Format_7 for the CCD models:

$$FPS_{In} = FPS_{CCD} = \frac{1}{T_{ChargeTrans} + T_{Dummy} + T_{Dump} + T_{Scan}} \quad (6.1)$$

It assumes that the maximum frame rate is the inverse of the sum of all events in a CCD, which take time such as:

- The time to transfer the charge to the vertical shift register (Charge transfer time)
- The time to shift out the dummy lines
- The time to dump the lines outside the AOI
- The time to shift out the lines of the AOI (Scanning time).

For different sensors, different values apply. Frame rates may be restricted by bandwidth limitation of the IEEE-1394 bus.

In some modes the IEEE-1394a bus limits the attainable frame rate. According to the 1394a specification on isochronous transfer, the largest data payload size of 4096 bytes per 125 μs cycle is possible with a bandwidth of 400 MB/s.

This leaves 20% (1024 bytes) free for asynchronous non-real time data transfer, such as loading LUT or shading correction image or changing shutter/gain.

The following formula establishes the relationship between the required Byte_Per_Packet size and certain variables for the image. It is valid only for Format_7.

$$BYTE\_PER\_PACKET = fps \cdot AoiWidth \cdot AoiHeight \cdot ByteDepth \cdot 125 \text{ μs} \quad (6.2)$$

If the value for "BYTE_PER_PACKET" is greater than 4096 (the maximum data payload), the sought-after frame rate cannot be attained. The attainable frame rate in fps can be calculated using the formula[1]:

$$fps \approx \frac{BYTE\_PER\_PACKET}{AoiWidth \cdot AoiHeight \cdot ByteDepth \cdot 125\mu s} \qquad (6.3)$$

ByteDepth is based on the following values:

$$\text{Mono8} \Rightarrow 8\,\text{bits/pixel} = 1\,\text{byte/pixel}$$
$$\text{Mono16} \Rightarrow 16\,\text{bits/pixel} = 2\,\text{bytes/pixel}$$
$$\text{YUV4:2:2} \Rightarrow 16\,\text{bits/pixel} = 2\,\text{bytes/pixel}$$
$$\text{YUV4:1:1} \Rightarrow 12\,\text{bits/pixel} = 1.5\,\text{bytes/pixel}$$

6.1 Example Formula for the b/w camera:
Mono16, 1392 × 1040 – 15 fps desired

$$BYTE\_PER\_PACKET = 15 \cdot 1392 \cdot 1040 \cdot 2 \cdot 125\,\mu s = 5428 > 4096$$

$$\Rightarrow fps_{reachable} \approx \frac{4096}{1392 \cdot 1040 \cdot 2 \cdot 125\,\mu s} = 11,32 \qquad (6.4)$$

6.8
Configuration of the Camera

All camera settings are made by writing specific values into the corresponding registers. This applies to both, values for general operating states such as video formats and modes, exposure times, etc., and to all extended features of the camera that are turned on and off and are controlled via corresponding registers. Fig. 6.54 gives an example.

6.8.1
Camera Status Register

The interoperability of cameras from different manufacturers is ensured by IIDC, formerly DCAM (Digital Camera Specification), published by the IEEE-1394 Trade Association.

IIDC is primarily concerned with setting memory addresses (e.g., CSR: Camera_Status_Register) and their meaning.

In principle all addresses in IEEE-1394 networks are 64 bits long.

The first 10 bits describe the Bus_Id, the next 6 bits the Node_Id.

Of the subsequent 48 bits, the first 16 are always FFFFh, leaving the description for the Camera_Status_Register in the last 32 bits.

---

[1] Provision: "BYTE_PER_PACKET" is divisible by 4

6.8 Configuration of the Camera

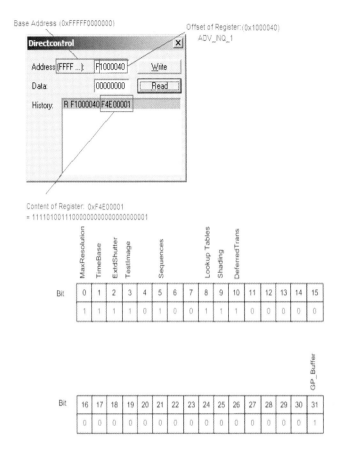

Fig. 6.54 Configuration of a camera.

If in the following, mention is made of a CSR F0F00600h, this means in full:

$$\text{Bus\_Id , Node\_Id , FFFF F0F00600h}$$

Writing and reading to and from the register can be done with programs such as *FireView* or by other programs that are developed using an API library (e.g., FirePackage).

Every register is 32 bit (Big Endian) and implemented as shown in Fig. 6.55.

This requires that, for example, to enable the *ISO_Enabled* mode (bit 0 in register 614h), the value 80000000 h must be written in the corresponding register.

Fig. 6.55 32-bit register.

Due to the IIDC standard, there is a variety of different APIs available, which make the integration of digital cameras with 1394 interface into customer's software easy.

Allied Vision Technologies, for example, offers three APIs suited for various requirements:

- FirePackage™ for demanding applications
- DirectFirePackage™ for WDM and DirectX/DirectShow
- Fire4Linux™ as an enhancement to libdc1394.

It should also be noted that the cameras are supported in major image processing libraries, such as,

- MVTEC Halcon™ and ActivVisionTools™,
- Stemmer's CommonVisionBlox™,
- Neurocheck™

and by manufacturers of machine vision components, such as, Matrox, National Instruments, and Cognex e.a.

6.9
Camera Noise[2]

In a CCD or CMOS camera various noise sources at different stages contribute to an overall total, appearing at the output of a digital camera.

Figure 6.56 details various noise sources:

Several parts of the camera or the sensor contribute different independent noises, which are discussed in detail below.

All noise sources are to be added in their quadrature.

[2] by Henning Haider, Allied Vision Technologies GmbH

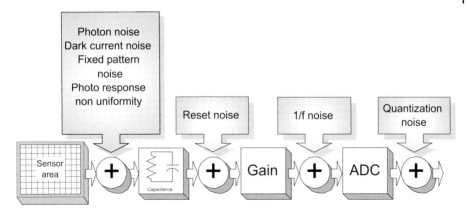

Fig. 6.56 Different noise components and their location.

6.9.1
Photon Noise

Photons generate electrons in the photodiode on a statistical base because of their quantum nature.

That means the count of electrons differs from image to image according to the so-called Poisson distribution.

Noise n in (count of electrons rms) is equal to the square root of the mean N of the generated count of electrons N_{photon}. The maximum of the generated photons is expressed as N_{well}, which is the full well capacity.

$$n_{photon} = \sqrt{N_{photon}} \quad (e^- \text{ rms}) \tag{6.5}$$

6.9.2
Dark Current Noise

Temperature in silicon material generates a flow of electrons, the so-called dark current, depending on the process and the temperature. For CCD sensors, the current is typically much lower than for CMOS sensors. Nevertheless it doubles per 7 °C to 9 °C and it adds up to the photon-created electrons in the pixels and the shift registers, thus limiting their capacity.

With CMOS sensors this can even limit longer integration times, because the sensor can reach saturation due to the dark current. Its nature is also statistical, so that the following formula applies:

Noise n in (electrons rms) is equal to the square root of the mean N of the dark current electrons:

$$n_{dark} = \sqrt{N_{dark}} \quad (e^- \text{ rms}) \qquad (6.6)$$

Fixed Pattern Noise (FPN)

Related with the dark current is its electrical behavior which is regionally different on the sensor. This introduces a structural spatial noise component, called fixed pattern noise, although not meant temporally, visible with low illumination conditions.

FPN is typically more dominant with CMOS sensors than with CCD, where mostly it can be ignored.

This noise n_{fpn} [%] is usually quantified in % of the mean dark level.

6.9.3
Photo Response Nonuniformity (PRNU)

Due to local differences in the pixel, the count of electrons for a given illumination is not exactly the same. Every pixel has a slightly different photon to electron response characteristic. Again, this is a structural spatial noise component, measured in % n_{prnu} [%] of the maximum signal level.

6.9.4
Reset Noise

Reading out a pixel requires the conversion of the charge into a voltage. This is done via destructive readout with a capacitance, which is reset after the actual pixel value is sampled. Again here the reset level varies from pixel to pixel, depending on the absolute temperature T and the capacitance C according to Eq. (6.7), with q being the elementary charge and k the Boltzmann constant:

$$n_{reset} = \frac{\sqrt{kTC}}{q} \quad (e^- \text{ rms}) \qquad (6.7)$$

This component can be ignored in the case of CCD sensors due to correlated double sampling in the analog front end, as discussed earlier.

6.9.5
1/f Noise (Amplifier Noise)

Every analog amplifier contains a noise component, which is proportional to $1/f$.

Nowadays, this component can also be neglected because of the use of CDS signal pre-processing.

6.9.6
Quantization Noise

Quantization noise is a side effect in the conversion of an analog signal to a digital number. It occurs due to uncertainty in switching from one step to the other, usually in the middle of one step.

Obviously, it can be made smaller by a larger amount of bits and quantization steps.

Assuming a sawtooth shape of the quantization error, in terms of noise electrons it can be expressed with $N_{electron}$ = count of signal electrons and N = #bits as:

$$n_{ADC} = \frac{N_{electron}}{2^N \cdot \sqrt{12}} (e^- \, rms) \tag{6.8}$$

Normally, quantization is smaller than the noise floor, so that it can be ignored.

6.9.7
Noise Floor

It can be seen from the above that the total noise of a camera has a component, which is dependent on the wanted signal, and one component independent of the wanted signal, called the noise floor. The noise floor sets the lower limit for the sensitivity of a camera.

The noise floor can be given as:

$$n_{floor} = \sqrt{n_{dark}^2 + n_{fpn}^2} = \sqrt{N_{dark} + n_{fpn}^2} \; (e^- \, rms) \tag{6.9}$$

6.9.8
Dynamic Range

The dynamic range of a camera is the quotient of the wanted signal to the ground floor.

$$DNR = \frac{N_{photon} - N_{dark}}{\sqrt{n_{dark}^2 + n_{fpn}^2}} = \frac{N_{photon} - N_{dark}}{\sqrt{N_{dark} + n_{fpn}^2}} \tag{6.10}$$

The DNR of a sensor is given, relative to the maximum signal, as:

$$DNR_{sensor} = \frac{N_{well} - N_{dark}}{\sqrt{N_{dark} + n_{fpn}^2}} \tag{6.11}$$

6.9.9
Signal to Noise Ratio

The signal to noise ratio of a camera is the quotient of the total signal to the total noise and could be given as:

$$SNR = \frac{N_{photon} + N_{dark}}{\sqrt{n_{photon}^2 + N_{dark} + n_{fpn}^2 + n_{prnu}^2 + n_{reset}^2 + n_{adc}^2}} \qquad (6.12)$$

Simplifying by eliminating the dark current component via clamping we get

$$SNR = \frac{N_{photon}}{\sqrt{N_{photon} + N_{dark} + n_{fpn}^2 + n_{prnu}^2 + n_{reset}^2 + n_{adc}^2}} \qquad (6.13)$$

In relation to the maximum of the sensor:

$$SNR = \frac{N_{well}}{\sqrt{N_{well} + N_{dark} + n_{fpn}^2 + n_{prnu}^2 + n_{reset}^2 + n_{adc}^2}} \qquad (6.14)$$

6.2 Example Sony ICX-285AL sensor (AVT Dolphin-F145B)
Although Sony does not publish full well capacity, the following specs can form the base for our DNR and SNR (at full illumination!) calculations, neglecting fixed pattern noise:

$$N_{well} \approx 18.000\, e^- \quad , \quad N_{dark} \approx 60\, e^-$$

$$DNR = \frac{18000 - 60}{\sqrt{60}} = 2316 = 67\, dB \approx 11.2\, bit$$

$$SNR = \frac{18000}{\sqrt{18000 + 60}} = 134 = 42.5\, dB \approx 7.1\, bit$$

6.3 Example Cypress (FillFactory) IBIS5A sensor (AVT Marlin-F131B)

$$N_{well} \approx 62500\, e^- \text{ (datasheet)} \quad , \quad n_{readout} \approx 40\, e^- \text{ (datasheet)}$$

$$DNR = \frac{62500}{40} = 1563 = 63.9\, dB \approx 10.6\, bit$$

$$SNR = \frac{62500}{\sqrt{62500 + 1600}} = 247 = 47.8\, dB \approx 7.9\, bit$$

6.10
Digital Interfaces

Today, more than a handful of rival digital interfaces is in the market.

Tab. 6.14 Digital interface comparison.

Interface	IEEE-1394a	USB 2.0	IEEE-1394b	Gigabit Ethernet (802.3AB)	Camera Link base
Maximum bit rate	400 MB/s	480 MB/s	800 MB/s	1000 MB/s	>2000 MB/s
Isochronous (video) mode	Yes	Yes	Yes	No	Yes
Bandwidth/ total usable bandwidth	Video: 32 MB/s (80%) Total: 40 MB/s	45 MB/s (90%)	Video: 64 MB/s (80%) Total: 80 MB/s	90 MB/s	255 MB/s (base) 680 MB/s (full)
Topology	Peer-to-peer	Master-slave, OTG(On the go)	Peer-to-peer	Networked, P2P	Master-slave
Single cable distance in in copper or other media	4.5 m, worst case; 10 m, typical camera application; 300 m GOF	5 m	10 m copper; 300 m GOF	25 m, 100 m (Cat5)	10 m
Maximum distance copper using repeaters	70 m	30 m	70 m	n.a.	30 m
Bus power	Up to 1.5 A and 36 V	Up to 0.5 A and 5 V	Up to 1.5 A and 36 V	None	None
Motherboard support	Many	Virtually all	Some	Some	None
PC load	Very low	Low	Very low	Middle	n.a.
OS support	Windows, Linux	Windows, Linux	Windows, Linux	Windows, Linux	Depending on vendor
Main applications	Multimedia electronics	PC-centric serial input	Multimedia electronics	Networking	High speed camera interface
Camera standard	IIDC V1.3	Video class(?)	IIDC V1.31	Proposal: GenIcam	
Devices per bus	63; 4(8) silmult. / card, accord. to 4(8) DMA's typ	127; 4 simult. / card, accord. to 4 DMA's typ	63; 4 silmult. / card, accord. to 4 DMA's typ	Dependant on software and available bandwidth	1 per interface

In principle any interface, able to deal with the amount of data of an uncompressed digital video stream, is a candidate.

The requirements can be summarized and prioritized as follows:

- Bandwidth requirement > 10 MB/s per camera in VGA, b/w @ 30 fps
- Secure protocol (HW/SW)
- Robust electrical interface, able to bridge distances of 3 to 30m (100m)
- No image loss
- Low CPU consumption for image acquisition and transfer
- Low latency
- Multi camera support
- Open (vendor independent) standardization for the protocol

Clearly, *1394a* was pioneering the terrain, and together with the coming 1394b it is having the biggest market share and the widest spread.

USB2.0 has proven that it also can handle the amount of data, needed to transmit raw, uncompressed digital images into the memory of the PC.

Recently, *GigaBit Ethernet* also claims its ability to transport video data over the network architecture with the increase of the network bandwidth.

All of the above are interfaces, which were not primarily designed for digital camera interfaces, but adopted for this use.

Camera Link today probably is the only open digital interface of relevance, which was developed specifically for use in digital image transfer.

It can be said that depending on the application, the user has the freedom of choice. Table 6.14 gives an overview of the important facts and figures.

References

1. www.fillfactory.com, Cypress/FillFactory CMOS sensors
2. www.framos.de, Datasheets of Sony sensors and components
3. http://www.kodak.com/global/en/digital/ccd/, KODAK sensors
4. www.micron.com/products/imaging/.com, MICRON CMOS sensors
5. www.photonfocus.com, Photonfocus CMOS sensors
6. http://www.sony.net/Products/SC-HP/, Sony semiconductor
7. http://www.cs.rit.edu/~ncs/color/, Information about color
8. http://www.tele.ucl.ac.be/PEOPLE/DOUXCHAMPS/ieee1394/cameras/index.html, FireWire™ camera list
9. http://en.wikipedia.org/wiki/Main_Page, WIKIPEDIA, the free encyclopedia
10. http://www-ise.stanford.edu/~tingchen/main.htm, Bayer demosaicing
11. www.alliedvisiontec.com, Homepage of Allied Vision Technologies
12. HOLST., G. C., *CCD Arrays, Cameras, and Displays*, SPIE Optical Engineering Press, Bellingham, **1998**.
13. THEUWIESEN, A. J. P., *Solid State Imaging with Charge Coupled Devices*, Kluwer, Dordrecht, The Netherlands **1995**.

7
Camera Computer Interfaces

Tony Iglesias, Anita Salmon, Johann Scholtz, Robert Hedegore, Julianna Borgendale, Brent Runnels, Nathan McKimpson, National Instruments

7.1
Overview

Machine vision acquisition architectures come in many different forms, but they all have the same end goal. That goal is to get image data from a physical sensor into a processing unit that can process the image and initiate an action. This goal is the same for personal computer (PC)-based machine vision systems, embedded compact vision systems, and smart cameras.

Of these architectures, PC-based machine vision systems are the most popular. PC-based systems provide the best performance for the price and the most flexibility. This chapter focuses on the PC; however, most of the concepts apply regardless to which physical architecture is used. All physical architectures have the same starting point and ending point. You have a sensor on one end and a processing unit on the other.

PC-based machine vision architectures typically consist of a camera connected to a computer through an interface card. Figure 7.1 shows the high level components of any vision system. The first item in the system is a camera or image sensor, which is responsible for detecting the image. The image data then travels across a camera bus and into an interface device. The interface device converts the data from the camera or sensor format, to a memory format that a PC can process. Driver software directs the transactions by acting on the interface card.

The focus of this chapter is on the camera bus, the interface card , the computer bus and the driver software. When designing your machine vision system you will need to make decisions and tradeoffs based on the strengths and weaknesses of each of these components.

The camera bus section discusses the most popular mechanisms for getting data from a camera to the interface card. Each camera bus has its own

Handbook of Machine Vision. Alexander Hornberg (Ed.)
Copyright © 2006 WILEY-VCH Verlag GmbH & Co. KGaA, Weinheim
ISBN: 3-527-40584-7

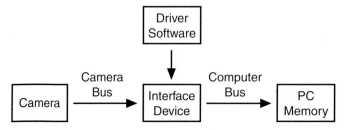

Fig. 7.1 Image Acquisition Architecture.

strengths and weaknesses. The section explores the very different approaches used in machine vision. The most basic approach is the analog approach, which uses a single wire to carry an analog signal that encodes image and data timing information. This section also discusses the brute-force mechanism of RS-422 and Low Voltage Differential Signaling (LVDS), used to transfer digital image data from the camera to the interface card, and Camera Link, the only digital bus specifically designed for machine vision. The camera buses section concludes by describing how machine vision component manufacturers are leveraging PC-standard buses like Universal Serial Bus (USB), IEEE 1394, and Gigabit Ethernet to lower the cost of machine vision. All of the approaches to getting image data from the camera into the computer memory discussed in the camera buses sections are very different. About the only thing they have in common is a cable, which allows the camera to be positioned some distance away from the PC.

The interface card connects the camera bus and the computer bus. The interface card is necessary because the camera does not directly output data in the same format that the computer uses internally. Even if the camera bus and the computer bus use the same protocol, an interface card will still be needed to physically convert from the cable to the motherboard. Machine vision interface cards vary in price, complexity, and performance. Interface cards that are specifically designed for machine vision, commonly referred to as frame grabbers, include many convenient features such as triggering support, image reconstruction, and preprocessing. General purpose interface cards such as USB and Gigabit Ethernet are so simple that the interface card consists of no more than a connector and a microchip, leaving the necessary machine vision features to either the camera or the host processor.

As the PC has evolved over the years, so have the computer buses. This section discusses the buses which have defined the decades such as ISA, PCI, and PCI Express as well as the intermediate buses such as EISA, 64-bit PCI, and PCI-X. This section also explores how each of these buses paved the way for PC-based machine vision.

The chapter concludes with a discussion on driver software. The driver software is responsible for getting images from the camera into the computer. The software does this by interacting directly with the interface device. It also serves as the interface between the programmer and the machine vision system. In addition to receiving images from the camera, the driver software also provides many convenient features such as software mechanisms for controlling the camera, for displaying images, and for transferring images to disk for later viewing or processing.

After reading this chapter, you will have a thorough understanding of the image acquisition component of a machine vision system. You will be able to evaluate your application needs and pick the right camera bus, interface card, and computer bus for your application. This knowledge will help to ensure that you meet your image acquisition form factor and performance goals without over-engineering your vision system.

7.2
Analog Camera Buses

Industrial analog cameras use coaxial cable to transmit an analog video signal from the camera to an image acquisition device or monitor. The analog video signal transmitted by the camera uses the same composite video formats used by TV stations to broadcast video signals around the world. For color video signals, there are two main video standards: National Television Systems Committee (NTSC) and Phase Alternative Line (PAL). NTSC is more common in North America and Japan, and PAL is more common in Europe. For monochrome video signals, the two main video standards are Electronic Industries Association (EIA) RS-170 and Consultative Committee for International Radio (CCIR). RS-170 is common in North America and CCIR is common in Europe.

7.2.1
Analog Video Signal

The analog video signal is generated when the image sensor in a camera is scanned. The image sensor is a matrix of photosensitive elements that accumulates an electric charge when exposed to light. Once the image sensor is exposed, the accumulated charge is transferred to an array of linked capacitors called a shift register. An amplifier at one end of the shift converts the charge on the capacitor to voltage, and drives the voltage onto a coaxial video cable.

7 Camera Computer Interfaces

Charges are sent to the amplifier as the image sensor is scanned. This creates a continuous analog voltage waveform on the video cable. The amplitude of the video signal at some point in time represents the Luminance, or brightness, of a particular pixel. Before each line of video, a signal known as the Back Porch is generated. The Back Porch signal is used as a reference by the receiver to restore any DC components from the AC-coupled video signal. The period when the DC components are removed is called the Clamping Interval.

Color information, or Chroma, can be transmitted along the same coaxial cable with a monochrome video signal. The Chroma signal is created by modulating two quadrature components onto a carrier frequency. The phase and amplitude of these components represent the color content of each pixel. The Chroma is added to the Luminance signal to create a color composite video signal.

7.2.2
Interlaced Video

According to the composite video format, an image frame is transmitted as two independent fields. One field contains all the odd lines, and the other contains all the even lines. When two independent fields are used to transmit an image it is called interlaced video transfer. Figure 7.2 shows a diagram of an interlaced video frame. In the RS-170 interlaced video format, frames are transmitted at 30 Hz, but the fields are updated every 60 Hz. The effect of interlaced video transfer is that the human eye perceives a 60 Hz video update rate when watching a monitor, as opposed to the true 30 Hz full image update rate.

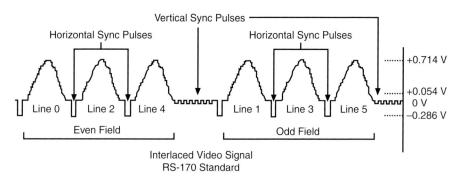

Fig. 7.2 Interlaced Video Frame.

7.2.3
Progressive Scan Video

Although interlaced video can deceive the human eye into seeing a higher video update rate, the reconstructed image of a moving object looks blurred when viewed as a singe frame. This is because the odd and even lines were exposed at different times. Progressive scan cameras are designed to eliminate motion blurring. Progressive scan cameras expose and scan the entire image sensor without interlacing even and odd lines. The result is improved performance for applications where the objects are constantly moving.

7.2.4
Timing Signals

In addition to video data, composite video signals also carry synchronization signals to designate the beginning of a field or line. A Horizontal Synchronization (Hsync) pulse is generated at the beginning of a line, and is seen as a negative-going pulse in the video stream. A Vertical Synchronization (Vsync) pulse is a negative-going pulse that designates the beginning of a field. Figure 7.3 illustrates a composite video signal.

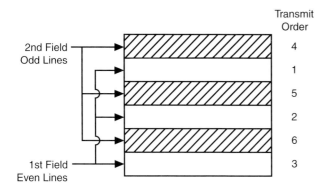

Fig. 7.3 Composite video signal.

The Color Burst is a timing signal used to decode color information from a video waveform. The Color Burst appears on the back porch, at the beginning of each line. It is a sinusoid whose frequency is equal to that of the Chroma carrier signal. The Color Burst is used as a phase reference to demodulate the quadrature-encoded Chroma signal.

The RS-170 video standard uses a 640 × 480 frame size. 780 discrete pixel values are transmitted between Hsync pulses, 640 of the pixel values transmitted are active image pixels that make up a full line of video. The remaining pixels are referred to as line blanking pixels and are unused. There are 525 lines per frame, 480 of the lines contain active image data. The remaining lines are called frame blanking lines and are unused. Table 7.1 lists the specifications for the most common analog video standards.

Tab. 7.1 Analog video standards.

Standard	Type	Frame Size (pixels)	Frame Rate (frames/sec)	Line Rate (lines/sec)
NTSC	Color	640×480	29.97	15,734
PAL	Color	768×576	25.00	15,625
RS-170	Black & White	640×480	30.00	15,750
CCIR	Black & White	768×576	25.00	15,625

7.2.5
Analog Image Acquisition

Analog image acquisition boards, commonly referred to as frame grabbers, typically use a Sync Separator to detect these timing signals and extract them from the video data stream. A Phase Lock Loop (PLL) synchronizes with the extracted Vsync and Hsync signals and generates a pixel clock, which is used to sample the video data. The pixel clock frequency of an RS-170 video signal is 12.27 MHz.

A DC Restore clamps to the Back Porch DC reference signal at the beginning of a line. An Analog to Digital Converter (ADC) then samples the analog video signal on each pixel clock cycle, generating discrete digital pixel values. The digital pixels are transferred across the computer bus and re-constructed on a frame in the system memory.

For color analog frame grabbers, the Chroma signal is separated from the Luminance signal with an analog filter. The two signals are sampled simultaneously by separate ADCs. Digital processing then converts the chrominance and luminance values into Red, Green, and Blue (RGB) intensity components. These values are then transferred across the computer bus into the system memory. Most color image processing algorithms use the RGB or Hue, Saturation, and Luminance (HSL) color formats.

7.2.6
S-Video

Combining the Chroma and Luminance signals on the same wire causes problems for applications that require high resolution, accuracy, and signal integrity. Sometimes, the higher frequency Chroma signal bleeds into the luminance spectrum and appears as noise in the acquired image.

To improve color composite video transmission, a new standard was developed called S-Video. S-Video uses the same timing and synchronization signals as composite video, but it uses two independent wires to transmit the Chroma and Luminance signals. S-Video is used in machine vision applications, but S-Video can also be seen on many consumer video devices today.

7.2.7
RGB

RGB cameras encode color information as Red, Green, and Blue components before transmitting the video signal across the cable. These components are transmitted on three different coaxial cables in parallel. Like the Luminance value of a composite video signal, a voltage on the coaxial wire represents the intensity value of each component. Vsync and Hsync timing signals are typically added to one of the three component signals for synchronization.

Since there is no high-frequency Chroma signal present in RGB video signals, the signal integrity of each component is better than that of a composite video signal. The video signal is transmitted in RGB color format, so there is no need for digital processing on the acquisition board. However, since there are three different analog signals, the acquisition hardware must have three analog to digital converters to sample a single video signal.

7.2.8
Analog Connectors

Composite video signals are transmitted on shielded, coaxial cables. A standard BNC connector is commonly used for machine vision applications. RGB video transfer also uses coaxial cables and BNC connectors for each component video signal. The S-Video standard defines a special connector and cable for video transfer. Figure 7.4A shows a BNC connector and Figure 7.4B shows an S-Video connector.

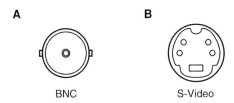

Fig. 7.4 A. BNC Connector, **B.** S-Video Connector

7.3
Parallel Digital Camera Buses

As higher-resolution camera sensors were developed, the limits of composite video were exceeded. As an alternative, camera manufacturers turned to digital video. In digital video, the analog video signal from the camera sensor is converted to a digital signal within the camera and sent to the image acquisition device in digital format.

To transmit digital video signals, custom-shielded cables were designed that bundle many conductors in parallel, each providing a separate stream of digital data. This allows for the guaranteed transmission of data at very high speeds with a high immunity to noise.

7.3.1
Digital Video Transmission

In composite video transmission, the Vsync and Hsync timing signals are embedded in the analog video signal and extracted using a PLL on the frame grabber. In most digital video transmission schemes, timing signals are sent on separate conductors in parallel with the video data signals. The Vsync signal is replaced with Frame Valid, which is asserted for the duration of a video frame. The Hsync signal is replaced with Line Valid, which is asserted for the duration of a video line. The Pixel Clock signal is generated by the camera and sent as a separate parallel signal, instead of being generated by the PLL on the frame grabber. All digital signals are transmitted from the camera at the same time as the Pixel Clock.

After the camera sensor is exposed, it must scanned, one pixel at a time. Analog voltages from the sensor are converted to digital pixel values and sent one after the other along the parallel cable. Typically, each data wire within a cable is dedicated to a particular bit of the pixel value. For example, a camera sensor that is sampled with a 10-bit ADC requires a cable with ten parallel data wires. Figure 7.5 shows the timing diagram for a parallel digital camera and the pixel locations on the image sensor. Two frames are read out. Each frame is two lines high and three pixels wide. The image sensor is scanned from upper-left to bottom-right.

7.3.2
Taps

As larger, faster sensors were developed, it became necessary to scan two, four, or even eight pixels at a time. Multiple pixels are sent down the parallel video cable in separate digital streams called taps. The scanning order and

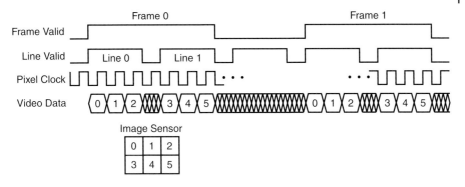

Fig. 7.5 Parallel digital timing diagram.

direction for camera sensors varies depending on the camera. Frame grabber companies rely on the camera manufacturer to indicate the scanning order and direction for multiple-tap cameras so images can be properly reconstructed in computer memory.

In Fig. 7.6, the image sensor is split into four quadrants, each corresponding to a different tap. Each tap is scanned from the middle of the sensor outward. Pixel 0 from each tap is transmitted at the same time, then pixel 1, and so on. If the bit depth of each pixel is 8 bits, this tap configuration requires a cable with 32 parallel wires, plus three additional wires for the timing signals. The frame grabber samples all 32 signals on every Pixel Clock cycle and stores the values in memory. Typically, the image is reconstructed in the frame grabber memory as a contiguous image, upper left to bottom right, before being sent across the computer bus.

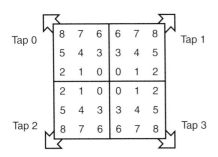

Fig. 7.6 Image Sensor Taps

Multiple taps can also be used to transmit color video signals. Pixels from color image sensors are typically represented by three independent 8-bit RGB components. Together, these three components represent any 24-bit color. Pixels are transmitted one per pixel clock on three different taps. Each tap carries an 8-bit RGB component value.

7.3.3
Differential Signaling

The earliest cameras transmitted digital video using Transistor-Transistor Logic (TTL) signaling. The limits of clock speed and cable length quickly forced manufacturers to consider a standard for differential signaling. RS-422 and RS-644 Low Voltage Differential Signaling (LVDS) were developed, and quickly adopted, as standards for digital video transmission. Differential signaling has better noise immunity, which allows for longer cable lengths, and can be run at much higher rates. However, a twisted pair of wires is now required for each digital data stream, doubling the number of wires in the cable and increasing cable cost.

7.3.4
Line Scan

Most image sensors are two-dimensional, having a matrix of pixels arranged in rows and columns. Light reflected from a three-dimensional object is projected onto the sensor and each pixel samples the light from a small region within the camera's field of view. Two-dimensional sensors are an ideal geometry for capturing images of stationary objects, but blurring occurs if the object is moving. This is because it usually takes many milliseconds to expose the entire sensor, during which time the object has moved. Special lighting and high-speed exposure can help the situation, but some applications use another kind of image sensor.

Applications that typically involve imaging a continuous object that is always moving in the same direction are known as web inspections. One example is a paper mill inspection system. Paper continuously rolls along a conveyor belt and is inspected by a vision system. The conveyor belt never stops, so the image sensor must be exposed very quickly. Furthermore, there are no breaks in the paper, so it must be imaged as one continuous object.

Line scan sensors are sensors designed specifically to solve web applications. These sensors are unique because they have only a single row of pixels. Since they are one-dimensional sensors, objects must be scanned one line at a time by moving the object past a stationary sensor. One line is exposed, sampled, and transmitted at a time. The frame grabber builds a two-dimensional image by stacking consecutive lines on top of each other in the system memory. The image height is theoretically infinite, since the image is continuous. Image blur is significantly reduced with a line scan camera because only one line is exposed at a time. Figure 7.7 illustrates a line scan sensor imaging a continuous paper reel moving along a conveyor.

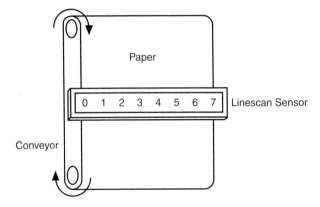

Fig. 7.7 Paper inspection using a line scan sensor.

7.3.5
Parallel Digital Connectors

Before any official standard for digital video could be put in place, camera vendors adopted their own proprietary formats for encoding analog camera sensor signals into digital video streams. Frame grabber manufacturers struggled to keep up with an ever-growing number of available formats. A list of compatible cameras and frame grabbers had to be maintained by each manufacturer, and custom cables had to be developed for a particular combination.

Since there was no early standard to mandate the transmission of digital video, camera and frame grabber manufacturers used an assortment of different high-density connectors for parallel digital connectivity. Custom cables were required to interface a specific camera with a particular frame grabber. Some commonly used connectors are shown in Figures 7.8A–F.

7.3.6
Camera Link

The cost of custom cables and the broad range of digital transmission formats were the driving force behind the development of the Automated Imaging Association (AIA) standard for high-speed transmission of digital video. The AIA standard is known as Camera Link. The Camera Link specification defines the cable, connector, and signal functionality. Any camera that complies with the Camera Link specification should work with any frame grabber that is also Camera Link compliant.

The Camera Link standard replaces expensive custom cables with a single, low-cost, standard cable with fewer wires. Special components on the camera are used to serialize 28 parallel TTL signals into four high-speed differential

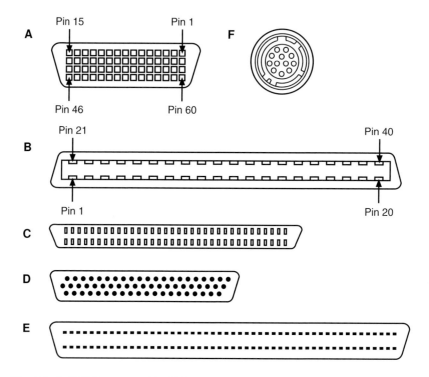

Fig. 7.8 **A.** DVI Connector, **B.** MDR Connector, **C.** VHDCI Connector, **D.** 62-pin High-Density DSUB Connector, **E.** 100-pin SCSI Connector, **F.** 12-pin Hirose Connector.

pairs, which are transmitted across the cable. A similar component is used on the frame grabber to de-serialize the data stream into parallel TTL signals. This reduces cable size and cost, and increases noise immunity and maximum cable length.

The Camera Link specification has come a long way in standardizing digital video transmission by defining the cable, the connector, and the signal functionality. However, Camera Link still leaves many aspects of the camera–frame grabber interface un-addressed. Tap configurations, for example, have not been defined in the initial release of the Camera Link specification. The frame grabber must be configured to properly reconstruct image data from a particular camera. The tap configuration of one camera may not even be supported by a particular frame grabber, although both are compliant with the Camera Link specification.

Although serial signals are defined on the cable pin-out, the specific serial commands for setting exposure, gain, and offset, for example, are not defined by the specification. The frame grabber driver software must be configured to accommodate a particular camera's serial commands. Control signals are also provided on the Camera Link cable for triggering and timing, but many manufacturers provide separate connectors for advanced triggering capabilities. Although the limited scope of the Camera Link specification does not provide Plug and Play compatibility, it gives camera and frame grabber companies an opportunity to differentiate their products by adding features or enhancing functionality.

7.3.7
Camera Link Signals

The Camera Link specification allows for three levels of support: Base, Medium, and Full configuration. A Base configuration camera uses one cable, one connector, and one serializer chip. Four TTL signals are used for timing, and 24 are used for data transfer, or three 8-bit taps. The serializer components run at up to 85 MHz, providing up to 255 MB/s of video throughput.

The timing signals include Frame Valid, Line Valid, Data Valid, and Reserved. Frame Valid and Line Valid are used the same way as in parallel digital transmission. The Data Valid signal asserts for each valid pixel. It can be deasserted to indicate that a particular clock cycle should not be sampled. The Reserved signal functionality is left to the camera manufacturer. Figure 7.9 shows the Camera Link timing signals for an image sensor with two lines, each three pixels wide. Notice that Data Valid deasserts for one clock cycle between Pixel 1 and Pixel 2, making that sample invalid.

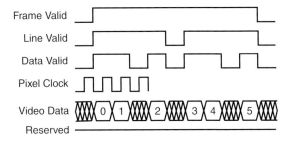

Fig. 7.9 Camera link timing diagram.

In addition to the data and timing signals, the TTL pixel clock is also converted into a fifth differential pair and transmitted along with the data. Camera configuration is supported through two serial data lines for communication to a Universal Asynchronous Receiver–Transmitter (UART). Four additional timing and triggering lines run from the frame grabber to the camera for precise exposure control. The serial and timing signals are also transmitted as differential pairs to provide higher speed and increased noise immunity. The connector and the cable pin-out is defined by the Camera Link specification for all the timing, data, serial, and trigger signals.

Medium configuration was added to the Camera Link specification as an extra layer of support to increase video throughput. 24 additional data signals are added for Medium configuration to support a total of six 8-bit taps. At 85 MHz, Medium configuration Camera Link supports a maximum throughput of 510 MB/s. Two serializer components are required to accommodate the additional data lines. Full configuration Camera Link uses three serializer components and two separate Camera Link cables for a total of ten 8-bit taps and a maximum throughput of 680 MB/s. Pixel data-bit positions for each configuration are defined by the Camera Link specification.

7.3.7.1 Camera Link Connectors

The Camera Link connector is defined in the specification as a 26-pin MDR connector from the 3M Corporation, shown in Fig. 7.10. This connector was designed specifically for high-speed LVDS signals. The camera and frame grabber connections are female, and the cable connections are male. The Camera Link specification defines a maximum cable length of 10 m. The wires within the cable are insulated and differential pairs are twisted and individually shielded to provide better noise immunity. An outer shield is added and tied to a chassis ground.

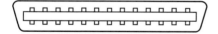

Fig. 7.10 Camera link connector.

7.4 Standard PC Buses

7.4.1 USB

The initial USB 1.0 specification was released in November 1995. The most recent USB 2.0 specification was released on April 27, 2000.

A USB is used to connect external devices to a computer. These external devices include printers, scanners, keyboards, mice, and joysticks, and audio\video devices. Prior to USB, a typical personal computer had the following:

- Two RS-232 serial ports for low bandwidth peripherals like modems
- One parallel port for high bandwidth peripherals such as printers

- Two PS/2 connectors for the keyboard and the mouse
- Other plug-in cards such as an audio capture or SCSI card

Each peripheral used a different connector. Older connectors took up a lot of space at the back of the computer. Older technologies also did not allow you to network peripherals.

A modern personal computer typically has between two and four USB ports. Multiple USB peripherals can be connected to a single USB port with an optional USB hub. A USB allows a theoretical maximum of 127 devices on a single port, but in practice a typical USB hub allows six devices per port.

Plug and Play is another benefit of USB. Plug and Play means that users can add and remove devices without restarting the computer. When a device is added, the device announces its presence to the host computer. At this point, the operating system launches the correct driver and any applications associated with the device.

USB has power on the cable. Low power devices, such as a mouse and keyboard, can draw power from the USB bus without needing an external power source. However, high power devices, such as scanners and printers, still require an external power source.

There is more than one basic connector for USB. Most peripherals that plug directly into a host computer use the flat, or Type A, USB connector shown in Fig. 7.11A. Peripherals that require a separate power cable use a Type A connector on the host side and a square, or Type B, connector on the peripheral side. Figure 7.11B shows a Type B USB connector. The different connector types prevent incorrect plugging of USB devices.

Fig. 7.11 **A.** USB Type A Connector, **B.** USB Type B Connector.

A USB defines four modes of data transfer – control, isochronous, bulk, and interrupt. All types of data transfers can occur simultaneously on the same network.

Control transfer is used for initial configuration of a device. Bulk transfer is used to move large amounts of data when the bus is not active. Interrupt transfer is used to poll peripherals that might need immediate service. Cyclic Redundancy Check (CRC) error checking is used to ensure that all the packets are transferred correctly.

Isochronous transfer guarantees timely delivery of data. No error correction mechanism is used, so the host must tolerate potentially corrupted data. Up to 90% of the available bandwidth can be allocated for isochronous transfers. Table 7.2 lists the specifications for USB.

Tab. 7.2 USB specifications.

Property	Specification
Throughput	1.5 Mbps
	12 Mbps
	480 Mbps
Connectors	Type A (flat shape found on many computers)
	Type B (square shape found on peripherals)
Cables	5 m (copper)
Max bus current	500 mA

7.4.1.1 USB for Machine Vision

Although the USB was developed for the consumer market, it has also been adopted by the industrial market. While USB 1.1 did not have sufficient bandwidth for anything more than a basic Web camera, USB 2.0 has sufficient bandwidth for streaming video. There are many USB 2.0 machine vision cameras in the market. USB 2.0 typically targets machine vision application with relatively low performance requirements and low cost cameras.

The one obstruction to the widespread adoption of USB for machine vision applications was the lack of a hardware specification for video acquisition devices. Each vendor had to implement their own hardware and software design. Different vendor designs prevented devices from different vendors from working together, and forced users to install device specific drivers.

The recently approved USB Video Class specification addresses the device compatibility issue. The new specification promotes interoperability between the hardware and software of different vendors while still allowing vendor-specific extensions.

7.4.2
IEEE 1394 (FireWire®)

The initial IEEE 1394 specification was released in December 1995.

Unlike USB, IEEE 1394 was never intended for basic computer peripherals. The initial speed of IEEE 1394 was 100 Mbps, compared to the 1.5 Mbps of a USB. This higher bandwidth is better suited for devices such as cameras and hard drives.

The Plug and Play feature is another benefit of IEEE 1394. With Plug and Play, users can add and remove devices without restarting the computer. When a device is added, the device announces its presence to the host computer. At this point, the operating system can launches the correct driver and any applications associated with the device.

IEEE 1394 has power on the cable. Low power devices can draw power off the IEEE 1394 bus without the need for an external power source. High power devices still require an external power source.

There are several different connectors available with IEEE 1394 devices. For IEEE 1394a there are 6-pin and 4-pin connectors. The 6-pin connector, shown in Figs. 7.12A, is commonly found on most IEEE 1394 devices and provides power over the cable. The smaller 4-pin connector, shown in Fig. 7.12B, is occasionally found on older laptops and does not provide power over the cable. There is also a latching 6-pin connector, shown in Fig. 7.12C. The latching 6-pin connector is suitable for industrial applications where cables should not be able to be easily unplugged. IEEE 1394b has a 9-pin connector, shown in Fig. 7.12D, which supersedes the 6-pin connector. The 9-pin connector can be used in beta or bilingual applications. IEEE 1394b connectors are dedicated to running at 800 Mbps while bilingual cables can interoperate with IEEE 1394a devices at 400 Mbps. Additionally, IEEE 1394b also specifies a fiber optic connector, shown in Fig. 7.12E, that is useful for applications that require longer cable lengths.

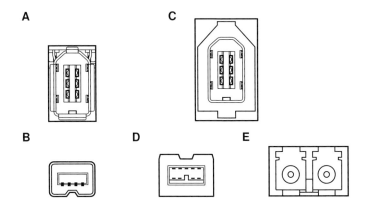

Fig. 7.12 **A.** IEEE 1394 6-pin Connector, **B.** IEEE 1394 4-pin Connector, **C.** IEEE 1394 Latched 6-pin Connector, **D.** IEEE 1394 9-pin Connector, **E.** IEEE 1394 Fiber Optic Connector.

IEEE 1394 defines two modes of data transfer – asynchronous and isochronous. Both types of data transfers can occur simultaneously on the same network. The IEEE 1394 specification assigns 20% of the available bandwidth to asynchronous communication and 80% to isochronous communication.

Asynchronous data transfer guarantees that all data is transferred correctly by means of a Data Acknowledge packet from the recipient. If a device is set up for asynchronous transfer, and it does not get a Data Acknowledge packet after sending data, the device retries the command. This mode of data transfer guarantees that data has been transmitted, but because retries are a possibility, there is no guarantee of bandwidth. Figure 7.13 illustrates asynchronous transfer.

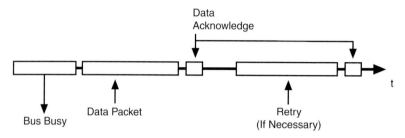

Fig. 7.13 Asynchronous transfer.

The following is an example of guaranteed data integrity, but not bandwidth. Assume there are ten devices on the bus. Nine of the devices are slow and need 100 KB/s. One is fast and requires 6 MB/s. The total amount of data that needs to be transported is approximately 7 MB/s. At first glance, you may think that this setup would not be a problem for a bus that can support 50 MB/s. However, in asynchronous transfer mode, IEEE 1394 uses round-robin bus access. Assume that size of each packet size is 1,024 bytes, and each device takes 25 μs to transfer its data. Every 250 μs, the fast device is allowed to send its 1,024 byte packet, giving a throughput of only 4 MB/s. In this scenario, all of the data from the device would not be transferred.

Isochronous data transfer guarantees bandwidth but not data integrity. As each device is connected to the bus, the devices request bandwidth on the bus. The amount of bandwidth the device requests is determined by the packet size of the device. The bus is divided into 125 μs cycles. Each cycle begins with an asynchronous cycle start packet. Devices that have requested isochronous bandwidth are guaranteed a single packet during each cycle. Once the packets are sent from each device, the remainder of the cycle is left for any other asynchronous transfers that need to take place. Overall, approximately 80% of each cycle is dedicated to isochronous data transfers. Because the recipient does not acknowledge that it has received a packet, there is no guarantee that the data made it across the bus. However, data received at one end is guaranteed to be in time with the data transmitted. In most cases, particularly those configurations that have only two devices on the bus, the transmission is successful. Figure 7.14 illustrates isochronous transfer.

IEEE 1394 devices may be capable of transferring data at certain rates, but since isochronous transfer splits this up into cycles, the actual throughput is calculated differently. Since every cycle of isochronous transfer is 125 μs, you

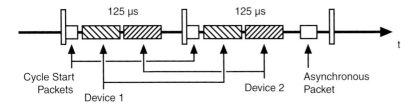

Fig. 7.14 Isochronous transfer.

will have 8,000 cycles in one second. Also, each device sends one packet per cycle. Therefore, the throughput in bytes per second can be determined by the packet size of the device in bytes multiplied by 8,000.

$$BytesPerSecond = (BytesPerPacket \times 8,000) \tag{7.1}$$

With IEEE 1394a, isochronous transfers theoretically account for 320 Mbps. In practice, the bus can transfer a maximum of 4,096 bytes every 125 µs, which is closer to 250 Mbps of bandwidth. With IEEE 1394b, the practical maximum doubles to 8,192 bytes every 125 µs, or 500 Mbps of bandwidth. There is almost zero transfer overhead, as isochronous headers are stripped of hardware and the payload is transferred directly into the system memory. Table 7.3 lists specifications for IEEE 1394.

Tab. 7.3 IEEE 1394 specifications.

Property	Specification
Throughput	400 Mbps (IEEE 1394a)
	800 Mbps (IEEE 1394b)
Connectors	IEEE 1394a 4-pin (no power on bus)
	IEEE 1394a 6-pin
	IEEE 1394a 6-pin latching
	IEEE 1394b 9-pin
	IEEE 1394b fiber optic
Cables	4.5 m (copper twisted pair)
	40 m (plastic fiber)
	100 m (glass fiber)
	100 m (copper Cat5)
Max bus current	1,500 mA (8–30 V)

7.4.2.1 IEEE 1394 for Machine Vision

From the early days of the IEEE 1394 bus, there has been a specification for consumer camcorders (DV) transferring data over IEEE 1394. The specification has led to widespread adoption of IEEE 1394 cameras in the consumer market.

The industrial market did not readily adopt the consumer camcorder specification. The main concern was that the video stream contained compressed images. While the DV specification requires less bandwidth to transfer data on the bus, it requires more processing time to decode the compressed images on the host computer. Additionally, images contain artifacts that are not suitable for the image processing necessary in machine vision applications.

To address the shortcomings of the consumer specification, the 1394 Trade Association formed a working group to define an industrial camera specification. The resulting 1394 Trade Association Industrial and Instrumentation specification for Digital Cameras (IIDC) defines a vendor agnostic hardware register map that allows basic query and control of the camera. Several video and external triggering modes are supported. The vendor agnostic nature of the specification promotes interoperability between different hardware and software.

The IIDC specification supports the following video modes:

- Monochrome (8 bits/pixel)
- Monochrome (16 bits/pixel)
- RGB (24 bits/pixel)
- RGB (48 bits/pixel)
- YUV 4:1:1 (12 bits/pixel)
- YUV 4:2:2 (16 bits/pixel)
- YUV 4:4:4 (24 bits/pixel)
- Bayer (8 bits/pixel)
- Bayer (16 bits/pixel)

It is important to note that IEEE 1394 is a big endian bus. If the host computer that is processing image data is little endian, then the acquisition requires an extra step in either software or hardware to correct the data for the incompatible data platform. In some cases additional steps must be taken to convert a frame for processing. An example of this is the Bayer images that must be converted from raw data to color data.

The IIDC specification uses asynchronous transfers for camera query and control and isochronous transfers for video transfer. IIDC specification scales well between a single fast acquisition and several slower acquisitions. An IEEE 1394b system is capable of supporting a single 640 × 480 8-bit camera running at 200 frames per second (fps), or three 640 × 480 8-bit cameras running at 66 fps. An IEEE 1394a system is capable of running at 100 fps on a single camera or at 33 fps on three identical cameras.

system memory. Onboard camera memory allows a camera to expose the next frame while transferring the current frame to the host computer.

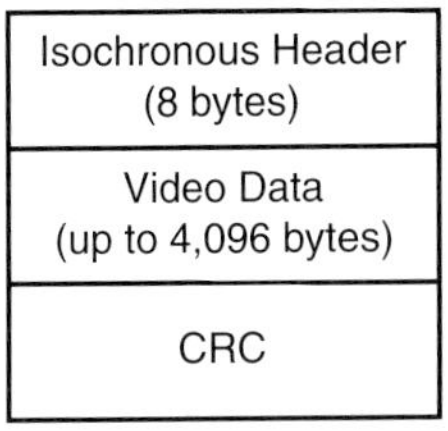

| Isochronous Header (8 bytes) |
| Video Data (up to 4,096 bytes) |
| CRC |

Fig. 7.18 IIDC Isochronous video packet.

Configuring the IEEE 1394 network topology is very important for proper acquisition of a machine vision application. Ideally, you want to create the shortest possible path for the data in order to minimize data loss. Longer cables and breaks in the cable can cause signal degradation and other noise-related issues. Figure 7.19 shows some of the common network topologies used with IEEE 1394.

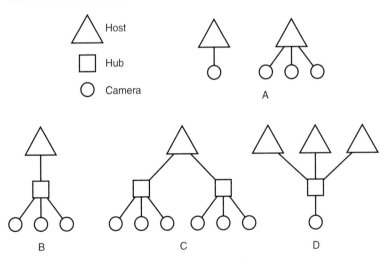

Fig. 7.19 Common IEEE 1394 network topologies.

The most basic setup, Fig. 7.19A, is one or more cameras directly connected to an IEEE 1394 interface card. This topology is probably the one most commonly used. Most IEEE 1394 interface cards have three ports, which is well-suited to the number of cameras that can be used simultaneously without violating the bandwidth limitations of the IEEE 1394.

An IEEE 1394 hub can be added to the topology above (Fig. 7.19B). The hub connects several cameras to a single IEEE 1394 port. Additionally, the hub extends the tethered range of cameras connected to the host computer.

Certain applications require more than three cameras on a single host computer (Fig. 7.19C). It is important to note that many IEEE 1394 host controllers cannot sustain more than four simultaneous isochronous receives. This limits a user to four simultaneously acquiring cameras per interface card. You can add another IEEE 1394 interface card to the system, but you should be cautious of exceeding the host computer bus bandwidth. Plugging several PCI IEEE 1394 adapters into a host computer and running each interface at the full IEEE 1394 bandwidth of 50 MB/s might saturate the PCI bandwidth limit of 133 MB/s.

Isochronous transfers are inherently broadcast transfers. To monitor a single camera from several host computers, connect a single camera to an IEEE 1394 hub (Fig. 7.19D). Connect the hub to several different host computers. One of the host computers will be the designated controller for starting and stopping acquisitions while receiving video data. The remaining computers will be listening stations to receive video data.

7.4.3
Gigabit Ethernet (IEEE 802.3z)

The original IEEE 802.3 Ethernet standard was published in 1985. Ethernet was originally developed by the Xerox Corporation in 1970 and originally ran at 10 Mbps. The 100 Mbps Fast Ethernet protocol was standardized in 1995, and the 1,000 Mbps Gigabit Ethernet protocol in 1998.

The RJ-45 Ethernet connector, shown in Fig. 7.20, is the standard connector found in most computer systems. The benefit of Gigabit Ethernet is that it is backwards compatible with original and fast Ethernet devices using existing connectors and cables. Gigabit Ethernet uses the same CSMA/CD protocol as the Ethernet.

Fig. 7.20 RJ-45 Ethernet connector.

Unlike a USB or the IEEE 1394, Ethernet was not originally intended to connect peripherals. Ethernet does not offer Plug and Play notification. Device discovery requires additional protocols or user intervention. Additionally, no power is provided on the bus. All these shortcomings are planned to be addressed in future revisions of various specifications.

The theoretical maximum bandwidth of the Gigabit Ethernet is 1,000 Mbps. With hardware limitations and software overheads, the practical maximum bandwidth is closer to 800 Mbps. Table 7.5 lists the specifications for the Gigabit Ethernet.

Tab. 7.5 Gigabit Ethernet specifications.

Property	Specification
Throughput	1,000 Mbps
Connectors	RJ-45
Cables	100 m (copper Cat5)
Max bus current	n/a

Ethernet has many derivative protocols like IP, UDP, and TCP, which offer portable transport layers. Each derivative layer is built on top of the previous one. For example, in order to transport user data via UDP, a UDP, IP, and Ethernet header must all be calculated and prefixed to the user data.

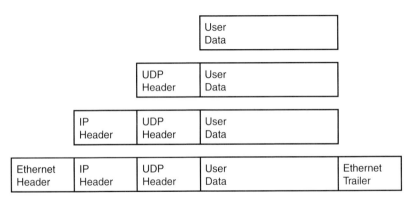

Fig. 7.21 Ethernet video packet.

7.4.3.1 Gigabit Ethernet for Machine Vision

While Ethernet was originally developed for networking computers, it has found some peripheral applications. The high bandwidth and ease of cabling makes the Gigabit Ethernet an attractive option for industrial cameras.

Currently, there is no accepted specification for Gigabit Ethernet industrial cameras; however, a few vendors have proprietary solutions. Many industry leaders are working to define an industrial Gigabit Ethernet camera specification. There are two main options under consideration for transmitting data over Ethernet: a proprietary protocol and an industry network protocol.

With a proprietary protocol, vendors would be free to develop protocols tailored to machine vision. As a benefit, such a specification could run with minimum overheads. The disadvantage is that a proprietary solution would not work with commercial hubs, switches, and routers. However, high speed video applications should not need hubs. Cameras should be directly connected to the host in order to minimize data loss.

With an industry protocol, vendors can build on top of any available protocols such as Ethernet, IP, UDP, or TCP. As a benefit, such a specification can leverage existing proven technology to implement a video stream. With reference designs in hardware and software, it could prove easier to develop than a proprietary specification. One potential problem, however, is that the layered protocols could add overheads in the form of headers that must be parsed and stripped from image data. Figure 7.22 shows an example of both a proprietary data packet and an industry standard data packet.

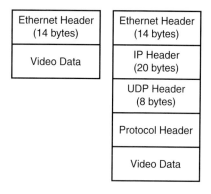

Fig. 7.22 Proprietary and industry protocol video packet.

Configuring the network topology is very important for proper image acquisition in a machine vision application. You should connect the network in such a way as to minimize network collisions and lost packets. Additionally, you should sufficiently shield cameras in an industrially noisy environment. Figure 7.23 shows some of the common topologies used with Ethernet networks.

The most basic configuration, shown in Fig. 7.23A, is to directly connect a camera to a host computer using a crossover cable. This topology will lead to the least amount of network collisions and lost packets. Having a separate network interface for camera communication and network traffic is essential.

Another configuration, shown in Fig. 7.23B, provides a network hub between several cameras and a host computer. The hub should be isolated from any external network. Network collisions can still occur between the connected cameras, but data loss due to lost packets should be minimal.

Yet another possible configuration, shown in Fig. 7.23C, involves connecting one or more cameras to an external network. One thing to consider with this configuration is that using an external network will induce more network collisions and lost packets. There is a good chance that packets may become stalled or fragmented before reaching the host computer based on surround-

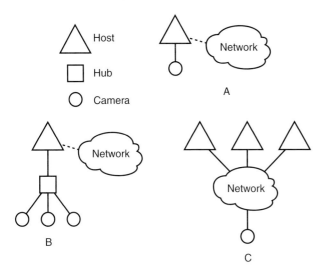

Fig. 7.23 Common ethernet topologies.

ing network traffic. One of the benefits to this configuration is that a camera can be placed far away and it does not need to be directly connected to the host computer. Also, a single camera can be shared between several monitoring stations.

7.5
Choosing a Camera Bus

The first machine vision systems used analog signals to send images from a camera to an acquisition device. For more than fifty years, analog video signal standards have helped ensure compatibility among a wide range of cameras, monitors, and frame grabbers. A camera from any manufacturer would likely work with a frame grabber from another manufacturer, provided the cameras use one of the standard video formats. The same video standards that promoted nearly universal compatibility also limit the performance of the camera for some applications. These common standards also diminished the ability of camera companies to differentiate their products. Since the development of analog video standards, several additional buses of interest for machine vision applications were developed for transmitting image data. Camera Link was developed specifically for machine vision applications, but others have been adapted from existing hardware and specifications. USB, IEEE 1394, and Gigabit Ethernet all utilize technologies well-established in other markets and, through the introduction of additional specifications in recent years, have all been adapted specifically for machine vision applications. Modern machine vision system developers have at least six different camera buses to choose from when designing a system. This section addresses the criteria to consider when choosing a camera bus and identifies which criteria are likely to be the most important for your application.

7.5.1
Bandwidth

The bandwidth, or throughput, of a camera bus is a measure of the rate at which image data is transferred from the camera to the acquisition device. Bandwidth is dictated by the frame rate, image resolution, and amount of data representing one pixel. 8-bit monochrome images are usually represented by one byte per pixel and 10-bit or larger monochrome images are usually represented by two bytes per pixel. Color images are usually transmitted as one to three bytes per pixel. Cameras that output data based on some color mosaic pattern, such as Bayer cameras, transmit data exactly in the same manner as monochrome cameras. Some color cameras output data in RGB format, usually as three bytes per pixel, although color scientific cameras sometimes use two bytes per color plane for a total of 6 bytes per pixel. Some IEEE 1394 cameras use the YUV color space, which requires from 1.5 bytes per pixel, for YUV 4:1:1, to 3 bytes per pixel, for YUV 4:4:4.

Analog cameras are suitable for low to medium bandwidth applications. Pixel clock rates are usually less than 40 MHz and most analog cameras transmit only one pixel per clock cycle. Typical acquisition rates include 640×480 at 60 fps, $1,024 \times 768$ at 30 fps, and $1,392 \times 1,040$ at 15 fps.

IEEE 1394 cameras offer similar or slightly higher throughput to analog cameras, but with much greater flexibility to choose between resolution and frame rate. The IIDC specification for IEEE 1394 cameras defines several standard frame rates that range from 1.875 fps to 240 fps as well as standard resolutions from 160×120 to $1,600 \times 1,200$. The specification also provides for a scalable image format, known as Format 7, which allows for almost any arbitrary resolution and frame rate. Format 7 images are limited only by the available bandwidth on the IEEE 1394 bus and the camera manufacturer's implementation decisions. For many cameras, frame rate varies inversely with image resolution along a roughly constant curve. The IEEE 1394a specification provides a maximum data rate of 400 Mbps, which is enough for a 640×480 8-bit monochrome acquisition at 100 fps. The IEEE 1394b specification doubles the available bandwidth to 800 Mbps and the maximum frame rate at 640×480 to 200 fps. Gigabit Ethernet offers a bandwidth that is similar to IEEE 1394b.

IEEE 1394a cameras are suitable for low to medium bandwidth applications and are available with diverse feature sets from many different manufacturers. IEEE 1394b and Gigabit Ethernet cameras raise the performance bar and provide the additional bandwidth necessary to solve a wider variety of vision applications; however, they are not yet widely available.

Camera Link is a high speed serial digital bus designed specifically for machine vision cameras. The specification defines a standard cable, a connector, a signal format, and a serial communication API for configuring cameras. Camera Link offers a three-tiered bandwidth structure (Base, Medium, and Full) to address a variety of applications. Base configuration cameras are allowed to acquire at up to 255 MB/s (3 bytes × 85 MHz), although the majority of cameras use roughly 100 MB/s or less (1 or 2 bytes × 50 MHz or less). A typical Base configuration camera might acquire 1 Mpixel images at 50 fps or more. Medium and Full configuration cameras acquire at up to 510 MB/s and 680 MB/s, respectively. A typical Full configuration camera can acquire $1,280 \times 1,024$ images at 500 fps, or larger 4 Mpixel images at more than 100 fps.

From a data transfer standpoint, low to medium bandwidth applications are often equally well served by multiple camera buses. Analog and IEEE 1394a cameras accommodate applications that require bandwidth of up to about 40 MB/s. Gigabit Ethernet and IEEE 1394b cameras currently fill the gap between 40 MB/s and about 100 MB/s. Parallel digital cameras (TTL, RS-422, and LVDS) offer performance similar to Base configuration Camera Link and once dominated the high speed image acquisition space. However, few new cameras are being developed with parallel digital connections. For applications beyond 100 MB/s, Medium and Full configuration Camera Link becomes a compelling, and sometimes the only, choice.

7.5.2
Resolution

Resolution is a measure of the ability of an imaging system to discern detail within a scene. Resolution is determined by many factors including optics, lighting, mechanical configuration, and the camera. For the sake of choosing a suitable imaging bus, only the camera itself has much impact. The resolution of a camera can be described by several measures, but generally the most noted specification is the number of pixels on the Charge Coupled Device (CCD) or Complementary Metal Oxide Semiconductor (CMOS) sensor. For area scan cameras, the width and height of the sensor are specified in pixels, while line scan cameras only specify the line width. It is assumed the height of a line scan image is arbitrarily large and limited only by the acquisition device. Most analog cameras, and some digital cameras, specify resolution in TV lines, a unit that takes into account factors other than just the number of pixels on the sensor. Ultimately the video signal will be digitized and the resolution of the camera described by the image resolution stored in digital form in the computer.

The majority of analog cameras adhere to one of a few well-established video standards. In North America and Japan, cameras use NTSC or the equivalent monochrome format, RS-170. Both are 525 lines per frame formats and once digitized, the resulting image is typically 640×480. Many other countries use the PAL or equivalent CCIR monochrome standards. PAL and CCIR are 625-line formats and the resulting digital image is typically 768×576. A wide variety of cameras are available for all of the standard video signal formats. Camera specifications such as sensitivity and signal to noise ratio impact the quality of the acquired image, however, the resolution is fundamentally limited by the signal format. For machine vision applications that require more detail, nonstandard analog cameras are available that offer higher resolution sensors. Sensors up to 2 Mpixel are available in analog cameras; however, the majority of high resolution analog cameras are limited to about 1 Mpixel. Since high resolution cameras do not follow a well-defined signal format, additional connections between the camera and frame grabber are sometimes required for synchronization signals, which increases the complexity of the cable. Analog cameras also use rectangular format sensors, although the aspect ratio will vary for nonstandard cameras.

Applications that require image resolutions greater than 1 Mpixel are generally better served by digital cameras. Digital cameras using Camera Link, IEEE 1394, or other buses are available in a wide variety of resolutions. Digital camera sensors cover all of the standard resolutions used in analog cameras, as well as many smaller and larger sizes. Sensors range in size from less than 128×128, for specialized scientific applications, to larger than 16 Mpixel. Several years ago, a system designer might choose the appropriate sensor for an application and be forced to use whichever bus the camera used. With a much larger selection of cameras currently in the market, many sensors are available with a choice of bus technology. Often the same camera model will be offered with the option of Camera Link or IEEE 1394 output. Parallel digital cameras (TTL, RS-422, and LVDS) are also available in a wide variety of resolutions, although the technology is being replaced by Camera Link.

Ultimately, the choice of a camera bus cannot be made based on resolution alone. Standard resolution cameras are offered with analog connections as well as for all of the digital buses. Only when the required sensor is more than about 1 Mpixel, do analog cameras start to be eliminated from the list of choices. Among the digital formats, there is not a tremendous difference in the resolutions available for Camera Link and IEEE 1394 formats. USB and Gigabit Ethernet cameras are not yet widely available for machine vision applications. Although with the eventual standardization of the Gigabit Ethernet format for imaging, Gigabit Ethernet cameras will likely offer many of the same sensors currently available in other digital formats.

7.5.3
Frame Rate

The maximum frame rate for standard analog cameras is limited by the signal format. NTSC and RS-170 cameras acquire images at a maximum of 30 fps, while PAL and CCIR cameras operate at 25 fps. Some analog cameras designed for high speed operations, acquire at twice the standard rate, or 60 fps and 50 fps for NTSC and PAL formats, respectively.

Digital cameras offer a much wider range of frame rates. Commonly, digital cameras acquire at about 15 to 60 fps at the full image size, although, many cameras are available that acquire beyond this range. At reduced image sizes, both IEEE 1394 and Camera Link can be configured to acquire at 1,000 fps or more. Parallel digital cameras also generally allow for higher frame rates at smaller image sizes, though not with the flexibility of modern cameras.

Choosing a camera bus based on frame rate alone is not practical. Although digital cameras usually offer higher frame rates, analog cameras, and particularly *double speed* cameras, are fast enough to solve many common applications. When the desired frame rate exceeds 60 fps, digital cameras are favored.

7.5.4
Cables

Cabling for analog cameras ranges from simple to complex. The most basic video connection to a standard analog camera requires only a single 75 Ω coaxial cable, often with standard BNC connectors. Nonstandard analog cameras sometimes require additional lines to carry the horizontal and vertical synchronization signals. Cameras in very compact packages often forego the BNC connector and use a standard 12-pin Hirose connector that combines video, timing, and power. The pin arrangement within this connector can vary slightly across manufacturers and even among similar cameras from the same manufacturer, though adapters are usually available. The recommended maximum cable length for analog video signals varies widely. Some sources suggest a length of 10 m or less for the best video quality, while other sources, particularly in security or broadcast video, say that runs of 100 m or more are acceptable with minimal loss in image quality.

IEEE 1394, USB, and Gigabit Ethernet cameras use standard, low cost cables that are widely available. Point to point connections for USB and IEEE 1394a are limited to less than 5 m, with longer distances possible using hubs or repeaters. IEEE 1394 also provides power over the same cable, which eliminates the need for a separate power cable. Gigabit Ethernet and IEEE 1394b cameras support longer single runs up to 100 m over fiber or Cat5 cable. Solutions are also available outside the specification that allow IEEE 1394a to be used across much longer distances over fiber optic cable.

Cables for parallel digital cameras are usually complex, costly, and customized. Since few standards exist for parallel digital cameras, there are a variety of connectors and pin-outs available and users generally do not expect to find an off-the-shelf cable. Due to the parallel data format, cables can have 40 or more wire pairs. Maximum cable length depends on several factors, but for moderate pixel clock rates, connections can be up to 10 m. At higher clock rates, the maximum length can be less than 5 m.

Camera Link uses a standardized cable that is relatively inexpensive. Camera Link cables have standard connectors that work with any Camera Link compliant camera and acquisition device. Base configuration cameras only require one cable. Medium and Full configuration cameras require two cables.

For longer distances and simplified cabling, Gigabit Ethernet is a good choice. At shorter distances, Camera Link and IEEE 1394 are also viable. Camera Link provides the bandwidth for high-speed applications, while IEEE 1394 offers the convenience of Plug and Play. Parallel digital cameras make sense for new applications when there is a compelling cost difference over an equivalent Camera Link camera, or when the camera has a unique feature not available elsewhere. Many scientific and infrared cameras continue to use parallel digital formats.

7.5.5
Line Scan

The majority of line scan cameras use parallel digital or Camera Link connections, with most new cameras offering Camera Link. A few Line scan cameras are also available with analog or IEEE 1394 connections. The trend in line scan sensors towards longer line lengths and higher rates makes Camera Link a good choice. Recently, Full configuration line scan cameras have been introduced with line lengths of more than 12,000 pixels and line rates greater than 20 kHz.

7.5.6
Reliability

Imaging reliability refers to the ability of an imaging system to guarantee acquisition of every image during an inspection process. In many applications such as manufacturing, inspection, and sorting, it is often important that the imaging system acquire and process every image to guarantee 100% inspection. If a fault occurs that causes the acquisition to miss an image, the user should be notified of the event. Events that can cause a system to miss an image include noise on the trigger lines or spurious trigger signals. Analog, parallel digital, and Camera Link image acquisition devices often provide trigger and timing functionality. In a typical system, an external signal from a part sensing device, such as an optical trigger, is connected to the frame grabber. The frame grabber receives the trigger signal and provides a signal to trigger the camera. Routing the trigger signals through the frame grabber provides advanced timing features, such as delayed or multiple pulse generation, or signal conditioning such as converting a 24 V isolated input to a TTL signal for the camera. The trigger signal also informs the frame grabber that an image is expected. If a trigger is received by the frame grabber and an image is not returned from the camera, the driver software can alert the application of a missed image. A missed image can be caused by an electrical problem or by sending triggers to the camera faster than it can accept them. Unlike analog, parallel digital, and Camera Link cameras, USB and IEEE 1394 cameras are usually triggered directly at the camera. The image acquisition system does not process the trigger and is not aware of the one to one relationship between trigger pulses and acquired images. If an IEEE 1394 camera receives a trigger at an invalid time, the camera ignores the trigger and the acquisition system may not be aware that a part was missed. This is one advantage of a frame grabber-based machine vision system.

Deterministic image transfer is also important for an image acquisition system. Determinism describes the confidence with which an image will be received within a specified time after a trigger event. The amount of time required to transfer an image will vary based on the camera resolution and the bus bandwidth, but the delay should be predictable. Camera Link, parallel digital, and analog cameras can usually guarantee deterministic timing due to the point-to-point nature of the bus and predictable camera timing. The transfer of data is controlled by the camera and there is no other data competing for bandwidth on the bus. USB and IEEE 1394 cameras also provide deterministic image transfer. Although various bus topologies are possible with hubs, each device is reserved a certain amount of bandwidth upon initialization. This allows the camera to send slices of data at known time intervals and transfer the image in a fixed time. Ethernet cameras cannot necessarily guarantee the transfer of an image. Network congestion can cause data packets to be dropped, requiring the data to be resent. This increases the transfer time in an unpredictable manner. In the worst case, some of the image data may never be received. For simple network topologies or point-to-point connections, data transfer should be reliable, but acquiring images over a larger network, where the data must contend with other traffic, can be challenging.

7.5.7
Summary of Camera Bus Specifications

Tab. 7.6 Comparison of common camera buses.

Property	Analog	Parallel Digital	Camera Link	IEEE 1394	USB	Ethernet
Throughput	40 MHz	TTL: 40 MB/s, RS422: 80 MB/s, LVDS: 200 MB/s	Base: 255 MB/s, Medium: 510 MB/s, Full: 680 MB/s	IEEE 1394a: 50 MB/s, IEEE 1394b: 100 MB/s	USB 1.1: 1.1 MB/s, USB 2.0: 60 MB/s	125 MB/s
Connectors	BNC, Hirose	AIA 68 pin, proprietary	MDR-26	IEEE 1394a: 4 or 6 pin, IEEE 1394b: 9 pin	Flat, 4 pin	RJ-45
Cable length	10 m	TTL: 2-10 m, RS422: 10 m, LVDS: 10 m	10 m	IEEE 1394a: 4.5 m per segment, 72 m with repeaters, IEEE 1394b: 4.5 m over copper, 40 m over plastic fiber, 100 m over glass fiber	5 m per segment, 30 m with repeaters	100 m per segment over Cat5e
Sensor resolution	640 × 480 to 1,920 × 1,080	16 Mpixel	11 Mpixel	5–8 Mpixel	6 Mpixel	Undetermined
Deterministic image transfer	Yes	Yes	Yes	Yes	Yes	No

7.5.8
Sample Use Cases

7.5.8.1 Manufacturing inspection

The inspection of manufactured goods is increasingly relying on machine vision systems. Verifying the quality of a product or process often includes measuring the size and shape of a part, checking for defects, or searching for the presence of components in an assembly. For example, a company that packages and sells window glass cleaner may need to verify the bottle cap and label placement as well as the fill level in the bottle. All of these tasks can be accomplished using machine vision. In this hypothetical example, the features to inspect are large and the field of view of the camera may only be a few inches diagonal. Feature size and field of view are the two system parameters that determine the necessary camera resolution. These inspections could likely be done with a standard resolution camera. The inspection rate for many applications will be limited by the mechanics of the manufacturing system to considerably less than 30 Hz, and a standard frame rate camera is sufficient. Cameras with this frame rate and resolution are available for any of the camera buses, and the decision will based on other criteria. Cost is almost always critical and analog, USB, and IEEE 1394 each can provide a cost effective solution. Triggering is more commonly found on analog and IEEE 1394 cameras, which may limit USB choices.

7.5.8.2 LCD inspection

Automated optical inspection is used to inspect LCD display panels during manufacturing to provide feedback for process control and to verify the quality of a finished product. Inspecting LCD displays typically requires detecting very small defects over a wide field of view. For example, defects as small as 5 μm must be detected on panels that are often 400 mm wide or more. Field of view and minimum feature size determine the necessary image resolution. Even with the high resolution cameras currently available, such as 11 Mpixel area scan cameras and 12 K line scan cameras, multiple images are required for full coverage. The throughput of the inspection system can be linked directly to the throughput of the acquisition. Camera Link provides the bandwidth necessary for high resolution and high speed inspections. High resolution IEEE 1394 and USB cameras offer similar sensor sizes and are useful where inspection time is limited by processing time or by other steps in the manufacturing process.

7.5.8.3 Security

Security and site monitoring applications present interesting challenges for a camera bus. Typically, several cameras are required and often they are located a considerable distance away from each other and from a monitoring/recording station. USB and IEEE 1394 cameras can be easily networked to accommodate multiple cameras, but the limited cable length severely restricts camera placement. Fiber optic extenders are available to extend the range to hundreds of meters, but adding an extender for each camera is a costly and complex solution. Since Camera Link and parallel digital cameras require point-to-point connections the number of cameras that can be monitored from one station is limited and security monitoring typically would not take advantage of the high bandwidth. Analog and Ethernet cameras are good candidates for this application. Both are capable of covering large distances. Ethernet cameras can be connected in a variety of arrangements, and the inherent networking ability allows many cameras to be controlled and viewed from one station. Likewise, several analog video signals can be connected to a single frame grabber, increasing the number of cameras that can be monitored. Connecting multiple cameras reduces the frame rate from each camera, however, high frame rates may not be required from all cameras simultaneously.

7.6
Computer Buses

Once you have acquired an image from a camera and sent the image data over the camera bus to an interface device, the data must still travel across a second bus, the internal computer bus, before it can be processed by the computer. The most efficient way to provide data to the processor is to first send the data to memory. Sending the data to memory allows the processor to continue working on other tasks while the data is being prepared for processing. Tight integration between the processor and main memory allows the processor to retrieve image data from memory much faster than retrieving it from another location. Using the wrong computer bus for your application can have significant impact on the overall performance of the system.

The following sections describe the most common computer buses for machine vision applications. There are many variations of these computer buses that offer different throughput, availability, and cost. The peak throughput, prevalence, and costs discussed in this section are for the most common and widely available computer buses. Refer to Fig. 7.24 for an overview of common computer buses. In most cases, the peak throughput of a bus will be much higher than the average throughput of the bus, due to the overheads of the bus protocol.

7.6.1
ISA/EISA

When personal computers first became widely available, there was no standard internal computer bus, forcing device manufacturers to make multiple versions of their devices with different interfaces for each brand of computer. As computers became common, developing custom devices became less practical, and a number of computer vendors began using a bus originally developed by IBM. The result of this was the Industry Standard Architecture (ISA) bus, introduced in the mid-1980s. Sometimes ISA buses are referred to by the original IBM name, AT. The ISA bus did not have an agreed-upon specification and initially suffered from compatibility problems. A few years later, a similar, more fully-featured bus was developed called the Extended Industry Standard Architecture (EISA) bus. Depending on the exact version, ISA and

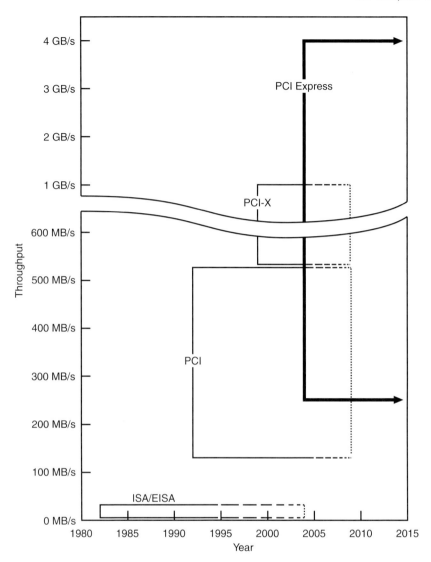

Fig. 7.24 Timeline of common computer buses.

EISA buses deliver a peak throughput of 4.7 to 33 MB/s. By the late 1990s, the superior performance and prevalence of the Peripheral Component Interconnect (PCI) bus had largely eclipsed the ISA and the EISA. It remained common, however, for computers to continue to offer EISA buses for use with legacy devices until the introduction of PCI Express. As computer manufacturers have added support for PCI Express, they have generally eliminated support for EISA. Even in cases where computers are still available with EISA, the low throughput and legacy status of the EISA bus does not make it practical for current machine vision systems.

7.6.2
PCI/CompactPCI/PXI

In the early 1990s, Intel developed a standardized bus called the Peripheral Component Interconnect (PCI) bus. PCI quickly gained acceptance in desktop, workstation, and server computers. PCI provides significantly improved performance over ISA/EISA buses. Initially PCI offered peak throughput of 132 MB/s and a data width of 32 bits. Later, a faster PCI bus was developed by increasing the frequency from 33 MHz to 66 MHz. A second variety increased the data width from 32 to 64 bits. By combining both of these changes, PCI offered a peak throughput of 528 MB/s. Due to higher development and implementation costs, the 64-bit/66 MHz PCI bus has not been as widely adopted as the original PCI implementation. 64-bit/66 MHz PCI buses are primarily limited to high-end computers that require additional performance.

The average throughput that can be achieved with PCI depends significantly on the configuration of the system. With only one device on the bus, the original implementation of PCI offers an average throughput of about 100 MB/s out of the 132 MB/s peak throughput. While the bus can achieve 132 MB/s for short periods of time, continuous use of the bus will yield about 100 MB/s due to some of the transfer time being used to prepare for the data transactions. PCI is a type of bus called a multi-drop bus because it can accomodate multiple devices. However, adding multiple devices to the PCI bus affects the available throughput. Refer to Fig. 7.25 for the diagram of a multi-drop bus configuration.

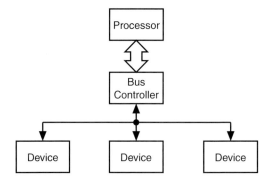

Fig. 7.25 Multi-drop bus configuration.

In a multi-drop bus all of the devices on the bus share the same physical wires to communicate with the processor and the memory. Because there is only one set of wires, only one device can *talk* at a time. This means that the 100 MB/s average throughput must be shared among all the devices on the bus. For example, an application that requires two input devices providing data at 80 MB/s each would not be possible on the 33 MHz/32-bit PCI bus because 2 × 80 MB/s = 160 MB/s which far exceeds the 100 MB/s that can be averaged on the PCI bus. In addition, when switching between multiple devices, a larger amount of throughput is required than the sum of the data rates of the two devices because some of the transfer time will be lost due to the time it takes to switch between the two devices. The exact amount of throughput lost to the overheads of switching will depend on your system configuration. Factors such as the amount and timing of data coming in on the two input devices, and the type of chipset used by the computer all affect the amount of overheads resulting from sharing the bus. Even with the drawback of sharing throughput among all devices, the PCI bus provides sufficient throughput for many machine vision applications.

As PCI became more widespread, extensions were built on top of the original specification to allow PCI to address the needs of specific markets. Two of these extensions, CompactPCI and PCI eXtensions for Instrumentation (PXI), leveraged the existing PCI standard to address the requirements of test, measurement, and automation applications. Timing and triggering capabilities were added to the standard and the mechanical design of the bus was changed to create a more rugged enclosure. The electrical interface and protocol from PCI were reused to provide consistency and reduce both development and implementation time and cost for markets not traditionally served by commodity computer technologies.

With the introduction of PCI Express, computer manufacturers have an incentive to reduce or eliminate traditional PCI device slots in new computers because PCI is no longer natively supported by some of the new chipsets required by the latest processors. Motherboard vendors must add additional circuitry to support PCI with these new chipsets, making the motherboard more costly. However, since PCI has a very large installed customer base, the need for a legacy PCI solution will continue for some time.

7.6.3
PCI-X

PCI-X was developed to provide additional data throughput, and address some design challenges associated with the 66 MHz version of the PCI bus. The initial PCI-X specification, released in 1999, allowed bus speeds of up to 133 MHz with a 64-bit data width to achieve up to 1 GB/s of peak throughput. Like the frequency and width extensions for PCI, PCI-X is primarily only found in server and workstation computers due to the higher implementation cost. Subsequent versions of PCI-X have continued to increase throughput, but they are not widely implemented. Many motherboard manufacturers that offer PCI-X operate the bus at a rate slower than 133 MHz, such as 100 MHz or 66 MHz, because running the bus at 133 MHz limits the motherboard to a single slot per PCI-X bus. Some computer vendors have designed multiple PCI-X buses into the same machine to address this limitation, but doing so further increases the cost of the computer. Running the bus at 100 MHz or 66 MHz reduces the peak throughput to either 800 MB/s or 533 MB/s and allows for two or four device slots per computer.

Like PCI, PCI-X is a multi-drop bus. Offering more than one device slot on the same PCI-X bus does not increase total throughput because the bus resources must be shared between the multiple devices. In order to gain more throughput, an entire second PCI-X bus would have to be connected to the computer chipset. Usually only the server chipsets support this level of functionality. With multiple devices installed in the same PCI-X bus, the percentage of the peak throughput that can be maintained by a device using PCI-X is about 75% of the peak bandwidth.

PCI-X does offer backward compatibility with PCI, which allows the use of devices using PCI bus frequencies and protocols in a PCI-X bus. Taking advantage of this functionality by using a PCI device in a PCI-X slot causes all of the devices to operate at the slower PCI rate. In this case, there is no value in selecting PCI-X over PCI.

Like PCI, the introduction of PCI Express has given computer manufacturers incentive to reduce or eliminate PCI-X slots on new motherboards. PCI-X is no longer natively supported by most new processor chipsets that support PCI Express. To support PCI-X devices on a PCI Express computer, vendors must add additional circuitry on top of the already expensive PCI-X implementation cost. Devices using PCI-X tend to be high-end devices, further adding to the speed of the transition. As with most technology products, the high-end of the device market moves quickly, rendering current high-end devices obsolete much faster than mid or low-level devices. While computers will likely continue to support PCI-X for some more time to allow users to switch from PCI and PCI-X devices over to PCI Express, it is not clear how long this transition period will last.

7.6.4
PCI Express/CompactPCI Express/PXI Express

PCI Express, sometimes referred to as PCIe, was designed to be scalable from desktop computers to servers. While the underlying hardware implementation of the bus was entirely redesigned, the software interface to PCI Express was kept backwards compatible with that of PCI. This was done to speed up the adoption of PCI Express and allow device manufacturers to leverage their existing software investment.

The most basic implementation of PCI Express uses one set of wires to receive data and a second set to transmit data. This is referred to as an x1 interface because one set of wires is used in each direction. Unlike PCI, where the 132 MB/s peak throughput is shared by traffic going both to and from the device, a PCI Express device offers a peak bandwidth of 250 MB/s per direction for the x1 interface. This results in a total throughput of 500 MB/s if the device accesses are evenly divided between sending and receiving data. For most devices, however, this is not the case. For example, the data of a video card is almost entirely being sent from the processor to the device and then out the video port. For a frame grabber, the data is almost entirely in the opposite direction – being sent to the processor as it is acquired from a camera. For this reason, throughput numbers for PCI Express are usually quoted as the throughput per direction, and for machine vision applications, this will be the most relevant figure.

Like the other buses, PCI Express can deliver its peak throughput only in short bursts. For the x1 link, the portion of this throughput that can be averaged by the input device can be above 200 MB/s. The amount of overheads on PCI Express, as a percentage of the peak throughput is generally larger than PCI. This can be particularly true under certain system configurations. One situation that can reduce the average throughput well below 200 MB/s is when only a few bytes of data are sent at a time. On PCI Express, data is transferred in packets. For each packet, additional information is sent along with the data such as the amount of data in the packet and a checksum. This information is used to ensure reliable transfer to the receiving device, but it uses some of the transfer time. Because the amount of overheads information that must be sent along with each packet is the same regardless of the amount of data in the packet, packets with small amounts of data dramatically reduce the achieved throughput. For most machine vision applications, however, the packet size does not dramatically impact performance.

Unlike the multi-drop buses of PCI and PCI-X, where the bus is shared by all devices, PCI Express is a point-to-point bus designed in a tree topology. Refer to Fig. 7.26 for the diagram of a PCI Express point-to-point configuration.

In a PCI Express bus, each device has its own interface to a chip on the motherboard. This interface combines the data from its downstream devices and sends it up the tree until it reaches the access point to the processor and main memory. As long as the available throughput of the connections going toward the processor are well-matched to the combined throughput of the branches, individual PCI Express devices can operate simultaneously without significantly impacting the performance of the other devices.

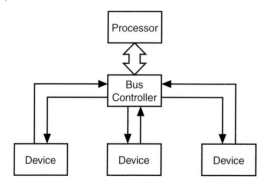

Fig. 7.26 Point-to-point bus configuration.

As was the case for increased throughput on the PCI bus, PCI Express also offers increased throughput by increasing the operating frequency and number of data bits transferred at a time. Unlike PCI, where the earliest computers only implemented the basic configuration, the initial PCI Express computers included slots that supported different levels of throughput. This was achieved through what PCI Express refers to as scalable widths, allowing the computer vendor to offer slots with different levels of performance. The most common widths in initial PCI Express computers were x1, x4, x8, and x16. Interfaces of different widths provide additional throughput in multiples of the 250 MB/s peak and 200 MB/s average offered by the x1 configuration. For example, x16 provides up to 16 times the peak throughput, or about 4 GB/s.

Because of the variety of PCI Express offerings, some care must be taken to avoid potential pitfalls when combining PCI Express devices and slots. When using a device of a particular width in a slot of a matched width, the situation is straightforward and performance will match the throughput of the common width. If, however, the device and slot are not the same width, some investigation might be necessary to avoid performance surprises. Refer to Fig. 7.27 for a diagram of allowable combinations.

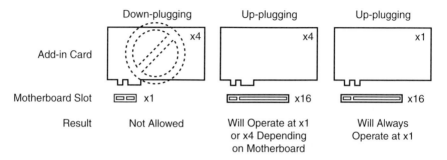

Fig. 7.27 PCI express width interoperability.

In PCI Express, a device may never be used in a slot of smaller width. For example, a x4 device will not fit in a x1 slot. It requires a x4 slot or larger. The opposite configuration, using a device in a larger slot, is called up-plugging and is permitted. In this case, however, the motherboard vendor has the freedom to choose whether it will support the device at its full data rate or only at the lowest x1 rate. This is done to allow motherboard vendors flexibility in controlling costs. Most commonly, it will be desktop computers that limit the device to x1 while workstations and servers support the device at its full rate. Even some desktop machines support the full data rate when a card is used in an up-plugging configuration. If you require the full performance of the device at greater than the x1 rate, verify with the computer manufacturer that the motherboard will support the device at its full rate in the intended slot.

In addition to the basic data transfer functionality that has been offered by previous buses, PCI Express also promises future capability to add support for a variety of advanced features such as regularly timed transfers. While many of these features were not implemented in the first computers to support PCI Express, they are likely to be added as the bus matures.

As with PCI, variations of PCI Express also serve markets with additional needs such as CompactPCI Express and PXI Express for the test, measurement, and automation markets.

7.6.5
Throughput

When selecting a machine vision system, one of the foremost considerations is how many parts can be inspected per minute. If the application only requires inspections for a limited period of time and does not require immediate output from the system during the acquisition, the system designer has more flexibility in system selection. In this scenario, the system can lag somewhat behind on transfer and processing during the acquisition as long as it has enough memory to queue up the incoming images until the acquisition is over.

In the case that a certain rate is required indefinitely, all components of the system must, on average operate at that rate. If any part of the system – the camera, camera bus, input device, computer bus, main memory, or processor – fails to keep pace with the incoming objects, objects will be missed.

Even when components on average operate at the required rate, there will often be periods of interruption when operation falls below average. When systems are designed, each component should be so chosen that its performance at least matches the required rate most of the time. In addition, the system should have some excess capacity to allow the system to catch up from periods of interruption. Refer to Fig. 7.28 for a depiction of such a system.

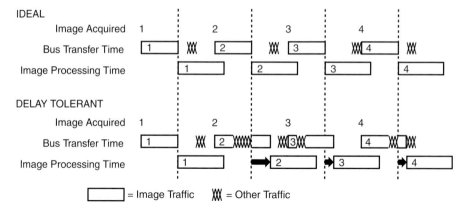

Fig. 7.28 Computer buses that tolerate deviation from the ideal.

The *Ideal* frame of Fig. 7.28 shows an example of a system where the computer bus is operating without interruption. As each image arrives, it is immediately transferred to the processor. The *Delay Tolerant* frame shows a more realistic view of a computer bus. Occasionally, another device on the computer bus, perhaps an Ethernet interface, interrupts image transfer, causing a delay in the data arriving at the processor's memory. Because this system was designed for headroom in addition to the transfer time required in an ideal situation, it is able to catch up and average the desired inspection rate. If, however, the interruption is for a period that is longer than the processor has memory for, an inspection can be lost. When only one buffer is available to store images in the processor memory, a loss of inspection happens as soon as a new image starts transferring while the processing on the previous image is not finished. In Fig. 7.28, a new image could not be sent to the image location in memory, because the processor was still working on the last image.

For the input device and its connection to the computer bus, there are two ways to address this issue. The first solution is to select a bus that either interrupts devices less often or a bus that is fast enough, relative to the required transfer rate to allow the device to quickly catch up once interruptions are over. The second solution is to choose an input device with large banks of memory to store up images, while waiting for access to the bus so that it can withstand the long interruptions of the bus.

When transferring data among devices on the same computer bus, each bus interrupts devices at different times. For example, all PCI and PCI-X devices must share the bus. If more than one device has data to provide at a time, the device must wait its turn, causing delays in starting the transfer of data. In PCI Express, each device has its own interface, and it is only delayed when the interface chips between the device and the computer exceed their larger interfaces to the processor. The primary method for ensuring that the inspections are not missed, however, is to select a computer bus that provides ample throughput to allow the bus to catch up once the device is interrupted. In this regard, the examples to follow will help you make this choice for your system.

The option of using large banks of memory is often used for acquisitions that only need to run for a pre-determined time limit. Using memory to store multiple images before using them is referred to as image buffering. Image buffering can be done on the camera, the input device, or in the main memory. Image buffering in the camera is relatively uncommon because camera manufacturers usually select the camera bus for the camera, to match the output requirements of the camera. When image buffering is implemented in the host memory, this is done because the computer bus is providing image data faster than the processor can process it. Buffering images in the main memory allows images to be queued up for the processor as they become available. This type of buffering will be discussed in the Aquisition Processing section. When buffering at the input device, additional memory in the input device itself is used to withstand delays in the availability of the computer bus.

When using image buffering, it is important to remember that for continuous application, buffering only delays the inevitable if the bus is not fast enough to catch up after interruptions. The system must stay close enough to the average transfer rate so that it never lags behind by more inspections than the number of buffers available. Otherwise, inspections will be missed.

7.6.6
Prevalence and Lifetime

A second important area of consideration during initial system specification is the prevalence and long-term availability of the selected computer bus. Selecting a dying bus limits supplier selection, resulting in difficulties if a decision is made later to deploy additional vision systems on the factory floor, and in the availability of replacement parts. As technologies age, the older technologies tend to become more expensive as they begin their phase-out. Once the technology is no longer available at a reasonable price, the task of adding to an existing line can require an entirely new re-specification process.

With the introduction of PCI Express, ISA/EISA is no longer commonly available in the latest computers and should not be selected for new systems. At this point, the timeline for the transition from PCI and PCI-X to PCI Express is unclear. If you are considering PCI or PCI-X and your system requires uncommon features in the computer or would require significant effort to replace the selected computer, ask your computer vendor for their plans to continue support of PCI and PCI-X.

7.6.7
Cost

Like the range in cost of the camera bus and its associated input device, the selected computer bus also has a cost impact as part of the overall cost of the computer. The first area of cost impact of the computer that offers the desired bus. Such buses, as PCI-X, are not available in desktop computers. If the system could otherwise use a desktop computer, selecting PCI-X would add cost to the system by forcing the use of a workstation or server-class computer. Other buses are available in all classes of computers, but if more than one input device is required in the system, a low-end computer might not provide enough slots for the devices, or it might not offer enough throughput for the combined requirements of the two devices. Again, this can force the purchase of a more expensive computer. Other buses, such as PCI and PCI Express, might be available in the low-cost computer, but not in the width or frequency required to meet the application's throughput needs.

The second area of cost compact is in the cost of the input device. High performance input devices for PCI-X and PCI Express often cost more than their PCI counterparts, although this is likely to change as PCI Express gains in popularity. The input device will also cost more, if it offers a large amount of on-board memory for image buffering, due to the selected bus being unable to maintain the required throughput. In some cases, it can be less expensive to buy a workstation-class computer and use PCI Express as the computer bus than to use a PCI device with memory for image buffering in a desktop computer. While all input devices include a small amount of unadvertised memory to deal with limited bus overheads, the amount of memory required to withstand the long periods of interruption typical when operating at a high inspection rate on a multi-drop bus can go into the megabytes or gigabytes, adding significantly to the cost of the device.

7.7
Choosing a Computer Bus

7.7.1
Determine Throughput Requirements

Determining whether or not a particular computer bus will meet an application's needs begins with gathering information about the system requirements. This section will discuss how to quantify requirements and evaluate the computer bus of the selected input device for appropriate throughput.

7.7 Choosing a Computer Bus | **473**

- *How many inspections per minute does the application require?*
  This will often be limited by the mechanics of the equipment on the factory floor.

- *Must inspections continue at this rate indefinitely or only for a limited period of time?*
  Is it acceptable to miss parts occasionally or must every part be inspected?

- *For the camera that will meet the imaging needs, what will be the width and height of the returned image?*
  If the application does not require the full width and height offered by the camera, can the image returned by the input device or camera be limited only to what is needed? This can reduce the transfer and processing burden.

- *How many pixels does the camera output simultaneously?*
  Sometimes this will be called a tap or the number of channels.

- *How many digitized bits are used to represent each pixel?*
  For digital cameras, this will be included in the camera's specification. Most commonly, a monochrome digital camera will be either 8- or 10-bit. For an RGB camera, this is commonly 8 bits each for red, green and blue. For analog cameras, the digitization of an analog video signal happens at the frame grabber. Some analog frame grabbers might support only 8-bit digitization while others support a selectable number of bits. For IEEE 1394 cameras, this value is often selectable when configuring the camera.

- *How many digitized bits will be transferred across the computer bus by the input device for each pixel?*
  For applications that involve immediate processing of the acquired image, the data will usually be transferred in a byte- or word-aligned format to allow more efficient access to the data by the processor. If using a 10-bit monochrome camera, it is likely that 16 bits (2 bytes) will be transferred, so that the processor will not have to use processing time to separate out the bits that belong to another pixel from the data that it gets from its memory. Likewise, for immediate processing of data from an RGB camera, transferring the three 8-bit color values is often done as 32 bits (4 bytes) so that the processor does not have to read memory twice if the data for a single pixel straddles multiple memory locations.

- *What bus does the proposed input device use?*
  For frame grabbers, this will be described as part of the product's specifications. For USB, IEEE 1394 or Ethernet ports, refer to the documen-

tation for the computer to determine if the port is integrated into the motherboard. Refer to the device documentation if the device is a plug-in device.

- *Are there any other devices sharing the computer bus?*
  On some computers, the integrated Ethernet port will often share the same PCI bus as the device slots. If the network is connected during your acquisitions, the bus throughput available for imaging can be dramatically reduced.

7.7.2
Applying the Throughput Requirements

With the above information, it is possible to determine the rate at which data must be transferred across the computer bus. While the examples in this section are intended to help you select a computer bus for a system that must run continuously, a similar process is applied to determine throughput needs of a system that will only acquire in bursts.
Consider the following example:

7.1 Example A particular application requires 480 inspections per minute that will be continuously acquired and processed. The user's camera evaluation determined that a monochrome camera is needed to produce images that are 1,280 pixels × 1,024 pixels with 10 bits used to represent each pixel. Two pixels are output by the camera at a time. Because the incoming data is 10-bits, the input device will transfer 2 bytes of data per pixel. Because of the continuous acquisition requirement, the selected computer bus must transfer an average of at least 480 images per minute. Each image will be made up of 1,280 × 1,024 = 1,310,720 pixels/image and the bus will transfer two bytes for each pixel, 1,310,720 pixels/image × 2 bytes/pixel = 2,621,440 bytes/image. So, with this information, we know how many bytes the bus will have to transfer in each acquisition interval. Combining this information with the number of images the system must transfer per second gives the data rate that the bus must average. Because bus throughput is specified in the amount of data per second, convert the inspections per minute to inspections per second by dividing the 480 images per minute by 60 s/min. This results in 8 images/s. By multiplying the amount of data per image by the number of images per second you get a data throughput of 2,621,440 bytes/image × 8 images/s = 20,971,520 bytes/s.
Bus throughput is usually given in megabits or megabytes per second, so we will need to convert to this format. One thing to remember is that although kilobytes are usually multiples of 1,024 bytes and megabytes are multiples of 1,024 kilobytes, when bus speeds are specified in megabytes per second, it is

usually in multiples of 1,000 kilobytes per second, which are in turn multiples of 1,000 bytes per second. So, to evaluate our calculated value in terms of the megabytes per second used when talking about computer buses, we will use 1,000 multiples which yield 20.97 MB/s.

When considering multi-drop buses, such as PCI and PCI-X the throughput used by other devices such as an Ethernet port that share the bus would need to be added to this calculated throughput requirement. Assuming that there are no other devices in the system, the throughput needs of this application can be solved by PCI, PCI-X or PCI Express. Lifetime and cost will be the factors that drive the decision.

7.8 Driver Software

The most commonly overlooked component of most machine vision systems is the software used to acquire images from your camera. The software which handles this responsibility is typically referred to as driver software. To get a clear understanding of driver software and how it pertains to your machine vision application, you should have a good understanding of the features of driver software, the acquisition modes possible with driver software, and how images are represented in memory and on disk.

To evaluate any driver software, you should consider the following dimensions: the Application Programming Interface (API), supported platforms, performance, and utility functions.

A typical driver software package has different layers to it each with a specific responsibility. Figure 7.29 shows each of these layers. For the purposes of this discussion, assume that the operating system you are using is a *dual mode* operating system (OS). Windows, Linux and MacOS are examples of dual mode operating systems. A dual mode OS means that the time critical, low level hardware interaction takes place in one partition, known as the kernel mode, while the less critical, high level user interaction takes place in another partition, called the user mode.

In this type of architecture, the lowest level component is the kernel level driver. The kernel level driver is in direct control of the hardware interface regardless of whether the hardware is a frame grabber, IEEE 1394 interface card, USB interface card or Gigabit Ethernet card. The kernel level driver manages detection of the hardware on system startup, directly reads and writes registers in the hardware during configuration and operation, and handles the low level hardware interactions such as servicing interrupts and initiating Di-

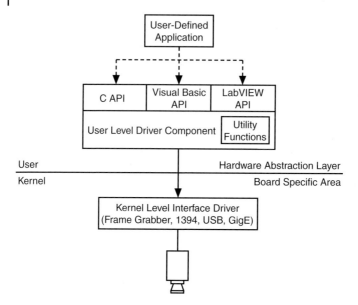

Fig. 7.29 Layers of typical driver software architecture.

rect Memory Access (DMA). The kernel level driver typically encapsulates the hardware interface device.

The next component is referred to as the user level driver. The user level driver interacts with the interface hardware through the kernel level driver and interacts with the user defined application through the Application Programming Interface (API). Typically, the user level driver still has fairly intimate knowledge of the interface hardware, but it also has knowledge of the interface hardware being used to connect to a camera. It also understands the concept of an image. The user level driver manages all the functions of the acquisition which are not time critical, and offloads the high priority tasks to the kernel level driver.

The user level driver will also expose a set of APIs to the user. The API is the set of functions which programmers have at their disposal for developing an application. An API is specific to a programming language such as C, Visual Basic, or LabVIEW and an application will use one of the exposed APIs to interface with the user level driver.

7.8.1
Application Programming Interface

The API is the highest layer of the driver software. The API is the set of functions that programmers have for controlling the interface hardware and the corresponding camera. The API is extremely important, because while the driver can be very powerful and efficient, it is useless unless it exposes those features easily to the programmer through the API.

Most driver software packages will expose more than one API. This is because of the variety of programming languages that programmers use throughout the industry. One of the first things you should look for in a driver software package is that it has support for the programming environment that you want to use. If there are going to be multiple programmers on the project, it is a good idea to make sure that all the necessary APIs are exposed.

Almost all driver software packages should expose a C API since it is one of the most powerful and well-adopted programming languages in the industry. If you are programming in Visual Basic, you should look for a driver software package that exposes an ActiveX interface. ActiveX is a software model which provides a simple-to-program interface to Visual Basic programmers. This software model (and API) is typically less powerful and less efficient than C. However, in many cases it is sufficient for machine vision applications and provides incredible advantages to developers because of the rapid development of user interfaces and applications.

While C and Visual Basic are well-established programming languages in machine vision, there are a couple of languages which are just starting to emerge. The first is VB.NET from Microsoft. This programming environment is intended to replace Visual Basic, although adoption has been slow. While many machine vision vendors are currently providing VB.NET compatibility through *wrappers*, none are providing native VB.NET controls. In the coming years, some software packages may emerge, but for now, Visual Basic is still more widely accepted.

Another exciting programming language for machine vision is LabVIEW. It is only since the late 1990s that LabVIEW has been used in machine vision applications, but the language has been used extensively since the mid-eighties in industries such as Test, Data Acquisition, and Instrument Control. LabVIEW is becoming popular because of the productivity and power it brings to machine vision. Designing a user interface in LabVIEW is very similar to the Visual Basic experience. However, writing code is a very intuitive experience. LabVIEW uses visual blocks to represent functions and "wires" to control program flow. The end result is code that looks very much like a flow chart, which mirrors the way many engineers think. Since LabVIEW has an optimized compiler, which generates raw machine code, it runs almost as efficiently as C-compiled software.

Regardless of which programming language you use, the two most important features of a good API are intuitive and identifiable function names and multi-dimensional scalability. Good function names are necessary for rapid code development and easy readability of the code. If the function name maps well to the concept the programmer can charge ahead without constantly referencing the documentation. The following code snippet provides an example of a good API.

```
/* Initialize variables */
int startNow = TRUE;
int continue = TRUE;
int waitForNext = TRUE;
int resize = TRUE;

/* Open the interface to the camera */
imgInterfaceOpen("img0",\&InterfaceID);

/* Start the image acquisition immediately */
imgGrabSetup(InterfaceID, startNow);

/* Continuously grab the next image and draw it to the screen */
while (continue == TRUE)      \{
   imgGrab (InterfaceID, myImage,waitForNext);
   imgDraw (myImage, windowNumber, resize);
\}

/* Close the interface when done */
imgClose (InterfaceID);
```

The first thing you will notice about this API is that it is very intuitive. The actual function names are very descriptive and someone reading the code can immediately understand what this section of code does even without extensive comments. In this case, we are setting up and initiating a continuous image acquisition with display. First, you initialize resources, next start the acquisition, then copy and display each image, and finally close the resources. Another defining aspect of the API is that each function is easily recognizable as part of this API since all functions have the same prefix. This may seem like an unimportant aspect, but since most applications require linking into multiple APIs, it can be very useful for debugging if you can quickly identify which library a particular function belongs to.

Another important feature of the API is scalability. It is very common for the needs of your application to change between the start of development and the final product. There are multiple dimensions in which you may want to scale your application. For example, you may start out with a low cost analog camera-based solution to your problem. However, as you start working on your processing, you may find that you need a higher resolution camera to resolve the features you need to inspect. With a good API, you should be able to change over to a high resolution Camera Link camera and Camera Link frame grabber without changing your code. You may have a multi-camera application with two frame grabbers and for cost reasons, want to port it to a single, multi-channel board. A major hurdle to porting to this new hardware is the need to rewrite the acquisition portion of your application. With a good API, this migration should not require much in the way of code changes.

Lastly, any good scalable API will offer both high level and low level options. For example, the above code sample was written with a high level API. It is very simple for doing common tasks, but does not provide the flexibility that may be needed for accomplishing advanced tasks.

7.8.2
Supported Platforms

Most machine vision applications today run on a PC. There are a number of operating systems in use today and most of them are appropriate for machine vision. Be sure that the hardware you buy has support for the platform that you intend to use for your application. The most prevalent platform is Windows. With the architectures of Windows 2000 and Windows XP, you can now develop very robust machine vision applications based on Windows.

A more recent operating system to emerge in machine vision is Linux. Linux has proven to be a very popular platform for vision in Europe and is catching on in the United States. Programmers like the open source nature of the OS because it provides transparency to the inner workings of the operating system which can allow developers to work around nuances or bugs that may exist at the lowest levels of the OS. Additionally, Linux is better suited for creating a streamlined environment for machine vision since it is fairly easy to include only the components you need for your solution and none of the additional drivers and utilities that you don't need.

There are also a number of real time operating systems emerging in the market. Real time operating systems such as VxWorks provide a light weight operating system that is extremely reliable and can run on embedded machines consuming very little resources. Many embedded vision systems and smart cameras are now taking advantage of low power processors and real time operating systems to provide machine vision in a highly reliable, small form factor machine vision solution.

7.8.3
Performance

Machine vision applications are typically some of the most demanding applications for a computer. The reason for this is the large size of the data set. A typical machine vision image has a resolution of 1,024 pixels by 1,024 lines, or a total of 1 million pixels. If each of these pixels is represented by 1 byte (8 bits), the total size of the image is 1 MB. You can imagine that an algorithm running on this image to find edges or patterns needs to execute arithmetic operations on each of these one million pixels. Many algorithms even require multiple passes over an image. If your application intends to run at 30 fps, it is not uncommon for your computer to perform tens of millions of instructions per second just to find your pattern.

Given the intense nature of image processing, it is important to minimize the overheads required in getting the image into the PC. This is the role of an efficient driver and hardware. PC architectures provide a mechanism for interface cards to move data into PC memory without using the processor. This mechanism is called Direct Memory Access (DMA). If your interface and driver support DMA, almost all of the overheads associated with copying data to the PC is handled by the hardware and driver and the CPU is free to process the resulting images.

Since copying the image data into the buffer is now handled without processor involvement, there needs to be some sort of notification to the processor that the image transfer is complete. Without this notification, the application running on the CPU would not know when it was safe to start processing the image. One way to implement this notification is with a polling mechanism. This mechanism consists of a loop which regularly checks a status bit of the hardware to see if the operation is complete. Polling is not ideal because it requires too much CPU usage to run the loop that reads the status bit. The CPU is too valuable a resource in most machine vision applications to be used in this manner. For this reason, most interfaces use a mechanism called an interrupt to let the processor know that an image is complete. When an interrupt asserts, the driver calls an interrupt service routine (ISR) to handle that interrupt. A well-written driver will minimize the amount of work that is done in an ISR so that it does not take valuable cycles away from the CPU until necessary. For example, if a frame grabber asserts an interrupt to signal that a frame has just been captured, an efficient ISR would simply increase a counter marking the count of frame which has arrived and perhaps store off the memory address into which that frame is stored. At a later time, the application can retrieve that image and perform a copy or other operations required to prepare the image for processing. In general, the ISR should defer as much as possible until the application requests that it be done.

Ideally, the basic acquisition of images from an interface should use less than 5% of the CPU. You will sometimes see more CPU usage if the camera outputs compressed data that the computer needs to convert. This is very common with color cameras which output images in a Bayer or YUV format, which need to be uncompressed for processing. If this is a concern, you may need to pay a little more for a camera that does not output compressed image data.

7.8.4 Utility Functions

Along with the basic components required for image acquisition, most drivers include utility functions for displaying and saving images. These features, while not a core part of the driver, are necessary components of a vision application. Display provides feedback that the images have been acquired properly and that the field of view and focus are properly set for your scene. The ability to save images allows you to share images with colleagues as well as record images that can be used to test processing algorithms. Later on in this chapter we will discuss image display and image files in more depth.

7.8.5 Acquisition Mode

One of the first things a programmer needs to determine is the correct acquisition mode for their application. There are two primary dimensions you will need to define in order to ensure that they pick the proper acquisition mode. The first question is whether your application will require a single shot or a continuous acquisition. A single shot acquisition runs once and then stops. A continuous acquisition will run indefinitely until you stop the acquisition. The second question is whether the application will require a single buffer or multiple buffers. Buffer refers to the memory space used to hold images and image corresponds to a single frame of data given by the camera. Figure 7.30 shows common names for each acquisition mode and a summary of when to use each one.

7.8.5.1 Snap

The most simple acquisition mode is the snap, which merely acquires a single image into a memory buffer, as shown in Fig. 7.31.

The image can be processed further after the acquisition is complete. The snap is useful when you only need to work with a single image. For example, if you want to take a test image of a part to save to disk and use it as a sample image for developing your image processing operation.

Another common use of the snap is to call the acquisition repeatedly in a software-timed loop. This method is inefficient and will not provide accurate timing, but can be useful when you only need to monitor a scene periodically. It is commonly used in monitoring applications, such as in observing a slowly-varying temperature on an analog meter.

	One Shot	Continuous
Single Buffer	**Snap** • Simple Programming • One Time Image Capture	**Grab** • Simple Programming • Minimal Processing Time Required • Suitable for Display Applications
Multi-Buffer	**Sequence** • High Speed Capture • Any Processing Done Offline • Typically Used for Capturing an Event	**Ring** • High Speed Capture • Processing Done Inline • Provides Buffer Protection During Processing

Fig. 7.30 Selecting the right acquisition mode.

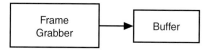

Fig. 7.31 Snap acquisition

7.8.5.2 Grab

In the grab acquisition mode, the frame grabber transfers each image into an acquisition buffer in the system memory. It continually overwrites the same buffer with new frames as long as the acquisition is in progress. The buffers are copied as necessary to a separate processing buffer where analysis or display may take place as shown in Fig. 7.32.

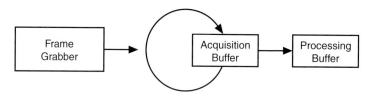

Fig. 7.32 Grab acquisition.

The grab acquisition is the simplest method of displaying a live image in real time. It is most useful in applications where a visual display is all you need. A typical application might involve monitoring a security gate to allow a guard to recognize and grant access to visitors from a remote location.

In some cases, the single acquisition buffer used in a grab is insufficient. If you are trying to acquire and process every single image, any operating system delays can cause frames to be missed. There is also a very short time available for the image to be copied into the processing buffer before the acquisition buffer is overwritten with the next frame. With dynamic scenes, you can clearly see this effect appear as a horizontal discontinuity in the image; one portion of the image comes from the latest frame, while the rest is from the previous frame. In these cases, a ring acquisition is recommended instead.

7.8.5.3 Sequence

A sequence acquisition uses multiple buffers, shown in Fig. 7.33, but writes to them only once. The image acquisition stops as soon as each buffer has been filled once.

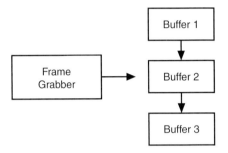

Fig. 7.33 Sequence acquisition.

Sequence acquisition is useful in cases where you need to capture a one-time event which will span over multiple frames. Any processing or display that needs to be done will typically be done after the acquisition is complete. Most applications that use this type of acquisition start the image acquisition on a trigger and then acquire at the maximum camera rate until all buffers are full. An example application would be monitoring a crash test.

7.8.5.4 Ring

A ring acquisition is the most complex, but also the most powerful acquisition mode. A ring uses and recycles multiple buffers in a continuous acquisition. Figure 7.34 shows a ring acquisition.

The interface copies images as they come from the camera into Buffer 1, then Buffer 2, then Buffer 3 and then back to Buffer 1, etc. The ring acquisition is the safest for robust machine vision applications.

This becomes apparent when we start thinking about adding processing into the equation. Inserting processing could cause problems if the algorithm takes more than one frame period to complete. Figure 7.35 illustrates a ring acquisition with processing.

The image processing for any image can begin as soon as a given image has been completely copied into a buffer. With a multi-buffered approach, processing can safely happen at the same time that the next image is being acquired. This would never be safe with a single-buffered approach.

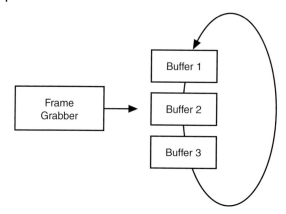

Fig. 7.34 Ring acquisition.

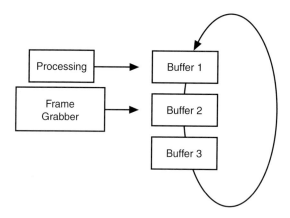

Fig. 7.35 Ring acquisition with processing.

In a single buffer approach, if the processing algorithm *blocks* and does not return until complete, the application may miss a frame. If the algorithm returns immediately, but runs in a separate thread, the next iteration of the acquisition loop may overwrite the image that is being processed. The multi-buffered approach ensures that your application can process one image while acquiring another.

The other benefit of ring acquisition is that if the processing for a particular image takes longer than average to complete, the acquisition can safely jump ahead to the next buffer. On average, however, the processing still does need to keep up with the acquisition. If not, the application will run out of buffers as the acquisition keeps pulling further and further ahead of the processing. However, multiple buffered acquisitions do allow for robustness when there is variability in processing time from image to image. This variability typically exists when the time that the algorithm takes to run is data driven. For example, if the algorithm counts objects and calculates statistics on each object, an image with one object will process much faster than an image with 20 objects. Sometimes the variability has nothing at all to do with the application itself, but with other tasks that the computer must attend to such as running an auto-save or reading a CD-ROM that a user has just inserted. A well-written application will have enough buffers in the acquisition list to accommodate the worst case scenarios.

7.8.6
Image Representation

Like everything else in a computer, images are stored as a series of bits. By understanding the bit structure, developers can better optimize their code for space and performance. The following sections discuss how images are stored both in memory and on disk.

7.8.6.1 Image Representation in Memory

An image is a 2D array of values representing light intensity. For the purposes of image processing, the term *image* refers to a digital image. An image is a function of the light intensity $f(x, y)$, where f is the brightness of the point (x, y) and x and y represent the spatial coordinates of a picture element, or pixel. By convention, the spatial reference of the pixel with the coordinates $(0, 0)$ is located at the top, left corner of the image. Notice in Fig. 7.36 that the value of x increases moving from left to right, and the value of y increases from top to bottom.

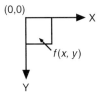

Fig. 7.36 Spatial reference of the $(0, 0)$ pixel.

In digital image processing, an imaging sensor converts an image into a discrete number of pixels. The imaging sensor assigns to each pixel a numeric location and a gray level or color value that specifies the brightness or color of the pixel.

A digitized image has three basic properties: resolution, definition, and number of planes. The spatial resolution of an image is determined by the number of rows and columns of pixels. An image composed of m columns and n rows has a resolution of $m \times n$. This image has m pixels along its horizontal axis and n pixels along its vertical axis.

The definition of an image indicates the number of shades that you can see in the image. The bit depth of an image is the number of bits used to encode the value of a pixel. For a given bit depth of n, the image has an image definition of 2^n, meaning a pixel can have 2^n different values. For example, if n equals 8 bits, a pixel can have 256 different values ranging from 0 to 255. If n equals 16 bits, a pixel can have 65,536 different values ranging from 0 to 65,535 or from 32,768 to 32,767. The manner in which you encode your image depends on the nature of the image acquisition device, the type of image processing you need to use, and the type of analysis you need to perform. For example, 8-bit encoding is sufficient if you need to obtain the shape information of objects in an image. However, if you need to precisely measure the light intensity of an image or region in an image, you must use 16-bit or floating-point encoding. Use color encoded images when your machine vision or image processing application depends on the color content of the objects you are inspecting or analyzing.

The number of planes in an image corresponds to the number of arrays of pixels that compose the image. A grayscale or pseudo-color image is composed of one plane, while a true-color image is composed of three planes – one each for the red, blue, and green components. In true-color images, the color component intensities of a pixel are coded into three different values. A color image is the combination of three arrays of pixels corresponding to the red, green, and blue components in an RGB image. HSL images are defined by their hue, saturation, and luminance values.

Most image processing libraries can manipulate both grayscale and color images. Grayscale images can be represented with either 1 or 2 bytes per pixel and color images can be represented as either RGB or as HSL values. Figure 7.37 shows how many bytes per pixel grayscale and color images use. For an identical spatial resolution, a color image occupies four times the memory space of an 8-bit grayscale image.

A grayscale image is composed of a single plane of pixels. Each pixel is encoded using either an 8-bit unsigned integer representing grayscale values between 0 and 255 or a 16-bit signed integer representing grayscale values between 0 and 65,535. Color image pixels are a composite of four values. RGB images store color information using 8 bits each for the red, green, and blue planes. HSL images store color information using 8 bits each for hue, saturation, and luminance. RGB U64 images store color information using 16 bits each for the red, green, and blue planes. In the RGB and HSL color models, an additional 8-bit value goes unused. This representation is known as 4×8-bit or 32-bit encoding. In the RGB U64 color model, an additional

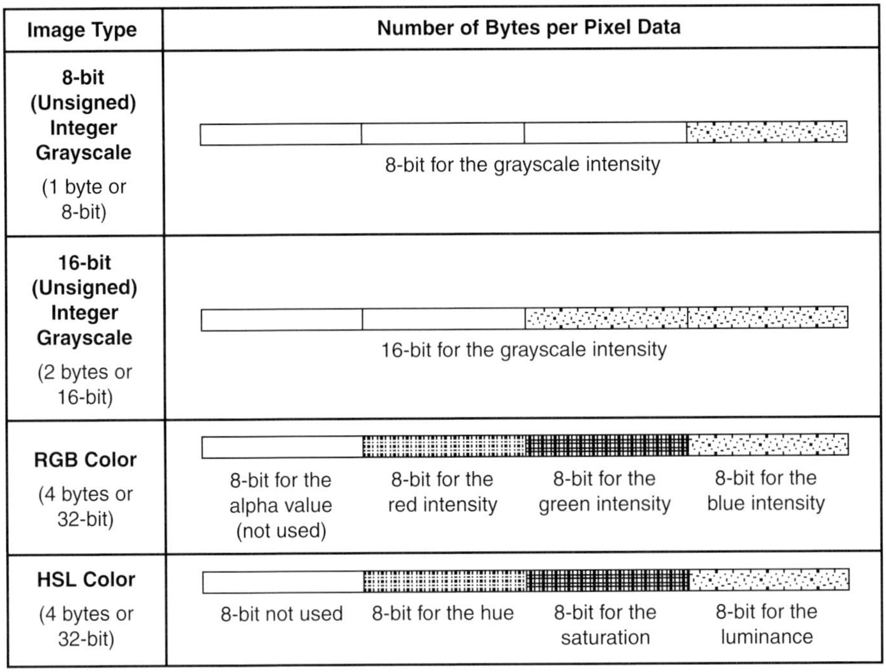

Image Type	Number of Bytes per Pixel Data
8-bit (Unsigned) Integer Grayscale (1 byte or 8-bit)	8-bit for the grayscale intensity
16-bit (Unsigned) Integer Grayscale (2 bytes or 16-bit)	16-bit for the grayscale intensity
RGB Color (4 bytes or 32-bit)	8-bit for the alpha value (not used) | 8-bit for the red intensity | 8-bit for the green intensity | 8-bit for the blue intensity
HSL Color (4 bytes or 32-bit)	8-bit not used | 8-bit for the hue | 8-bit for the saturation | 8-bit for the luminance

Fig. 7.37 Bytes per pixel.

16-bit value goes unused. This representation is known as 4 × 16-bit or 64-bit encoding. Figure 7.38 shows how RGB and HSL store color information.

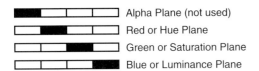

Alpha Plane (not used)
Red or Hue Plane
Green or Saturation Plane
Blue or Luminance Plane

Fig. 7.38 RGB and HSL pixel representation.

Figure 7.39 depicts an example of how an image is represented in the system memory. Please note that this is only one type of representation and the actual representation may vary slightly depending on your software package. In addition to the image pixels, the stored image includes additional rows and columns of pixels called the image border and the left and right alignments. Specific processing functions involving pixel neighborhood operations use image borders. The alignment regions ensure that the first pixel of the image is 8-byte aligned in memory. The size of the alignment blocks depend on the image width and border size. Aligning the image increases processing speed by as much as 30%. The line width is the total number of pixels in a horizontal line of an image, which includes the sum of the horizontal resolution, the image borders, and the left and right alignments. The horizontal resolution and line width may be the same length if the horizontal resolution is a multiple of 8 bytes and the border size is 0.

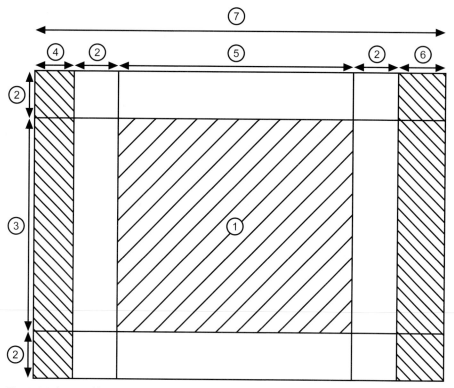

Fig. 7.39 Internal image representation: 1 image, 2 image border, 3 vertical resolution, 4 left alignment, 5 right alignment, 7 line width.

Many image processing functions process a pixel by using the values of its neighbors. A neighbor is a pixel whose value affects the value of a nearby pixel when an image is processed. Pixels along the edge of an image do not have neighbors on all four sides. If you need to use a function that processes pixels based on the value of their neighboring pixels, specify an image border that surrounds the image to account for these outlying pixels. You define the image border by specifying a border size and the values of the border pixels.

The size of the border should accommodate the largest pixel neighborhood required by the function you are using. The size of the neighborhood is specified by the size of a 2D array. For example, if a function uses the eight adjoining neighbors of a pixel for processing, the size of the neighborhood is 3×3, indicating an array with three columns and three rows. Set the border size to be greater than or equal to half the number of rows or columns of the 2D array rounded down to the nearest integer value. For example, if a function uses a 3 × 3 neighborhood, the image should have a border size of at least 1. If a function uses a 5 × 5 neighborhood, the image should have a border size of at least 2.

7.8.6.2 Bayer Color Encoding

Bayer encoding is a method you can use to produce color images using a single imaging sensor, instead of three individual sensors for the red, green, and blue components of light. This technology greatly reduces the cost of cameras. In some higher end cameras, the decoding algorithm is implemented in the camera firmware. This type of implementation guarantees that the decoding works in real time. Other cameras simply output the raw Bayer pattern so that the image you get in memory needs to be decoded by the software to get the expected RGB format.

The Bayer Color Filter Array (CFA) is a primary color, mosaic pattern of 50% green, 25% red, and 25% blue pixels. Green pixels comprise half of the total pixels because the human eye gets most of its sharpness information from green light.

Figure 7.40 describes how the Bayer CFA is used in the image acquisition process.

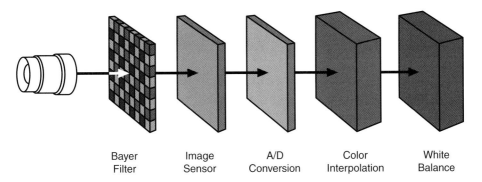

Fig. 7.40 Bayer encoding.

Light travels through the camera lens onto an image sensor that provides one value for each sensor cell. The sensor is an array of tiny, light-sensitive diodes called photosites. The sensor converts light into electrical charges. The sensor is covered by the Bayer CFA so that only one color value reaches any given pixel. The raw output is a mosaic of red, green, and blue pixels of differing intensities. When the image is captured, the accumulated charge for each cell is read, and analog values are converted to digital pixel values using an analog-to-digital converter.

Color interpolation, sometimes referred to as demosaicing, fills in the missing colors. A decoding algorithm determines a value for the RGB components of each pixel in the array by averaging the color values of selected neighboring pixels and producing an estimate of color and intensity. After the interpolation process is complete, the white balancing process further enhances the image by adjusting the red and blue signals to match the green signal in the white areas of the image.

7.2 Example Several decoding algorithms perform color decoding, including nearest neighbor, linear, cubic, and cubic spline interpolations. The following example provides a simple explanation of the interpolation process. Determine the value of the pixel in the center of the following group:

R	G	R
G	B	G
R	G	R

These pixels have the following values:

200	050	220
060	100	062
196	058	198

Neighboring pixels are used to determine the RGB values for the center pixel. The blue component is taken directly from the pixel value, and the green and red components are the average of the surrounding green and red pixels, respectively.

$$R = (200 + 220 + 196 + 198)/4 = 203.5 \sim 204$$
$$G = (50 + 60 + 62 + 58)/4 = 57.5 \sim 58$$
$$B = 100$$

The final RGB value for the pixel is $(204, 58, 100)$. This process is repeated for each pixel in the image.

White balancing is a method you can use to adjust for different lighting conditions and optical properties of the filter. While the human eye compensates for light with a color bias based on its memory of white, a camera captures the real state of light. Optical properties of the Bayer filter may result in mismatched intensities between the red, green, and blue components of the image.

To adjust image colors more closely to the human perception of light, white balancing assumes that if a white area can be made to look white, the remaining colors will be accurate as well. White balancing involves identifying the portion of an image that is closest to white, adjusting this area to white, and correcting the balance of colors in the remainder of the image based on the white area. You should perform white balancing every time lighting conditions change. Setting the white balance incorrectly can cause color inconsistencies in the image.

The white level defines the brightness of an image after white balancing. The values for the red, green, and blue gains are determined by dividing the white level by the mean value of each component color. The maximum white level is 255. If the white level is too high or too low, the image will appear too light or too dark. You can adjust the white level to fine-tune the image brightness.

7.8.6.3 Image Representation on Disk

An image file is composed of a header followed by pixel values. Depending on the file format, the header contains image information about the horizontal and vertical resolution, pixel definition, and the original palette. Image files may also store supplementary information such as calibration, overlays, or timestamps. The following are common image file formats:

Bitmap (BMP) – the bitmap file format represents most closely the image representation in memory. Bitmap files are uncompressed so the size on disk will be very near the size in memory. One benefit of the bitmap format is that you do not lose any data when you save your image.

Tagged Image File Format (TIFF) – a popular file format for bitmapped graphics that stores the information defining graphical images in discrete blocks called tags. Each tag describes a particular attribute of the image. TIFF files support compressed and noncompressed images.

Portable Network Graphics (PNG) – a very flexible file format which support lossless compression as well as support for 16-bit monochrome images and 64-bit color images. PNG also offers the capability of storing additional image information such as calibration, overlays, or timestamps.

Joint Photographic Experts Group (JPEG) – JPEG images offer support for grayscale and color images, and a high level of compression. However, this compression is lossy. Once the image is saved, you can never recover all of the image information.

Standard formats for 8-bit grayscale and RGB color images are BMP, TIFF, PNG, and JPEG. The standard formats for 16-bit grayscale and 64-bit RGB is PNG.

7.8.7
Image Display

Displaying images is an important component of machine vision applications because it gives you the ability to visualize your data. Image processing and image visualization are distinct and separate elements. Image processing refers to the creation, acquisition, and analysis of images. Image visualization refers to how image data is presented and how you can interact with the visualized images. A typical imaging application uses many images in memory that the application never displays.

Use display functions to visualize your image data, to retrieve generated events and the associated data from an image display environment, to select regions of interest from an image interactively, and to annotate the image with additional information.

Display functions display images, set attributes of the image display environment, assign color palettes to image display environments, close image display environments, and set up and use an image browser in image display environments. Some Region of Interest (ROI) functions, a subset of the display functions, interactively define ROIs in image display environments. These ROI functions configure and display different drawing tools, detect draw events, retrieve information about the region drawn on the image display environment, and move and rotate ROIs. Nondestructive overlays display important information on top of an image without changing the values of the image pixels.

7.8.7.1 Understanding Display Modes

One of the key components of displaying images is the display mode that the video adaptor operates. The display mode indicates how many bits specify the color of a pixel on the display screen. Generally, the display mode available from a video adaptor ranges from 8- to 32-bits per pixel, depending on the amount of video memory available on the video adaptor and the screen resolution you choose. If you have an 8-bit display mode, a pixel can be one of 256 different colors. If you have a 16-bit display mode, a pixel can be one of 65,536 colors. In 24-bit or 32-bit display mode, the color of a pixel on the screen is encoded using 3 or 4 bytes, respectively. In these modes, information is stored using 8 bits each for the red, green, and blue components of the pixel. These modes offer the possibility to display about 16.7 million colors.

Understanding your display mode is important to understanding how your software displays the different image types on a screen. Image processing functions often use grayscale images. Because display screen pixels are made of red, green, and blue components, the pixels of a grayscale image cannot be rendered directly. In 24-bit or 32-bit display mode, the display adaptor uses 8 bits to encode a grayscale value, offering 256 gray shades. This color resolution is sufficient to display 8-bit grayscale images. However, higher bit depth images, such as 16-bit grayscale images, are not accurately represented in 24- or 32-bit display mode. To display a 16-bit grayscale image, either ignore the least significant bits or use a mapping function to convert 16 bits to 8 bits.

Mapping functions evenly distribute the dynamic range of the 16-bit image to an 8-bit image. The following techniques describe the common ways in which software convert 16-bit images to 8-bit images and display those images using mapping functions.

Full Dynamic – the minimum intensity value of the 16-bit image is mapped to 0 and the maximum intensity value is mapped to 255. All other values in the image are mapped between 0 and 255 using the following equation:

$$z = \frac{x - y}{v - y} \cdot 255 \qquad (7.2)$$

Where
- z is the 8-bit pixel value
- x is the 16-bit value
- y is the minimum intensity value
- v is the maximum intensity value

The full dynamic mapping method is a good general purpose method because it ensures the display of the complete dynamic range of the image. Because the minimum and maximum pixel values in an image are used to determine the full dynamic range of that image, the presence of noisy or defective pixels with minimum or maximum values can affect the appearance of the displayed image.

Given Range – this technique is similar to the Full Dynamic method, except that the minimum and maximum values to be mapped to 0 and 255 are user defined. You can use this method to enhance the contrast of some regions of the image by finding the minimum and maximum values of those regions and computing the histogram of those regions. A histogram of this region shows the minimum and maximum intensities of the pixels. Those values are used to stretch the dynamic range of the entire image.

Downshifts – this technique is based on shifts of the pixel values. This method applies a given number of right shifts to the 16-bit pixel value and displays the least significant bit. This technique truncates some of the lowest bits, which are not displayed. The downshifts method is very fast, but it reduces the real dynamic of the sensor to 8-bit sensor capabilities. It requires knowledge of the bit-depth of the imaging sensor that has been used. For example, an image acquired with a 12-bit camera should be visualized using four right shifts in order to display the eight most significant bits acquired with the camera.

7.8.7.2 Palettes

At the time a grayscale image is displayed on the screen, the software converts the value of each pixel of the image into red, green, and blue intensities for the corresponding pixel displayed on the screen. This process uses a color table, called a palette, which associates a color to each possible grayscale value of an image.

With palettes, you can produce different visual representations of an image without altering the pixel data. Palettes can generate effects, such as photonegative displays or color-coded displays. In the latter case, palettes are useful for detailing particular image constituents in which the total number of colors is limited.

Displaying images in different palettes helps emphasize regions with particular intensities, identify smooth or abrupt gray-level variations, and convey details that might be difficult to perceive in a grayscale image.

For example, the human eye is much more sensitive to small intensity variations in a bright area than in a dark area. Using a color palette may help you distinguish these slight changes.

A palette is a pre-defined or user-defined array of RGB values. It defines for each possible gray-level value a corresponding color value to render the pixel. The gray-level value of a pixel acts as an address that is indexed into the table, returning three values corresponding to an RGB intensity. This set of RGB values defines a palette in which varying amounts of red, green, and blue are mixed to produce a color representation of the value range. Color palettes are composed of 256 RGB elements. A specific color is the result of applying a value between 0 and 255 for each of the three color components. If the red, green, and blue components have an identical value, the result is a gray level pixel value. A gray palette associates different shades of gray with each value to produce a continuous linear gradation of gray from black to white. You can set up the palette to assign the color black to the value 0 and white to 255, or vice versa. Other palettes can reflect linear or nonlinear gradations going from red to blue, light brown to dark brown, and so on.

7.8.7.3 Nondestructive Overlays

A nondestructive overlay enables you to annotate the display of an image. You can overlay text, lines, points, complex geometric shapes, and bitmaps on top of your image without changing the underlying pixel values in your image; only the display of the image is affected. Figure 7.41 shows how you can use the overlay to depict the orientation of each particle in the image.

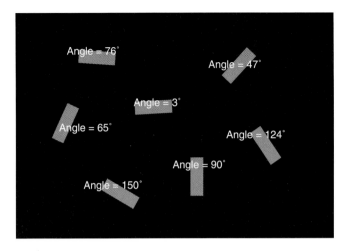

Fig. 7.41 Nondestructive overlay.

You can use nondestructive overlays for many purposes, such as the following:

- Highlighting the location in an image where objects have been detected
- Adding quantitative or qualitative information to the displayed image – like the match score from a pattern matching function
- Displaying ruler grids or alignment marks

Overlays do not affect the results of any analysis or processing functions – they affect only the display. The overlay is associated with an image, so there are no special overlay data types. You only need to add the overlay to your image.

7.9
Features of a Machine Vision System

A machine vision system is composed of a camera connected through a camera bus to an interface device, which interacts with the host PC memory through the computer bus. Machine vision systems use certain features to acquire the pixel data from a camera sensor, manipulate the pixel data, and deliver the image of interest to the host PC memory. While the features are common across machine vision applications, they can be implemented in a variety of ways and in any of the major system components. Machine vision systems use image reconstruction, timing and triggering, memory handling,

LUTs, ROIs, color space conversion, and shading correction to manipulate the image data in preparation for image processing. This section provides an introduction to the purpose, functionality, and implementation options for these features.

7.9.1
Image Reconstruction

There is a disparity between the way camera sensors output pixel data and the way that image processing software expects to receive the image data. Camera sensors output individual pixel data rather than the data for a full image. Therefore, you need a way to properly interpret and orient the data so that a recognizable image is available for image processing or display. Image reconstruction refers to the reordering of the sensor data to create the acquired image in the host memory.

Sensors output image data in what is called a tap, or channel. A tap is defined as a group of data lines that acquire one pixel each. A camera that latches only one pixel on the active edge of the pixel clock is known as a single tap camera. Multi-tap cameras acquire multiple pixels on separate data lines that are available on the same active edge of the pixel clock. Using multiple taps greatly increases the camera's acquisition speed and frame rate.

Camera sensors output data in a variety of configurations. A single tap camera scans data beginning with the top left pixel and moves to the right before proceeding down to the next line, continuing in this manner until the last line is completed for that frame. The camera then starts acquiring the pixel at the top left corner of the next frame. In contrast, the data from each tap of a multi-tap camera can be independently acquired using one of several different configurations. Figure 7.42 depicts several of the possible tap configurations machine vision sensors use.

Image reconstruction can occur at any point in the machine vision system. Since the sensor does not output pixel data in order, the pixel data must be directed to the correct location in memory to reconstruct the image.

Image reconstruction requires a large amount of memory to appropriately reorder the image data. To minimize size and cost, cameras rarely provide the memory needed to store pixel data during image reconstruction. Image reconstruction can be implemented with software on the host PC to reorder the pixel data in memory; however, this increases the amount of memory required in the system. Most commonly, a frame grabber interface device provides on-board memory and logic to perform inline reconstruction of the image.

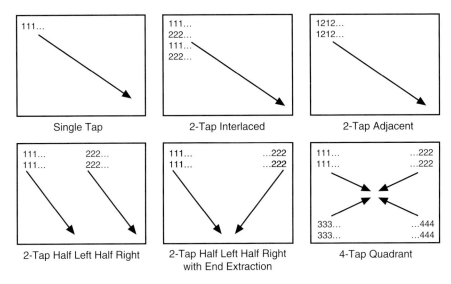

Fig. 7.42 Tap configurations.

7.9.2
Timing and Triggering

Most machine vision systems rely on timing and triggering features to control and synchronize various parts of the system. Timing is the ability to precisely determine when a measurement is taken. Triggering offers the flexibility to couple timing signals between several channels and devices to provide precise control of signal generation and acquisition timing. In machine vision systems, timing and triggering provide a method to coordinate a vision action or function with events external to the computer, such as sending a pulse to strobe lighting or receiving a pulse from a position sensor to signal the presence of an item on an assembly line.

In the simplest system configurations, a trigger is not used, and the camera acquires images continuously. However, if your application only requires certain images, you can use a trigger to acquire a single frame (or line), start an acquisition, or stop an acquisition. These types of triggers are commonly generated by external devices, such as data acquisition boards, position encoders, strobe lights, or limit switches. Triggering can capture repetitive vision events that occur at unknown intervals by starting an acquisition upon each occurrence of the event.

Several triggering configurations are available to acquire images in a machine vision application. Some cameras are designed to connect directly to the output trigger of an external device. This connectivity is simple; however, this configuration provides limited functionality. Since the trigger is passed directly from the external device to the camera, the way that the external device outputs the trigger pulse needs to be compatible with the way that the camera uses the trigger. Without this compatibility, the camera may not acquire the desired image. Alternatively, the output trigger of an external device can be connected to a frame grabber. The frame grabber receives the external device trigger, programmatically determines the appropriate action for that trigger, and sends out a conditioned trigger to the camera. In response to a trigger, the frame grabber can either take no action (trigger disabled), send a pulse to the camera to start a continuous acquisition, send a pulse to the camera to acquire a single image in response to each trigger received, or perform other trigger conditioning. Figure 7.43 depicts a possible configuration for triggering in a machine vision system. In this configuration, the proximity sensor outputs a trigger when an item moves past the sensor. The trigger is routed to the frame grabber, which sends out a trigger to acquire an image with the camera.

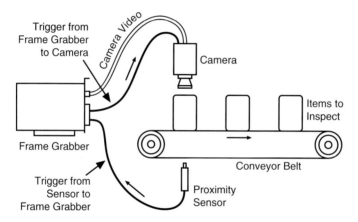

Fig. 7.43 Trigger configuration example.

In many cases, trigger conditioning is required to provide camera control and to ensure that the desired images are acquired for your application. There are many types of trigger conditioning to accomplish different types of control for the system. A delay can be inserted between the times when the frame grabber receives a trigger from the external device and sends a trigger out to another device. For example, a delay could be used in a manufacturing conveyor belt system that uses a proximity sensor located at a distance from the camera. The frame grabber receives a trigger when the object passes by the proximity sensor, applies a delay, and sends a trigger to the camera to start the acquisition when the object will be in front of the camera. For some applications and cameras, the width of the trigger pulse can define the duration of the exposure for the camera's sensor. For advanced triggering in systems, pattern generation provides synchronization of multiple events using multiple trigger lines, such as camera exposure, lighting, and plunger. A frame grabber can also be programmed to ignore triggers received from the external device. This feature is most useful to for ignoring pulses that occur while the camera is still outputting the previous image data.

Ignoring triggers is very useful in applications that use asynchronous reset, a feature provided by many high-end cameras. The term *asynchronous* refers to the fact that the response of the camera to an external input (the trigger) is immediate. This causes the camera to *reset* and capture an image as soon as it receives a trigger. If the frame grabber receives a trigger before the camera completes the transmission of the current frame, the camera resets without completing the current frame, and the application receives a partial frame. For this reason, certain triggers should be ignored to avoid resetting the camera while transmitting data from the previous frame. Additionally, the frame grabber can keep a count of the ignored triggers so that you can determine whether any trigger events did not result in an acquired image.

Applications that use line scan cameras also benefit from triggering in a machine vision system. In contrast to area scan cameras that provide images of a set image size, line scan cameras can theoretically provide a continuous image. For this reason, they are commonly used in document scanning and web inspection applications. The stationary line scan camera acquires a single line of pixels with each exposure, and the object of interest is moved in front of the camera at a rate compatible with the exposure frequency. Each acquired line is combined with previous lines to build the entire image of the object. A lack of synchronization between the acquisition and the object speed can distort the image. If the object speeds up, the image is stretched. If the belt slows down, the image is compressed. Quadrature encoder triggers account for changes in speed of the object of interest. Frame grabbers can provide additional trigger conditioning by triggering the acquisition at different multiples of the quadrature encoder clicks, giving extra control over the acquisition.

Variable height acquisition (VHA) is often used in line scan applications where you do not know the exact size of the object you are acquiring. In VHA mode, you can use an external sensor that indicates whether an object is within the field of view to trigger your image acquisition from this object's present/not present signal. The advantage of this mode is that the acquired image has a *variable height*. This guarantees that your object will fit in one image because the acquisition is based on whether the object is present. In contrast, area scan cameras might acquire only a part of an object because the image size is smaller than the object. Since VHA is an advanced function, it is commonly implemented using triggering with frame grabbers and not with cameras.

Triggering serves an essential role in the integration of machine vision systems. Triggers allow for coordination between vision actions, such as image acquisition, and other events in the system. Frame grabbers provide many benefits in the control and manipulation of triggers through routing and trigger conditioning that are not available with cameras or software alone.

7.9.3
Memory Handling

Machine vision applications are memory intensive due to the amount of data involved in image acquisition and processing. For this reason, you need a location to buffer the acquired image data before transferring or processing the data. Without buffering, you could only acquire at the maximum rate supported by the bottleneck of the system (commonly the camera bus or the computer bus bandwidth). If you exceed this maximum rate, you would have a data overflow and frames would be lost. Memory for image buffering can be provided by the camera, interface device, or system memory.

By allocating a sufficient number of buffers in the system memory, you should be able to prevent losing frames even with an arbitrarily large delay caused by limited bus throughput. Unfortunately, that is not exactly true. To illustrate this scenario, compare the architectures of buffering in the system memory versus buffering in onboard memory on a frame grabber. Figure 7.44 depicts an acquisition into the system memory and Fig. 7.45 shows an acquisition into frame grabber memory.

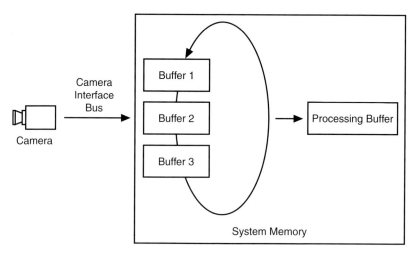

Fig. 7.44 Acquiring an image into the system memory.

By populating memory on the frame grabber, the user can acquire images from a high speed event into the onboard memory and then transfer them to the system memory across the PCI bus more slowly. If there is a delay associ-

7.9 Features of a Machine Vision System

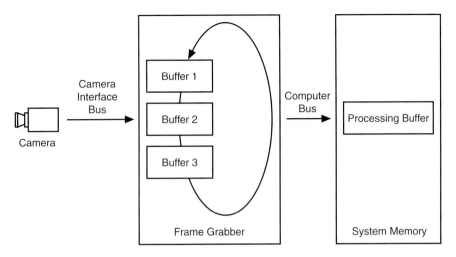

Fig. 7.45 Acquiring an image into memory on a frame grabber.

ated with the PCI bus, the operating system, or image processing, the onboard buffers will continue to be filled until resources are freed to catch up with the acquisition. Cameras can include memory for buffering; however, buffering is most commonly handled by the frame grabber. Most frame grabbers provide some amount of onboard memory, either in the form of a small FIFO buffer designed to hold a few video lines (4 KB), or a larger memory space sufficient to hold several images (16–80 MB). Even if the acquisition is configured to go to the system memory, the onboard memory acts as a buffer to absorb the various delays that might be encountered.

In addition to the buffering advantages, onboard memory allows burst acquisitions at very high rates. Some very high-speed applications, such as tracking a bullet, require extremely high frame rates for short periods of time. The PCI bus can only sustain 100 MB/sec, and the frame grabber can be designed to acquire at a much greater rate from the camera (for example, 32 bits/pixel at 50 MHz resulting in 200 MB/s of data). Onboard memory allows you to store the image data from a high speed event on the frame grabber before transferring the data to the system memory across the PCI bus more slowly. As a result, you can bypass the PCI bus limitation for short duration acquisitions, and guarantee that no frames will be lost until the onboard memory is filled. Using onboard memory is also beneficial when the system memory is limited. In this case, you could acquire the images onto the onboard memory and transfer one buffer at a time into the system memory for processing. This configuration reduces the required system memory because you only hold a single buffer on the host PC at a given time.

Onboard memory can also assist in minimizing the PCI bus traffic used by the machine vision application. Unnecessary traffic across the PCI bus can be avoided by evaluating the processing results of each buffer and discontinuing the transfers as soon as the information of interest has been obtained. Alternatively, the acquisition can store images into onboard memory at the full frame rate and only transfer images across the PCI bus when the user requests them. This solution is commonly used in monitoring applications that don't require every image but do require PCI bandwidth for other data. This provides the most recent image, and the PCI bus is only busy with image transfers when the user requests an image. This results in continuous onboard acquisition that allows the most immediate response to image requests without taking up unnecessary PCI bandwidth.

7.9.4
Additional Features

While all machine vision systems take advantage of image reconstruction, timing and triggering, and memory handling features, there are several other features available to help manipulate the acquired image data to prepare for image processing. These features are commonly used as pre-processing operations to prepare the image for analysis. These operations include look-up tables, regions of interest, color space conversion, and shading correction.

7.9.4.1 **Look-up Tables**
A look-up table (LUT) is a commonly used feature in machine vision systems. LUT transformations are basic image processing functions that improve the contrast and brightness of an image by modifying the dynamic intensity of regions with poor contrast. LUTs efficiently transform one value into another by receiving a pixel value, relating that value to the index of a location in the table, and reassigning the pixel value to the value stored in that table location. Common LUT transformations applied in machine vision applications include data inversion, binarization, contrast enhancement, gamma correction, or other nonlinear transfer functions. Table 7.7 presents an example of a data inversion LUT for an 8-bit grayscale image type.

Commonly, the LUT index value is directly related to the pixel intensity value. In a data inversion LUT, the pixel intensity values are reversed; the lowest intensity pixel value (0, or black) is converted to the highest intensity pixel value (255, or white), and the highest intensity pixel value is converted to the lowest pixel intensity value. When constructing an LUT, a specific output value is determined for each index using an equation or other logic. Figure 7.46 illustrates the results of applying various LUT transformations to an image.

Tab. 7.7 Sample data inversion look-up table.

LUT Index	LUT Output
0	255
1	254
2	253
...	...
253	2
254	1
255	0

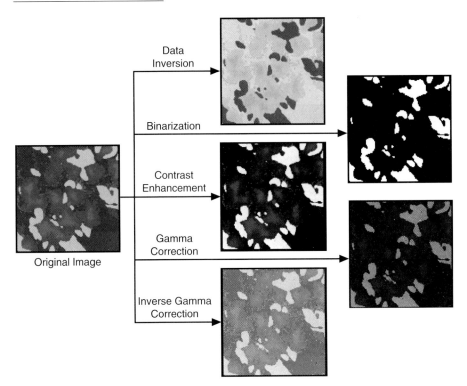

Fig. 7.46 Sample transformations using LUTs.

Many machine vision applications require manipulation of the image to separate the objects of interest from the background. Binarization, also known as thresholding, segments an image into two regions: a particle region and a background region. Binarization is accomplished by selecting a pixel value range, known as the gray-level interval or threshold interval. Binarization works by setting all image pixels that fall within the threshold interval to white and setting all other image pixels to black. The result is a binary image that can easily undergo additional binary image processing algorithms. Since certain detail may be lost in the image, binarization is commonly used in applications where the outline or shape of objects in the image is of interest for processing.

Often, the brightness of an image does not make full use of the available dynamic range. Contrast enhancement is a simple image processing technique that increases the intensity dynamic of an image. Also known as an LUT table, this transfer function increases the contrast by evenly distributing a specified gray-level range over the full gray scale of intensities. For processing, the increased contrast helps to differentiate between the lighter and darker features in the image without losing image details, as happens with binarization.

Gamma correction, or Power of Y transformation, changes the intensity range of an image by simultaneously adjusting the contrast and brightness of an image. This transformation uses a gamma coefficient, Y, to control the adjustment. For Gamma correction transformations, the higher the gamma coefficient, the higher the intensity correction. Gamma correction enhances high intensity pixel values by expanding high gray-level ranges while compressing low gray-level ranges. This decreases the overall brightness of an image and increases the contrast in bright areas at the expense of the contrast in dark areas. Inverse gamma correction, or Power of 1/Y, enhances the low intensity pixel values by expanding low gray-level ranges while compressing high gray-level ranges. This increases the overall brightness of an image and increases the contrast in dark areas at the expense of the contrast in bright areas. Gamma correction is commonly used in applications where the intensity range is too large to be stored or displayed.

Data inversion, binarization, contrast enhancement, and gamma correction are just a sample of the transformations that can be performed with LUTs. LUTs are highly customizable, allowing you to create user-defined LUTs to perform other operations such as gain or offset transformations. LUTs provide an efficient way to perform data transformations because the output values are pre-calculated and stored in the table. Retrieving the table value and processing takes less time than performing the calculation to determine the output value for each pixel in an image. LUTs can be implemented in the hardware to do inline processing of the image, or in the software to process the image after the image is acquired.

7.9.4.2 Region of Interest

A region of interest (ROI) is a user-defined subset of an image. The region of interest identifies the areas of interest for machine vision application, and removes the uninteresting image data that is beyond the ROI. Figure 7.47 illustrates the process and the result of defining a region of interest.

ROIs can be applied by the camera, the interface device, or the host PC. For cameras and interface devices, an ROI is a hardware-programmable rectangular portion of the acquisition window defining the specific area of the image

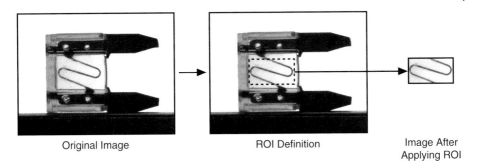

Original Image   ROI Definition   Image After Applying ROI

Fig. 7.47 Region of interest.

to acquire. In an ROI acquisition with hardware, only the selected region is transferred across the PCI bus. As a result, defining an ROI increases the sustained frame rate for the system because the image size is smaller and there is less data per acquisition to transfer across the bus. At the host PC, ROIs are the regions of an image in which you want to focus your image processing and analysis. These regions can be defined using standard contours, such as ovals or rectangles, or freehand contours. In software, the user can define one or more regions to be used for analysis. For image processing, applying an ROI reduces the time needed to perform the algorithms because there is less data to process.

7.9.4.3 Color Space Conversion

A color space is a subspace within a 3D coordinate system where each color is represented by a point. You can use color spaces to facilitate the description of colors between persons, machines, or software programs. Most color spaces are geared toward displaying images with hardware, such as color monitors and printers, or toward applications that manipulate color information, such as computer graphics and image processing.

A number of different color spaces are used in various industries and applications. Humans perceive color according to parameters such as brightness, hue, and intensity, while computers perceive color as a combination of red, green, and blue. The printing industry uses cyan, magenta, and yellow to specify color. Every time you process color images, you must define a color space. The RGB and HSL color spaces are the most common color spaces used in machine vision applications. Table 7.8 lists and describes common color spaces.

The RGB color space is the most commonly used color space. The human eye receives color information in separate red, green, and blue components through cones, the color receptors present in the human eye. These three colors are known as additive primary colors. In an additive color system, the

Tab. 7.8 Common color spaces.

Color Space	Description
RGB	Based on red, green, and blue. Used by computers to display images.
HSL	Based on hue, saturation, and luminance. Used in image processing applications.
CIE	Based on brightness, hue, and colorfulness. Defined by the Commission Internationale de l'Eclairage (International Commission on Illumination)as the different sensations of color that the human brain perceives.
CMY	Based on cyan, magenta, and yellow. Used by the printing industry.
YIQ	Separates the luminescence information (Y) from the color information (I and Q). Used for TV broadcasting.

human brain processes the three primary light sources and combines them to compose a single color *image*. The three primary color components can combine to reproduce the most possible colors. The RGB space simplifies the design of computer monitors, but it is not ideal for all applications. In the RGB color space, the red, green, and blue color components are all necessary to describe a color. Therefore, RGB is not as intuitive as other color spaces.

The HSL color space describes color using the hue component, which makes HSL the best choice for many image processing applications, such as color matching. The HSL color space was developed to put color in terms that are easy for humans to quantify. Hue, saturation, and luminance are characteristics that distinguish one color from another in the HSL space. Hue corresponds to the dominant wavelength of the color. The hue component is a color, such as orange, green, or violet. You can visualize the range of hues as a rainbow. Saturation refers to the amount of white added to the hue and represents the relative purity of a color. A color without any white is fully saturated. The degree of saturation is inversely proportional to the amount of white light added. Colors such as pink, composed of red and white, and lavender, composed of purple and white, are less saturated than red and purple. Brightness embodies the chromatic notion of luminance, or the amplitude or power of light. Chromaticity is the combination of hue and saturation. The relationship between chromaticity and brightness characterizes a color. Systems that manipulate hue use the HSL color space. Overall, two principle factors – the de-coupling of the intensity component from the color information, and the close relationship between chromaticity and human perception of color – make the HSL space ideal for developing machine vision applications.

Variables, such as the lighting conditions, influence which color space to use in your image processing. If you do not expect the lighting conditions to vary considerably during your application, and you can easily define the colors you are looking for using red, green, and blue, use the RGB space. Since the RGB space reproduces an image as you would expect to see it, use the RGB space if you want to display color images, but not process them. If you expect the lighting conditions to vary considerably during your color machine vision application, use the HSL color space. The HSL color space provides more accurate color information than the RGB space when running color processing functions, such as color matching or advanced image processing algorithms.

A machine vision application might require conversion between two color spaces. There are standard ways to convert RGB to grayscale and to convert one color space to another. For example, the transformation from RGB to grayscale is linear. Other transformations from one color space to another, such as the conversion of RGB color space to HSL space, are nonlinear because some color spaces represent colors that cannot be represented in the other color space. Color space conversion often occurs in software on the host PC, or in real time on some color frame grabbers, to prepare the image for processing or display.

7.9.4.4 Shading Correction

Differences in the lighting, camera, and geometry of an object can cause shadows in the acquired image. For example, there might be a bright spot in the center of an image, or there could be a gradient from one side of the image to the other. These variations in pixel intensities can hinder the processing of the image. To demonstrate the impact of shading in image processing, Fig. 7.48 shows an acquired image and the binarization of that image.

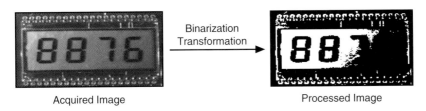

Fig. 7.48 Effects of shading on image processing

The acquired image is brighter on the left than on the right. When a binarization transformation is applied to the image, some of the image information is lost due to this uneven illumination. Shading correction is an image pre-processing function to compensate for uneven illumination, sensor non-linearity, or other factors contributing to the shading of the image. There are several techniques and algorithms that can be used to correct image shading. While certain methods are simple to implement in hardware, cameras do not often perform shading correction. Shading correction is most commonly provided by a frame–grabber interface device or as a pre-processing algorithm executed in the software of the host PC.

7.9.5
Summary

Machine vision systems can employ a variety of features to manipulate image data in preparation for image processing and analysis. Image reconstruction, timing and triggering, memory handling, LUTs, ROIs, color space conversion, and shading correction are common features used in machine vision applications. The host PC, interface device, and camera can provide this functionality; however, there are differences in the implementations that make certain solutions preferable to others.

The primary trade-offs between the implementations are the differences in memory and CPU requirements. Implementing features such as image reconstruction, memory handling, and look-up tables in the camera or interface device requires populating memory on the hardware. While onboard memory increases the cost of the hardware, it also provides a higher performance solution. These types of machine vision system features are less expensive to implement with software on the host PC; however, these solutions are more processor intensive. For these reasons, the best implementation of these features is very system- and application-dependent. Some applications have heavy processing loads, which necessitate efficient methods of image data manipulation. In contrast, other systems have plenty of processing power but require a low cost solution.

Frame grabber hardware provides the most efficient implementation of these features for machine vision systems. Since frame grabber interface devices are specifically meant for machine vision applications, they can be designed to provide many of the image data manipulation features useful in machine vision applications. It is highly efficient for the frame grabber to implement these features in line with the translation of image data from the camera interface bus protocol to the PC memory protocol. As a result, more system processing power is available for use in image processing and analysis.

Cameras that communicate with the host via a standard computer bus, such as IEEE 1394 or USB, cannot rely on the interface device to provide these features. Since these buses are not specific to machine vision applications, they do not implement the features required by machine vision applications. Therefore, the implementation of these features becomes the responsibility of the camera or the host software. Many cameras include some amount of onboard memory to buffer for potential bus delays, and some cameras provide region of interest control. However, to reduce cost, most cameras leave the majority of machine vision features up to the host computer and the software. The resulting system development effort directly relates to the features and ease of use provided by your vision software package. Robust vision software packages will ease your development by providing you with the functions and features needed to quickly get your system up and running.

8
Machine Vision Algorithms

Dr. Carsten Steger, MVTec Software GmbH

In the previous chapters, we have examined the different hardware components that are involved in delivering an image to the computer. Each of the components plays an essential role in the machine vision process. For example, illumination is often crucial to bring out the objects we are interested in. Triggered frame grabbers and cameras are essential if the image is to be taken at the right time with the right exposure. Lenses are important for acquiring a sharp and aberration-free image. Nevertheless, none of these components can "see," i.e., extract the information we are interested in from the image. This is analogous to human vision. Without our eyes we cannot see. Yet, even with eyes we could not see anything without our brain. The eye is merely a sensor that delivers data to the brain for interpretation. To extend this analogy a little further, even if we are myopic we can still see – only worse. Hence, we can see that the processing of the images delivered to the computer by sensors is truly the core of machine vision. Consequently, in this chapter we will discuss the most important machine vision algorithms.

8.1
Fundamental Data Structures

Before we can delve into the study of the machine vision algorithms, we need to examine the fundamental data structures that are involved in machine vision applications. Therefore, in this section we will take a look at the data structures for images, regions, and subpixel-precise contours.

8.1.1
Images

An image is the basic data structure in machine vision, since this is the data that an image acquisition device typically delivers to the computer's memory. As we saw in Chapter 5, a pixel can be regarded as a sample of the energy that

Handbook of Machine Vision. Alexander Hornberg (Ed.)
Copyright © 2006 WILEY-VCH Verlag GmbH & Co. KGaA, Weinheim
ISBN: 3-527-40584-7

falls on the sensor element during the exposure, integrated over the spectral distribution of the light and the spectral response of the sensor. Depending on the camera type, typically the spectral response of the sensor will comprise the entire visible spectrum and optionally a part of the near infrared spectrum. In this case, the camera will return one sample of the energy per pixel, i.e., a single-channel gray value image. RGB cameras, on the other hand, will return three samples per pixel, i.e., a three-channel image. These are the two basic types of sensors that are encountered in machine vision applications. However, in other application areas, e.g., remote sensing, images with a very large number samples per pixel, which sample the spectrum very finely, are possible. For example, the HYDICE sensor collects 210 spectral samples per pixel [64]. Therefore, to handle all possible applications, an image can be considered as a set of an arbitrary number of channels. Intuitively, an image channel can simply be regarded as a two-dimensional array of numbers. This is also the data structure that is used to represent images in a programming language. Hence, the gray value at the pixel (r,c) can be interpreted as an entry of a matrix: $g = f_{r,c}$. In a more formalized manner, we can regard an image channel f of width w and height h as a function from a rectangular subset $R = \{0,\ldots,h-1\} \times \{0,\ldots,w-1\}$ of the discrete 2D plane $\mathbb{Z}^2$ (i.e., $R \subset \mathbb{Z}^2$) to a real number: $f : R \mapsto \mathbb{R}$, with the gray value g at the pixel position (r,c) defined by $g = f(r,c)$. Likewise, a multi-channel image can be regarded as a function $f : R \mapsto \mathbb{R}^n$, where n is the number of channels.

In the above discussion, we have assumed that the gray values are given by real numbers. In almost all cases, the image acquisition device will not only discretize the image spatially, but will also discretize the gray values to a fixed number of gray levels. In most cases, the gray values will be discretized to 8 bits (one byte), i.e., the set of possible gray values will be $\mathbb{G}_8 = \{0,\ldots,255\}$. In some cases, a higher bit depth will be used, e.g., 10, 12, or even 16 bits. Consequently, to be perfectly accurate, a single channel image should be regarded as a function $f : R \mapsto \mathbb{G}_b$, where $\mathbb{G}_b = \{0,\ldots,2^b-1\}$ is the set of discrete gray values with b bits. However, in many cases this distinction is unimportant, so we will regard an image as a function to the set of real numbers.

Up to now, we have regarded an image as a function that is sampled spatially because this is the manner in which we receive the image from an image acquisition device. For theoretical considerations, it is sometimes convenient to regard the image as a function in an infinite continuous domain, i.e., $f : \mathbb{R}^2 \mapsto \mathbb{R}^n$. We will use this convention occasionally in this chapter. It will be obvious from the context which of the two conventions is used.

8.1.2
Regions

One of the tasks in machine vision is to identify regions in the image that have certain properties, e.g., by performing a threshold operation (see Section 8.4). Therefore, we at least need a representation for an arbitrary subset of the pixels in an image. Furthermore, for morphological operations, we will see in Section 8.6.1 that it will be essential that regions can also extend beyond the image borders to avoid artifacts. Therefore, we define a region as an arbitrary subset of the discrete plane: $R \subset \mathbb{Z}^2$.

The choice of the letter R is intentionally identical to the R that is used in the previous section to denote the rectangle of the image. In many cases, it is extremely useful to restrict the processing to a certain part of the image that is specified by a region of interest (ROI). In this context, we can regard an image as a function from the region of interest to a set of numbers: $f : R \mapsto \mathbb{R}^n$. The region of interest is sometimes also called the domain of the image because it is the domain of the image function f. We can even unify both views: we can associate a rectangular ROI with every image that uses the full number of pixels. Therefore, from now on, we will silently assume that every image has an associated region of interest, which will be denoted by R.

In Section 8.4.2, we will also see that often we will need to represent multiple objects in an image. Conceptually, this can simply be achieved by considering sets of regions.

From an abstract point of view, it is therefore simple to talk about regions in the image. It is not immediately clear, however, how to best represent regions. Mathematically, we can describe regions as sets, as in the above definition. An equivalent definition is to use the characteristic function of the region:

$$\chi_R(r,c) = \begin{cases} 1 & (r,c) \in R \\ 0 & (r,c) \notin R \end{cases} \qquad (8.1)$$

This definition immediately suggests to use binary images to represent regions. A binary image has a gray value of 0 for points that are not included in the region and 1 (or any other number different from 0) for points that are included in the region. As an extension to this, we could represent multiple objects in the image as label images, i.e., as images in which the gray value encodes to which region the point belongs. Typically, a label of 0 would be used to represent points that are not included in any region, while numbers > 0 would be used to represent different regions.

The representation of regions as binary images has one obvious drawback: it needs to store (sometimes very many) points that are not included in the region. Furthermore, the representation is not particularly efficient: we need to store at least one bit for every point in the image. Often, the representation actually uses one byte per point because it is much easier to access bytes than bits. This representation is also not particularly efficient for runtime purposes: to determine which points are included in the region, we need to perform a test for every point in the binary image. In addition, it is a little awkward to store regions that extend to negative coordinates as binary images, which also leads to cumbersome algorithms. Finally, the representation of multiple regions as label images leads to the fact that overlapping regions cannot be represented, which will cause problems if morphological operations are performed on the regions. Therefore, a representation that only stores the points included in a region in an efficient manner would be very useful.

Figure 8.1 shows a small example region. We first note that either horizontally or vertically, there are extended runs in which adjacent pixels belong to the region. This is typically the case for most regions. We can use this property and only store the necessary data for each run. Since images are typically stored line by line in memory, it is better to use horizontal runs. Therefore, the minimum amount of data for each run is the row coordinate of the run and the start and end columns of the run. This method of storing a region is called a run-length representation or a run-length encoding. With this representation, the example region can be stored with just four runs, as shown in Fig. 8.1. Consequently, the region can also be regarded as the union of all of its runs:

$$R = \bigcup_{i=1}^{n} \mathbf{r}_i \qquad (8.2)$$

Here, $\mathbf{r}_i$ denotes a single run, which can also be regarded as a region. Note that the runs are stored sorted in lexicographic order according to their row and start column coordinates. This means that there is an order on the runs $\mathbf{r}_i = (r_i, cs_i, ce_i)$ in R defined by: $\mathbf{r}_i \prec \mathbf{r}_j \Leftrightarrow r_i < r_j \vee r_i = r_j \wedge cs_i < cs_j$. This order is crucial for the execution speed of algorithms that use run-length encoded regions.

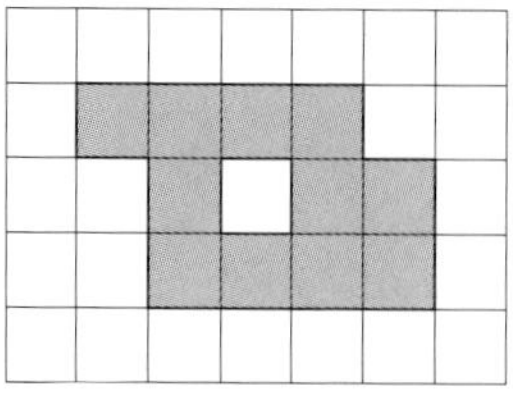

Run	Row	Start Column	End Column
1	1	1	4
2	2	2	2
3	2	4	5
4	3	2	5

Fig. 8.1 Run-length representation of a region.

In the above example, the binary image could be stored with 35 bytes if one byte per pixel is used or with 5 bytes if one bit per pixel is used. If the coordinates of the region are stored as two-byte integers the region can be represented with 24 bytes in the run-length representation. This is already a saving, albeit a small one, compared to binary images stored with one byte per pixel, but no saving if the binary image is stored as compactly as possible with one bit per pixel. To get an impression of how much this representation really saves, we note that we are roughly storing the boundary of the region in the run-length representation. On average, the number of points on the boundary of the region will be proportional to the square root of the area of the region. Therefore, we can typically expect a very significant saving from the run-length representation compared to binary images, which must at least store every pixel in the surrounding rectangle of the region. For example, a full rectangular ROI of a $w \times h$ image can be stored with h runs instead of $w \times h$ pixels in a binary image (i.e., wh or $\lceil w/8 \rceil h$ bytes, depending on whether one byte or one bit per pixel is used). Similarly, a circle with the diameter d can be stored with d runs as opposed to at least $d \times d$ pixels. We can see that the run-length representation often leads to an enormous reduction in memory consumption. Furthermore, since this representation only stores the points actually contained in the region, we do not need to perform a test whether a point lies in the region or not. These two features can save a significant amount of execution time. Also, with this representation it is straightforward to have regions with negative coordinates. Finally, to represent multiple regions, lists or arrays of run-length encoded regions can be used. Since in this case each region is treated separately, overlapping regions do not pose any problems.

8.1.3
Subpixel-Precise Contours

The data structures we have considered so far have been pixel-precise. Often, it is important to extract subpixel-precise data from an image because the application requires an accuracy that is higher than the pixel resolution of the image. The subpixel data can, for example, be extracted with subpixel thresholding (see Section 8.4.3) or subpixel edge extraction (see Section 8.7.3). The results of these operations can be described with subpixel-precise contours. Figure 8.2 displays several example contours. As we can see, the contours can basically be represented as a polygon, i.e., an ordered set of control points (r_i, c_i), where the ordering defines which control points are connected to each other. Since the extraction typically is based on the pixel grid, the distance between the control points of the contour is approximately one pixel on average. In the computer, the contours are simply represented as arrays of floating point row and column coordinates. From Fig. 8.2 we can also see that there is a rich topology associated with the contours. For example, contours can be closed (contour 1) or open (contours 2–5). Closed contours are usually represented by having the first contour point identical to the last contour point or by a special attribute that is stored with the contour. Furthermore, we can see that several contours can meet in a junction point, e.g., contours 3, 4, and 5. It is sometimes useful to explicitly store this topological information with the contours.

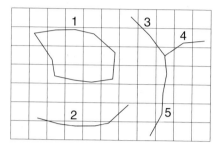

Fig. 8.2 Different subpixel-precise contours. Contour 1 is a closed contour, while contours 2–5 are open contours. Contours 3, 4, and 5 meet in a junction point.

8.2
Image Enhancement

In the preceding chapters, we have seen that we have various means at our disposal to obtain a good image quality. The illumination, lenses, cameras, and image acquisition devices all play a crucial role here. However, although we try very hard to select the best possible hardware setup, sometimes the image quality is not sufficient. Therefore, in this section we will take a look at several common techniques for image enhancement.

8.2.1
Gray Value Transformations

Despite our best efforts in controlling the illumination, in some cases it is necessary to modify the gray values of the image. One of the reasons for this may be a weak contrast. With controlled illumination, this problem usually only occurs locally. Therefore, we may only need to increase the contrast locally. Another possible reason for adjusting the gray values may be that the contrast or brightness of the image has changed from the settings that were in effect when we have set up our application. For example, illuminations typically age and produce a weaker contrast after some time.

A gray value transformation can be regarded as a point operation. This means that the transformed gray value $t_{r,c}$ only depends on the gray value $g_{r,c}$ in the input image at the same position: $t_{r,c} = f(g_{r,c})$. Here, $f(g)$ is a function that defines the gray value transformation to apply. Note that the domain and range of $f(g)$ typically are $\mathbb{G}_b$, i.e., they are discrete. Therefore, to increase the transformation speed, gray value transformations can be implemented as a look-up table (LUT) by storing the output gray value for each possible input gray value in a table. If we denote the LUT as f_g, we have $t_{r,c} = f_g[g_{r,c}]$, where the [] operator denotes the table look-up.

The most important gray value transformation is a linear gray value scaling: $f(g) = ag + b$. If $g \in \mathbb{G}_b$, we need to ensure that the output value is also in $\mathbb{G}_b$. Hence, we must clip and round the output gray value as follows: $f(g) = \min(\max(\lfloor ag + b + 0.5 \rfloor, 0), 2^b - 1)$. For $|a| > 1$ the contrast is increased, while for $|a| < 1$ the contrast is decreased. If $a < 0$ the gray values are inverted. For $b > 0$ the brightness is increased, while for $b < 0$ the brightness is decreased.

Figure 8.3(a) shows a small part of an image of a printed circuit board. The entire image was acquired such that the full range of gray values is used. Three components are visible in the image. As we can see, the contrast of the components is not as good as it could be. Figures 8.3(b)–(e) show the effect of applying a linear gray value transformation with different values for a and b. As we can see from Fig. 8.3(e), the component can be seen more clearly for $a = 2$.

The parameters of the linear gray value transformation must be selected appropriately for each application and must be adapted to changed illumination conditions. Since this can be quite cumbersome, we ideally would like to have a method that selects a and b automatically based on the conditions in the image. One obvious method to do this is to select the parameters such that the maximum range of the gray value space $\mathbb{G}_b$ is used. This can be done as follows: let $g_{\min}$ and $g_{\max}$ be the minimum and maximum gray value in the ROI under consideration. Then, the maximum range of gray values will be used if $a = (2^b - 1)/(g_{\max} - g_{\min})$ and $b = -ag_{\min}$. This transformation can be thought of as a normalization of the gray values. Figure 8.3(f) shows the effect of the gray value normalization of the image in Fig. 8.3(a). As we can see, the contrast is not much better than in the original image. This happens because there are specular reflections on the solder, which have the maximum gray value, and because there are very dark parts in the image with a gray value of almost 0. Hence, there is not much room to improve the contrast.

The problem with the gray value normalization is that a single pixel with a very bright or dark gray value can prevent us from using the desired gray value range. To get a better understanding of this point, we can take a look at the gray value histogram of the image. The gray value histogram is defined as the frequency with which a particular gray value occurs. Let n be the number of points in the ROI under consideration and n_i be the number of pixels that have the gray value i. Then, the gray value histogram is a discrete function

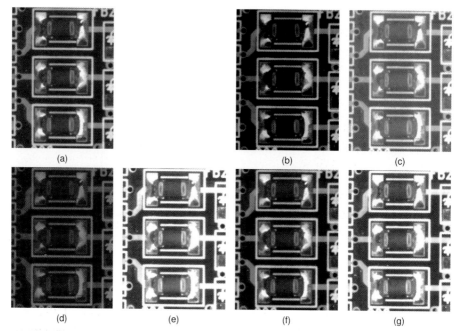

Fig. 8.3 Examples for linear gray value transformations. (a) Original image. (b) Decreased brightness ($b = -50$). (c) Increased brightness ($b = 50$). (d) Decreased contrast ($a = 0.5$). (e) Increased contrast ($a = 2$). (f) Gray value normalization. (g) Robust gray value normalization ($p_l = 0$, $p_u = 0.8$).

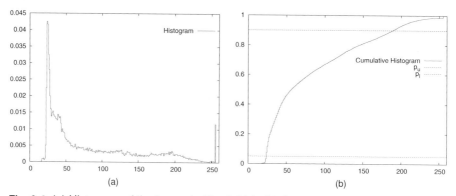

Fig. 8.4 (a) Histogram of the image in Fig. 8.3(a). (b) Corresponding cumulative histogram with probability thresholds p_u and p_l superimposed.

with domain $\mathbb{G}_b$ that has the values

$$h_i = \frac{n_i}{n} \tag{8.3}$$

In probabilistic terms, the gray value histogram can be regarded as the probability density of the occurrence of gray value i. We can also compute the cumulative histogram of the image as follows:

$$c_i = \sum_{j=0}^{i} h_j \tag{8.4}$$

This corresponds to the probability distribution of the gray values. Figure 8.4 shows the histogram and cumulative histogram of the image in Fig. 8.3(a). We can see the specular reflections on the solder create a peak in the histogram at gray value 255. Furthermore, we can see that the smallest gray value in the image is 16. This explains why the gray value normalization did not increase the contrast significantly. We can also see that the dark part of the gray value range contains the most information about the components, while the bright part contains the information corresponding to the specular reflections as well as the printed rectangles on the board. Therefore, to get a more robust gray value normalization, we can simply ignore a part of the histogram that includes a fraction p_l of the darkest gray values and a fraction $1 - p_u$ of the brightest gray values. This can be easily done based on the cumulative histogram by selecting the smallest gray value for which $c_i \geq p_l$ and the largest gray value for which $c_i \leq p_u$. Conceptually, this corresponds to intersecting the cumulative histogram with the lines $p = p_l$ and $p = p_u$. Figure 8.4(b) shows two example probability thresholds superimposed on the cumulative histogram. For the example image in Fig. 8.3(a), it is best to ignore only the bright gray values that correspond to the reflections and print on the board to get a robust gray value normalization. Figure 8.3(g) shows the result that is obtained with $p_l = 0$ and $p_u = 0.8$. As we can see, the contrast of the components is significantly improved.

The robust gray value normalization is an extremely powerful method that is used, for example, as a feature extraction method for optical character recognition (see Section 8.11), where it can be used to make the OCR features invariant to illumination changes. However, it requires transforming the gray values in the image, which is computationally expensive. If we want to make an algorithm robust to illumination changes it is often possible to adapt the parameters to the changes in the illumination, e.g., as described in Section 8.4.1 for the segmentation of images.

8.2.2
Radiometric Calibration

Many image processing algorithms rely on the fact that there is a linear correspondence between the energy that the sensor collects and the gray value in the image: $G = aE + b$, where E is the energy that falls on the sensor and G is the gray value in the image. Ideally, $b = 0$, which means that twice as much energy on the sensor leads to twice the gray value in the image. However, $b = 0$ is not necessary for measurement accuracy. The only requirement is that the correspondence is linear. If the correspondence is nonlinear the accuracy of the results returned by these algorithms typically will degrade. Examples for this are the subpixel-precise threshold (see Section 8.4.3), the gray value features (see Section 8.5.2), and, most notably, subpixel-precise edge extraction (see Section 8.7, in particular Section 8.7.4). Unfortunately, sometimes the gray value correspondence is nonlinear, i.e., either the camera or frame grabber produces a nonlinear response to the energy. If this is the case, and we want to perform accurate measurements, we must determine the nonlinear response and invert it. If we apply the inverse response to the images, the resulting images will have a linear response. The process of determining the inverse response function is known as radiometric calibration.

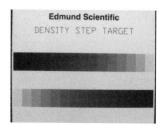

Fig. 8.5 Example of a calibrated density target that is traditionally used for radiometric calibration in laboratory settings.

In laboratory settings, traditionally calibrated targets (typically gray ramps such as the target shown in Fig. 8.5) are used to perform the radiometric calibration. Consequently, the corresponding algorithms are called the chart based. The procedure is to measure the gray values in different patches and to compare them to the known reflectance of the patches. This yields a small number of measurements (e.g., 15 independent measurements in the target in Fig. 8.5), through which a function is fitted, e.g., a gamma response function that includes gain and offset, given by

$$f(g) = (a + bg)^\gamma \qquad (8.5)$$

There are several problems with this approach. First, it requires a very even illumination throughout the entire field of view in order to be able to determine the gray values of the patches correctly. While this may be achievable in laboratory settings, it is much harder to achieve in a production environment, where the calibration often must be performed. Furthermore, effects like vignetting may lead to an apparent light drop off toward the border, which also prevents the extraction of the correct gray values. This problem is always present, independent of the environment. Another problem is that the format of the calibration targets is not standardized, and hence it is difficult to implement a general algorithm for finding the patches on the targets and to determine their correspondence to the true reflectances. In addition, the reflectances on the density targets are often tailored for photographic applications, which means that the reflectances are specified as a linear progression in density, which is related logarithmically to the reflectance. For example, the target in Fig. 8.5 has a linear density progression, i.e., a logarithmic gray value progression. This means that the samples for the curve fitting are not evenly distributed, which can cause the fitted response to be less accurate in the parts of the curve that contain the samples with the larger spacing. Finally, the range of functions that can be modeled for the camera response is limited to the single function that is fitted through the data.

Because of the above problems, a radiometric calibration algorithm that does not require any calibration target is highly desirable. These algorithms are called chart-less radiometric calibration. They are based on taking several images of the same scene with different exposures. The exposure can be varied by changing the aperture of the lens or by varying the shutter time of the camera. Since the aperture can be set less accurately than the shutter time, and since the shutter time of most industrial cameras can be controlled very accurately in software, varying the shutter time is the preferred method of acquiring images with different exposures. The advantages of this approach are that no calibration targets are required and that they do not require an even illumination. Furthermore, the range of possible gray values can be covered with multiple images instead of a single image, as required by the algorithms that use calibration targets. The only requirement on the image content is that there should be no gaps in the histograms of different images within the gray value range that each image covers. Furthermore, with a little extra effort, even overexposed (i.e., saturated) images can be handled.

To derive an algorithm for chart-less calibration, let us examine what two images with different exposures tell us about the response function. We know that the gray value G in the image is a nonlinear function r of the energy E that falls onto the sensor during the exposure e [63]:

$$G = r(eE) \qquad (8.6)$$

Note that e is proportional to the shutter time and proportional to the aperture of the lens, i.e., proportional to $(1/k)^2$, where k is the f-stop of the lens. As described above, in industrial applications we typically leave the aperture constant and vary the shutter time. Therefore, we can think of e as the shutter time.

The goal of the radiometric calibration is to determine the inverse response $q = r^{-1}$. The inverse response can be applied to an image via a LUT to achieve a linear response.

Now, let us assume that we have acquired two images with different exposures e_1 and e_2. Hence, we know $G_1 = r(e_1 E)$ and $G_2 = r(e_2 E)$. By applying the inverse response q to both equations, we obtain $q(G_1) = e_1 E$ and $q(G_2) = e_2 E$. We can now divide both equations to eliminate the unknown energy E, and obtain:

$$\frac{q(G_1)}{q(G_2)} = \frac{e_1}{e_2} = e_{1,2} \tag{8.7}$$

As we can see, q only depends on the gray values in the images and the ratio $e_{1,2}$ of the exposures, and not the exposures e_1 and e_2 themselves. Equation (8.7) is the defining equation for all chart-less radiometric calibration algorithms.

One way to determine q based on (8.7) is to discretize q in a look-up table. Thus, $q_i = q(G_i)$. To derive a linear algorithm to determine q, we can take logarithms on both sides of (8.7) to obtain $\log(q_1/q_2) = \log e_{1,2}$, i.e., $\log(q_1) - \log(q_2) = \log e_{1,2}$ [63]. If we set $Q_i = \log(q_i)$ and $E_{1,2} = \log e_{1,2}$, each pixel in the image pair yields one linear equation for the inverse response function Q:

$$Q_1 - Q_2 = E_{1,2} \tag{8.8}$$

Hence, we obtain a linear equation system $AQ = E$, where Q is a vector of the LUT for the logarithmic inverse response function, while A is a matrix with 256 columns for byte images. A and E have as many rows as pixels in the image, e.g., 307200 for a 640 × 480 image. Therefore, this equation system is much too large to be solved in acceptable time. To derive an algorithm that solves the equation system in acceptable time, we note that each row of the equation system has the following form:

$$(\,0 \;\ldots\; 0 \;\; 1 \;\; 0 \;\ldots\; 0 \;\; -1 \;\; 0 \;\ldots\; 0\,)Q = E_{1,2} \tag{8.9}$$

The indices of the 1 and -1 entries in the above equation are determined by the gray values in the first and second image. Note that each pair of gray values that occurs multiple times leads to several identical rows in A. Also note that $AQ = E$ is an overdetermined equation system, which can be solved through the normal equations $A^\top A Q = A^\top E$. This means that each row that occurs k times in A will have the weight k in the normal equations. The same behavior is obtained by multiplying the row (8.9) that corresponds to the gray value pair by $\sqrt{k}$ and to include that row only once in A. This typically reduces the number of rows in A from several hundred thousand to a few thousand, and thus makes the solution of the equation system feasible.

The simplest method to determine k is to compute the 2D histogram of the image pair. The 2D histogram determines how often gray value i occurs in the first image, while gray value j occurs in the second image at the same position. Hence, for byte images the 2D histogram is a 256×256 image in which the column coordinate indicates the gray value in the first image, while the row coordinate indicates the gray value in the second image. It is obvious that the 2D histogram contains the required values of k. We will see examples for the 2D histograms below.

Note that the discussion so far has assumed that the calibration is performed from a single image pair. It is, however, very simple to include multiple images in the calibration since additional images provide the same type of equations as in (8.9), and can thus simply be added to A. This makes it much easier to cover the entire range of gray values. Thus, we can start with a fully exposed image and successively reduce the shutter time until we reach an image in which the smallest possible gray values are assumed. We could even start with a slightly overexposed image to ensure that the highest gray values are assumed. However, in this case we have to take care that the overexposed (saturated) pixels are excluded from A because they violate the defining Eq. (8.7). This is a very tricky problem to solve in general since some cameras exhibit a bizarre saturation behavior. For many cameras it is sufficient to exclude pixels with the maximum gray value from A.

Despite the fact that A has many more rows than columns, the solution Q is not uniquely determined because we cannot determine the absolute value of the energy E that falls onto the sensor. Hence, the rank of A is at most 255 for byte images. To solve this problem, we could arbitrarily require $q(255) = 255$, i.e., scale the inverse response function such that the maximum gray value range is used. Since the equations are solved in a logarithmic space it is slightly more convenient to require $q(255) = 1$ and to scale the inverse response to the full gray value range later. With this, we obtain one additional equation of the form

$$(0 \quad \ldots \quad 0 \quad k\,)Q = 0 \qquad (8.10)$$

To enforce the constraint $q(255) = 1$, the constant k must be chosen such that (8.10) has the same weight as the sum of all other Eqs. (8.9), i.e., $k = \sqrt{wh}$, where w and h are the width and height of the image, respectively.

Even with this normalization, we still face some practical problems. One problem is that if the images contain very little noise the equations in (8.9) can become decoupled, and hence do not provide a unique solution for Q. Another problem is that if the possible range of gray values is not completely covered by the images there are no equations for the range of gray values that are not covered. Hence, the equation system will become singular. Both problems can be solved by introducing smoothness constraints for Q, which couple the equations and enable an extrapolation of Q into the range of gray values that is not covered by the images. The smoothness constraints require that the second derivative of Q should be small. Hence, for byte images they lead to 254 equations of the form

$$(0 \ \ldots \ 0 \ s \ -2s \ s \ 0 \ \ldots \ 0)Q = 0 \tag{8.11}$$

The parameter s determines the amount of smoothness that is required. Like for (8.10), s must be chosen such that (8.11) has the same weight as the sum of all other equations, i.e., $s = c\sqrt{wh}$, where c is a small number. Empirically, $c = 4$ works well for a wide range of cameras. The approach of tabulating the inverse response q has two slight drawbacks. First, if the camera has a resolution of more than 8 bits the equation system and 2D histograms will become very large. Second, the smoothness constraints will lead to straight lines in the logarithmic representation of q, i.e., exponential curves in the normal representation of q in the range of gray values that is not covered by the images. Therefore, sometimes it may be preferable to model the inverse response as a polynomial, e.g., as in [66]. This model also leads to linear equations for the coefficients of the polynomial. Since polynomials are not very robust in the extrapolation into areas in which no constraints exist we also have to add smoothness constraints in this case by requiring that the second derivative of the polynomial is small. Because this is done in the original representation of q the smoothness constraints will extrapolate straight lines into the gray value range that is not covered.

Let us now consider two cameras: One with a linear response and one with a strong gamma response, i.e., with a small γ in (8.5), and hence with a large γ in the inverse response q. Figure 8.6 displays the 2D histograms of two images taken with each camera with an exposure ratio of 0.5. Note that in both cases the values in the 2D histogram correspond to a line. The only difference is the slope of the line. A different slope, however, could also be caused by a different exposure ratio. Hence, we can see that it is quite important to know the exposure ratios precisely if we want to perform the radiometric calibration.

To conclude this section, we give two examples of the radiometric calibration. The first camera is a linear camera. Here, five images were acquired with shutter times of 32, 16, 8, 4, and 2 ms, as shown in Fig. 8.7(a). The calibrated inverse response curve is shown in Fig. 8.7(b). Note that the response is linear, but the camera has set a slight offset in the amplifier, which prevents very small gray values from being assumed. The second camera is a camera with a gamma response. In this case, six images were taken with shutter times of 30, 20, 10, 5, 2.5, and 1.25 ms. The calibrated inverse response curve is shown in Fig. 8.7(d). Note the strong gamma response of the camera. The 2D histograms in Fig. 8.6 were computed from the second and third brightest images in both sequences.

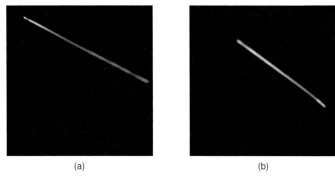

Fig. 8.6 (a) 2D histogram of two images taken with an exposure ratio of 0.5 with a linear camera. (b) 2D histogram of two images taken with an exposure ratio of 0.5 with a camera with a strong gamma response curve. For better visualization, the 2D histograms are displayed with a square root LUT. Note that in both cases the values in the 2D histogram correspond to a line. Hence, linear responses cannot be distinguished from gamma responses without knowing the exact exposure ratio.

8.2.3
Image Smoothing

Every image contains some degree of noise. For the purposes of this chapter, noise can be regarded as random changes in the gray values, which occur for various reasons, e.g., because of the randomness of the photon flux. In most cases, the noise in the image will need to be suppressed by using image smoothing operators.

In a more formalized manner, noise can be regarded as a stationary stochastic process [72]. This means that the true gray value $g_{r,c}$ is disturbed by a noise term $n_{r,c}$ to get the observed gray value: $\hat{g}_{r,c} = g_{r,c} + n_{r,c}$. We can regard the noise $n_{r,c}$ as a random variable with mean 0 and variance σ^2 for every pixel. We can assume a mean of 0 for the noise because any mean different from 0 would constitute a systematic bias of the observed gray values, which we could not detect anyway. Stationary means that the noise does not depend on the position in the image, i.e., is identically distributed for each pixel. In particular, σ^2 is assumed constant throughout the image. The last assumption is a convenient abstraction that does not necessarily hold because the variance of the noise sometimes depends on the gray values in the image. However, we will assume that the noise is always stationary.

Figure 8.8 shows an image of an edge from a real application. The noise is clearly visible in the bright patch in Fig. 8.8(a) and in the horizontal gray value profile in Fig. 8.8(b). Figures 8.8(c) and (d) show the actual noise in the image.

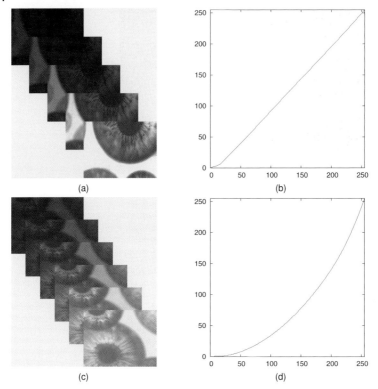

Fig. 8.7 (a) Five images taken with a linear camera with shutter times of 32, 16, 8, 4, and 2 ms. (b) Calibrated inverse response curve. Note that the response is linear, but the camera has set a slight offset in the amplifier, which prevents very small gray values from being assumed. (c) Six images taken with a camera with a gamma response with shutter times of 30, 20, 10, 5, 2.5, and 1.25 ms. (d) Calibrated inverse response curve. Note the strong gamma response of the camera.

How the noise has been calculated is explained below. It can be seen that there is slightly more noise in the dark patch of the image.

With the above discussion in mind, noise suppression can be regarded as a stochastic estimation problem, i.e., given the observed noisy gray values $\hat{g}_{r,c}$ we want to estimate the true gray values $g_{r,c}$. An obvious method to reduce the noise is to acquire multiple images of the same scene and to simply average these images. Since the images are taken at different times, we will refer to this method as temporal averaging or the temporal mean. If we acquire n images, the temporal average is given by

$$g_{r,c} = \frac{1}{n} \sum_{i=1}^{n} \hat{g}_{r,c;i} \qquad (8.12)$$

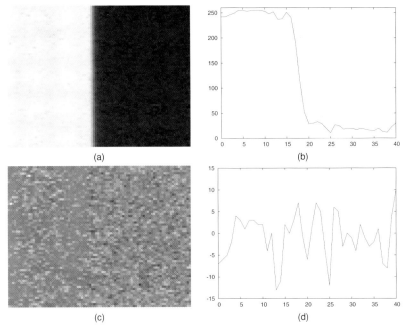

Fig. 8.8 (a) An image of an edge. (b) Horizontal gray value profile through the center of the image. (c) The noise in (a) scaled by a factor of 5. (d) Horizontal gray value profile of the noise.

where $\hat{g}_{r,c;i}$ denotes the noisy gray value at position (r,c) in image i. This approach is frequently used in X-ray inspection systems, which inherently produce quite noisy images. From probability theory [72], we know that the variance of the noise is reduced by a factor of n by this estimation: $\sigma_m^2 = \sigma^2/n$. Consequently, the standard deviation of the noise is reduced by a factor of $\sqrt{n}$. Figure 8.9 shows the result of acquiring 20 images of an edge and computing the temporal average. Compared to Fig. 8.8(a), which shows one of the 20 images, the noise has been reduced by a factor of $\sqrt{20} \approx 4.5$, as can be seen from Fig. 8.9(b). Since this temporally averaged image is a very good estimate for the true gray values, we can subtract it from any of the images that were used in the averaging to obtain the noise in that image. This is how the image in Fig. 8.8(c) was computed.

One of the drawbacks of the temporal averaging is that we have to acquire multiple images to reduce the noise. This is not very attractive if the speed of the application is important. Therefore, other means for reducing the noise are required in most cases. Ideally, we would like to use only one image to estimate the true gray value. If we turn to the theory of stochastic processes again, we see that the temporal averaging can be replaced with a spatial averaging if the stochastic process, i.e., the image, is ergodic [72]. This is precisely

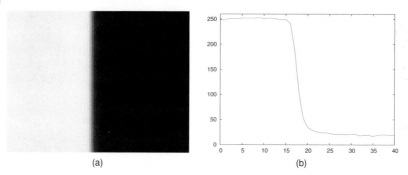

Fig. 8.9 (a) An image of an edge obtained by averaging 20 images of the edge. (b) Horizontal gray value profile through the center of the image.

the definition of ergodicity, and we will assume for the moment that it holds for our images. Then, the spatial average or spatial mean can be computed over a window (also called the mask) of $(2n+1) \times (2m+1)$ pixels as follows:

$$g_{r,c} = \frac{1}{(2n+1)(2m+1)} \sum_{i=-n}^{n} \sum_{j=-m}^{m} \hat{g}_{r-i,c-j} \qquad (8.13)$$

This spatial averaging operation is also called a mean filter. Like for temporal averaging, the noise variance is reduced by a factor that corresponds to the number of measurements that are used to calculate the average, i.e., by $(2n+1)(2m+1)$. Figure 8.10 shows the result of smoothing the image of Fig. 8.8 with a 5×5 mean filter. The standard deviation of the noise is reduced by a factor of 5, which is approximately the same as the temporal averaging in Fig. 8.9. However, we can see that the edge is no longer as sharp as for temporal averaging. This happens, of course, because the images are not ergodic in general, only in areas of constant intensity. Therefore, in contrast to the temporal mean, the spatial mean filter blurs edges.

In Eq. (8.13), we have ignored that the image has finite extent. Therefore, if the mask is close to the image border it will partially stick out of the image, and consequently will access undefined gray values. To solve this problem, several approaches are possible. A very simple approach is to calculate the filter only for pixels for which the mask lies completely within the image. This means that the output image is smaller than the input image, which is not very helpful if multiple filtering operations are applied in sequence. We could also define that the gray values outside the image are 0. For the mean filter, this would mean that the result of the filter would become progressively darker as the pixels get closer to the image border. This is also not desirable. Another approach would be to use the closest gray value on the image border for pixels outside the image. This approach would still create unwanted edges at the

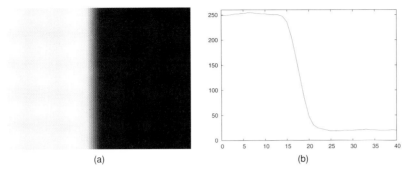

Fig. 8.10 (a) An image of an edge obtained by smoothing the image of Fig. 8.8(a) with a 5 × 5 mean filter. (b) Horizontal gray value profile through the center of the image.

image border. Therefore, typically the gray values are mirrored at the image border. This creates the least amount of artifacts in the result.

As was mentioned above, noise reduction from a single image is preferable for speed reasons. Therefore, let us take a look at the number of operations involved in the calculation of the mean filter. If the mean filter is implemented based on (8.13) the number of operations will be $(2n+1)(2m+1)$ for each pixel in the image, i.e., the calculation will have the complexity $O(whmn)$, where w and h are the width and height of the image, respectively. For $w = 640$, $h = 480$, and $m = n = 5$ (i.e., a 11 × 11 filter), the algorithm will perform 37,171,200 additions and 307,200 divisions. This is quite a substantial amount of operations, so we should try to reduce the operation count as much as possible. One way to do this is to use the associative law of the addition of real numbers as follows:

$$g_{r,c} = \frac{1}{(2n+1)(2m+1)} \sum_{i=-n}^{n} \left(\sum_{j=-m}^{m} \hat{g}_{r-i,c-j} \right) \qquad (8.14)$$

This may seem like a trivial observation, but if we look closer we can see that the term in parentheses only needs to be computed once and can be stored, e.g., in a temporary image. Effectively, this means that we are first computing the sums in the column direction of the input image, save them in a temporary image, and then compute the sums in the row direction of the temporary image. Hence, the double sum in (8.13) of complexity $O(nm)$ is replaced with two sums of total complexity $O(n+m)$. Consequently, the complexity drops from $O(whmn)$ to $O(wh(m+n))$. With the above numbers, now only 6,758,400 additions are required. The above transformation is so important that it has its own name. Whenever a filter calculation allows decomposition into separate row and column sums the filter is called separable. It is obviously of great advantage if a filter is separable, and often is the best speed improvement that can be achieved. In this case, however, it is not the best we can do. Let us take a look at the column sum, i.e., the part in the parentheses in (8.14), and let the result of the column sum be denoted by $t_{r,c}$. Then, we have

$$t_{r,c} = \sum_{j=-m}^{m} \hat{g}_{r,c-j} = t_{r,c-1} + \hat{g}_{r,c+m} - \hat{g}_{r,c-m-1} \tag{8.15}$$

i.e., the sum at position (r, c) can be computed based on the already computed sum at position $(r, c-1)$ with just two additions. The same holds, of course, also for the row sums. The result of this is that we only need to compute the complete sum only once for the first column or row, and can then update it very efficiently. With this, the total complexity is $O(wh)$. Note that the mask size does not influence the runtime in this implementation. Again, since this kind of transformation is so important it has a special name. Whenever a filter can be implemented with this kind of updating scheme based on previously computed values it is called a recursive filter. For the above example, the mean filter requires just 1,238,880 additions for the entire image. This is more than a factor of 30 faster for this example than the naive implementation based on (8.13). Of course, the advantage becomes even bigger for larger mask sizes.

In the above discussion, we have called the process of spatial averaging a mean filter without defining what is meant by the word filter. We can define a filter as an operation that takes a function as input and produces a function as output. Since images can be regarded as functions (see Section 8.1.1), for our purposes a filter transforms an image into another image.

The mean filter is an instance of a linear filter. Linear filters are characterized by the following property: applying a filter to a linear combination of two input images yields the same result as applying the filter to the two images and then computing the linear combination. If we denote the linear filter by h, and the two images by f and g, we have: $h\{af(p) + bg(p)\} = ah\{f(p)\} + bh\{g(p)\}$, where $p = (r, c)$ denotes a point in the image and the $\{\ \}$ operator denotes the application of the filter. Linear filters can be computed by a convolution. For a one-dimensional function on a continuous domain, the convolution is given by

$$f * h = (f * h)(x) = \int_{-\infty}^{\infty} f(t)h(x-t)\,dt \tag{8.16}$$

Here, f is the image function and the filter h is specified by another function called the convolution kernel of the filter mask. Similarly, for two-dimensional functions we have:

$$f * h = (f * h)(r, c) = \int_{-\infty}^{\infty} \int_{-\infty}^{\infty} f(u, v)h(r-u, c-v)\,du\,dv \tag{8.17}$$

For functions with discrete domains, the integrals are replaced by sums

$$f * h = \sum_{i=-\infty}^{\infty} \sum_{j=-\infty}^{\infty} f_{i,j} h_{r-i,c-j} \tag{8.18}$$

The integrals and sums are formally taken over an infinite domain. Of course, to be able to compute the convolution in a finite amount of time, the filter $h_{r,c}$ must be 0 for sufficiently large r and c. For example, the mean filter is given by

$$h_{r,c} = \begin{cases} \dfrac{1}{(2n+1)(2m+1)} & |r| \leq n \wedge |c| \leq m \\ 0 & \text{otherwise} \end{cases} \tag{8.19}$$

The notion of separability can be extended for arbitrary linear filters. If $h(r,c)$ can be decomposed as $h(r,c) = s(r)t(c)$ (or $h_{r,c} = s_r t_c$), h is called separable. As for the mean filter, we can factor out s in this case to get a more efficient implementation:

$$\begin{aligned} f * h &= \sum_{i=-\infty}^{\infty} \sum_{j=-\infty}^{\infty} f_{i,j} h_{r-i,c-j} = \sum_{i=-\infty}^{\infty} \sum_{j=-\infty}^{\infty} f_{i,j} s_{r-i} t_{c-j} \\ &= \sum_{i=-\infty}^{\infty} s_{r-i} \left(\sum_{j=-\infty}^{\infty} f_{i,j} t_{c-j} \right) \end{aligned} \tag{8.20}$$

Obviously, separable filters have the same speed advantage as the separable implementation of the mean filter. Therefore, separable filters are preferred over nonseparable filters. There is also a definition for recursive linear filters, which we cannot cover in detail. The interested reader is referred to [21]. Recursive linear filters have the same speed advantage as the recursive implementation of the mean filter, i.e., the runtime does not depend on the filter size. Unfortunately, many interesting filters cannot be implemented as recursive filters. They usually can only be approximated by a recursive filter.

Although the mean filter produces good results, it is not the optimum smoothing filter. To see this, we note that noise primarily manifests itself as high-frequency fluctuations of the gray values in the image. Ideally, we would like a smoothing filter to remove these high-frequency fluctuations. To see how well the mean filter performs this task, we can examine how the mean filter responds to certain frequencies in the image. The theory how to do this is provided by the Fourier transform. The interested reader is referred to [13] for details. Figure 8.11(a) shows the frequency response of a 3×3 mean filter. In this plot, the row and column coordinates of 0 correspond to a frequency of 0, which represents the medium gray value in the image, while the row and column coordinates of ± 1 represent the highest possible frequencies in the image. For example, the frequencies with column coordinate 0 and row coordinate ± 1 correspond to a grid with alternating one pixel wide vertical bright and dark lines. From Fig. 8.11(a), we can see that the 3 × 3 mean filter removes certain frequencies completely. These are the points for which the response has a value of 0. They occur for relatively high frequencies. However, we can also see that the highest frequencies are not removed completely. To illustrate this, Fig. 8.11(b) shows an image with one pixel wide lines spaced three pixels apart. From Fig. 8.11(c), we can see that this frequency is completely removed by the 3 × 3 mean filter: the output image has a constant gray value. If we change the spacing of the lines to two pixels, as in Fig. 8.11(d), we can see from Fig. 8.11(e) that this higher frequency is not removed completely. This is an undesirable behavior since it means that noise is not removed completely by the mean filter. Note also that the polarity of the lines has been reversed by the mean filter, which also is undesirable. Furthermore, from Fig. 8.11(a) we can see that the frequency response of the mean filter is not rotationally symmetric, i.e., it is anisotropic. This means that diagonal structures are smoothed differently than horizontal or vertical structures.

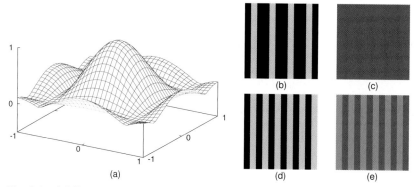

Fig. 8.11 (a) Frequency response of the 3 × 3 mean filter.
(b) Image with one pixel wide lines spaced three pixels apart.
(c) Result of applying the 3 × 3 mean filter to the image in (b). Note that all the lines have been smoothed out. (d) Image with one pixel wide lines spaced two pixels apart. (e) Result of applying the 3 × 3 mean filter to the image in (d). Note that the lines have not been completely smoothed out although they have a higher frequency than the lines in (b). Note also that the polarity of the lines has been reversed.

Because the mean filter has the above drawbacks, the question arises which smoothing filter is optimal. One way to approach this problem is to define certain natural criteria that the smoothing filter should fulfill and then to search for the filters that fulfill the desired criteria. The first natural criterion is that the filter should be linear. This is natural because we can imagine an image being composed of multiple objects in an additive manner. Hence, the filter output should be a linear combination of the input. Furthermore, the filter should be position invariant, i.e., it should produce the same results no matter where an object is in the image. This is automatically fulfilled for linear filters. Also, we would like the filter to be rotation invariant, i.e., isotropic, so that it produces the same result independent of the orientation of the objects in the image. As we saw above, the mean filter does not fulfill this criterion. We would also like to control the amount of smoothing (noise reduction) that is being performed. Therefore, the filter should have a parameter t that can be used to control the smoothing, where higher values of t indicate more smoothing. For the mean filter, this corresponds to the mask sizes m and n. Above, we saw that the mean filter does not suppress all high frequencies, i.e., noise, in the image. Therefore, a criterion that describes the noise suppression of the filter in the image should be added. One such criterion is that the larger t gets, the more local maxima in the image should be eliminated. This is a quite intuitive criterion, as can be seen in Fig. 8.8(a), where many local maxima can be detected because of noise. Note that because of linearity we only need to require maxima to be eliminated. This automatically implies that local minima are eliminated as well. Finally, we sometimes would like to execute the smoothing filter several times in succession. If we do this, we would also have a simple means to predict the result of the combined filtering. Therefore, first filtering with t and then with s should be identical to a single filter operation with $t + s$. It can be shown that among all smoothing filters, the Gaussian filter is the only filter that fulfills all of the above criteria [61]. Other natural criteria for a smoothing filter have been proposed [99, 5, 31], which also single out the Gaussian filter as the optimal smoothing filter. In one dimension, the Gaussian filter is given by

$$g_\sigma(x) = \frac{1}{\sqrt{2\pi}\sigma} e^{-\frac{x^2}{2\sigma^2}} \qquad (8.21)$$

This is the function that also defines the probability density of a normally distributed random variable. In two dimensions, the Gaussian filter is given by

$$g_\sigma(r,c) = \frac{1}{2\pi\sigma^2} e^{-\frac{r^2+c^2}{2\sigma^2}} = \frac{1}{\sqrt{2\pi}\sigma} e^{-\frac{r^2}{2\sigma^2}} \frac{1}{\sqrt{2\pi}\sigma} e^{-\frac{c^2}{2\sigma^2}} = g_\sigma(r)g_\sigma(c) \qquad (8.22)$$

Hence, the Gaussian filter is separable. Therefore, it can be computed very efficiently. In fact, it is the only isotropic separable smoothing filter. Unfortunately, it cannot be implemented recursively. However, some recursive approximations have been proposed [22, 101]. Figure 8.12 shows plots of the 1D and 2D Gaussian filter with $\sigma = 1$. The frequency response of a Gaussian filter is also a Gaussian function, albeit with σ inverted. Therefore, Fig. 8.12(b) also gives a qualitative impression of the frequency response of the Gaussian filter. It can be seen that the Gaussian filter suppresses high frequencies much better than the mean filter.

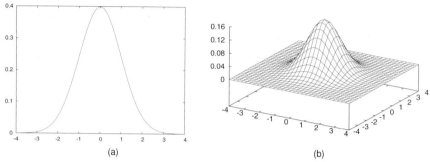

Fig. 8.12 (a) One-dimensional Gaussian filter with $\sigma = 1$. (b) Two-dimensional Gaussian filter with $\sigma = 1$.

Like the mean filter, any linear filter will change the variance of the noise in the image. It can be shown that for a linear filter $h(r,c)$ or $h_{r,c}$ the noise variance is multiplied by the following factor [72]:

$$\int_{-\infty}^{\infty}\int_{-\infty}^{\infty} h(r,c)^2 \, dr \, dc \quad \text{or} \quad \sum_{i=-\infty}^{\infty}\sum_{j=-\infty}^{\infty} h_{r,c}^2 \qquad (8.23)$$

For a Gaussian filter, this factor is $1/(4\pi\sigma^2)$. If we compare this to a mean filter with a square mask with parameter n, we see that to get the same noise reduction with the Gaussian filter, we need to set $\sigma = (2n+1)/(2\sqrt{\pi})$. For example, a 5×5 mean filter has the same noise reduction effect as a Gaussian filter with $\sigma \approx 1.41$.

Figure 8.13 compares the results of the Gaussian filter with those of the mean filter of an equivalent size. For small filter sizes ($\sigma = 1.41$ and 5×5), there is hardly any noticeable difference between the results. However, if larger filter sizes are used it becomes clear that the mean filter turns the edge into a ramp, leading to a badly defined edge that is also visually quite hard to locate, whereas the Gaussian filter produces a much sharper edge. Hence, we can see that the Gaussian filter produces better results, and consequently is usually the preferred smoothing filter if the quality of the results is the primary concern. If speed is the primary concern the mean filter is preferable.

We close this section with a nonlinear filter that can also be used for noise suppression. The mean filter is a particular estimator for the mean value of a sample of random values. From probability theory, we know that other estimators are also possible, most notably the median of the samples. The median is defined as the value for which 50% of the values in the probability distribution of the samples are smaller than the median and 50% are larger. From a practical point of view, if the sample set contains n values g_i, $i = 0, \ldots, n-1$, we sort the values g_i in ascending order to get s_i, and then select the value $\text{median}(g_i) = s_{n/2}$. Hence, we can obtain a median filter by calculating the

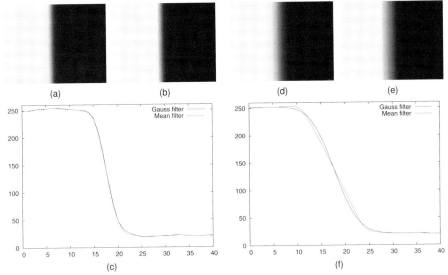

Fig. 8.13 Images of an edge obtained by smoothing the image of Fig. 8.8(a) with a Gaussian filter with $\sigma = 1.41$ (a) and with a mean filter of size 5×5 (b) and corresponding gray value profiles (c). Note that both filters return very similar results in this example. Result of a Gaussian filter with $\sigma = 3.67$ (d) and a 13×13 mean filter (e) and corresponding profiles (f). Note that the mean filter turns the edge into a ramp, leading to a badly defined edge, whereas the Gaussian filter produces a much sharper edge.

median instead of the mean inside a window around the current pixel. Let W denote the window, e.g., a $(2n+1) \times (2m+1)$ rectangle as for the mean filter. Then the median filter is given by

$$g_{r,c} = \underset{(i,j) \in W}{\operatorname{median}} \hat{g}_{r-i,c-j} \qquad (8.24)$$

The median filter is not separable. However, with sophisticated algorithms it is possible to obtain a runtime complexity that is comparable to a separable linear filter: $O(whn)$ [47, 23]. The properties of the median filter are quite difficult to analyze. We can note, however, that it performs no averaging of the input gray values, but simply selects one of them. This can lead to surprising results. For example, the result of applying a 3×3 median filter to the image in Fig. 8.11(b) would be a completely black image. Hence, the median filter would remove the bright lines because they cover less than 50% of the window. This property can sometimes be used to remove objects completely from the image. On the other hand, applying a 3×3 median filter to the image in Fig. 8.11(d) would swap the bright and dark lines. This result is as undesirable as the result of the mean filter on the same image. On the edge image of

Fig. 8.8(a), the median filter produces quite good results, as can be seen from Fig. 8.14. In particular, it should be noted that the median filter preserves the sharpness of the edge even for large filter sizes. However, it cannot be predicted if and how much the position of the edge is changed by the median filter, which is possible for the linear filters. Furthermore, we cannot estimate how much noise is removed by the median filter, in contrast to the linear filters. Therefore, for high-accuracy measurements the Gaussian filter should be used.

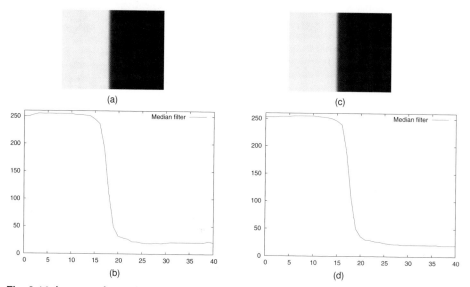

Fig. 8.14 Images of an edge obtained by smoothing the image of Fig. 8.8(a) with a median filter of size 5 × 5 (a) and corresponding gray value profile (b). Result of a 13 × 13 median filter (c) and corresponding profile (d). Note that the median filter preserves the sharpness of the edge to a great extent.

Finally, it should be mentioned that the median is a special case of the more general class of rank filters. Instead of selecting the median $s_{n/2}$ of the sorted gray values, the rank filter would select the sorted gray value at a particular rank r, i.e., s_r. We will see other cases of rank operators in Section 8.6.2.

8.3
Geometric Transformations

In many applications, it cannot be ensured that the objects to be inspected are always in the same position and orientation in the image. Therefore, the inspection algorithm must be able to cope with these position changes. Hence, one of the problems is to detect the position and orientation, also called the pose, of the objects to be examined. This will be the subject of later sections of this chapter. For the purposes of this section, we assume that we know the pose already. In this case, the simplest procedure to adapt the inspection to a particular pose is to align the ROIs appropriately. For example, if we know that an object is rotated by 45°, we could simply rotate the ROI by 45° before performing the inspection. In some cases, however, the image must be transformed (aligned) to a standard pose before the inspection can be performed. For example, the segmentation of the OCR is much easier if the text is either horizontal or vertical. Another example is the inspection of objects for defects based on a reference image. Here, we also need to align the image of the object to the pose in the reference image or vice versa. Therefore, in this section we will examine different geometric image transformations that are useful in practice.

8.3.1
Affine Transformations

If the position and rotation of the objects cannot be kept constant with the mechanical setup we need to correct the rotation and translation of the object. Sometimes the distance of the object to the camera changes, leading to an apparent change in size of the object. These transformations are part of a very useful class of transformations called affine transformations, which are transformations that can be described by the following equation:

$$\begin{pmatrix} \tilde{r} \\ \tilde{c} \end{pmatrix} = \begin{pmatrix} a_{11} & a_{12} \\ a_{21} & a_{22} \end{pmatrix} \begin{pmatrix} r \\ c \end{pmatrix} + \begin{pmatrix} t_r \\ t_c \end{pmatrix} \qquad (8.25)$$

Hence, an affine transformation consists of a linear part given by a 2×2 matrix and a translation. The above notation is a little cumbersome, however, since we always have to list the translation separately. To circumvent this, we can use a representation where we extend the coordinates with a third coordinate of 1, which enables us to write the transformation as a simple matrix multiplication

$$\begin{pmatrix} \tilde{r} \\ \tilde{c} \\ 1 \end{pmatrix} = \begin{pmatrix} a_{11} & a_{12} & a_{13} \\ a_{21} & a_{22} & a_{23} \\ 0 & 0 & 1 \end{pmatrix} \begin{pmatrix} r \\ c \\ 1 \end{pmatrix} \qquad (8.26)$$

Note that the translation is represented by the elements a_{13} and a_{23} of the matrix A. This representation with an added redundant third coordinate is called homogeneous coordinates. Similarly, the representation with two coordinates in (8.25) is called inhomogeneous coordinates. We will see the true power of the homogeneous representation below. Any affine transformation can be constructed from the following basic transformations, where the last row of the matrix has been omitted:

$$\begin{pmatrix} 1 & 0 & t_r \\ 0 & 1 & t_c \end{pmatrix} \quad \text{Translation} \tag{8.27}$$

$$\begin{pmatrix} s_r & 0 & 0 \\ 0 & s_c & 0 \end{pmatrix} \quad \text{Scaling in the row and column direction} \tag{8.28}$$

$$\begin{pmatrix} \cos\alpha & -\sin\alpha & 0 \\ \sin\alpha & \cos\alpha & 0 \end{pmatrix} \quad \text{Rotation by an angle of } \alpha \tag{8.29}$$

$$\begin{pmatrix} \cos\alpha & 0 & 0 \\ \sin\alpha & 1 & 0 \end{pmatrix} \quad \text{Skew of the row axis by an angle of } \theta \tag{8.30}$$

The first three basic transformations need no further explanation. The skew (or slant) is a rotation of only one axis, in this case the row axis. It is quite useful to rectify slanted characters in the OCR.

8.3.2
Projective Transformations

An affine transformation enables us to correct almost all relevant pose variations that an object may undergo. However, sometimes affine transformations are not general enough. If the object in question may rotate in three dimensions it will undergo a general perspective transformation, which is quite hard to correct because of the occlusions that may occur. If, however, the object is planar we can model the transformation of the object by a two-dimensional perspective transformation, which is a special two-dimensional projective transformation [42, 27]. Projective transformations are given by

$$\begin{pmatrix} \tilde{r} \\ \tilde{c} \\ \tilde{w} \end{pmatrix} = \begin{pmatrix} h_{11} & h_{12} & h_{13} \\ h_{21} & h_{22} & h_{23} \\ h_{31} & h_{32} & h_{33} \end{pmatrix} \begin{pmatrix} r \\ c \\ w \end{pmatrix} \tag{8.31}$$

Note the similarity to the affine transformation in (8.26). The only changes that were made are that the transformation is now described by a full 3×3 matrix and that we have replaced the 1 in the third coordinate with a variable w. This representation is actually the true representation in homogeneous coordinates. It can also be used for affine transformations, which are special projective transformations. With this third coordinate, it is not obvious how we are able to obtain a transformed 2D coordinate, i.e., how to compute the corresponding inhomogeneous point. First, it must be noted that in homogeneous coordinates all points $p = (r, c, w)^\top$ are only defined up to a scale factor, i.e., the vectors p and λp ($\lambda \neq 0$) represent the same 2D point [42, 27]. Consequently, the projective transformation given by the matrix H is also only defined up to a scale factor, and hence has only eight independent parameters. To obtain an inhomogeneous 2D point from the homogeneous representation, we must divide the homogeneous vector by w. This requires $w \neq 0$. Such points are called finite points. Conversely, points with $w = 0$ are called points at infinity because they can be regarded as lying infinitely far away in a certain direction [42, 27].

Since a projective transformation has eight independent parameters it can be uniquely determined from four corresponding points [42, 27]. This is how the projective transformations will usually be determined in machine vision applications. We will extract four points in an image, which typically represent a rectangle, and will rectify the image so that the four extracted points will be transformed to the four corners of the rectangle, i.e., to their corresponding points. Unfortunately, because of space limitations we cannot give the details of how the transformation is computed from the point correspondences. The interested reader is referred to [42, 27].

8.3.3
Image Transformations

After we have taken a look at how coordinates can be transformed with affine and projective transformations, we can consider how an image should be transformed. Our first idea might be to go through all the pixels in the input image, to transform their coordinates, and to set the gray value of the transformed point in the output image. Unfortunately, this simple strategy does not work. This can be seen by checking what happens if an image is scaled by a factor of 2: only one quarter of the pixels in the output image would be set. The correct way to transform an image is to loop through all the pixels in the output image and to calculate the position of the corresponding point in the input image. This is the simplest way to ensure that all relevant pixels in the output image are set. Fortunately, calculating the positions in the original image is simple: we only need to invert the matrix that describes the affine or projective transformation, which results again in an affine or projective transformation.

When the image coordinates are transformed from the output image to the input image, typically not all pixels in the output image transform back to coordinates that lie in the input image. This can be taken into account by computing a suitable ROI for the output image. Furthermore, we see that the resulting coordinates in the input image will typically not be integer coordinates. An example of this is given in Fig. 8.15, where the input image is transformed by an affine transformation consisting of a translation, rotation, and scaling. Therefore, the gray values in the output image must be interpolated.

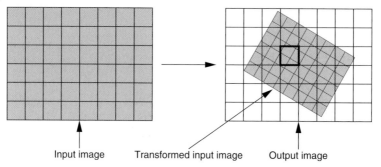

Input image   Transformed input image   Output image

Fig. 8.15 An affine transformation of an image. Note that integer coordinates in the output image transform to noninteger coordinates in the original image, and hence must be interpolated.

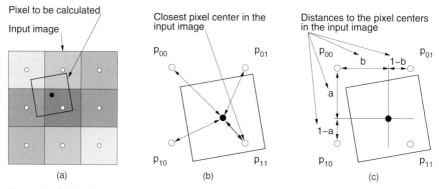

Fig. 8.16 (a) A pixel in the output image is transformed back to the input image. Note that the transformed pixel center lies on a noninteger position between four adjacent pixel centers. (b) Nearest neighbor interpolation determines the closest pixel center in the input image and uses its gray value in the output image. (c) Bilinear interpolation determines the distances to the four adjacent pixel centers and weights their gray values using the distances.

The interpolation can be done in several ways. Figure 8.16(a) displays a pixel in the output image that has been transformed back to the input image. Note that the transformed pixel center lies on a noninteger position between four adjacent pixel centers. The simplest and fastest interpolation method is to calculate the closest of the four adjacent pixel centers, which only involves rounding the floating point coordinates of the transformed pixel center, and to use the gray value of the closest pixel in the input image as the gray value of the pixel in the output image, as shown in Fig. 8.16(b). This interpolation method is called the nearest neighbor interpolation. To see the effect of this interpolation, Figs. 8.17(a) displays an image of a serial number of a bank note, where the characters are not horizontal. Figures 8.17(c) and (d) display the result of rotating the image such that the serial number is horizontal using this interpolation. Note that because the gray value is taken from the closest pixel center in the input image, the edges of the characters have a jagged appearance, which is undesirable.

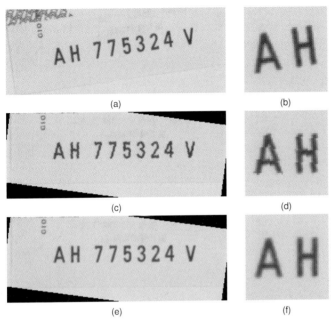

Fig. 8.17 (a) Image showing a serial number of a bank note. (b) Detail of (a). (c) Image rotated such that the serial number is horizontal using the nearest neighbor interpolation. (d) Detail of (c). Note the jagged edges of the characters. (e) Image rotated using bilinear interpolation. (f) Detail of (e). Note the smooth edges of the characters.

The reason for the jagged appearance in the result of the nearest neighbor interpolation is that essentially we are regarding the image as a piecewise constant function: every coordinate that falls within a rectangle of extent ± 0.5 in each direction is assigned the same gray value. This leads to discontinuities in the result, which cause the jagged edges. This behavior is especially noticeable if the image is scaled by a factor > 1. To get a better interpolation, we can use more information than the gray value of the closest pixel. From Fig. 8.16(a), we can see that the transformed pixel center lies in a square of four adjacent pixel centers. Therefore, we can use the four corresponding gray values and weight them appropriately. One way to do this is to use bilinear interpolation, as shown in Fig. 8.16(c). First, we compute the horizontal and vertical distances of the transformed coordinate to the adjacent pixel centers. Note that these are numbers between 0 and 1. Then, we weight the gray values according to their distances to get the bilinear interpolation:

$$\tilde{g} = b(ag_{11} + (1-a)g_{01}) + (1-b)(ag_{10} + (1-a)g_{00}) \quad (8.32)$$

Figures 8.17(e) and (f) display the result of rotating the image of Fig. 8.17(a) using bilinear interpolation. Note that the edges of the characters now have a very smooth appearance. This much better result more than justifies the longer computation time (typically a factor of more than 5).

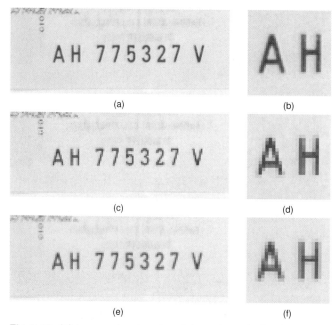

Fig. 8.18 (a) Image showing a serial number of a bank note. (b) Detail of (a). (c) The image of (a) scaled down by a factor of 3 using bilinear interpolation. (d) Detail of (c). Note the different stroke widths of the vertical strokes of the letter H. This is caused by aliasing. (e) Result of scaling the image down by integrating a smoothing filter (in this case a mean filter) into the image transformation. (f) Detail of (e).

To conclude the discussion on interpolation, we discuss the effects of scaling an image down. In the bilinear interpolation scheme, we would interpolate from the closest four pixel centers. However, if the image is scaled down adjacent pixel centers in the output image will not necessarily be close in the input image. Imagine a larger version of the image of Fig. 8.11(b) (one pixel wide vertical lines spaced three pixels apart) being scaled down by a factor of 4 using the nearest neighbor interpolation: we would get an image with one pixel wide lines that are four pixels apart. This is certainly not what we would expect. For bilinear interpolation, we would get similar unexpected results. If we scale down an image we are essentially subsampling it. As a consequence, we may obtain an image that contains a content that was not present in the original image. This effect is called aliasing. Another example for aliasing can be seen in Fig. 8.18. The image in Fig. 8.18(a) is scaled down by a factor of 3 in Fig. 8.18(c) using bilinear interpolation. Note that the stroke widths of the vertical strokes of the letter H, which are equally wide in Fig. 8.18(a), now appear to be substantially different. This is undesirable. To improve the image transformation, the image must be smoothed before it is scaled down, e.g., using a mean or a Gaussian filter. Alternatively, the smoothing can be integrated into the gray value interpolation. Figure 8.18(e) shows the result of integrating a mean filter into the image transformation. Because of the smoothing, the strokes of the H now have the same width again.

In the above examples, we have seen the usefulness of the affine transformations for rectifying text. Sometimes, an affine transformation is not sufficient for this purpose. Figures 8.19(a) and (b) show two images of license plates on cars. Because the position of the camera with respect to the car could not be controlled in this example, the images of the license plates show severe perspective distortions. Figures 8.19(c) and (d) show the result of applying projective transformations to the images that cut out the license plates and rectify them. Hence, the images in Figs. 8.19(c) and (d) would result if we would have looked at the license plates perpendicularly from in front of the car. Obviously, it is now much easier to segment and read the characters on the license plates.

Fig. 8.19 (a), (b) Images of license plates. (c), (d) Result of a projective transformation that rectifies the perspective distortion of the license plates.

8.3.4
Polar Transformations

We conclude this section with another very useful geometric transformation: the polar transformation. This transformation is typically used to rectify parts of images that show objects that are circular or that are contained in circular rings in the image. An example is shown in Fig. 8.20(a). Here, we can see the inner part of a CD that contains a ring with a bar code and some text. To read the bar code, one solution is to rectify the part of the image that contains the bar code. For this purpose, the polar transformation can be used, which converts the image into polar coordinates (d, ϕ), i.e., into the distance d to the center of the transformation and the angle ϕ of the vector to the center of the transformation. Let the center of the transformation be given by (m_r, m_c). Then, the polar coordinates of a point (r, c) are given by

$$d = \sqrt{(r - m_r)^2 + (c - m_c)^2}$$
$$\phi = \arctan\left(-\frac{r - m_r}{c - m_c}\right) \tag{8.33}$$

In the calculation of the arc tangent function, the correct quadrant must be used, based on the sign of the two terms in the fraction in the argument of arctan. Note that the transformation of a point into polar coordinates is quite expensive to compute because of the square root and the arc tangent. Fortunately, to transform an image, like for affine and projective transformations the inverse of the polar transformation is used, which is given by

$$r = m_r - d \sin \phi$$
$$c = m_c + d \cos \phi \tag{8.34}$$

Here, the sines and cosines can be tabulated because they only occur for a finite number of discrete values, and hence only need to be computed once. Therefore, the polar transformation of an image can be computed efficiently. Note that by restricting the range of d and ϕ, we can transform arbitrary circular sectors. Figure 8.20(b) shows the result of transforming a circular ring that contains the bar code in Fig. 8.20(a). Note that because of the polar transformation the bar code is straight and horizontal, and consequently can be read easily.

8.4
Image Segmentation

In the preceding sections, we have looked at operations that transform an image into another image. These operations do not give us information about the

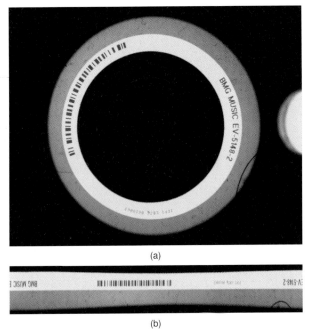

Fig. 8.20 (a) Image of the center of a CD showing a circular bar code. (b) Polar transformation of the ring that contains the bar code. Note that the bar code is now straight and horizontal.

objects in the image. For this purpose, we need to segment the image, i.e., extract regions from the image that correspond to the objects we are interested in. More formally, the segmentation is an operation that takes an image as input and returns one or more regions or subpixel-precise contours as output.

8.4.1
Thresholding

The simplest segmentation algorithm is to threshold the image. The threshold operation is defined by

$$S = \{(r,c) \in R \mid g_{\min} \leq f_{r,c} \leq g_{\max}\} \tag{8.35}$$

Hence, the threshold operation selects all points in the ROI R of the image that lie within a specified range of gray values into the output region S. Often, $g_{\min} = 0$ or $g_{\max} = 2^b - 1$ is used. If the illumination can be kept constant the thresholds $g_{\min}$ and $g_{\max}$ are selected when the system is set up and are never modified. Since the threshold operation is based on the gray values themselves, it can be used whenever the object to be segmented and the background have significantly different gray values.

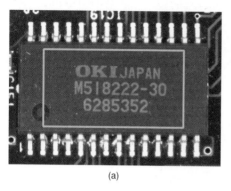

(a)   (b)

OKI JAPAN
M518222-30
6285352

OKI JAPAN
M518222-30
6285352

(c)   (d)

Fig. 8.21 (a), (b) Images of prints on ICs with a rectangular ROI overlaid in light gray. (c), (d) Result of thresholding the images in (a), (b) with $g_{min} = 90$ and $g_{max} = 255$.

Figures 8.21(a) and (b) show two images of ICs on a printed circuit board (PCB) with a rectangular ROI overlaid in light gray. The result of thresholding the two images with $g_{min} = 90$ and $g_{max} = 255$ is shown in Figs. 8.21(c) and (d). Since the illumination is kept constant the same threshold works for both images. Note also that there are some noisy pixels in the segmented regions. They can be removed, e.g., based on their area (see Section 8.5) or based on morphological operations (see Section 8.6).

The constant threshold only works well as long as the gray values of the object and the background do not change. Unfortunately, this occurs less frequently than one would wish, e.g., because of a changing illumination. Even if the illumination is kept constant, different gray value distributions on similar objects may prevent us from using a constant threshold. Figure 8.22 shows an example for this. In Figs. 8.22(a) and (b) two different ICs on the same PCB are shown. Despite the identical illumination the prints have a substantially different gray value distribution, which will not allow us to use the same threshold for both images. Nevertheless, the print and the background can be separated easily in both cases. Therefore, ideally we would like to have a method

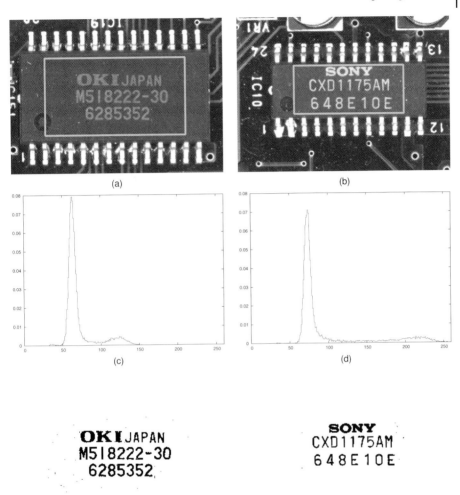

Fig. 8.22 (a), (b) Images of prints on ICs with a rectangular ROI overlaid in light gray. (c), (d) Gray value histogram of the images in (a), (b) within the respective ROI. (e), (f) Result of thresholding the images in (a), (b) with a threshold selected automatically based on the gray value histogram.

that is able to determine the thresholds automatically. This can be done based on the gray value histogram of the image. Figures 8.22(c) and (d) show the histograms of the images in Figs. 8.22(a) and (b). It is obvious that there are two relevant peaks (maxima) in the histograms in both images. The one with the smaller gray value corresponds to the background, while the one with the higher gray value corresponds to the print. Intuitively, a good threshold corresponds to the minimum between the two peaks in the histogram. Unfortunately, neither the two maxima nor the minimum is well defined because of random fluctuations in the gray value histogram. Therefore, to robustly select the threshold that corresponds to the minimum, the histogram must be smoothed, e.g., by convolving it with a one-dimensional Gaussian filter. Since it is not clear which σ to use, a good strategy is to smooth the histogram with progressively larger values of σ until two unique maxima with a unique minimum in between are obtained. The result of using this approach to select the threshold automatically is shown in Figs. 8.22(e) and (f). As can be seen, for both images suitable thresholds have been selected. This approach to select the thresholds is not the only approach. Further approaches are described, e.g., in [39, 49]. All these approaches have in common that they are based on the gray value histogram of the image. One example for such a different approach is to assume that the gray values in the foreground and background each have a normal (Gaussian) probability distribution, and to jointly fit two Gaussian densities to the histogram. The threshold is then defined as the gray value for which the two Gaussian densities have equal probabilities.

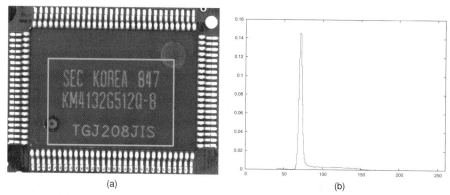

Fig. 8.23 (a) Image of a print on an IC with a rectangular ROI overlaid in light gray. (b) Gray value histogram of the image in (a) within the ROI. Note that there are no significant minima and only one single significant maximum in the histogram.

While calculating the thresholds from the histogram often works extremely well, it fails whenever the assumption that there are two peaks in the histogram is violated. One such example is shown in Fig. 8.23. Here, the print is so noisy that the gray values of the print are extremely spread out, and consequently there is no discernible peak for the print in the histogram. Another reason for the failure of the desired peak to appear is an inhomogeneous illumination. This typically destroys the relevant peaks or moves them so that they are in the wrong location. An uneven illumination often even prevents us from using a threshold operation altogether because there are no fixed thresholds that work throughout the entire image. Fortunately, often the objects of interest can be characterized by being locally brighter or darker than their local background. The prints on the ICs we have examined so far are a good example for this. Therefore, instead of specifying global thresholds we would like to specify by how much a pixel must be brighter or darker than its local background. The only problem we have is how to determine the gray value of the local background. Since a smoothing operation, e.g., the mean, Gaussian, or median filter (see Section 8.2.3), calculates an average gray value in a window around the current pixel we can simply use the filter output as an estimate of the gray value of the local background. The operation of comparing the image to its local background is called a dynamic thresholding operation. Let the image be denoted by $f_{r,c}$ and the smoothed image be denoted by $s_{r,c}$. Then, the dynamic thresholding operation for bright objects is given by

$$S = \{(r,c) \in R \mid f_{r,c} - s_{r,c} \geq g_{\text{diff}}\} \tag{8.36}$$

while the dynamic thresholding operation for dark objects is given by

$$S = \{(r,c) \in R \mid f_{r,c} - s_{r,c} \leq -g_{\text{diff}}\} \tag{8.37}$$

Figure 8.24 gives an example of how the dynamic thresholding works. In Fig. 8.24(a), a small part of a print on an IC with a one pixel wide horizontal ROI is shown. Figure 8.24(b) displays the gray value profiles of the image and the image smoothed with a 9 × 9 mean filter. It can be seen that the text is substantially brighter than the local background estimated by the mean filter. Therefore, the characters can be segmented easily with the dynamic thresholding operation.

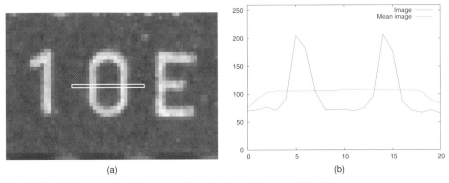

Fig. 8.24 (a) Image showing a small part of a print on an IC with a one pixel wide horizontal ROI. (b) Gray value profiles of the image and the image smoothed with a 9 × 9 mean filter. Note that the text is substantially brighter than the local background estimated by the mean filter.

In the dynamic thresholding operation, the size of the smoothing filter determines the size of the objects that can be segmented. If the filter size is too small the local background will not be estimated well in the center of the objects. As a rule of thumb, the diameter of the mean filter must be larger than the diameter of the objects to be recognized. The same holds for the median filter. An analogous relation exists for the Gaussian filter. Furthermore, in general if larger filter sizes are chosen for the mean and Gaussian filters the filter output will be more representative of the local background. For example, for light objects the filter output will become darker within the light objects. For the median filter, this is not true since it will completely eliminate the objects if the filter mask is larger than the diameter of the objects. Hence, the gray values will be representative of the local background if the filter is sufficiently large. If the gray values in the smoothed image are more representative of the local background we can typically select a larger threshold g_{diff}, and hence can suppress noise in the segmentation better. However, the filter mask cannot be chosen arbitrarily large because neighboring objects might adversely influence the filter output. Finally, it should be noted that the dynamic thresholding operation not only returns a segmentation result for objects that are brighter or darker than their local background. It also returns a segmentation result at the bright or dark region around edges.

(a) (b)

Fig. 8.25 (a) Image of a print on an IC with a rectangular ROI overlaid in light gray. (b) Result of segmenting the image in (a) with a dynamic thresholding operation with $g_{\text{diff}} = 5$ and a 31 × 31 mean filter.

Figure 8.25(a) again shows the image of Fig. 8.23(a), which could not be segmented with an automatic threshold. In Fig. 8.25(b), the result of segmenting the image with a dynamic thresholding operation with $g_{\text{diff}} = 5$ is shown. The local background was obtained with a 31 × 31 mean filter. Note that the difficult print is segmented very well with the dynamic thresholding.

8.4.2
Extraction of Connected Components

The segmentation algorithms in the previous section return one region as the segmentation result (recall the definitions in (8.35)–(8.37)). Typically, the segmented region contains multiple objects that should be returned individually. For example, in the examples in Figs. 8.21–8.25 we are interested in obtaining each character as a separate region. Typically, the objects we are interested in are characterized by forming a connected set of pixels. Hence, to obtain the individual regions we must compute the connected components of the segmented region.

To be able to compute the connected components, we must define when two pixels should be considered connected. On a rectangular pixel grid, there are only two natural options to define the connectivity. The first possibility is to define two pixels as being connected if they have an edge in common, i.e., if the pixel is directly above, below, left, or right of the current pixel, as shown in Fig. 8.26(a). Since each pixel has four connected pixels this definition is called the 4-connectivity or 4-neighborhood. Alternatively, the definition can be also extended to include the diagonally adjacent pixels, as shown in Fig. 8.26(b). This definition is called the 8-connectivity or 8-neighborhood.

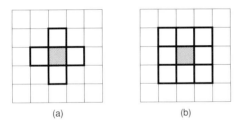

Fig. 8.26 The two possible definitions of connectivity on rectangular pixel grids: (a) 4-connectivity, (b) 8-connectivity.

While these definitions are easy to understand, they cause problematic behavior if the same definition is used on both the foreground and background. Figure 8.27 shows some of the problems that occur if 8-connectivity is used for the foreground and background. In Fig. 8.27(a), there is clearly a single line in the foreground, which divides the background into two connected components. This is what we would intuitively expect. However, as Fig. 8.27(b) shows, if the line is slightly rotated we still obtain a single connected component in the foreground. However, now the background is also a single component. This is quite counterintuitive. Figure 8.27(c) shows another peculiarity. Again, the foreground region consists of a single connected component. Intuitively, we would say that the region contains a hole. However, the background also is a single connected component, indicating that the region contains no hole. The only remedy for this problem is to use opposite connectivities on the foreground and background. If, for example, 4-connectivity is used for the background in the examples in Fig. 8.27 all of the above problems are solved. Likewise, if 4-connectivity is used for the foreground and 8-connectivity for the background the inconsistencies are avoided.

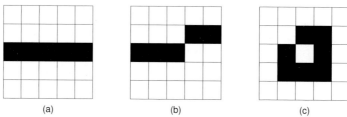

(a) (b) (c)

Fig. 8.27 Some peculiarities when the same connectivity, in this case 8-connectivity, is used for the foreground and background. (a) The single line in the foreground clearly divides the background into two connected components. (b) If the line is very slightly rotated there is still a single line, but now the background is a single component, which is counterintuitive. (c) The single region in the foreground intuitively contains one hole. However, the background is a single connected component, indicating that the region has no hole, which also is counterintuitive.

To compute the connected components on the run-length representation of a region, a classical depth-first search can be performed [82]. We can repeatedly search for the first unprocessed run and then search for overlapping runs in the adjacent rows of the image. The used connectivity determines whether two runs overlap. For 4-connectivity, the runs must at least have one pixel in the same column, while for the 8-connectivity the runs must at least touch diagonally. An example for this procedure is shown in Fig. 8.28. The run-length representation of the input region is shown in Fig. 8.28(a), the search tree for the depth-first search using the 8-connectivity is shown in Fig. 8.28(b), and the resulting connected components are shown in Fig. 8.28(c). For the 8-connectivity, three connected components result. If the 4-connectivity were used four connected components would result.

It should be noted that the connected components can also be computed from the representation of a region as a binary image. The output of this operation is a label image. Therefore, this operation is also called labeling or component labeling. For a description of algorithms that compute the connected components from a binary image see [39, 49].

To conclude this section, Figs. 8.29(a) and (b) show the result of computing the connected components of the regions in Figs. 8.22(e) and (f). As can be seen, each character is a connected component. Furthermore, the noisy segmentation results are also returned as separate components. Thus, it is easy to remove them from the segmentation, e.g., based on their area.

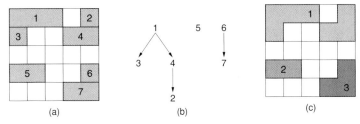

Fig. 8.28 (a) Run-length representation of a region containing seven runs. (b) Search tree when performing a depth-first search for the connected components of the region in (a) using 8-connectivity. The numbers indicate the runs. (c) Resulting connected components.

Fig. 8.29 (a), (b) Result of computing the connected components of the regions in Figs. 8.22(e) and (f). The connected components are visualized by using eight different gray values cyclically.

8.4.3
Subpixel-Precise Thresholding

All the thresholding operations we have discussed so far have been pixel precise. In most cases, this precision is sufficient. However, some applications require a higher accuracy than the pixel grid. Therefore, an algorithm that returns a result with subpixel precision is sometimes required. Obviously, the result of this subpixel-precise thresholding operation cannot be a region, which is only pixel precise. The appropriate data structure for this purpose therefore is a subpixel-precise contour (see Section 8.1.3). This contour will represent the boundary between regions in the image that have gray values above the gray value threshold g_{sub} and regions that have gray values below g_{sub}. To obtain this boundary, we must convert the discrete representation of the image into a continuous function. This can be done, for example, with bilinear interpolation (see Eq. (8.32) in Section 8.3.3). Once we have obtained a continuous representation of the image, the subpixel-precise thresholding operation conceptually consists of intersecting the image function $f(r,c)$ with the constant function $g(r,c) = g_{\text{sub}}$. Figure 8.30 shows the bilinearly interpolated image $f(r,c)$ in a 2 × 2 block of the four closest pixel centers. The closest pixel centers are lying at the corners of the graph. The bottom of the graph shows the intersection curve of the image $f(r,c)$ in this 2 × 2 block with the constant gray value $g_{\text{sub}} = 100$. Note that this curve is a part of a hyperbola. Since this hyperbolic curve would be quite cumbersome to represent, we can simply substitute it with a straight line segment between the two points where the hyperbola leaves the 2 × 2 block. This line segment constitutes one segment of the subpixel contour we are interested in. Each 2 × 2 block in the image typically contains between zero and two of these line segments. If the 2 × 2 block contains an intersection of two contours four line segments may occur. To obtain meaningful contours, these segments need to be linked. This can be done by repeatedly selecting the first unprocessed line segment in the image as the first segment of the contour and then to trace the adjacent line segments until the contour closes, reaches the image border, or reaches an intersection point. The result of this linking step typically are closed contours that enclose a region in the image in which the gray values are either larger or smaller than the threshold. Note that if such a region contains holes, one contour will be created for the outer boundary of the region and one for each hole.

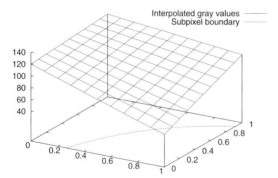

Fig. 8.30 The graph shows gray values that are interpolated bilinearly between four pixel centers, lying at the corners of the graph, and the intersection curve with the gray value $g_{\text{sub}} = 100$ at the bottom of the graph. This curve (a part of a hyperbola) is the boundary between the region with gray values > 100 and gray values < 100.

Figure 8.31(a) shows an image of a PCB that contains a ball grid array (BGA) of solder pads. To ensure good electrical contact, it must be ensured that the pads have the correct shape and position. This requires high accuracy, and since in this application typically the resolution of the image is small compared to the size of the balls and pads the segmentation must be performed with subpixel accuracy. Figure 8.31(b) shows the result of performing a subpixel-precise thresholding operation on the image in Fig. 8.31(a). To see enough details of the results, the part that corresponds to the white rectangle in Fig. 8.31(a) is displayed. The boundary of the pads is extracted with very good accuracy. Figure 8.31(c) shows even more detail: the left pad in the center row of Fig. 8.31(b), which contains an error that must be detected. As can be seen, the subpixel-precise contour correctly captures the erroneous region of the pad. We can also easily see the individual line segments in the subpixel-precise contour and how they are contained in the 2 × 2 pixel blocks. Note that each block lies between four pixel centers. Therefore, the contour's line segments end at the lines that connect the pixel centers. Note also that in this part of the image there is only one block in which two line segments are contained: at the position where the contour enters the error on the pad. All the other blocks contain one or no line segments.

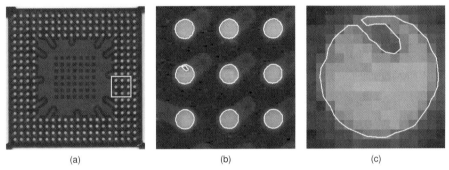

Fig. 8.31 (a) Image of a PCB with BGA solder pads. (b) Result of applying a subpixel-precise threshold to the image in (a). The part that is being displayed corresponds to the white rectangle in (a). (c) Detail of the left pad in the center row of (b).

8.5
Feature Extraction

In the previous sections, we have seen how to extract regions or subpixel-precise contours from an image. While the regions and contours are very useful, they often are not sufficient because they contain the raw description of the segmented data. Often, we must select certain regions or contours from the segmentation result, e.g., to remove unwanted parts of the segmentation. Furthermore, often we are interested in gauging the objects. In other applications, we might want to classify the objects, e.g., in the OCR, to determine the type of the object. All these applications require that we determine one or more characteristic quantities from the regions or contours. The quantities we determine are called features. They typically are real numbers. The process of determining the features is called feature extraction. There are different kinds of features. Region features are features that can be extracted from the regions themselves. In contrast, gray value features also use the gray values in the image within the region. Finally, contour features are based on the coordinates of the contour.

8.5.1
Region Features

By far the simplest region feature is the area of the region:

$$a = |R| = \sum_{(r,c) \in R} 1 = \sum_{i=1}^{n} ce_i - cs_i + 1 \qquad (8.38)$$

Hence, the area a of the region is simply the number of points $|R|$ in the region. If the region is represented as a binary image the first sum has to be used to compute the area, whereas if a run-length representation is used the second sum can be used. Recall from (8.2) that a region can be regarded as the union of its runs and the area of a run is extremely simple to compute. Note that the second sum contains much fewer terms than the first sum, as discussed in Section 8.1.2. Hence, the run-length representation of a region will lead to a much faster computation of the area. This is true for almost all region features.

Figure 8.32 shows the result of selecting all regions with an area ≥ 20 from the regions in Figs. 8.29(a) and (b). Note that all the characters have been selected, while all of the noisy segmentation results have been removed. These regions could now be used as input for the OCR.

(a)

(b)

Fig. 8.32 (a), (b) Result of selecting regions with an area ≥ 20 from the regions in Figs. 8.29(a) and (b). The connected components are visualized by using eight different gray values cyclically.

The area is a special case of a more general class of features called the moments of the region. The moment of order (p,q), $p \geq 0$, $q \geq 0$, is defined as

$$m_{p,q} = \sum_{(r,c) \in R} r^p c^q \qquad (8.39)$$

Note that $m_{0,0}$ is the area of the region. As for the area, simple formulas to compute the moments solely based on the runs can be derived. Hence, the moments can be computed very efficiently in the run-length representation.

The moments in (8.39) depend on the size of the region. Often, it is desirable to have features that are invariant to the size of the objects. To obtain such features, we can simply divide the moments by the area of the region if $p + q \geq 1$ to get normalized moments:

$$n_{p,q} = \frac{1}{a} \sum_{(r,c) \in R} r^p c^q \qquad (8.40)$$

The most interesting feature that can be derived from the normalized moments is the center of gravity of the region, which is given by $(n_{1,0}, n_{0,1})$. It can be used to describe the position of the region. Note that the center of gravity is a subpixel-precise feature, even though it is computed from pixel-precise data.

The normalized moments depend on the position in the image. Often, it is useful to make the features invariant to the position of the region in the image. This can be done by calculating the moments relative to the center of gravity of the region. These central moments are given by ($p + q \geq 2$):

$$\mu_{p,q} = \frac{1}{a} \sum_{(r,c) \in R} (r - n_{1,0})^p (c - n_{0,1})^q \qquad (8.41)$$

Note that they are also normalized. The second central moments ($p + q = 2$) are particularly interesting. They enable us to define an orientation and an extent for the region. This is done by assuming that the moments of order 1 and 2 of the region were obtained from an ellipse. Then, from these five moments the five geometric parameters of the ellipse can be derived. Figure 8.33 displays the ellipse parameters graphically. The center of the ellipse is identical to the center of gravity of the region. The major and minor axes r_1 and r_2 and the angle of the ellipse with respect to the column axis are given by

$$\begin{aligned} r_1 &= \sqrt{2\left(\mu_{2,0} + \mu_{0,2} + \sqrt{(\mu_{2,0} - \mu_{0,2})^2 + 4\mu_{1,1}^2}\right)} \\ r_2 &= \sqrt{2\left(\mu_{2,0} + \mu_{0,2} - \sqrt{(\mu_{2,0} - \mu_{0,2})^2 + 4\mu_{1,1}^2}\right)} \\ \theta &= -\frac{1}{2} \arctan \frac{2\mu_{1,1}}{\mu_{0,2} - \mu_{2,0}} \end{aligned} \qquad (8.42)$$

For a derivation of these results see [39] (note that there the diameters are used instead of the radii). From the ellipse parameters, we can derive another very useful feature: the anisometry r_1/r_2. It is scale invariant and describes how elongated a region is.

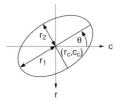

Fig. 8.33 The geometric parameters of an ellipse.

The ellipse parameters are extremely useful to determine orientations and sizes of regions. For example the angle θ can be used to rectify rotated text. Figure 8.34(a) shows the result of thresholding the image in Fig. 8.17(a). The segmentation result is treated as a single region, i.e., the connected components have not been computed. Figure 8.34(a) also displays the ellipse parameters by overlaying the major and minor axis of the equivalent ellipse. Note that the major axis is slightly longer than the region because the equivalent ellipse does not need to have the same area as the region. It only needs to have the same moments of order 1 and 2. The angle of the major axis is a very good estimate for the rotation of the text. In fact, it has been used to rectify the images in Figs. 8.17(b) and (c). Figure 8.34(a) shows the axes of the characters after the connected components have been computed. Note how well the orientation of the regions corresponds with our intuition.

While the ellipse parameters are extremely useful, they have two minor shortcomings. First, the orientation can only be determined if $r_1 \neq r_2$. Our first thought might be that this only applies to circles, which have no meaningful orientation anyway. Unfortunately, this is not true. There is a much larger class of objects for which $r_1 = r_2$. All objects that have a fourfold rotational symmetry like squares have $r_1 = r_2$. Hence, their orientation cannot be determined with the ellipse parameters. The second slight problem is that since the underlying model is an ellipse the orientation θ can only be determined modulo π (180°). This problem can be solved by determining the point in the region that has the largest distance from the center of gravity and use it to select θ or $\theta + \pi$ as the correct orientation.

In the above discussion, we have used various transformations to make the moment-based features invariant to certain transformations, e.g., translation and scaling. Several approaches have been proposed to create moment-based features that are invariant to a larger class of transformations, e.g., translation,

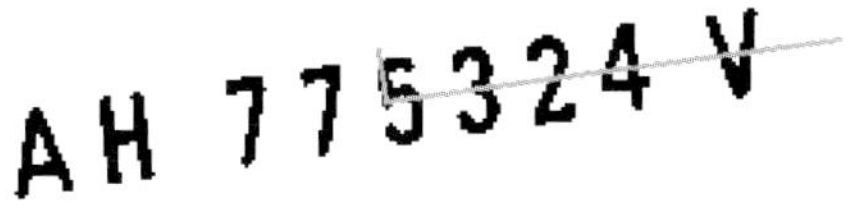

AH 775324 V

(a)

AH 775324 V

(b)

Fig. 8.34 Result of thresholding the image in Fig. 8.17(a) overlaid with a visualization of the ellipse parameters. The light gray lines represent the major and minor axes of the regions. Their intersection is the center of gravity of the regions. (a) The segmentation is treated as a single region. (b) The connected components of the region are used. The angle of the major axis in (a) has been used to rotate the images in Figs. 8.17(b) and (c).

rotation, and scaling [46] or even general affine transformations [32, 62]. They are primarily used to classify objects.

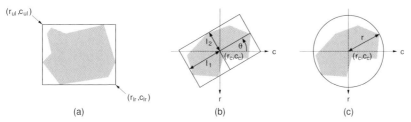

Fig. 8.35 (a) The smallest axis-parallel enclosing rectangle of a region. (b) The smallest enclosing rectangle of arbitrary orientation. (c) The smallest enclosing circle.

Apart from the moment-based features, there are several other useful features that are based on the idea of finding an enclosing geometric primitive for the region. Figure 8.35(a) displays the smallest axis-parallel enclosing rectangle of a region. This rectangle is often also called the bounding box of the region. It can be calculated very easily based on the minimum and maximum row and column coordinates of the region. Based on the parameters of the rectangle, other useful quantities like the width and height of the region and their ratio can be calculated. The parameters of the bounding box are particularly useful if we want to quickly find out whether two regions can intersect. Since the smallest axis-parallel enclosing rectangle sometimes is not very tight we can also define a smallest enclosing rectangle of arbitrary orientation, as shown in Fig. 8.35(b). Its computation is much more complicated than the computation of the bounding box, however, so we cannot give details here. An efficient implementation can be found in [93]. Note that an arbitrarily oriented rectangle has the same parameters as an ellipse. Hence, it also enables us to define the position, size, and orientation of a region. Note that in contrast to the ellipse parameters, a useful orientation for squares is returned. The final useful enclosing primitive is an enclosing circle, as shown in Fig. 8.35(c). Its computation is also quite complex [98]. It also enables us to define the position and size of a region.

The computation of the smallest enclosing rectangle of arbitrary orientation and the smallest enclosing circle is based on first computing the convex hull of the region. The convex hull of a set of points, and in particular a region, is the smallest convex set that contains all the points. A set is convex if for any two points in the set the straight line between them is completely contained in the set. The convex hull of a set of points can be computed efficiently [7, 71]. The convex hull of a region is often useful to construct ROIs from regions that have been extracted from the image. Based on the convex hull of the region, another useful feature can be defined: the convexity, which is defined as the ratio of the area of the region to the area of its convex hull. It is a feature between 0 and 1 that measures how compact the region is. A convex region has a convexity of 1. The convexity can, for example, be used to remove unwanted segmentation results, which often are highly nonconvex.

Another useful feature of a region is its contour length. To compute it, we need to trace the boundary of the region to get a linked contour of the boundary pixels [39]. Once the contour has been computed, we simply need to sum the Euclidean distances of the contour segments, which are 1 for horizontal and vertical segments and $\sqrt{2}$ for diagonal segments. Based on the contour length l and the area a of the region, we can define another measure for the compactness of a region: $c = l^2/(4\pi a)$. For circular regions, this feature is 1, while all other regions have larger values. The compactness has similar uses as the convexity.

8.5.2
Gray Value Features

We have already seen some gray value features in Section 8.2.1: the minimum and maximum gray value within the region:

$$g_{\min} = \min_{(r,c) \in R} g_{r,c} \qquad g_{\max} = \max_{(r,c) \in R} g_{r,c} \qquad (8.43)$$

They are used for the gray value normalization in Section 8.2.1. Another obvious feature is the mean gray value within the region:

$$\bar{g} = \frac{1}{a} \sum_{(r,c) \in R} g_{r,c} \qquad (8.44)$$

Here, a is the area of the region, given by (8.38). The mean gray value is a measure for the brightness of the region. A single measurement within a reference region can be used to measure additive brightness changes with respect to the conditions when the system was set up. Two measurements within different reference regions can be used to measure linear brightness changes, and hence to compute a linear gray value transformation (see Section 8.2.1) that compensates the brightness change, or to adapt segmentation thresholds. The mean gray value is a statistical feature. Another statistical feature is the variance of the gray values

$$s^2 = \frac{1}{a-1} \sum_{(r,c) \in R} (g_{r,c} - \bar{g})^2 \qquad (8.45)$$

and the standard deviation $s = \sqrt{s^2}$. Measuring the mean and standard deviation within a reference region can also be used to construct a linear gray value transformation that compensates brightness changes. The standard deviation can be used to adapt segmentation thresholds. Furthermore, the standard deviation is a measure for the amount of texture that is present within the region.

The gray value histogram (8.3) and the cumulative histogram (8.4), which we have already encountered in Section 8.2.1, are also gray value features. From the histogram, we have already used a feature for the robust contrast normalization: the α-quantile

$$g_\alpha = \min\{g : c_g \geq \alpha\} \qquad (8.46)$$

where c_g is defined in (8.4). It was used to obtain the robust minimum and maximum gray values in Section 8.2.1. The quantiles were called p_l and p_u there. Note that for $\alpha = 0.5$ we obtain the median gray value. It has similar uses as the mean gray value.

In the previous section, we have seen that the region's moments are extremely useful features. They can be extended to gray value features in a natural manner. The gray value moment of order (p, q), $p \geq 0$, $q \geq 0$, is defined as

$$m_{p,q} = \sum_{(r,c) \in R} g_{r,c} r^p c^q \qquad (8.47)$$

This is the natural generalization of the region moments because we obtain the region moments from the gray value moments by using the characteristic function χ_R (8.1) of the region as the gray values. Like for the region moments the moment $a = m_{0,0}$ can be regarded as the gray value area of the region. It is actually the "volume" of the gray value function $g_{r,c}$ within the region. Like for the region moments, normalized moments can be defined by

$$n_{p,q} = \frac{1}{a} \sum_{(r,c) \in R} g_{r,c} r^p c^q \qquad (8.48)$$

The moments $(n_{1,0}, n_{0,1})$ define the gray value center of gravity of the region. With this, central gray value moments can be defined by

$$\mu_{p,q} = \frac{1}{a} \sum_{(r,c) \in R} g_{r,c} (r - n_{1,0})^p (c - n_{0,1})^q \qquad (8.49)$$

Like for the region moments, based on the second central moments we can define the ellipse parameters major and minor axis and the orientation. The formulas are identical to (8.42). Furthermore, the anisometry can also be defined identically as for the regions.

All the moment-based gray value features are very similar to their region-based counterparts. Therefore, it is interesting to look at their differences. Like we saw, the gray value moments reduce to the region moments if the characteristic function of the region is used as the gray values. The characteristic function can be interpreted as a membership of a pixel to the region. A membership of 1 means that the pixel belongs to the region, while 0 means the pixel does not belong to the region. This notion of belonging to the region is crisp, i.e., for every pixel a hard decision must be made. Suppose now that instead of making a hard decision for every pixel we could make a "soft" or "fuzzy" decision about whether a pixel belongs to the region and that we encode the degree of belonging to the region by a number $\in [0, 1]$. We can interpret the degree of belonging as a fuzzy membership value, as opposed to the crisp binary membership value. With this, the gray value image can be regarded as a fuzzy set [65]. The advantage of regarding the image as a fuzzy set is that we do not have to make a hard decision whether a pixel belongs to the object or not. Instead, the fuzzy membership value determines what percentage of the pixel belongs to the object. This enables us to measure the position and size of the objects much more accurately, especially for small objects, because in the transition zone between the foreground and background there will be some mixed pixels that allow us to capture the geometry of the object more accurately. An example of this is shown in Fig. 8.36. Here, a synthetically generated subpixel-precise ideal circle of radius 3 is shifted in subpixel increments. The gray values represent a fuzzy membership, scaled to values between 0 and 200 for display purposes. The figure displays a pixel-precise region, thresholded with a value of 100, which corresponds to a membership above 0.5, as well as two circles that have a center of gravity and area that were obtained from the region and gray value moments. The gray value moments were computed in the entire image. It can be seen that the area and center of gravity are computed much more accurately by the gray value moments because the decision of whether a pixel belongs to the foreground or not has been avoided. In this example, the gray value moments result in an area error that is always smaller than 0.25% and a position error smaller than 1/200 pixel. In contrast, the area error for the region moments can be up to 13.2% and the position error can be up to 1/6 pixel. Note that both types of moments yield subpixel-accurate measurements. We can see that on ideal data it is possible to obtain an extremely high accuracy with the gray value moments, even for very small objects. On real data the accuracy will necessarily be somewhat lower. It should also be noted that the accuracy advantage of the gray value moments primarily occurs for small objects. Because the gray value moments must access every pixel within the region, whereas the region moments can be computed solely based on the run-length representation of the region, the region moments can be computed much faster. Hence, the gray moments are typically only used for relatively small regions.

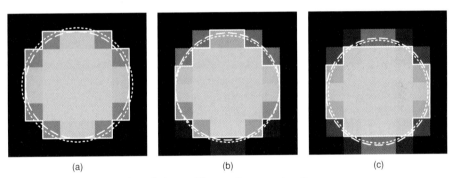

(a) (b) (c)

Fig. 8.36 Subpixel-precise circle position and area using the gray value and region moments. The image represents a fuzzy membership, scaled to values between 0 and 200. The solid line is the result of segmenting with a membership of 100. The dotted line is a circle that has the same center of gravity and area as the segmented region. The dashed line is a circle that has the same gray value center of gravity and gray value area as the image. (a) Shift: 0, error in the area for the region moments: 13.2%, for the gray value moments: -0.05%. (b) Shift: 5/32 pixel, Error in the row coordinate for the region moments: -0.129, for the gray value moments: 0.003. (c) Shift: 1/2 pixel, error in the area for the region moments: -8.0%, for the gray value moments: -0.015%. Note that the gray value moments yield a significantly better accuracy for this small object.

The only question we need to answer is how we can define the fuzzy membership value of a pixel. If we assume that the camera has a fill factor of 100% and the gray value response of the image acquisition device and camera are linear the gray value difference of a pixel to the background gray value is proportional to the portion of the object that is covered by the pixel. Consequently, we can define a fuzzy membership relation as follows: every pixel that has a gray value below the background gray value g_{min} has a membership value of 0. Conversely, every pixel that has a gray value above the foreground gray value g_{max} has a membership value of 1. In between, the membership values are interpolated linearly. Since this procedure would require floating point images, the membership is typically scaled to an integer image with b bits, typically 8 bits. Consequently, the fuzzy membership relation is a simple linear gray value scaling, as defined in Section 8.2.1. If we scale the fuzzy membership image in this manner the gray value area needs to be divided by the maximum gray value, e.g., 255, to obtain the true area. The normalized and central gray value moments do not need to be modified in this manner since they are, by definition, invariant to a scaling of the gray values.

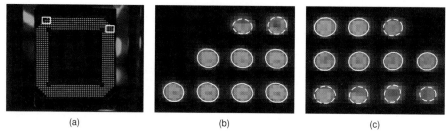

(a)   (b)   (c)

Fig. 8.37 (a) Image of a BGA device. The two rectangles correspond to the image parts shown in (b) and (c). The results of inspecting the balls for correct size (gray value area $\geq$ 20) and correct gray value anisometry ($\leq$ 1.25) are visualized in (b) and (c). Correct balls are displayed as solid ellipses, while defective balls are displayed as dashed ellipses.

Figure 8.37 displays a real application where the above principles are used. In Fig. 8.37(a) a ball grid array (BGA) device with solder balls is displayed, along with two rectangles that indicate the image parts shown in Figs. 8.37(b) and (c). The image in Fig. 8.37(a) is first transformed into a fuzzy membership image using $g_{min} = 40$ and $g_{max} = 120$ with 8 bits resolution. The individual balls are segmented and then inspected for correct size and shape by using the gray value area and the gray value anisometry. The erroneous balls are displayed with dashed lines. To aid the visual interpretation, the ellipses representing the segmented balls are scaled such that they have the same area as the gray value area. This is done because the gray value ellipse parameters typically return an ellipse with a different area than the gray value area, analogously to the region ellipse parameters (see the discussion following (8.42) in Section 8.5.1). As can be seen, all the balls that have an erroneous size or shape, indicating partially missing solder, have been correctly detected.

8.5.3
Contour Features

Many of the region features we have discussed in Section 8.5.1 can be transferred to subpixel-precise contour features in a straightforward manner. For example, the length of the subpixel-precise contour is even easier to compute because the contour is already represented explicitly by its control points (r_i, c_i), $i = 1, \ldots, n$. It is also simple to compute the smallest enclosing axis-parallel rectangle (the bounding box) of the contour. Furthermore, the convex hull of the contour can be computed like for regions [7, 71]. From the convex hull, we can also derive the smallest enclosing circles [98] and the smallest enclosing rectangles of arbitrary orientation [93].

In the previous two sections, we have seen that the moments are extremely useful features. An interesting question, therefore, is whether they can be defined for contours. In particular, it is interesting whether a contour has an area. Obviously, for this to be true the contour must enclose a region, i.e., it must be closed and must not intersect itself. To simplify the formulas, let us assume that a closed contour is specified by $(r_1, c_1) = (r_n, c_n)$. Let the subpixel-precise region that the contour encloses be denoted by R. Then, the moment of order (p, q) is defined as

$$m_{p,q} = \iint_{(r,c) \in R} r^p c^q \, dr \, dc \tag{8.50}$$

Like for regions, we can define normalized and central moments. The formulas are identical to (8.40) and (8.41) with the sums being replaced by integrals. It can be shown that these moments can be computed solely based on the control points of the contour [86]. For example, the area and center of gravity of the contour are given by

$$\begin{aligned} a &= \frac{1}{2} \sum_{i=1}^{n} r_{i-1} c_i - r_i c_{i-1} \\ n_{1,0} &= \frac{1}{6a} \sum_{i=1}^{n} (r_{i-1} c_i - r_i c_{i-1})(r_{i-1} + r_i) \\ n_{0,1} &= \frac{1}{6a} \sum_{i=1}^{n} (r_{i-1} c_i - r_i c_{i-1})(c_{i-1} + c_i) \end{aligned} \tag{8.51}$$

Analogous formulas can be derived for the second order moments. Based on them, we can again compute the ellipse parameters major axis, minor axis, and orientation. The formulas are identical to (8.42). The moment-based contour features can be used for the same purposes as the corresponding region and gray value features. By performing an evaluation similar to that in Fig. 8.36, it can be seen that the contour center of gravity and the ellipse parameters are equally accurate as the gray value center of gravity. The accuracy of the contour area is slightly worse than the gray value area because we have approximated the hyperbolic segments with line segments. Since the true contour is a circle, the line segments always lie inside the true circle. Nevertheless, subpixel-thresholding and the contour moments could also have been used to detect the erroneous balls in Fig. 8.37.

8.6
Morphology

In Section 8.4 we have discussed how to segment regions. We have already seen that segmentation results often contain unwanted noisy parts. Furthermore, sometimes the segmentation will contain parts in which the shape of the object we are interested in has been disturbed, e.g., because of reflections. Therefore, we often need to modify the shape of the segmented regions to obtain the desired results. This is the subject of the field of mathematical morphology, which can be defined as a theory for the analysis of spatial structures [84]. For our purposes, mathematical morphology provides a set of extremely useful operations that enable us to modify or describe the shape of objects. Morphological operations can be defined on regions and gray value images. We will discuss both types of operations in this section.

8.6.1
Region Morphology

All region morphology operations can be defined in terms of six very simple operations: union, intersection, difference, complement, translation, and transposition. We will take a brief look at these operations first.

The union of two regions R and S is defined as the set of points that lie in R or in S:

$$R \cup S = \{p \mid p \in R \vee p \in S\} \qquad (8.52)$$

One important property of the union is that it is commutative: $R \cup S = S \cup R$. Furthermore, it is associative: $(R \cup S) \cup T = R \cup (S \cup T)$. While this may seem like a trivial observation, it will enable us to derive very efficient implementations for the morphological operations below. The algorithm to compute the union of two binary images is obvious: we simply need to compute the logical *or* of the two images. The runtime complexity of this algorithm is obviously $O(wh)$, where w and h are the width and height of the binary image, respectively. In the run-length representation, the union can be computed with a lower complexity: $O(n+m)$, where n and m are the number of runs in R and

S. The principle of the algorithm is to merge the runs of the two regions while observing the order of the runs (see Section 8.1.2) and then to pack overlapping runs into single runs.

The intersection of two regions is defined as the set of points that lie in R and in S:

$$R \cap S = \{p \mid p \in R \land p \in S\} \tag{8.53}$$

Like the union, the intersection is commutative and associative. Again, the algorithm on binary images is obvious: we compute the logical *and* of the two images. For the run-length representation, again an algorithm that has complexity $O(n+m)$ can be found.

The difference of two regions is defined as the set of points that lie in R but not in S:

$$R \setminus S = \{p \mid p \in R \land p \notin S\} = R \cap \overline{S} \tag{8.54}$$

The difference is not commutative and not associative. Note that it can be defined in terms of the intersection and the complement of a region R, which is defined as all the points that do not lie in R:

$$\overline{R} = \{p \mid p \notin R\} \tag{8.55}$$

Since the complement of a finite region is infinite it is impossible to represent it as a binary image. Therefore, for the representation of regions as binary images it is important to define the operations without the complement. It is, however, possible to represent it as a run-length-encoded region by adding a flag that indicates whether the region or its complement is being stored. This can be used to define a more general set of morphological operations. There is an interesting relation between the number of connected components of the background $|C(\overline{R})|$ and the number of holes of the foreground $|H(R)|$: $|C(\overline{R})| = 1 + |H(R)|$. As discussed in Section 8.4.2, complimentary connectivities must be used for the foreground and the background for this relation to hold.

Apart from the set operations, there are two basic geometric transformations that are used in morphological operations. The translation of a region by a vector t is defined as

$$R_t = \{p \mid p - t \in R\} = \{q \mid q = p + t \text{ for } p \in R\} \tag{8.56}$$

Finally, the transposition of a region is defined as a mirroring about the origin:

$$\check{R} = \{-p \mid p \in R\} \tag{8.57}$$

Note that this is the only operation where a special point (the origin) is singled out. All the other operations do not depend on the origin of the coordinate system, i.e., they are translation invariant.

With these building blocks, we can now take a look at the morphological operations. They typically involve two regions. One of these is the region we want to process, which will be denoted by R below. The other region has a special meaning. It is called the structuring element and will be denoted by S. The structuring element is the means by which we can describe the shapes we are interested in.

The first morphological operation we consider is the Minkowski addition, which is defined by

$$R \oplus S = \{r + s \mid r \in R, s \in S\} = \bigcup_{s \in S} R_s = \bigcup_{r \in R} S_r = \{t \mid R \cap (\check{S})_t \neq \emptyset\} \quad (8.58)$$

It is interesting to interpret the formulas. The first formula says that to get the Minkowski addition of R with S, we take every point in R and every point in S and compute the vector sum of the points. The result of the Minkowski addition is the set of all points thus obtained. If we single out S this can also be interpreted as taking all points in S, translating the region R by the vector corresponding to the point s from S, and to compute the union of all the translated regions. Thus, we obtain the second formula. By symmetry, we can also translate S by all points in R to obtain the third formula. Another way to look at the Minkowski addition is the fourth formula: it tells us that we move the transposed structuring element around in the plane. Whenever the translated transposed structuring element and the region have at least one point in common we copy the translated reference point into the output. Figure 8.38 shows an example for the Minkowski addition.

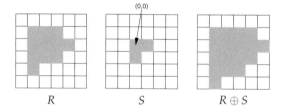

Fig. 8.38 Example for the Minkowski addition $R \oplus S$.

While the Minkowski addition has a simple formula, it has one small drawback. Its geometric criterion is that the transposed structuring element has at least one point in common with the region. Ideally, we would like to have an operation that returns all translated reference points for which the structuring element itself has at least one point in common with the region. To achieve this, we only need to use the transposed structuring element in the Minkowski addition. This operation is called a dilation, and is defined by

$$R \oplus \check{S} = \{t \mid R \cap S_t \neq \emptyset\} = \bigcup_{s \in S} R_{-s} \quad (8.59)$$

Figure 8.39 shows an example for the dilation. Note that the result of the Minkowski addition and dilation are different. This is true whenever the structuring element is not symmetric with respect to the origin. If the structuring element is symmetric the Minkowski addition and dilation are identical. Please be aware that this is assumed in many discussions about and implementations of the morphology. Therefore, the dilation is often defined without the transposition, which is technically incorrect.

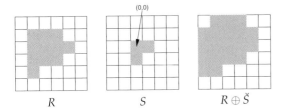

Fig. 8.39 Example for the dilation $R \oplus \check{S}$.

The implementation of the Minkowski addition for binary images is straightforward. As suggested by the second formula in (8.58), it can be implemented as a nonlinear filter with logical *or* operations. The runtime complexity is proportional to the size of the image times the number of pixels in the structuring element. The second factor can be reduced to roughly the number of boundary pixels in the structuring element [23]. Also, for binary images represented with one bit per pixel very efficient algorithms can be developed for special structuring elements [10]. Nevertheless, in both cases the runtime complexity is proportional to the number of pixels in the image. To derive an implementation for the run-length representation of the regions, we first need to examine some algebraic properties of the Minkowski addition. It is commutative: $R \oplus S = S \oplus R$. Furthermore, it is distributive with respect to the union: $(R \cup S) \oplus T = R \oplus T \cup S \oplus T$. Since a region can be regarded as the union of its runs we can use the commutativity and distributivity to transform the Minkowski addition as follows:

$$R \oplus S = \left(\bigcup_{i=1}^{n} \mathbf{r}_i \right) \oplus \left(\bigcup_{j=1}^{m} \mathbf{s}_j \right) = \bigcup_{j=1}^{m} \left(\left(\bigcup_{i=1}^{n} \mathbf{r}_i \right) \oplus \mathbf{s}_j \right) = \bigcup_{i=1}^{n} \bigcup_{j=1}^{m} \mathbf{r}_i \oplus \mathbf{s}_j \quad (8.60)$$

Therefore, the Minkowski addition can be implemented as the union of nm dilations of single runs, which are trivial to compute. Because the union of the runs can also be computed easily, the runtime complexity is $O(mn)$, which is better than for binary images.

As we have seen above, the dilation and Minkowski addition enlarge the input region. This can be used, for example, to merge separate parts of a region into a single part, and thus to obtain the correct connected components of objects. One example for this is shown in Fig. 8.40. Here, we want to segment each character as a separate connected component. If we compute the connected components of the thresholded region in Figure 8.40(b) we can see that the characters and their dots are separate components (Fig. 8.40(c)). To solve this problem, we first need to connect the dots with their characters. This can be achieved using a dilation with a circle of diameter 5 (Fig. 8.40(d)). With this, the correct connected components are obtained (Fig. 8.40(e)). Unfortunately, they have the wrong shape because of the dilation. This can be corrected by intersecting the components with the originally segmented region. Figure 8.40(f) shows that with these simple steps we have obtained one component with the correct shape for each character.

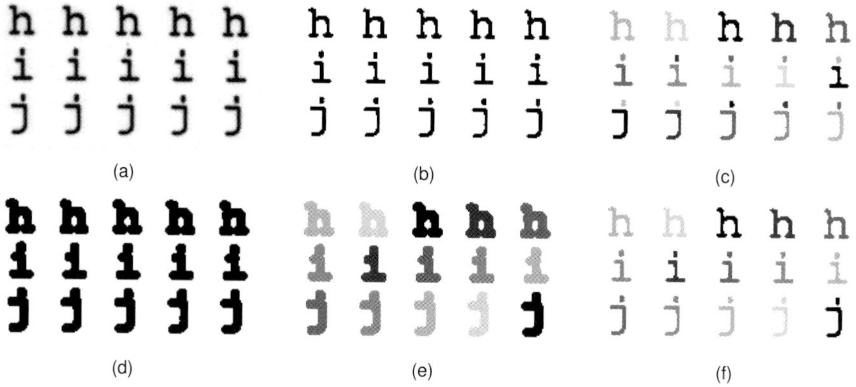

Fig. 8.40 (a) Image of a print of several characters. (b) Result of thresholding (a). (c) Connected components of (b) displayed with six different gray values. Note that the characters and their dots are separate connected components, which is undesirable. (d) Result of dilating the region in (b) with a circle of diameter 5. (e) Connected components of (d). Note that each character is now a single connected component. (f) Result of intersecting the connected components in (e) with the original segmentation in (b). This transforms the connected components into the correct shape.

The dilation is also very useful for constructing ROIs based on regions that were extracted from the image. We will see an example for this in Section 8.7.3.

The second type of morphological operations is the Minkowski subtraction. It is defined by

$$R \ominus S = \bigcap_{s \in S} R_s = \{r \mid \forall s \in S : r - s \in R\} = \{t \mid (\check{S})_t \subseteq R\} \qquad (8.61)$$

The first formula is similar to the second formula in (8.58) with the union having been replaced by an intersection. Hence, we can still think about moving the region R by all vectors s from S. However, now the points must be contained in all translated regions (instead of at least one translated region). This is what the second formula in (8.61) expresses. Finally, if we look at the third formula we see that we can also move the transposed structuring element around in the plane. If it is completely contained in the region R we add its reference point to the output. Again, note the similarity to the Minkowski addition, where the structuring element had to have at least one point in common with the region. For the Minkowski subtraction it must lie completely within the region. Figure 8.41 shows an example for the Minkowski subtraction.

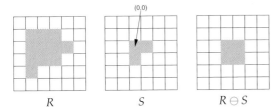

Fig. 8.41 Example for the Minkowski subtraction $R \ominus S$.

The Minkowski subtraction has the same small drawback as the Minkowski addition: its geometric criterion is that the transposed structuring element must completely lie within the region. As for the dilation, we can use the transposed structuring element in the Minkowski subtraction. This operation is called an erosion and is defined by

$$R \ominus \check{S} = \bigcap_{s \in S} R_{-s} = \{t \mid S_t \subseteq R\} \qquad (8.62)$$

Figure 8.42 shows an example for the erosion. Again, note that the Minkowski subtraction and erosion only produce identical results if the structuring element is symmetric with respect to the origin. Please be aware that this is often silently assumed and the erosion is defined as a Minkowski subtraction, which is technically incorrect.

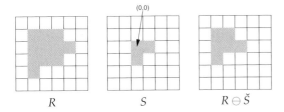

Fig. 8.42 Example for the erosion $R \ominus \check{S}$.

As we have seen from the small examples, the Minkowski subtraction and erosion shrink the input region. This can, for example, be used to separate objects that are attached to each other. Figure 8.43 shows an example for this. Here, the goal is to segment the individual globular objects. The result of thresholding the image is shown in Fig. 8.43(b). If we compute the connected components of this region an incorrect result is obtained because several objects touch each other (Fig. 8.43(c)). The solution is to erode the region with a circle of diameter 15 (Fig. 8.43(d)) before computing the connected components (Fig. 8.43(e)). Unfortunately, the connected components have the wrong shape. Here, we cannot use the same strategy that we used for the dilation (intersecting the connected components with the original segmentation) because the erosion has shrunk the region. To approximately get the original shape back, we can dilate the connected components with the same structuring element that we used for the erosion (Figure 8.43(f)).

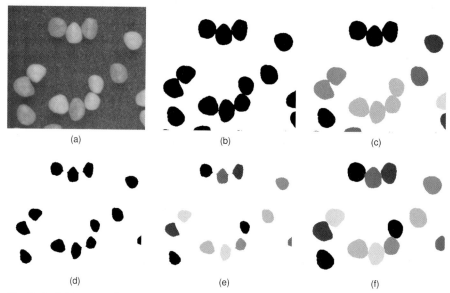

Fig. 8.43 (a) Image of several globular objects. (b) Result of thresholding (a). (c) Connected components of (b) displayed with six different gray values. Note that several objects touch each other and hence are in the same connected component. (d) Result of eroding the region in (b) with a circle of diameter 15. (e) Connected components of (d). Note that each object is now a single connected component. (f) Result of dilating the connected components in (e) with a circle of diameter 15. This transforms the correct connected components into approximately the correct shape.

We can see another use of the erosion if we remember its definition: it returns the translated reference point of the structuring element S for every translation for which S_t completely fits into the region R. Hence, the erosion acts like a template matching operation. An example of this use of the erosion is shown in Fig. 8.44. In Fig. 8.44(a), we can see an image of a print of several letters with the structuring element used for the erosion overlaid in white. The structuring element corresponds to the center line of the letter "e." The reference point of the structuring element is its center of gravity. The result of eroding the thresholded letters (Fig. 8.44(b)) with the structuring element is shown in Fig. 8.44(c). Note that all letters "e" have been correctly identified. In Figs. 8.44(d)–(f), the experiment is repeated with another set of letters. The structuring element is the center line of the letter "o." Note that the erosion correctly finds the letters "o." However, additionally the circular parts of the letters "p" and "q" are found since the structuring element completely fits into them.

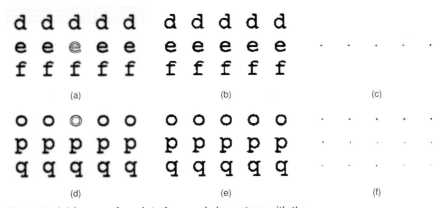

Fig. 8.44 (a) Image of a print of several characters with the structuring element used for the erosion overlaid in white. (b) Result of thresholding (a). (c) Result of the erosion of (b) with the structuring element in (a). Note that the reference point of all letters "e" has been found. (d) A different set of characters with the structuring element used for the erosion overlaid in white. (e) Result of thresholding (d). (f) Result of the erosion of (e) with the structuring element in (d). Note that the reference point of the letter "o" has been identified correctly. In addition, the circular parts of the letters "p" and "q" have been extracted.

An interesting property of the Minkowski addition and subtraction as well as the dilation and erosion is that they are dual to each other with respect to the complement operation. For the Minkowski addition and subtraction we have: $R \oplus S = \overline{\overline{R} \ominus S}$ and $R \ominus S = \overline{\overline{R} \oplus S}$. The same identities hold for the dilation and erosion. Hence, a dilation of the foreground is identical to an erosion of the background and vice versa. We can make use of the duality whenever we want to avoid computing the complement explicitly, and hence to speed up some operations. Note that the duality only holds if the complement can be infinite. Hence, it does not hold for binary images, where the complemented region needs to be clipped to a certain image size.

One extremely useful application of the erosion and dilation is the calculation of the boundary of a region. The algorithm to compute the true boundary as a linked list of contour points is quite complicated [39]. However, an approximation to the boundary can be computed very easily. If we want to compute the inner boundary, we simply need to erode the region appropriately and to subtract the eroded region from the original region:

$$\partial R = R \setminus (R \ominus S) \quad (8.63)$$

By duality, the outer boundary (the inner boundary of the background) can be computed with a dilation:

$$\partial R = (R \oplus S) \setminus R \quad (8.64)$$

To get a suitable boundary, the structuring element S must be chosen appropriately. If we want to obtain an 8-connected boundary we must use the structuring element S_8 in Fig. 8.45. If we want a 4-connected boundary we must use S_4.

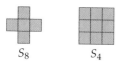

Fig. 8.45 The structuring elements for computing the boundary of a region with 8-connectivity (S_8) and 4-connectivity (S_4).

Figure 8.46 displays an example of computation of the inner boundary of a region. A small part of the input region is shown in Fig. 8.46(a). The boundary of the region computed by (8.63) with S_8 is shown in Fig. 8.46(b), while the result with S_4 is shown in Fig. 8.46(c). Note that the boundary is only approximately 8- or 4-connected. For example, in the 8-connected boundary there are occasional 4-connected pixels. Finally, the boundary of the region as computed by an algorithm that traces around the boundary of the region and links the boundary points into contours is shown in Fig. 8.46(d). Note that this is the true boundary of the region. Also note that since only part of the region is displayed there is no boundary at the bottom of the displayed part.

As we have seen above, the erosion can be used as a template matching operation. However, it is sometimes not selective enough and returns too many matches. The reason for this is that the erosion does not take into account the background. For this reason, an operation that explicitly models the background is needed. This operation is called the hit-or-miss transform. Since the foreground and background should be taken into account it uses a structuring element that consists of two parts: $S = (S^f, S^b)$ with $S^f \cap S^b = \emptyset$. With this, the hit-or-miss transform is defined as

$$R \otimes S = (R \ominus \check{S}^f) \cap (\overline{R} \ominus \check{S}^b) = (R \ominus \check{S}^f) \setminus (R \oplus \check{S}^b) \quad (8.65)$$

8.6 Morphology

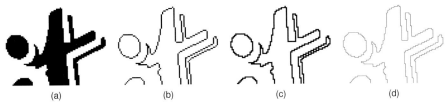

Fig. 8.46 (a) Detail of a larger region. (b) 8-connected boundary of (a) computed by (8.63). (c) 4-connected boundary of (a). (d) Linked contour of the boundary of (a).

Hence the hit-or-miss transform returns those translated reference points for which the foreground structuring element S^f completely lies within the foreground and the background structuring element S^b completely lies within the background. The second equation is especially useful from an implementation point of view since it avoids having to compute the complement. The hit-or-miss transform is dual to itself if the foreground and background structuring elements are exchanged: $R \otimes S = \overline{R} \otimes S'$, where $S' = (S^b, S^f)$.

Figure 8.47 shows the same image as Fig. 8.44(d). The goal here is to match only the letters "o" in the image. To do so, we can define a structuring element that crosses the vertical strokes of the letters "p" and "q" (and also "b" and "d"). One possible structuring element for this purpose is shown in Fig. 8.47(b). With the hit-or-miss transform, we are able to remove the found matches for the letters "p" and "q" from the result, as can be seen from Fig. 8.47(c).

Fig. 8.47 (a) Image of a print of several characters. (b) The structuring element used for the hit-or-miss transform. The black part is the foreground structuring element and the light gray part is the background structuring element. (c) Result of the hit-or-miss transform of the thresholded image (see Fig. 8.44(e)) with the structuring element in (b). Note that only the reference point of the letter "o" has been identified, in contrast to the erosion (see Fig. 8.44(f)).

We now turn our attention to operations in which the basic operations we have discussed so far are executed in succession. The first such operation is the opening, defined by

$$R \circ S = (R \ominus \check{S}) \oplus S = \bigcup_{S_t \subseteq R} S_t \qquad (8.66)$$

Hence, the opening is an erosion followed by a Minkowski addition with the same structuring element. The second equation tells us that we can visualize the opening by moving the structuring element around the plane. Whenever the structuring element completely lies within the region we add the entire translated structuring element to the output region (and not just the translated reference point as in the erosion). The opening's definition causes the location of the reference point to cancel out, which can be seen from the second equation. Therefore, the opening is translation invariant with respect to the structuring element. In contrast to the erosion and dilation, the opening is idempotent, i.e., applying it multiple times has the same effect as applying it once: $(R \circ S) \circ S = R \circ S$.

Like the erosion, the opening can be used as a template matching operation. In contrast to the erosion and hit-or-miss transform, it returns all points of the input region into which the structuring element fits. Hence it preserves the shape of the object to find. An example of this is shown in Fig. 8.48, where the same input images and structuring elements as in Fig. 8.44 are used. Note that the opening has found the same instances of the structuring elements as the erosion but has preserved the shape of the matched structuring elements. Hence, in this example it also finds the letters "p" and "q." To find only the letters "o," we could combine the hit-or-miss transformation with a Minkowski addition to get a hit-or-miss opening: $R \odot S = (R \otimes S) \oplus S^f$.

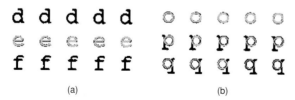

(a)   (b)

Fig. 8.48 (a) Result of applying an opening with the structuring element in Fig. 8.44(a) to the segmented region in Fig. 8.44(b). (b) Result of applying an opening with the structuring element in Fig. 8.44(d) to the segmented region in Fig. 8.44(e). The result of the opening is overlaid in light gray onto the input region, displayed in black. Note that the opening finds the same instances of the structuring elements as the erosion but preserves the shape of the matched structuring elements.

Another very useful property of the opening results if structuring elements like circles or rectangles are used. If an opening with these structuring elements is performed parts of the region that are smaller than the structuring element are removed from the region. This can be used to remove unwanted appendages from the region and to smooth the boundary of the region by removing small protrusions. Furthermore, small bridges between object parts can be removed, which can be used to separate objects. Finally, the opening can be used to suppress small objects. Figure 8.49 shows an example of using the opening to remove unwanted appendages and small objects from the segmentation. In Fig. 8.49(a), an image of a ball-bonded die is shown. The goal is to segment the balls on the pads. If the image is thresholded (Fig. 8.49(b)) the wires that are attached to the balls are also extracted. Furthermore, there are extraneous small objects in the segmentation. By performing an opening with a circle of diameter 31, the wires and small objects are removed, and only smooth region parts that correspond to the balls are retained.

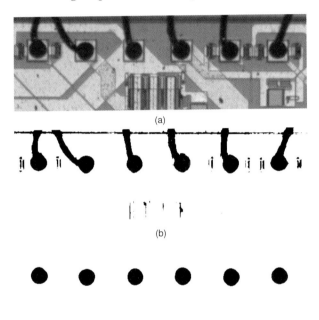

Fig. 8.49 (a) Image of a ball-bonded die. The goal is to segment the balls. (b) Result of thresholding (a). The segmentation includes the wires that are bonded to the pads. (c) Result of performing an opening with a circle of diameter 31. The wires and the other extraneous segmentation results have been removed by the opening and only the balls remain.

The second interesting operation in which the basic morphological operations are executed in succession is the closing, defined by

$$R \bullet S = (R \oplus \check{S}) \ominus S = \overline{\bigcup_{S_t \subseteq \overline{R}} S_t} \qquad (8.67)$$

Hence, the closing is a dilation followed by a Minkowski subtraction with the same structuring element. There is, unfortunately, no simple formula that tells us how the closing can be visualized. The second formula is actually defined by the duality of the opening and the closing: a closing on the foreground is identical to an opening on the background and vice versa: $R \bullet S = \overline{\overline{R} \circ S}$ and $R \circ S = \overline{\overline{R} \bullet S}$. Like the opening, the closing is translation invariant with respect to the structuring element. Furthermore, it is also idempotent.

Since the closing is dual to the opening it can be used to merge objects that are separated by gaps that are smaller than the structuring element. If structuring elements like circles or rectangles are used the closing can be used to close holes and to remove indentations that are smaller than the structuring element. The second property enables us to smooth the boundary of the region.

Figure 8.50 shows how the closing can be used to remove indentations in a region. In Fig. 8.50(a), a molded plastic part with a protrusion is shown. The goal is to detect the protrusion because it is a production error. Since the actual object is circular if the entire part were visible the protrusion could be detected by performing an opening with a circle that is almost as large as the object and then subtracting the opened region from the original segmentation. However, only a part of the object is visible, so the erosion in the opening would create artifacts or remove the object entirely. Therefore, by duality we can pursue the opposite approach: we can segment the background and perform a closing on it. Figure 8.50(b) shows the result of thresholding the background. The protrusion is now an indentation in the background. The result of performing a closing with a circle of diameter 801 is shown in Fig. 8.50(c). The diameter of the circle was set to 801 because it is large enough to completely fill the indentation and to recover the circular shape of the object. If much smaller circles were used, e.g., with a diameter of 401, the indentation would not be filled completely. To detect the error itself, we can compute the difference between the closing and the original segmentation. To remove some noisy pixels that are caused because the boundary of the original segmentation is not as smooth as the closed region, the difference can be post-processed with an opening, e.g., with a 5×5 rectangle, to remove the noisy pixels. The resulting error region is shown in Fig. 8.50(d).

The operations we have discussed so far have been mostly concerned with the region as a two-dimensional object. The only exception has been the calculation of the boundary of a region, which reduces a region to its one-dimensional outline, and hence gives a more condensed description of the region. If the objects are mostly linear, i.e., are regions that have a much greater length than width, a more salient description of the object would be obtained if we could somehow capture its one pixel wide center line. This center line is called the skeleton or medial axis of the region. Several definitions of a skeleton can be given [84]. One intuitive definition can be obtained if we imagine that we try to fit circles that are as large as possible into the region. More pre-

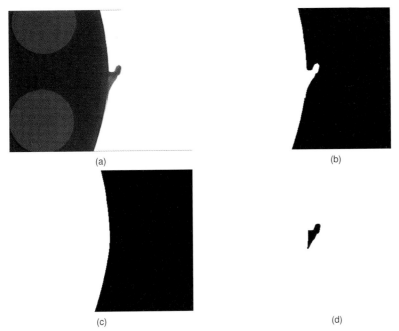

Fig. 8.50 (a) Image of size 768×576 showing a molded plastic part with a protrusion. (b) Result of thresholding the background of (a). (c) Result of a closing on (b) with a circle of diameter 801. Note that the protrusion (the indentation in the background) has been filled in and the circular shape of the plastic part has been recovered. (d) Result of computing the difference between (c) and (b) and performing an opening with a 5×5 rectangle on the difference to remove small parts. The result is the erroneous protrusion of the mold.

cisely, a circle C is maximal in the region R if there is no other circle in R that is a superset of C. The skeleton then is defined as the set of the centers of the maximal circles. Consequently, a point on the skeleton has at least two different points on the boundary of the region to which it has the same shortest distance. Algorithms to compute the skeleton are given in [84, 55]. They basically can be regarded as sequential hit-or-miss transforms that find points on the boundary of the region that cannot belong to the skeleton and delete them. The goal of the skeletonization is to preserve the homotopy of the region, i.e., the number of connected components and holes. One set of structuring elements for computing an 8-connected skeleton is shown in Fig. 8.51 [84]. These structuring elements are used sequentially in all four possible orientations to find pixels with the hit-or-miss transform that can be deleted from the region. The iteration is continued until no changes occur. It should be noted that the skeletonization is an example of an algorithm that can be implemented more efficiently on binary images than on the run-length representation.

Fig. 8.51 The structuring elements for computing an 8-connected skeleton of a region. These structuring elements are used sequentially in all four possible orientations to find pixels that can be deleted.

Figure 8.52(a) shows a part of an image of a printed circuit board with several conductors. The image is threshold (Fig. 8.52(b)), and the skeleton of the thresholded region is computed with the above algorithm (Fig. 8.52(c)). Note that the skeleton contains several undesirable branches on the upper two conductors. For this reason, many different skeletonization algorithms have been proposed. One algorithm that produces relatively few unwanted branches is described in [24]. The result of this algorithm is shown in Fig. 8.52(d). Note that there are no undesirable branches in this case.

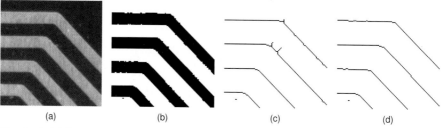

Fig. 8.52 (a) Image showing a part of a PCB with several conductors. (b) Result of thresholding (a). (c) 8-connected skeleton computed with an algorithm that uses the structuring elements in Fig. 8.51 [84]. (d) Result of computing the skeleton with an algorithm that produces fewer skeleton branches [24].

The final region morphology operation we will discuss is the distance transform, which returns an image instead of a region. This image contains for each point in the region R the shortest distance to a point outside the region (i.e., to $\overline{R}$). Consequently, all points on the inner boundary of the region have a distance of 1. Typically, the distance of the other points is obtained by considering paths that must be contained in the pixel grid. Thus, the chosen connectivity defines which paths are allowed. If the 4-connectivity is used the corresponding distance is called the city-block distance. Let (r_1, c_1) and (r_2, c_2) be two points. Then the city-block distance is given by $d_4 = |r_2 - r_1| + |c_2 - c_1|$. Figure 8.53(a) shows the city-block distance between two points. In the example, the city-block distance is 5. On the other hand, if 8-connectivity is used the corresponding distance is called the chessboard distance. It is given by $d_8 = \max\{|r_2 - r_1|, |c_2 - c_1|\}$. In the example in Fig. 8.53(b), the chessboard distance between the two points is 3. Both of these distances are approximations to the Euclidean distance, given by $d_e = \sqrt{(r_2 - r_1)^2 + (c_2 - c_1)^2}$. For the example in Fig. 8.53(c), the Euclidean distance is $\sqrt{13}$.

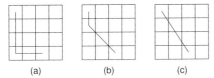

Fig. 8.53 (a) City-block distance between two points. (b) Chessboard distance. (c) Euclidean distance.

Algorithms to compute the distance transform are described in [11]. They work by initializing the distance image outside the region with 0 and within the region with a suitably chosen maximum distance, i.e., $2^b - 1$, where b is the number of bits in the distance image, e.g., $2^{16} - 1$. Then, two sequential line-by-line scans through the image are performed, one from the top left to the bottom right corner, and the second in the opposite direction. In each case, a small mask is placed at the current pixel and the minimum over the elements in the mask of the already computed distances plus the elements in the mask is computed. The two masks are shown in Fig. 8.54. If $d_1 = 1$ and $d_2 = \infty$ are used (i.e., d_2 is ignored) the city-block distance is computed. For $d_1 = 1$ and $d_2 = 1$, the chessboard distance results. Interestingly, if $d_1 = 3$ and $d_2 = 4$ is used and the distance image is divided by 3, a very good approximation to the Euclidean distance results, which can be computed solely with integer operations. This distance is called the chamfer-3-4 distance [11]. With slight modifications, the true Euclidean distance can be computed [19]. The principle is to compute the number of horizontal and vertical steps to reach the boundary using masks similar to the ones in Fig. 8.54, and then to compute the Euclidean distance from the number of steps.

d_2	d_1	d_2
d_1	0	

0	d_1	
d_2	d_1	d_2

Fig. 8.54 Masks used in the two sequential scans to compute the distance transform. The left mask is used in the left-to-right, top-to-bottom scan. The right mask is used in the scan in the opposite direction.

The skeleton and the distance transform can be combined to compute the width of linear objects very efficiently. In Fig. 8.55(a), a PCB with conductors that have several errors is shown. The protrusions on the conductors are called spurs, while the indentations are called mouse bites [67]. They are deviations from the correct conductor width. Figure 8.55(b) shows the result of computing the distance transform with the chamfer-3-4 distance on the segmented conductors. The errors are clearly visible in the distance transform.

To extract the width of the conductors, we need to calculate the skeleton of the segmented conductors (Fig. 8.55(c)). If the skeleton is used as the ROI for the distance image each point on the skeleton will have the corresponding distance to the border of the conductor. Since the skeleton is the center line of the conductor this distance is the width of the conductor. Hence, to detect errors, we simply need to threshold the distance image within the skeleton. Note that in this example it is extremely useful that we have defined that images can have an arbitrary region of interest. Figure 8.55(d) shows the result of drawing circles at the centers of gravity of the connected components of the error region. All major errors have been detected correctly.

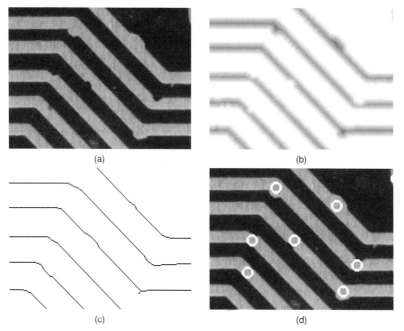

Fig. 8.55 (a) Image showing a part of a PCB with several conductors that have spurs and mouse bites. (b) Distance transform of the result of thresholding (a). The distance image is visualized inverted (dark gray values correspond to large distances). (c) Skeleton of the segmented region. (d) Result of extracting too narrow or too wide parts of the conductors by using (c) as the ROI for (b) and thresholding the distances. The errors are visualized by drawing circles at the centers of gravity of the connected components of the error region.

8.6.2
Gray Value Morphology

Because morphological operations are very versatile and useful the question whether they can be extended to gray value images arises quite naturally. This can indeed be done. In analogy to the region morphology, let $g(r,c)$ denote the image that should be processed and let $s(r,c)$ be an image with ROI S. Like in the region morphology, the image s is called the structuring element. The gray value Minkowski addition is then defined as

$$g \oplus s = (g \oplus s)_{r,c} = \max_{(i,j) \in S} \{g_{r-i,c-j} + s_{i,j}\} \qquad (8.68)$$

This is a natural generalization because the Minkowski addition for regions is obtained as a special case if the characteristic function of the region is used as the gray value image. If, additionally, an image with gray value 0 within the ROI S is used as the structuring element the Minkowski addition becomes:

$$g \oplus s = \max_{(i,j) \in S} \{g_{r-i,c-j}\} \qquad (8.69)$$

For characteristic functions, the maximum operation corresponds to the union. Furthermore, $g_{r-i,c-j}$ corresponds to the translation of the image by the vector (i,j). Hence, (8.69) is equivalent to the second formula in (8.58).

Like in the region morphology, the dilation can be obtained by transposing the structuring element. This results in the following definition:

$$g \oplus \check{s} = (g \oplus \check{s})_{r,c} = \max_{(i,j) \in S} \{g_{r+i,c+j} + s_{i,j}\} \qquad (8.70)$$

The typical choice for the structuring element in the gray value morphology is the flat structuring element that was already used above: $s(r,c) = 0$ for $(r,c) \in S$. With this, the gray value dilation has a similar effect as the region dilation: it enlarges the foreground, i.e., parts in the image that are brighter than their surroundings, and shrinks the background, i.e., parts in the image that are darker than their surroundings. Hence, it can be used to connect disjoint parts of a bright object in the gray value image. This is sometimes useful if the object cannot be segmented easily using region operations alone. Conversely, the dilation can be used to split dark objects.

The Minkowski subtraction for gray value images is given by

$$g \ominus s = (g \ominus s)_{r,c} = \min_{(i,j) \in S} \{g_{r-i,c-j} - s_{i,j}\} \qquad (8.71)$$

As above, by transposing the structuring element we obtain the gray value erosion:

$$g \ominus \check{s} = (g \ominus \check{s})_{r,c} = \min_{(i,j) \in S} \{g_{r+i,c+j} - s_{i,j}\} \qquad (8.72)$$

Like the region erosion, the gray value erosion shrinks the foreground and enlarges the background. Hence, the erosion can be used to split touching bright objects and to connect disjoint dark objects. In fact, the dilation and erosion, as well as the Minkowski addition and subtraction, are dual to each other, like for regions. For the duality, we need to define what the complement of an image should be. If the images are stored with b bits the natural definition for the complement operation is $\overline{g}_{r,c} = 2^b - 1 - g_{r,c}$. With this, it can be easily shown that the erosion and dilation are dual: $g \oplus s = \overline{\overline{g} \ominus s}$ and $g \ominus s = \overline{\overline{g} \oplus s}$. Therefore, all properties that hold for one operation for bright objects hold for the other operation for dark objects and vice versa.

Note that the dilation and erosion can also be regarded as two special rank filters (see Section 8.2.3) if flat structuring elements are used. They select the minimum and maximum gray value within the domain of the structuring element, which can be regarded as the filter mask. Therefore, the dilation and erosion are sometimes referred to as the maximum and minimum filters (or max and min filters).

Efficient algorithms to compute the dilation and erosion are given in [23]. Their runtime complexity is $O(whn)$, where w and h are the dimensions of the image, while n is roughly the number of points on the boundary of the domain of the structuring element for flat structuring elements. For rectangular structuring elements, algorithms with a runtime complexity of $O(wh)$, i.e., with a constant number of operations per pixel, can be found [35]. This is similar to the recursive implementation of a linear filter.

With these building blocks, we can define a gray value opening like for regions as an erosion followed by a Minkowski addition

$$g \circ s = (g \ominus \check{s}) \oplus s \tag{8.73}$$

and the closing as a dilation followed by a Minkowski subtraction

$$g \bullet s = (g \oplus \check{s}) \ominus s \tag{8.74}$$

The gray value opening and closing have similar properties as their region counterparts. In particular, with the above definition of the complement for images they are dual to each other: $g \circ s = \overline{\overline{g} \bullet s}$ and $g \bullet s = \overline{\overline{g} \circ s}$. Like the region operations, they can be used to fill in small holes or, by duality, to remove small objects. Furthermore, they can be used to join or separate objects and to smooth the inner and outer boundaries of objects in the gray value image.

Figure 8.56 shows how the gray value opening and closing can be used to detect errors on the conductors on a PCB. We have already seen in Fig. 8.55 that some of these errors can be detected by looking at the width of the conductors with the distance transform and the skeleton. This technique is very

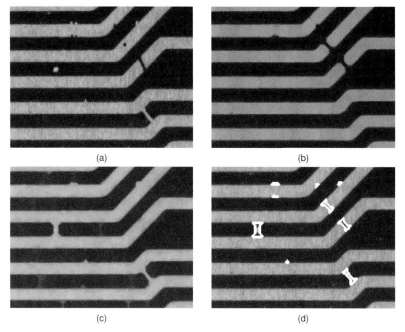

Fig. 8.56 (a) Image showing a part of a PCB with several conductors that have spurs, mouse bites, pin holes, spurious copper, and open and short circuits. (b) Result of performing a gray value opening with an octagon of diameter 11 on (a). (c) Result of performing a gray value closing with an octagon of diameter 11 on (a). (d) Result of segmenting the errors in (a) by using a dynamic threshold operation with the images of (b) and (c).

useful because it enables us to detect relatively large areas with errors. However, small errors are harder to detect with this technique because the distance transform and skeleton are only pixel-precise, and consequently the diameter of the conductor can only be determined reliably with a precision of two pixels. Smaller errors can be detected more reliably with the gray value morphology. Figure 8.56(a) shows a part of a PCB with several conductors that have spurs, mouse bites, pin holes, spurious copper, and open and short circuits [67]. The result of performing a gray value opening and closing with an octagon of diameter 11 is shown in Figs. 8.56(b) and (c). Because of the horizontal, vertical, and diagonal layout of the conductors, using an octagon as the structuring element is preferable. It can be seen that the opening smooths out the spurs, while the closing smooths out the mouse bites. Furthermore, the short circuit and spurious copper are removed by the opening, while the pin hole and open circuit are removed by the closing. To detect these errors, we can require that the opened and closed images should not differ too much. If there were no errors the differences would solely be caused by the texture on the conductors. Since the gray values of the opened image are always smaller than those of the closed image we can use the dynamic threshold operation for bright objects (8.36) to perform the required segmentation. Every pixel that has a gray value difference greater than g_{diff} can be considered as an error. Figure 8.56(d) shows the result of segmenting the errors using a dynamic threshold $g_{\text{diff}} = 60$. This detects all the errors on the board.

We conclude this section with an operator that computes the range of gray values that occur within the structuring element. This can be obtained easily by calculating the difference between the dilation and erosion:

$$g \diamond s = (g \oplus š) - (g \ominus š) \qquad (8.75)$$

Since this operator produces similar results as a gradient filter (see Section 8.7) it is sometimes called the morphological gradient.

Figure 8.57 shows how the gray range operator can be used to segment punched serial numbers. Because of the scratches, texture, and illumination it is difficult to segment the characters in Fig. 8.57(a) directly. In particular, the scratch next to the upper left part of the "2" cannot be separated from the "2" without splitting several of the other numbers. The result of computing the gray range within a 9×9 rectangle is shown in Fig. 8.57(b). With this, it is easy to segment the numbers (Fig. 8.57(c)) and to separate them from other segmentation results (Fig. 8.57(d)).

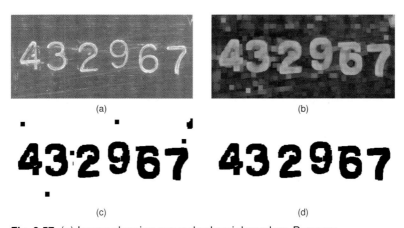

Fig. 8.57 (a) Image showing a punched serial number. Because of the scratches, texture, and illumination it is difficult to segment the characters directly. (b) Result of computing the gray range within a 9×9 rectangle. (c) Result of thresholding (b). (d) Result of computing the connected components of (c) and selecting the characters based on their size.

8.7
Edge Extraction

In Section 8.4 we have discussed several segmentation algorithms. They have in common that they are based on thresholding the image, either with pixel or subpixel accuracy. It is possible to achieve very good accuracies with these approaches, as we saw in Section 8.5. However, the accuracy of the measurements we can derive from the segmentation result in most cases critically depends on choosing the correct threshold for the segmentation. If the threshold is chosen incorrectly the extracted objects typically become larger or smaller because of the smooth transition from the foreground to the background gray value. This problem is especially grave if the illumination can change since in this case the adaptation of the thresholds to the changed illumination must be very accurate. Therefore, a segmentation algorithm that is robust with respect to illumination changes is extremely desirable. From the above discussion, we see that the boundary of the segmented region or subpixel-precise contour moves if the illumination changes or the thresholds are chosen inappropriately. Therefore, the goal of a robust segmentation algorithm must be to find the boundary of the objects as robustly and accurately as possible. The best way to describe the boundaries of the objects robustly is by regarding them as edges in the image. Therefore, in this section we will examine methods to extract edges.

8.7.1
Definition of Edges in 1D and 2D

To derive an edge extraction algorithm, we need to define what edges actually are. For the moment, let us make the simplifying assumption that the gray values in the object and in the background are constant. In particular, we assume that the image contains no noise. Furthermore, let us assume that the image is not discretized, i.e., is continuous. To illustrate this, Fig. 8.58(b) shows an idealized gray value profile across the part of a workpiece, that is indicated in Fig. 8.58(a).

From the above example, we can see that edges are areas in the image in which the gray values change significantly. To formalize this, let us regard the image for the moment as a one-dimensional function $f(x)$. From elementary calculus we know that the gray values change significantly if the first derivative of $f(x)$ differs significantly from 0: $|f'(x)| \gg 0$. Unfortunately, this alone is insufficient to define a unique edge location because there are typically many connected points for which this condition is true since the transition between the background and foreground gray value is smooth. This can be seen in Fig. 8.59(a), where the first derivative $f'(x)$ of the ideal gray

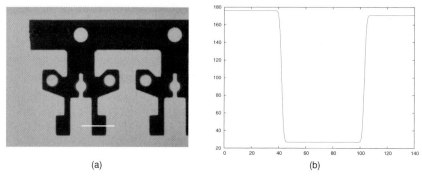

Fig. 8.58 (a) An image of a back-lit workpiece with a horizontal line that indicates the location of the idealized gray value profile in (b).

value profile in Fig. 8.58(b) is displayed. Note, for example, that there is an extended range of points for which $|f'(x)| \geq 20$. Therefore, to obtain a unique edge position we must additionally require that the absolute value of the first derivative $|f'(x)|$ is locally maximal. This is called nonmaximum suppression.

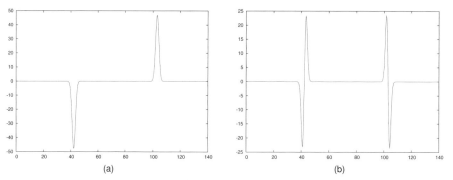

Fig. 8.59 (a) First derivative $f'(x)$ of the ideal gray value profile in Fig. 8.58(b). (b) Second derivative $f''(x)$.

From elementary calculus we know that at the points where $|f'(x)|$ is locally maximal the second derivative vanishes: $f''(x) = 0$. Hence, edges are given by the locations of inflection points of $f(x)$. To remove flat inflection points, we would additionally have to require $f'(x)f'''(x) < 0$. However, this restriction is seldom observed. Therefore, in 1D an alternative and equivalent definition to the maxima of the absolute value of the first derivative is to define edges as the locations of the zero crossings of the second derivative. Figure 8.59(b) displays the second derivative $f''(x)$ of the ideal gray value profile in Fig. 8.58(b). Clearly, the zero crossings are in the same positions as the maxima of the absolute value of the first derivative in Fig. 8.59(a).

From Fig. 8.59(a), we can also see that in 1D we can easily associate a polarity with an edge based on the sign of $f'(x)$. We speak of a positive edge if $f'(x) > 0$ and of a negative edge if $f'(x) < 0$.

We now turn to edges in continuous 2D images. Here, the edge itself is a curve $s(t) = (r(t), c(t))$, which is parameterized by a parameter t, e.g., its arc length. At each point of the edge curve, the gray value profile perpendicular to the curve is a 1D edge profile. With this, we can adapt the first 1D edge definition above for the 2D case: we define an edge as the points in the image where the directional derivative in the direction perpendicular to the edge is locally maximal. From differential geometry we know that the direction $n(t)$ perpendicular to the edge curve $s(t)$ is given by $n(t) = s'(t)^\perp \parallel s''(t)$. Unfortunately, the edge definition seemingly requires us to know the edge position $s(t)$ already to obtain the direction perpendicular to the edge, and hence looks like a circular definition. Fortunately, the direction $n(t)$ perpendicular to the edge can be determined easily from the image itself. It is given by the gradient vector of the image, which points into the direction of steepest ascent of the image function $f(r, c)$. The gradient of the image is given by the vector of its first partial derivatives:

$$\nabla f = \nabla f(r,c) = \left(\frac{\partial f(r,c)}{\partial r}, \frac{\partial f(r,c)}{\partial c} \right) = (f_r, f_c) \tag{8.76}$$

In the last equation, we have used a subscript to denote the partial derivative with respect to the subscripted variable. We will use this convention throughout this section. The Euclidean length $\|\nabla f\|_2 = \sqrt{f_r^2 + f_c^2}$ of the gradient vector is the equivalent of the absolute value of the first derivative $|f'(x)|$ in 1D. We will also call the length of the gradient vector its magnitude. It is also often called the amplitude. The gradient direction is, of course, directly given by the gradient vector. We can also convert it to an angle by calculating $\phi = -\arctan(f_r/f_c)$. Note that ϕ increases in the mathematically positive direction (counterclockwise) starting at the column axis. This is the usual convention. With the above definitions, we can define edges in 2D as the points in the image where the gradient magnitude is locally maximal in the direction of the gradient. To illustrate this definition, Fig. 8.60(a) shows a plot of the gray values of an idealized corner. The corresponding gradient magnitude is shown in Fig. 8.60(b). The edges are the points at the top of the ridge in the gradient magnitude.

In 1D, we have seen that the second edge definition (the zero crossings of the second derivative) is equivalent to the first definition. Therefore, it is natural to ask whether this definition can be adapted for the 2D case. Unfortunately, there is no direct equivalent for the second derivative in 2D since there are three partial derivatives of order two. A suitable definition for the second

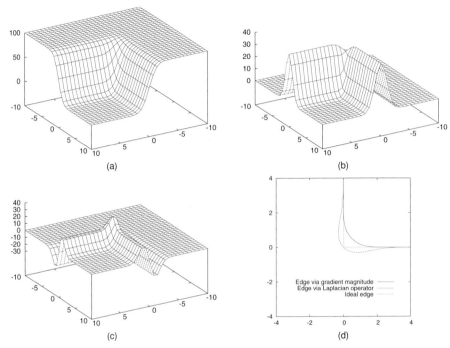

Fig. 8.60 (a) Image of an idealized corner, e.g., one of the corners at the bottom of the workpiece in Fig. 8.58(a). (b) Gradient magnitude of (a). (c) Laplacian of (a). (d) Comparison of the edges that result from the two definitions in 2D.

derivative in 2D is the Laplacian operator (Laplacian for short), defined by

$$\Delta f = \Delta f(r,c) = \frac{\partial^2 f(r,c)}{\partial r^2} + \frac{\partial^2 f(r,c)}{\partial c^2} = f_{rr} + f_{cc} \quad (8.77)$$

With this, the edges can be defined as the zero crossings of the Laplacian: $\Delta f(r,c) = 0$. Figure 8.60(c) shows the Laplacian of the idealized corner in Fig. 8.60(a). The results of the two edge definitions are shown in Fig. 8.60(d). It can be seen that, unlike for the 1D edges, the two definitions do not result in the same edge positions. The edge positions are only identical for straight edges. Whenever the edge is significantly curved the two definitions return different results. It can be seen that the definition via the maxima of the gradient magnitude always lies inside the ideal corner, whereas the definition via the zero crossings of the Laplacian always lies outside the corner and passes directly through the ideal corner. The Laplacian edge also is in a different position than the true edge for a larger part of the edge. Therefore, in 2D the definition via the maxima of the gradient magnitude is usually preferred. However, in some applications the fact that the Laplacian edge passes through the corner can be used to measure objects with sharp corners more accurately.

8.7.2
1D Edge Extraction

We now turn our attention to edges in real images, which are discrete and contain noise. In this section, we will discuss how to extract edges from 1D gray value profiles. This is a very useful operation that is used frequently in machine vision applications because it is extremely fast. It is typically used to determine the position or diameter of an object.

The first problem we have to address is how to compute the derivatives of the discrete 1D gray value profile. Our first idea might be to use the differences of consecutive gray values on the profile: $f'_i = f_i - f_{i-1}$. Unfortunately, this definition is not symmetric. It would compute the derivative at the "half pixel" positions $f_{i-\frac{1}{2}}$. A symmetric way to compute the first derivative is given by

$$f'_i = \frac{1}{2}(f_{i+1} - f_{i-1}) \qquad (8.78)$$

This formula is obtained by fitting a parabola through three consecutive points of the profile and computing the derivative of the parabola in the center point. The parabola is uniquely defined by the three points. With the same mechanism, we can also derive a formula for the second derivative:

$$f''_i = \frac{1}{2}(f_{i+1} - 2f_i + f_{i-1}) \qquad (8.79)$$

Note that the above methods to compute the first and second derivatives are linear filters, and hence can be regarded as the following two convolution masks: $\frac{1}{2} \cdot (\ 1 \ \ 0 \ \ -1\ )$ and $\frac{1}{2} \cdot (\ 1 \ \ -2 \ \ 1\ )$. Note that the -1 is the last element in the first derivative mask because the elements of the mask are mirrored in the convolution (see (8.16)).

Figure 8.61(a) displays the true gray value profile taken from the horizontal line in the image in Fig. 8.58(a). Its first derivative, computed with (8.78), is shown in Fig. 8.61(b). We can see that the noise in the image causes a very large number of local maxima in the absolute value of the first derivative, and consequently also a large number of zero crossings in the second derivative. The salient edges can be easily selected by thresholding the absolute value of the first derivative: $|f'_i| \geq t$. For the second derivative, the edges cannot be selected as easily. In fact, we have to resort to calculating the first derivative as well to be able to select the relevant edges. Hence, the edge definition via the first derivative is preferable because it can be done with one filter operation instead of two, and consequently the edges can be extracted much faster.

The gray value profile in Fig. 8.61(a) already contains relatively little noise. Nevertheless, in most cases it is desirable to suppress the noise even further. If the object we are measuring has straight edges in the part in which we are

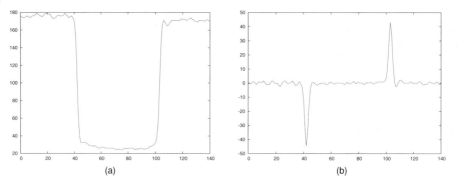

Fig. 8.61 (a) Gray value profile taken from the horizontal line in the image in Fig. 8.58(a). (b) First derivative f'_i of the gray value profile.

performing the measurement we can use the gray values perpendicular to the line along which we are extracting the gray value profile and average them in a suitable manner. The simplest way to do this is to compute the mean of the gray values perpendicular to the line. If, for example, the line along which we are extracting the gray value profile is horizontal we can calculate the mean in the vertical direction as follows:

$$f_i = \frac{1}{2m+1} \sum_{j=-m}^{m} f_{r+j,c+i} \qquad (8.80)$$

This acts like a mean filter in one direction. Hence, the noise variance is reduced by a factor of $2m+1$. Of course, we could also use a one-dimensional Gaussian filter to average the gray values. However, since this would require larger filter masks for the same noise reduction, and consequently would lead to longer execution times, in this case the mean filter is preferable.

Fig. 8.62 Creation of the gray value profile from an inclined line. The line is visualized by the heavy solid line. The circles indicate the points that are used to compute the profile. Note that they do not lie on the pixel centers. The direction in which the 1D mean is computed is visualized by the dashed lines.

If the line along which we want to extract the gray value profile is horizontal or vertical the calculation of the profile is simple. If we want to extract the profile from inclined lines or from circles or ellipses the computation is slightly more difficult. To enable meaningful measurements for distances, we must sample the line with a fixed distance, typically one pixel. Then, we need to generate lines perpendicular to the curve along which we want to extract the profile. This procedure is shown for an inclined line in Fig. 8.62. Because of this, the points from which we must extract the gray values typically do not lie on pixel centers. Therefore, we will have to interpolate them. This can be done with the techniques discussed in Section 8.3.3, i.e., with nearest neighbor or bilinear interpolation.

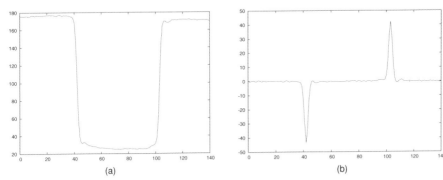

Fig. 8.63 (a) Gray value profile taken from the horizontal line in the image in Fig. 8.58(a) and averaged vertically over 21 pixels. (b) First derivative f'_i of the gray value profile.

Figure 8.63 shows a gray value profile and its first derivative obtained by vertically averaging the gray values along the line shown in Fig. 8.58(a). The size of the 1D mean filter was 21 pixels in this case. If we compare this with Fig. 8.61, which shows the profile obtained from the same line without averaging, we can see that the noise in the profile has been reduced significantly. Because of this, the salient edges are even easier to select than without the averaging.

Unfortunately, the averaging perpendicular to the curve along which the gray value profile is extracted sometimes is insufficient to smooth the profiles enough to enable us to extract the relevant edges easily. One example is shown in Fig. 8.64. Here, the object to be measured has a significant amount of texture, which is not as random as noise and consequently does not average out completely. Note that on the right side of the profile there is a negative edge with an amplitude almost as large as the edges we want to extract. Another reason for the noise not to cancel out completely may be that we cannot choose the size of the averaging large enough, e.g., because the object's boundary is curved.

To solve these problems, we must smooth the gray value profile itself to suppress the noise even further. This is done by convolving the profile with a smoothing filter: $f_s = f * h$. We then can extract the edges from the smoothed profile via its first derivative. This would involve two convolutions: one for

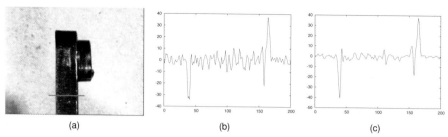

Fig. 8.64 (a) An image of a relay with a horizontal line that indicates the location of the gray value profile. (b) First derivative of the gray value profile without averaging. (c) First derivative of the gray value profile with vertical averaging over 21 pixels.

the smoothing filter and one for the derivative filter. Fortunately, the convolution has a very interesting property that we can use to save one convolution. The derivative of the smoothed function is identical to the convolution of the function with the derivative of the smoothing filter: $(f * h)' = f * h'$. We can regard h' as an edge filter.

Like for the smoothing filters, the natural question to ask is which edge filter is optimal. This problem was addressed by Canny [15]. He proposed three criteria that an edge detector should fulfill. First, it should have a good detection quality, i.e., it should have a low probability of falsely detecting an edge point and also a low probability of erroneously missing an edge point. This criterion can be formalized as maximizing the signal-to-noise ratio of the output of the edge filter. Second, the edge detector should have good localization quality, i.e., the extracted edges should be as close as possible to the true edges. This can be formalized by minimizing the variance of the extracted edge positions. Finally, the edge detector should return only a single edge for each true edge, i.e., it should avoid multiple responses. This criterion can be formalized by maximizing the distance between the extracted edge positions. Canny then combined these three criteria into one optimization criterion and solved it using the calculus of variations. To do so, he assumed that the edge filter has a finite extent (mask size). Since adapting the filter to a particular mask size involves solving a relatively complex optimization problem Canny looked for a simple filter that could be written in closed form. He found that the optimal edge filter can be approximated very well with the first derivative of the Gaussian filter:

$$g'_\sigma(x) = \frac{-x}{\sqrt{2\pi}\sigma^3} e^{-\frac{x^2}{2\sigma^2}} \qquad (8.81)$$

One drawback of using the true derivative of the Gaussian filter is that the edge amplitudes become progressively smaller as σ is increased. Ideally, the edge filter should return the true edge amplitude independent of the smoothing. To achieve this for an idealized step edge, the output of the filter must be multiplied by $\sqrt{2\pi}\sigma$.

Note that the optimal smoothing filter would be the integral of the optimal edge filter, i.e., the Gaussian smoothing filter. It is interesting to note that, like the criteria in Section 8.2.3, Canny's formulation indicates that the Gaussian filter is the optimal smoothing filter.

Since the Gaussian filter and its derivatives cannot be implemented recursively (see Section 8.2.3) Deriche used Canny's approach to find optimal edge filters that can be implemented recursively [20]. He derived the following two filters:

$$d'_\alpha(x) = -\alpha^2 x e^{-\alpha|x|}$$
$$e'_\alpha(x) = -2\alpha \sin(\alpha x) e^{-\alpha|x|} \qquad (8.82)$$

The corresponding smoothing filters are:

$$d_\alpha(x) = \frac{\alpha}{4}(\alpha|x| + 1)e^{-\alpha|x|}$$
$$e_\alpha(x) = \frac{\alpha}{2}(\sin(\alpha|x|) + \cos(\alpha|x|))e^{-\alpha|x|} \qquad (8.83)$$

Note that in contrast to the Gaussian filter, where larger values for σ indicate more smoothing, smaller values for α indicate more smoothing in the Deriche filters. The Gaussian filter has similar effects as the first Deriche filter for $\sigma = \sqrt{\pi}/\alpha$. For the second Deriche filter, the relation is $\sigma = \sqrt{\pi}/(2\alpha)$. Note that the Deriche filters are significantly different from the Canny filter. This can also be seen from Fig. 8.65, which compares the Canny and Deriche smoothing and edge filters with equivalent filter parameters.

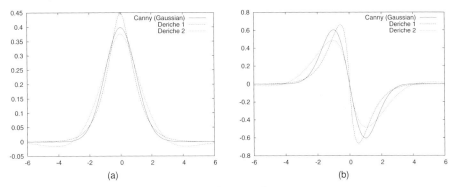

Fig. 8.65 Comparison of the Canny and Deriche filters. (a) Smoothing filters. (b) Edge filters.

Figure 8.66 shows the result of using the Canny edge detector with $\sigma = 1.5$ to compute the smoothed first derivative of the gray value profile in Fig. 8.64(a). Like in Fig. 8.64(c), the profile was obtained by averaging over

21 pixels vertically. Note that the amplitude of the unwanted edge on the right side of the profile has been reduced significantly. This enables us to select the salient edges more easily.

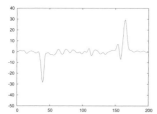

Fig. 8.66 Result of applying the Canny edge filter with $\sigma = 1.5$ to the gray value profile in Fig. 8.64(a) with vertical averaging over 21 pixels.

To extract the edge position, we need to perform the nonmaximum suppression. If we are only interested in the edge positions with pixel accuracy we can proceed as follows. Let the output of the edge filter be denoted by $e_i = |f * h'|_i$, where h' denotes one of the above edge filters. Then, the local maxima of the edge amplitude are given by the points for which $e_i > e_{i-1} \wedge e_i > e_{i+1} \wedge e_i \geq t$, where t is the threshold to select the relevant edges.

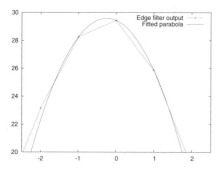

Fig. 8.67 Principle of extracting edge points with subpixel accuracy. The local maximum of the edge amplitude is detected. Then, a parabola is fitted through the three points around the maximum. The maximum of the parabola is the subpixel-accurate edge location. The edge amplitude was taken from the right edge in Fig. 8.66.

Unfortunately, extracting the edges with pixel accuracy is often not accurate enough. To extract edges with subpixel accuracy, we note that around the maximum the edge amplitude can be approximated well with a parabola. Figure 8.67 illustrates this by showing a zoomed part of the edge amplitude around the right edge in Fig. 8.66. If we fit a parabola through three points around the maximum edge amplitude and calculate the maximum of the parabola we can obtain the edge position with subpixel accuracy. If an ideal camera system is assumed this algorithm is as accurate as the precision with which the floating point numbers are stored in the computer [88].

We conclude the discussion of the 1D edge extraction by showing the results of the edge extraction on the two examples we have used so far. Figure 8.68(a) shows the edges that have been extracted along the line shown in Fig. 8.58(a) with the Canny filter with $\sigma = 1.0$. From the two zoomed parts around the extracted edge positions, we can see that by coincidence both edges lie very close to the pixel centers. Figure 8.68(b) displays the result of extracting edges along the line shown in Fig. 8.64(a) with the Canny filter with $\sigma = 1.5$. In this case, the left edge is almost exactly in the middle of two pixel centers. Hence, we can see that the algorithm is successful in extracting the edges with subpixel precision.

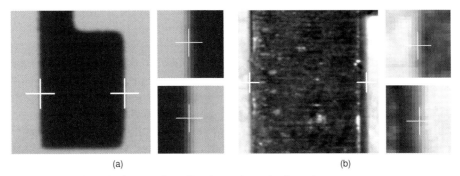

(a) (b)

Fig. 8.68 (a) Result of extracting 1D edges along the line shown in Fig. 8.58(a). The two small images show a zoomed part around the edge positions. In this case, they both lie very close to the pixel centers. The distance between the two edges is 60.95 pixels. (b) Result of extracting 1D edges along the line shown in Fig. 8.64(a). Note that the left edge, shown in detail in the upper right image, is almost exactly in the middle between two pixel centers. The distance between the two edges is 125.37 pixels.

8.7.3
2D Edge Extraction

As discussed in Section 8.7.1, there are two possible definitions for edges in 2D, which are not equivalent. Like in the 1D case, the selection of salient edges will require us to perform a thresholding based on the gradient magnitude. Therefore, the definition via the zero crossings of the Laplacian requires us to compute more partial derivatives than the definition via the maxima of the gradient magnitude. Consequently, we will concentrate on the maxima of the gradient magnitude for the 2D case. We will add some comments on the zero crossings of the Laplacian at the end of this section.

As in the 1D case, the first question we need to answer is how to compute the partial derivatives of the image that are required to calculate the gradient. Similar to (8.78), we could use finite differences to calculate the partial derivatives. In 2D, they would be $f_{r;i,j} = \frac{1}{2}(f_{i+1,j} - f_{i-1,j})$ and $f_{c;i,j} = \frac{1}{2}(f_{i,j+1} - f_{i,j-1})$. However, as we have seen above, typically the image must be smoothed to obtain good results. For time-critical applications, the filter masks should be as small as possible, i.e., 3×3. All 3×3 edge filters can be brought into the following form by scaling the coefficients appropriately (note that the filter masks are mirrored in the convolution):

$$\begin{pmatrix} 1 & 0 & -1 \\ a & 0 & -a \\ 1 & 0 & -1 \end{pmatrix} \quad \begin{pmatrix} 1 & a & 1 \\ 0 & 0 & 0 \\ -1 & -a & -1 \end{pmatrix} \quad (8.84)$$

If we use $a = 1$ we obtain the Prewitt filter. Note that it performs a mean filter perpendicular to the derivative direction. For $a = \sqrt{2}$ the Frei filter is obtained, while for $a = 2$ we obtain the Sobel filter, which performs an approximation to a Gaussian smoothing perpendicular to the derivative direction. Of the above three filters, the Sobel filter returns the best results because it uses the best smoothing filter. Interestingly enough, 3×3 filters are still actively investigated. For example, Ando [2] has recently proposed an edge filter that tries to minimize the artifacts that invariably are obtained with small filter masks. In our notation, his filter would correspond to $a = 2.435101$. Unfortunately, like the Frei filter it requires floating point calculations, which makes it unattractive for time-critical applications.

The 3×3 edge filters are primarily used to quickly find edges with moderate accuracy in images of relatively good quality. Since speed is important and the calculation of the gradient magnitude via the Euclidean length (the 2-norm) of the gradient vector ($\|\nabla f\|_2 = \sqrt{f_r^2 + f_c^2}$) requires an expensive square root calculation the gradient magnitude is typically computed by one of the following norms: the 1-norm $\|\nabla f\|_1 = |f_r| + |f_c|$ or the maximum norm $\|\nabla f\|_\infty = \max(|f_r|, |f_c|)$. Note that the first norm corresponds to the cityblock distance in the distance transform, while the second norm corresponds to the chessboard distance (see Section 8.6.1). Furthermore, the nonmaximum suppression also is relatively expensive and is often omitted. Instead, the gradient magnitude is simply thresholded. Because this results in edges that are wider than one pixel the thresholded edge regions are skeletonized. Note that this implicitly assumes that the edges are symmetric.

Figure 8.69 shows an example where this simple approach works quite well because the image is of good quality. Figure 8.69(a) displays the edge amplitude around the leftmost hole of the workpiece in Fig. 8.58(a) computed with the Sobel filter and the 1-norm. The edge amplitude is thresholded (Fig. 8.69(b)) and the skeleton of the resulting region is computed (Fig. 8.69(c)).

Since the assumption that the edges are symmetric is fulfilled in this example the resulting edges are in the correct location.

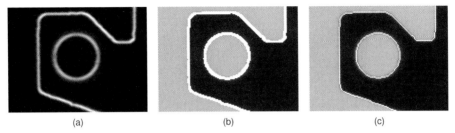

Fig. 8.69 (a) Edge amplitude around the leftmost hole of the workpiece in Fig. 8.58(a) computed with the Sobel filter and the 1-norm. (b) Thresholded edge region. (c) Skeleton of (b).

This approach fails to produce good results on the more difficult image of the relay in Fig. 8.64(a). As can be seen from Fig. 8.70(a), the texture on the relay causes many areas with high gradient magnitude, which are also present in the segmentation (Fig. 8.70(b)) and the skeleton (Fig. 8.70(c)). Another interesting thing to note is that the vertical edge at the right corner of the top edge of the relay is quite blurred and asymmetric. This produces holes in the segmented edge region, which are exacerbated by the skeletonization.

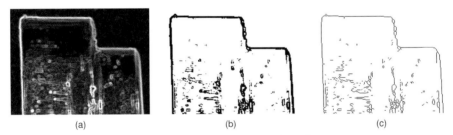

Fig. 8.70 (a) Edge amplitude around the top part of the relay in Fig. 8.64(a) computed with the Sobel filter and the 1-norm. (b) Thresholded edge region. (c) Skeleton of (b).

Because the 3×3 filters are not robust against noise and other disturbances, e.g., textures, we need to adapt the approach to optimal 1D edge extraction described in the previous section to the 2D case. In 2D, we can derive the optimal edge filters by calculating the partial derivatives of the optimal smoothing filters since the properties of the convolution again allow us to move the derivative calculation into the filter. Consequently, Canny's optimal edge filters in 2D are given by the partial derivatives of the Gaussian filter. Because the Gaussian filter is separable, so are its derivatives: $g_r = \sqrt{2\pi}\sigma g'_\sigma(r)g_\sigma(c)$ and $g_c = \sqrt{2\pi}\sigma g_\sigma(r)g'_\sigma(c)$ (see the discussion following (8.81) for the factors of $\sqrt{2\pi\sigma}$). To adapt the Deriche filters to the 2D case, the separability of the filters is postulated. Hence, the optimal 2D Deriche filters are given by $d'_\alpha(r)d_\alpha(c)$ and $d_\alpha(r)d'_\alpha(c)$ for the first Deriche filter and $e'_\alpha(r)e_\alpha(c)$ and $e_\alpha(r)e'_\alpha(c)$ for the second Deriche filter (see (8.82)).

The advantage of the Canny filter is that it is isotropic, i.e., rotation invariant (see Section 8.2.3). Its disadvantage is that it cannot be implemented recursively. Therefore, the execution time depends on the amount of smoothing specified by σ. The Deriche filters, on the other hand, can be implemented recursively, and hence their runtime is independent of the smoothing parameter α. However, they are anisotropic, i.e., the edge amplitude they calculate depends on the angle of the edge in the image. This is undesirable because it makes the selection of the relevant edges harder. Lanser has shown that the anisotropy of the Deriche filters can be corrected [58]. We will refer to the isotropic versions of the Deriche filters as the Lanser filters.

Figure 8.71 displays the result of computing the edge amplitude with the second Lanser filter with $\alpha = 0.5$. Compared to the results with the Sobel filter, the Lanser filter was able to suppress the noise and texture significantly better. This can be seen from the edge amplitude image (Fig. 8.71(a)) as well as the thresholded edge region (Fig. 8.71(b)). Note, however, that the edge region still contains a hole for the vertical edge that starts at the right corner of the topmost edge of the relay. This happens because the edge amplitude has only been thresholded and the important step of the nonmaximum suppression has been omitted to compare the results of the Sobel and Lanser filters.

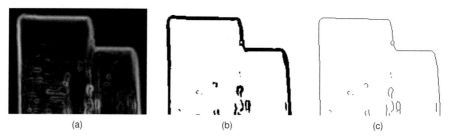

Fig. 8.71 (a) Edge amplitude around the top part of the relay in Fig. 8.64(a) computed with the second Lanser filter with $\alpha = 0.5$. (b) Thresholded edge region. (c) Skeleton of (b).

As we saw in the above examples, thresholding the edge amplitude and then skeletonizing the region sometimes does not yield the desired results. To obtain the correct edge locations, we must perform the nonmaximum suppression (see Section 8.7.1). In the 2D case, this can be done by examining the two neighboring pixels that lie closest to the gradient direction. Conceptually, we can think of transforming the gradient vector into an angle. Then, we divide the angle range into eight sectors. Figure 8.72 shows two examples for this. Unfortunately, with this approach diagonal edges often still are two pixels wide. Consequently, the output of the nonmaximum suppression still must be skeletonized.

Angle 15°

Angle 35°

Fig. 8.72 Examples for the pixels that are examined in the non-maximum suppression for different gradient directions.

Figure 8.73 shows the result of applying the nonmaximum suppression to the edge amplitude image in Fig. 8.71(a). From the thresholded edge region in Fig. 8.73(b) it can be seen that the edges are now in the correct locations. In particular, the incorrect hole in Fig. 8.71 is no longer present. We can also see that the few diagonal edges are sometimes two pixels wide. Therefore, their skeleton is computed and displayed in Fig. 8.73(c).

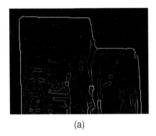
(a)

(b)

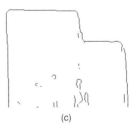
(c)

Fig. 8.73 (a) Result of applying the nonmaximum suppression to the edge amplitude image in Fig. 8.71(a). (b) Thresholded edge region. (c) Skeleton of (b).

Up to now, we have been using simple thresholding to select the salient edges. This works well as long as the edges we are interested in have roughly the same contrast or have a contrast that is significantly different from the contrast of noise, texture, or other irrelevant objects in the image. In many applications, however, we face the problem that if we select the threshold so high that only the relevant edges are selected they are often fragmented. If, on the other hand, we set the threshold so low that the edges we are interested in are not fragmented, we end up with many irrelevant edges. These two situations are illustrated in Figs. 8.74(a) and (b). A solution for this problem was proposed by Canny [15]. He devised a special thresholding algorithm for segmenting edges: the hysteresis thresholding. Instead of a single threshold, it uses two thresholds. Points with an edge amplitude greater than the higher threshold are immediately accepted as safe edge points. Points with an edge amplitude smaller than the lower threshold are immediately rejected. Points with an edge amplitude between the two thresholds are only accepted if they are connected to safe edge points via a path in which all points have an edge amplitude above the lower threshold. We can also think about this operation as selecting the edge points with an amplitude above the upper threshold first, and then extending the edges as far as possible while remaining above the lower threshold. Figure 8.74(c) shows that the hysteresis thresholding enables us to select only the relevant edges without fragmenting them or missing edge points.

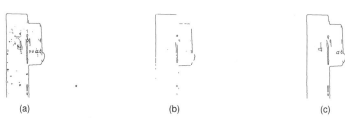

Fig. 8.74 (a) Result of thresholding the edge amplitude for the entire relay image in Fig. 8.64(a) with a threshold of 60. This causes many irrelevant texture edges to be selected. (b) Result of thresholding the edge amplitude with a threshold of 140. This only selects relevant edges. However, they are severely fragmented and incomplete. (c) Result of hysteresis thresholding with a low threshold of 60 and a high threshold of 140. Only the relevant edges are selected, and they are complete.

Like in the 1D case, the pixel-accurate edges we have extracted so far are often not accurate enough. We can use a similar approach as for 1D edges to extract edges with subpixel accuracy: we can fit a 2D polynomial to the edge amplitude and extract its maximum in the direction of the gradient vector [89, 88]. The fitting of the polynomial can be done with convolutions with special filter masks (so-called facet model masks) [39, 41]. To illustrate this, Fig. 8.75(a) shows a 7×7 part of an edge amplitude image. The fitted 2D polynomial obtained from the central 3×3 amplitudes is shown in Fig. 8.75(b), along with an arrow that indicates the gradient direction. Furthermore, contour lines of the polynomial are shown. They indicate that the edge point is offset by approximately a quarter of a pixel in the direction of the arrow.

The above procedure gives us one subpixel-accurate edge point per non-maximum suppressed pixel. These individual edge points must be linked into subpixel-precise contours. This can be done by repeatedly selecting the first unprocessed edge point to start the contour and then to successively find adjacent edge points until the contour closes, reaches the image border, or reaches an intersection point.

Figure 8.76 illustrates the subpixel edge extraction along with a very useful strategy to increase the processing speed. The image in Fig. 8.76(a) is the same workpiece as in Fig. 8.58(a). Because the subpixel edge extraction is rel-

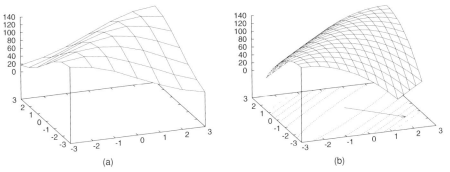

Fig. 8.75 (a) 7 × 7 part of an edge amplitude image. (b) Fitted 2D polynomial obtained from the central 3 × 3 amplitudes in (a). The arrow indicates the gradient direction. The contour lines in the plot indicate that the edge point is offset by approximately a quarter of a pixel in the direction of the arrow.

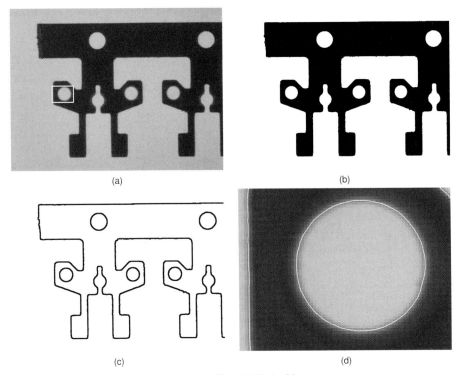

Fig. 8.76 (a) Image of the workpiece in Fig. 8.58(a) with a rectangle that indicates the image part shown in (d). (b) Thresholded workpiece. (c) Dilation of the boundary of (b) with a circle of diameter 5. This is used as the ROI for the subpixel edge extraction. (d) Subpixel-accurate edges of the workpiece extracted with the Canny filter with $\sigma = 1$.

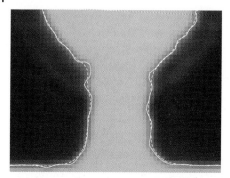 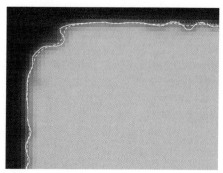

Fig. 8.77 Comparison of the subpixel-accurate edges extracted via the maxima of the gradient magnitude in the gradient direction (dashed lines) and the edges extracted via the subpixel-accurate zero crossings of the Laplacian. In both cases, a Gaussian filter with $\sigma = 1$ was used. Note that since the Laplacian edges must follow the corners they are much more curved than the gradient magnitude edges.

atively costly we want to reduce the search space as much as possible. Since the workpiece is back-lit we can threshold it easily (Fig. 8.76(b)). If we calculate the inner boundary of the region with (8.63) the resulting points are close to the edge points we want to extract. We only need to dilate the boundary slightly, e.g., with a circle of diameter 5 (Fig. 8.76(c)), to obtain a region of interest for the edge extraction. Note that the ROI is only a small fraction of the entire image. Consequently, the edge extraction can be done an order of magnitude faster than in the entire image, without any loss of information. The resulting subpixel-accurate edges are shown in Fig. 8.76(d) for the part of the image indicated by the rectangle in Fig. 8.76(a). Note how well they capture the shape of the hole.

We conclude this section with a look at the second edge definition via the zero crossings of the Laplacian. Since the zero crossings are just a special threshold we can use the subpixel-precise thresholding operation, defined in Section 8.4.3, to extract edges with subpixel accuracy. To make this as efficient as possible, we must first compute the edge amplitude in the entire ROI of the image. Then, we threshold the edge amplitude and use the resulting region as the ROI for the computation of the Laplacian and for the subpixel-precise thresholding. The resulting edges for two parts of the workpiece image are compared to the gradient magnitude edge in Fig. 8.77. Note that since the Laplacian edges must follow the corners they are much more curved than the gradient magnitude edges, and hence are more difficult to process further. This is another reason why the edge definition via the gradient magnitude is usually preferred.

Despite the above arguments, the property that the Laplacian edge exactly passes through corners in the image can be used advantageously in some applications. Figure 8.78(a) shows an image of a screw for which the depth of the thread must be measured. Figures 8.78(b)–(d) display the results of extracting the border of the screw with subpixel-precise thresholding, the gradient magnitude edges with a Canny filter with $\sigma = 0.7$, and the Laplacian edges with a Gaussian filter with $\sigma = 0.7$. Note that in this case the most suitable results are obtained with the Laplacian edges.

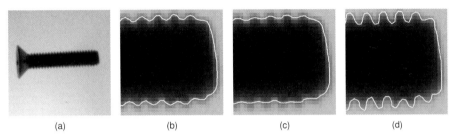

Fig. 8.78 (a) Image of a screw for which the depth of the thread must be measured. (b) Result of performing a subpixel-precise thresholding operation. (c) Result of extracting the gradient magnitude edges with a Canny filter with $\sigma = 0.7$. (d) Result of extracting the Laplacian edges with a Gaussian filter with $\sigma = 0.7$. Note that for this application the Laplacian edges return the most suitable result.

8.7.4 Accuracy of Edges

In the previous two sections, we have seen that edges can be extracted with subpixel resolution. We have used the terms "subpixel-accurate" and "subpixel-precise" to describe these extraction mechanisms without actually justifying the use of the words: accurate and precise. Therefore, in this section we will examine whether the edges we can extract are actually subpixel-accurate and subpixel precise.

Since the words accuracy and precision are often confused or used interchangeably, let us first define what we mean by them. By precision, we denote how close on average an extracted value is to its mean value. Hence, precision measures how repeatable we can extract the value. By accuracy, we denote how close on average the extracted value is to its true value [40]. Note that the precision does not tell us anything about the accuracy of the extracted value. It could, for example, be offset by a systematic bias, but still be very precise. Conversely, the accuracy does not necessarily tell us how precise the extracted value is. The measurement could be quite accurate, but not very repeatable. Figure 8.79 shows different situations that can occur. Also note that the accuracy and precision are statements about the average distribution of the extracted values. From a single value, we cannot tell whether the measurements are accurate or precise.

If we adopt a statistical point of view, the extracted values can be regarded as random variables. With this, the precision of the values is given by the variance of the values: $V[x] = \sigma_x^2$. If the extracted values are precise they have a small variance. On the other hand, the accuracy can be described by the difference of the expected value $E[x]$ of the values to the true value T: $|E[x] - T|$. Since we typically do not know anything about the true probability distribution of the extracted values, and consequently cannot determine $E[x]$ and $V[x]$, we must estimate them with the empirical mean and variance of the extracted values.

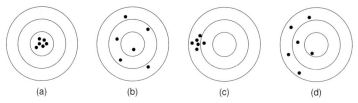

Fig. 8.79 Comparison of accuracy and precision. The center of the circles indicates the true value of the feature. The dots indicate the outcome of the measurements of the feature. (a) Accurate and precise. (b) Accurate but not precise. (c) Not accurate but precise. (d) Neither accurate nor precise.

The accuracy and precision of edges is analyzed extensively in [87, 88]. The precision of ideal step edges extracted with the Canny filter is derived analytically. If we denote the true edge amplitude by a and the noise variance in the image by σ_n^2, it can be shown that the variance of the edge positions σ_e^2 is given by

$$\sigma_e^2 = \frac{3}{8} \frac{\sigma_n^2}{a^2} \tag{8.85}$$

Even though this result was derived analytically for continuous images, it also holds in the discrete case. This result has also been verified empirically in [87, 88]. Note that it is quite intuitive. The larger the noise in the image is, the less precisely the edges can be located. Furthermore, the larger the edge amplitude is, the higher the precision of the edges is. Note also that, possibly contrary to our intuition, increasing the smoothing does not increase the precision. This happens because the noise reduction achieved by the larger smoothing cancels out exactly with the weaker edge amplitude that results from the smoothing. From (8.85), we can see that the Canny filter is subpixel precise ($\sigma_e \leq 1/2$) if the signal-to-noise ratio $a^2/\sigma_n^2 \geq 3/2$. This can, of course, be easily achieved in practice. Consequently, we see that we were justified in calling the Canny filter subpixel precise.

The same derivation can also be performed for the Deriche and Lanser filters. For continuous images, the following variances result:

$$\sigma_e^2 = \frac{5}{64} \frac{\sigma_n^2}{a^2} \quad \text{and} \quad \sigma_e^2 = \frac{3}{16} \frac{\sigma_n^2}{a^2} \qquad (8.86)$$

Note that the Deriche and Lanser filters are more precise than the Canny filter. Like for the Canny filter, the smoothing parameter α has no influence on the precision. In the discrete case, this is, unfortunately, no longer true because of discretization of the filter. Here, less smoothing (larger values of α) leads to slightly worse precisions than predicted by (8.86). However, for practical purposes we can assume that the smoothing for all edge filters we have discussed has no influence on the precision of the edges. Consequently, if we want to control the precision of the edges we must maximize the signal-to-noise ratio by using proper lighting, cameras, and frame grabbers. Furthermore, if analog cameras are used the frame grabber should have a line jitter that is as small as possible.

For ideal step edges, it is also easy to convince oneself that the expected position of the edge under noise corresponds to its true position. This happens because both the ideal step edge and the above filters are symmetric with respect to the true edge positions. Therefore, the edges that are extracted from noisy ideal step edges must be distributed symmetrically around the true edge position. Consequently, their mean value is the true edge position. This is also verified empirically for the Canny filter in [87, 88]. Of course, it can also be verified for the Deriche and Lanser filters.

While it is easy to show that edges are very accurate for ideal step edges, we must also perform experiments on real images to test the accuracy on real data. This is important because some of the assumptions that are used in the edge extraction algorithms may not hold in practice. Because these assumptions are seldom stated explicitly we should carefully examine them here. Let us focus on straight edges because, as we have seen from the discussion in Section 8.7.1, especially Fig. 8.60, sharply curved edges will necessarily lie in incorrect positions. See also [8] for a thorough discussion for the positional errors of the Laplacian edge detector for ideal corners of two straight edges with varying angles. Because we concentrate on straight edges we can reduce the edge detection to the 1D case, which is simpler to analyze. From Section 8.7.1 we know that 1D edges are given by inflection points of the gray value profiles. This implicitly assumes that the gray value profile, and consequently its derivatives, are symmetric with respect to the true edge. Furthermore, to obtain subpixel positions the edge detection implicitly assumes that the gray values at the edge change smoothly and continuously as the edge moves in subpixel increments through a pixel. For example, if an edge covers 25% of a pixel we would assume that the gray value in the pixel is a mixture of 25% of the foreground gray value and 75% of the background gray value. We will see whether these assumptions hold in real images below.

To test the accuracy of the edge extraction on real images, it is instructive to repeat the experiments in [87, 88] with a different camera. In [87, 88], a print of an edge is mounted on an xy-stage and shifted in 50 µm increments, which corresponds to approximately 1/10 pixel, for a total of 1 mm. The goals are to determine whether the shifts of 1/10 pixel can be detected reliably and to obtain information about the absolute accuracy of the edges. Figure 8.80(a) shows an image used in this experiment. We are not going to repeat the test whether the subpixel shifts can be detected reliably here. The 1/10 pixel shifts can be detected with a very high confidence (more than 99.99999%). What is more interesting is to look at the absolute accuracy. Since we do not know the true edge position we must get an estimate for it. Because the edge was shifted in linear increments in the test images, such an estimate can be obtained by fitting a straight line through the extracted edge positions and subtracting the line from the measured edge positions.

Figure 8.80(b) displays the result of extracting the edge in Fig. 8.80(a) along a horizontal line with the Canny filter with $\sigma = 1$. The edge position error is shown in Fig. 8.80(c). We can see that there are errors of up to approximately 1/22 pixel. What causes these errors? As we discussed above, for ideal cameras no error occurs, so one of the assumptions must be violated. In this case, the assumption that the gray value is a mixture of the foreground and background gray values, which is proportional to the area of the pixel covered by the object is violated. This happens because the camera did not have a fill factor of 100%, i.e., the light sensitive area of a pixel on the sensor was much smaller than the total area of the pixel. Consider what happens when the edge moves across the pixel and the image is perfectly focused. In the light sensitive area of the pixel the gray value changes as expected when the edge moves across the pixel because the sensor integrates the incoming light. However, when the edge enters the light-insensitive area the gray value no longer changes [60]. Consequently, the edge does not move in the image. In the real image, the focus is not perfect. Hence, the light is spread slightly over adjacent sensor elements. Therefore, the edges do not jump as they would in a perfectly focused image, but shift continuously. Nevertheless, the poor fill factor causes errors in the edge positions. This can be seen very clearly from Fig. 8.80(c). Recall that the shift of 50 µm corresponds to 1/10 pixel. Consequently, the entire shift of 1 mm corresponds to two pixels. This is why we see a sine wave with two periods in Fig. 8.80(c). Each period corresponds exactly to one pixel. That these effects are caused by the fill factor can also be seen if the lens is defocused. In this case, the light is spread over more sensor elements. This helps to create an artificially increased fill factor, which causes smaller errors.

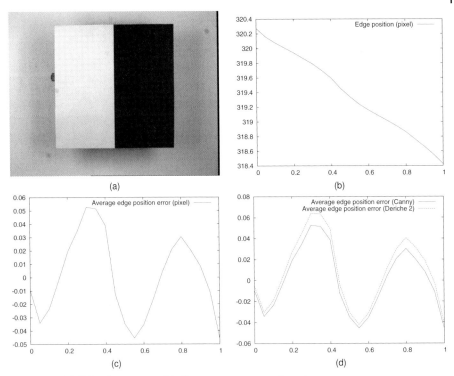

Fig. 8.80 (a) Edge image used in the accuracy experiment. (b) Edge position extracted along a horizontal line in the image with the Canny filter. The edge position is given in pixels as a function of the true shift in millimeters. (c) Error of the edge positions obtained by fitting a line through the edge positions in (b) and subtracting the line from (b). (d) Comparison of the errors obtained with the Canny and the second Deriche filters.

From the above discussion, it would appear that the edge position can be extracted with an accuracy of 1/22 pixel. To check whether this is true, let us repeat the experiment with the second Deriche filter. Figure 8.80(d) shows the result of extracting the edges with $\alpha = 1$ and computing the errors with the line fitted through the Canny edge positions. The last part is done to make the errors comparable. We can see to our surprise that the Deriche edge positions are systematically shifted in one direction. Does this mean that the Deriche filter is less accurate than the Canny filter? Of course, it does not, since on ideal data both filters return the same result. It shows us that another assumption must be violated. In this case, it is the assumption that the edge profile is symmetric with respect to the true edge position. This is the only reason why the two filters, which are symmetric themselves, can return a different result.

There are many reasons why edge profiles may become asymmetric. One reason is that the gray value response of the camera and image acquisition device are nonlinear. Figure 8.81 illustrates that an originally symmetric edge profile becomes asymmetric by a nonlinear gray value response function. It can be seen that the edge position accuracy is severely degraded by the nonlinear response. To correct the nonlinear response of the camera, it must be calibrated radiometrically with the methods described in Section 8.2.2.

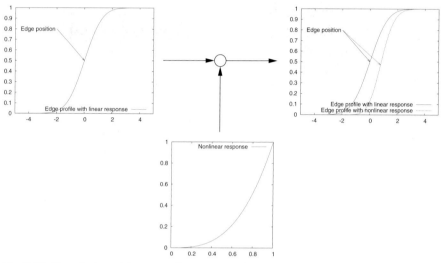

Fig. 8.81 Result of applying a nonlinear gray value response curve to an ideal symmetric edge profile. The ideal edge profile is shown in the upper left graph, the nonlinear response in the bottom graph. The upper right graph shows the modified gray value profile along with the edge positions on the profiles. Note that the edge position is affected substantially by the nonlinear response.

Unfortunately, even if the camera has a linear response or is calibrated radiometrically, other factors may cause the edge profiles to become asymmetric. In particular, lens aberrations like coma, astigmatism, and chromatic aberrations may cause asymmetric profiles. Since lens aberrations cannot be corrected easily with image processing algorithms they should be as small as possible.

While all the error sources discussed above influence the edge accuracy, we have so far neglected the largest source of errors. If the camera is not calibrated geometrically extracting edges with subpixel accuracy is pointless because the lens distortions alone are sufficient to render any subpixel position meaningless. Let us, for example, assume that the lens has a distortion that is smaller than 1% in the entire field of view. At the corners of the image, this means that the edges are offset by 4 pixels for a 640×480 image. We can see that extracting edges with subpixel accuracy is an exercise in futility if the lens distortions are not corrected, even for this relatively small distortion. This is illustrated in Fig. 8.82, where the results of correcting the lens distortions after calibrating the camera with the approach in [59] are shown. Note that despite the fact that the application used a very high quality lens, the lens distortions cause an error of approximately 3 pixels.

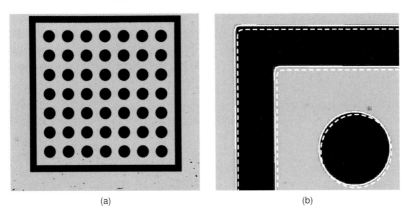

Fig. 8.82 (a) Image of a calibration target. (b) Extracted subpixel-accurate edges (solid lines) and edges after the correction of the lens distortions (dashed lines). Note that the lens distortions cause an error of approximately 3 pixels.

Another detrimental influence on the accuracy of the extracted edges is caused by the perspective distortions in the image. They happen whenever we cannot mount the camera perpendicularly to the objects we want to measure. Figure 8.83(a) shows the result of extracting the 1D edges along the ruler markings on a caliper. Because of the severe perspective distortions the distances between the ruler markings vary greatly throughout the image. If the camera is calibrated, i.e., its interior orientation and the exterior orientation of the plane in which the objects to measure lie have been determined (e.g., with the approach in [59]) the measurements in the image can be converted into measurements in world coordinates in the plane determined by the calibration. This is done by intersecting the optical ray that corresponds to each edge point in the image with the plane in the world. Figure 8.83(b) displays the results of converting the measurements in Fig. 8.83(a) into millimeters with this approach. Note that the measurements are extremely accurate even in the presence of severe perspective distortions.

From the above discussion, we can see that extracting edges with subpixel accuracy relies on careful selection of the hardware components. First, the gray value response of the camera and image acquisition device should be linear. To ensure this, the camera should be calibrated radiometrically. Furthermore, lenses with very small aberrations like coma and astigmatism should be chosen. Furthermore, monochromatic light should be used to avoid the effects of chromatic aberrations. In addition, the fill factor of the camera should

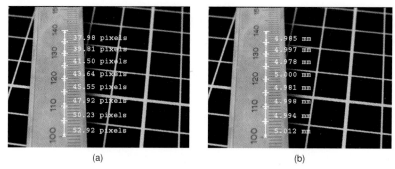

Fig. 8.83 Result of extracting 1D edges along the ruler markings on a caliper. (a) Pixel distances between the markings. (b) Distances converted to world units using camera calibration.

be as large as possible to avoid the effects of "blind spots." Finally, the camera should be calibrated geometrically to obtain meaningful results. All these requirements for the hardware components are, of course, also valid for other subpixel algorithms, e.g., the subpixel-precise thresholding (see Section 8.4.3), the gray value moments (see Section 8.5.2), and the contour features (see Section 8.5.3).

8.8
Segmentation and Fitting of Geometric Primitives

In Sections 8.4 and 8.7, we have seen how to segment images by thresholding and edge extraction. In both cases, the boundary of objects is either returned explicitly or can be derived by some post-processing (see Section 8.6.1). Therefore, for the purposes of this section we can assume that the result of the segmentation is a contour with the points of the boundary, which may be subpixel accurate. This approach often creates an enormous amount of data. For example, the subpixel-accurate edge of the hole in the workpiece in Fig. 8.76(d) contains 172 contour points. However, we are typically not interested in such a large amount of information. For example, in the application in Fig. 8.76(d), we would probably be content with knowing the position and radius of the hole, which can be described with just three parameters. Therefore, in this section we will discuss methods to fit geometric primitives to contour data. We will only examine the most relevant geometric primitives: lines, circles, and ellipses. Furthermore, we will examine how contours can be segmented automatically into parts that correspond to the geometric primitives. This will enable us to reduce the amount of data that needs to be processed substantially, while also providing us with a symbolic description of the data. Furthermore, the fitting of the geometric primitives will enable us to reduce the influence of incorrectly or inaccurately extracted points (so-called outliers). We will start by examining the fitting of the geometric primitives in Sections 8.8.1–8.8.3. In each case, we will assume that the contour or part of the contour we are examining corresponds to the primitive we are trying to fit, i.e., we are assuming that the segmentation into different primitives has already been performed. The segmentation itself will be discussed in Section 8.8.4.

8.8.1
Fitting Lines

If we want to fit lines we first need to think about the representation of lines. In images, lines can occur in any orientation. Therefore, we have to use a representation that enables us to represent all lines. For example, the common representation $y = mx + b$ does not allow us to do this. One representation that can be used is the Hessian normal form of the line, given by

$$\alpha r + \beta c + \gamma = 0 \tag{8.87}$$

This is actually an over-parameterization since the parameters (α, β, γ) are homogeneous [42, 27]. Therefore, they are only defined up to a scale factor. The scale factor in the Hessian normal form is fixed by requiring that $\alpha^2 + \beta^2 = 1$. This has the advantage that the distance of a point to the line can simply be obtained by substituting its coordinates into (8.87).

To fit a line through a set of points (r_i, c_i), $i = 1, \ldots, n$, we can minimize the sum of the squared distances of the points to the line:

$$\varepsilon^2 = \sum_{i=1}^{n} (\alpha r_i + \beta c_i + \gamma)^2 \tag{8.88}$$

While this is correct in principle, it does not work in practice because we can achieve a zero error if we select $\alpha = \beta = \gamma = 0$. This is caused by the over-parameterization of the line. Therefore, we must add the constraint $\alpha^2 + \beta^2 = 1$ as a Lagrange multiplier, and hence must minimize the following error:

$$\varepsilon^2 = \sum_{i=1}^{n} (\alpha r_i + \beta c_i + \gamma)^2 - \lambda(\alpha^2 + \beta^2 - 1)n \tag{8.89}$$

The solution to this optimization problem is derived in [39]. It can be shown that (α, β) is the eigenvector corresponding to the smaller eigenvalue of the following matrix:

$$\begin{pmatrix} \mu_{2,0} & \mu_{1,1} \\ \mu_{1,1} & \mu_{0,2} \end{pmatrix} \tag{8.90}$$

With this, γ is given by $\gamma = -(\alpha n_{1,0} + \beta n_{0,1})$. Here, $\mu_{2,0}$, $\mu_{1,1}$, and $\mu_{0,2}$ are the second-order central moments of the point set (r_i, c_i), while $n_{1,0}$ and $n_{0,1}$ are the normalized first-order moments (the center of gravity) of the point set. If we replace the area a of a region with the number n of points and sum over the points in the point set instead of the points in the region, the formulas to compute these moments are identical to the region moments (8.40) and (8.41) in Section 8.5.1. It is interesting to note that the vector (α, β) thus obtained, which is the normal vector of the line, is the minor axis that would be obtained from the ellipse parameters of the point set. Consequently, the major axis of the ellipse is the direction of the line. This is a very interesting connection between the ellipse parameters and the line fitting, because the results were derived using different approaches and models.

Figure 8.84(b) illustrates the line fitting procedure for an oblique edge of the workpiece shown in Fig. 8.84(a). Note that by fitting the line we were able to reduce the effects of the small protrusion on the workpiece. As mentioned above, by inserting the coordinates of the edge points into the line equation (8.87) we can easily calculate the distances of the edge points to the line. Therefore, by thresholding the distances the protrusion can be detected easily.

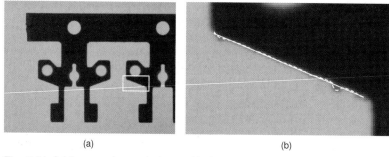

(a) (b)

Fig. 8.84 (a) Image of a workpiece with the part shown in (b) indicated by the white rectangle. (b) Extracted edge within a region around the inclined edge of the workpiece (dashed line) and straight line fitted to the edge (solid line).

As can be seen from the above example, the line fit is robust to small deviations from the assumed model (small outliers). However, Fig. 8.85 shows that large outliers severely affect the quality of the fitted line. In this example, the line is fitted through the straight edge as well as the large arc caused by the relay contact. Since the line fit must minimize the sum of the squared distances of the contour points the fitted line has a direction that deviates from that of the straight edge.

The least-squares line fit is not robust to large outliers since points that lie far from the line have a very large weight in the optimization because of the squared distances. To reduce the influence of far away points, we can introduce a weight w_i for each point. The weight should be $\ll 1$ for far away points. Let us for the moment assume that we had a way to compute these

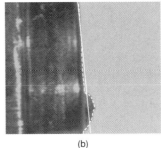

Fig. 8.85 (a) Image of a relay with the part shown in (b) indicated by the light gray rectangle. (b) Extracted edge within a region around the vertical edge of the relay (dashed line) and straight line fitted to the edge (solid line). To provide a better visibility of the edge and line, the contrast of the image has been reduced in (b).

weights. Then, the minimization becomes

$$\varepsilon^2 = \sum_{i=1}^{n} w_i(\alpha r_i + \beta c_i + \gamma)^2 - \lambda(\alpha^2 + \beta^2 - 1)n \qquad (8.91)$$

The solution of this optimization problem is again given by the eigenvector corresponding to the smaller eigenvalue of a moment matrix like in (8.90) [57]. The only difference is that the moments are computed by taking the weights w_i into account. If we interpret the weights as gray values the moments are identical to the gray value center of gravity and the second-order central gray value moments (see (8.48) and (8.49) in Section 8.5.2). Like above, the fitted line corresponds to the major axis of the ellipse obtained from the weighted moments of the point set. Hence, there is an interesting connection to the gray value moments.

The only remaining problem is how to define the weights w_i. Since we want to give smaller weights to points with large distances the weights must be based on the distances $\delta_i = |\alpha r_i + \beta c_i + \gamma|$ of the points to the line. Unfortunately, we do not know the distances without fitting the line, so this seems an impossible requirement. The solution is to fit the line in several iterations. In the first iteration, $w_i = 1$ is used, i.e., a normal line fit is performed to calculate the distances δ_i. They are used to define weights for the following iterations by using a weight function $w(\delta)$ [57]. In practice, one of the following two weight functions can be used. They both work very well. The first weight function was proposed by Huber [48, 57]. It is given by

$$w(\delta) = \begin{cases} 1 & |\delta| \leq \tau \\ \tau/|\delta| & |\delta| > \tau \end{cases} \qquad (8.92)$$

The parameter τ is the clipping factor. It defines which points should be regarded as outliers. We will see how it is computed below. For now, please note that all points with a distance $\leq \tau$ receive a weight of 1. This means that for small distances the squared distance is used in the minimization. Points with a distance $> \tau$, on the other hand, receive a progressively smaller weight. In fact, the weight function is chosen such that points with large distances use the distance itself and not the squared distance in the optimization. Sometimes, these weights are not small enough to suppress outliers completely. In this case, the Tukey weight function can be used [68, 57]. It is given by

$$w(\delta) = \begin{cases} (1-(\delta/\tau)^2)^2 & |\delta| \leq \tau \\ 0 & |\delta| > \tau \end{cases} \tag{8.93}$$

Again, τ is the clipping factor. Note that this weight function completely disregards points that have a distance $> \tau$. For distances $\leq \tau$, the weight changes smoothly from 1 to 0.

In the above two weight functions, the clipping factor specifies which points should be regarded as outliers. Since the clipping factor is a distance it could simply be set manually. However, this would ignore the distribution of the noise and outliers in the data, and consequently would have to be adapted for each application. It is more convenient to derive the clipping factor from the data itself. This is typically done based on the standard deviation of the distances to the line. Since we expect outliers in the data we cannot use the normal standard deviation, but must use a standard deviation that is robust to outliers. Typically, the following formula is used to compute the robust standard deviation:

$$\sigma_\delta = \frac{\text{median } |\delta_i|}{0.6745} \tag{8.94}$$

The constant in the denominator is chosen such that for normally distributed distances the standard deviation of the normal distribution is computed. The clipping factor is then set to a small multiple of σ_δ, e.g., $\tau = 2\sigma_\delta$.

In addition to the Huber and Tukey weight functions, other weight functions can be defined. Several other possibilities are discussed in [42].

Figure 8.86 displays the result of fitting a line robustly to the edge of the relay using the Tukey weight function with a clipping factor of $\tau = 2\sigma_\delta$ with five iterations. If we compare this to the standard least-squares line fit in Fig. 8.85(b) we see that with the robust fit the line is now fitted to the straight line part of the edge and outliers caused by the relay contact have been suppressed.

It should also be noted that the above approach to outlier suppression by weighting down the influence of points with large distances can sometimes fail because the initial fit, which is a standard least-squares fit, can produce a solution that is dominated by outliers. Consequently, the weight function

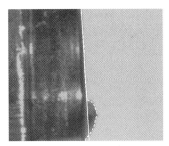

Fig. 8.86 Straight line (solid line) fitted robustly to the vertical edge (dashed line). In this case, the Tukey weight function with a clipping factor of $\tau = 2\sigma_\delta$ with five iterations was used. Compared to Fig. 8.85(b), the line is now fitted to the straight line part of the edge.

will drop inliers. In this case, other robust methods must be used. The most important approach is the random sample consensus (RANSAC) algorithm, proposed by Fischler and Bolles in [29]. Instead of dropping outliers successively it constructs a solution (e.g., a line fit) from the minimum number of points (e.g., two for lines), which are selected randomly, and then checks how many points are consistent with the solution. The process of randomly selecting points, constructing the solution, and checking the number of consistent points is continued until a certain probability of having found the correct solution, e.g., 99%, is achieved. At the end, the solution with the largest number of consistent points is selected.

8.8.2
Fitting Circles

Fitting circles or circular arcs to a contour uses the same idea as fitting lines: we want to minimize the sum of the squared distances of the contour points to the circle:

$$\varepsilon^2 = \sum_{i=1}^{n} \left(\sqrt{(r_i - \alpha)^2 + (c_i - \beta)^2} - \rho \right)^2 \quad (8.95)$$

Here, (α, β) is the center of the circle and ρ is its radius. Unlike the line fitting, this leads to a nonlinear optimization problem, which can only be solved iteratively using nonlinear optimization techniques. Details can be found in [39, 50, 1].

Figure 8.87(a) shows the result of fitting circles to the edges of the holes of a workpiece, along with the extracted radii in pixels. In Fig. 8.87(b), details of the upper right hole are shown. Note how well the circle fits the extracted edges.

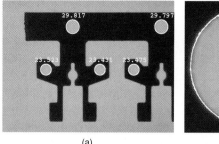

Fig. 8.87 (a) Image of a workpiece with circles fitted to the edges of the holes in the workpiece. (b) Details of the upper right hole with the extracted edge (dashed line) and the fitted circle (solid line).

Like the least-squares line fit, the least-squares circle fit is not robust to outliers. To make the circle fit robust, we can use the same approach that we used for the line fitting: we can introduce a weight that is used to reduce the influence of outliers. Again, this requires that we perform a normal least-squares fit first and then use the distances that result from it to calculate the weights in later iterations. Since it is possible that large outliers prevent this algorithm from converging to the correct solution, a RANSAC approach might be necessary in extreme cases.

Figure 8.88 compares the standard circle fitting with the robust circle fitting using the BGA example of Fig. 8.31. With the standard fitting (Fig. 8.88(b)), the circle is affected by the error in the pad, which acts like an outlier. This is corrected with the robust fitting (Fig. 8.88(c)).

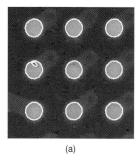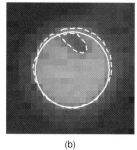

Fig. 8.88 (a) Image of a BGA with pads extracted by subpixel-precise thresholding (see also Fig. 8.31). (b) Circle fitted to the left pad in the center row of (a). The fitted circle is shown as a solid line while the extracted contour is shown as a dashed line. The fitted circle is affected by the error in the pad, which acts like an outlier. (c) Result of robustly fitting a circle. The fitted circle corresponds to the true boundary of the pad.

To conclude this section, we should spend some thoughts on what happens when a circle is fitted to a contour that only represents a part of a circle (a circular arc). In this case, the accuracy of the parameters becomes progressively worse as the angle of the circular arc becomes smaller. An excellent analysis of this effect is performed in [50]. This effect is obvious from the geometry of the problem. Simply think about a contour that only represents a 5° arc. If the contour points are disturbed by noise we have a very large range of radii and centers that lead to almost the same fitting error. On the other hand, if we fit to a complete circle the geometry of the circle is much more constrained. This effect is caused by the geometry of the fitting problem and not by a particular fitting algorithm, i.e., it will occur for all fitting algorithms.

8.8.3
Fitting Ellipses

To fit an ellipse to a contour, we would like to use the same principles as for lines and circles: minimize the distance of the contour points to the ellipse. This requires us to determine the closest point to each contour point on the ellipse. While this can be determined easily for lines and circles, it requires finding the roots of a fourth-degree polynomial for ellipses. Since this is quite complicated and expensive, ellipses are often fitted by minimizing a different kind of distance. The principle is similar to the line fitting approach: we write down an implicit equation for ellipses (for lines, the implicit equation is given by (8.87)), and then substitute the point coordinates into the implicit equation to get a distance measure for the points to the ellipse. For the line fitting problem, this procedure returns the true distance to the line. For the ellipse fitting, it only returns a value that has the same properties as a distance, but is not the true distance. Therefore, this distance is called the algebraic distance. Ellipses are described by the following implicit equation:

$$\alpha r^2 + \beta rc + \gamma c^2 + \delta r + \zeta c + \eta = 0 \qquad (8.96)$$

Like for lines, the set of parameters is a homogeneous quantity, i.e., only defined up to scale. Furthermore, (8.96) also describes hyperbolas and parabolas. Ellipses require $\beta^2 - 4\alpha\gamma < 0$. We can solve both problems by requiring $\beta^2 - 4\alpha\gamma = -1$. An elegant solution to fitting ellipses by minimizing the algebraic error with a linear method was proposed by Fitzgibbon. The interested reader is referred to [30] for details. Unfortunately, minimizing the algebraic error can result in biased ellipse parameters. Therefore, if the ellipse parameters should be determined with maximum accuracy the geometric error should be used. A nonlinear approach for fitting ellipses based on the geometric error is proposed in [1]. It is significantly more complicated than the linear approach in [30].

Like the least-squares line and circle fits, fitting ellipses via the algebraic or geometric distance is not robust to outliers. We can again introduce weights to create a robust fitting procedure. If the ellipses are fitted with the algebraic distance this again results in a linear algorithm in each iteration of the robust fit [57]. In applications with a very large number of outliers or with very large outliers a RANSAC approach might be necessary.

The ellipse fitting is very useful in camera calibration, where often circular marks are used on the calibration targets [60, 59, 43]. Since circles project to ellipses fitting ellipses to the edges in the image is the natural first step in the calibration process. Figure 8.89(a) displays an image of a calibration target. The ellipses fitted to the extracted edges of the calibration marks are shown in Fig. 8.89(b). In Fig. 8.89(c), a detailed view of the center mark with the fitted ellipse is shown. Since the subpixel edge extraction is very accurate there is hardly any visible difference between the edge and the ellipse, and consequently the edge is not shown in the figure.

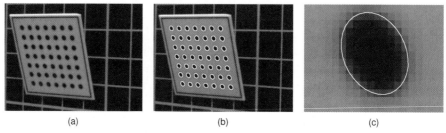

Fig. 8.89 (a) Image of a calibration target. (b) Ellipses fitted to the extracted edges of the circular marks of the calibration target. (c) Detail of the center mark of the calibration target with the fitted ellipse.

To conclude this section, we should note that if ellipses are fitted to contours that only represent a part of an ellipse the same comments that were made for circular arcs at the end of the last section apply: the accuracy of the parameters will become worse as the angle that the arc subtends becomes smaller. The reason for this behavior lies in the geometry of the problem and not in the fitting algorithms we use.

8.8.4
Segmentation of Contours into Lines, Circles, and Ellipses

So far, we have assumed that the contours to which we are fitting the geometric primitives correspond to a single primitive of the correct type, e.g., a line segment. Of course, a single contour may correspond to multiple primitives of different types. Therefore, in this section we will discuss how contours can be segmented into different primitives.

8.8 Segmentation and Fitting of Geometric Primitives

We will start by examining how a contour can be segmented into lines. To do so, we would like to find a polygon that approximates the contour sufficiently well. Let us call the contour points $p_i = (r_i, c_i)$, $i = 1, \ldots, n$. Approximating the contour by a polygon means we want to find a subset p_{i_j}, $j = 1, \ldots, m$, $m \leq n$, of the control points of the contour that describes the contour reasonably well. Once we have found the approximating polygon, each line segment $(p_{i_j}, p_{i_{j+1}})$ of the polygon is a part of the contour that can be approximated well with a line. Hence, we can fit lines to each line segment afterwards to obtain a very accurate geometric representation of the line segments.

The question we need to answer is how do we define whether a polygon approximates the contour sufficiently well. A large number of different definitions has been proposed over the years. A very good evaluation of many polygonal approximation methods has been carried out by Rosin in [75] and [76]. In both cases, it was established that the algorithm proposed by Ramer [74], which curiously enough is one of the oldest algorithms, is the best overall method.

The Ramer algorithm performs a recursive subdivision of the contour until the resulting line segments have a maximum distance to the respective contour segments that is lower than a user-specified threshold $d_{\max}$. Figure 8.90 illustrates how the Ramer algorithm works. We start out by constructing a single line segment between the first and last contour point. If the contour is closed we construct two segments: one from the first point to the point with index $n/2$ and the second one from $n/2$ to n. We then compute the distances of all the contour points to the line segment and find the point with the maximum distance to the line segment. If its distance is larger than the threshold we have specified we subdivide the line segment into two segments at the point with the maximum distance. Then, this procedure is applied recursively to the new segments until no more subdivisions occur, i.e., until all segments fulfill the maximum distance criterion.

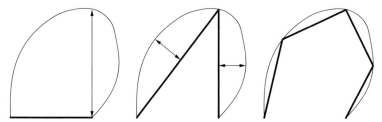

Fig. 8.90 Example of the recursive subdivision that is performed in the Ramer algorithm. The contour is displayed as a thin line, while the approximating polygon is displayed as a thick line.

Figure 8.91 illustrates the use of the polygonal approximation in a real application. In Fig. 8.91(a), a back-lit cutting tool is shown. In the application, the dimensions and angles of the cutting tool must be inspected. Since the tool consists of straight edges the obvious approach is to extract edges with subpixel accuracy (Fig. 8.91(b)) and to approximate them with a polygon using the Ramer algorithm. From Fig. 8.91(c) we can see that the Ramer algorithm splits the edges correctly. We can also see the only slight drawback of the Ramer algorithm: it sometimes places the polygon control points into positions that are slightly offset from the true corners. In this application, this poses no problem since to achieve maximum accuracy we must fit lines to the contour segments robustly anyway (Fig. 8.91(d)). This enables us to obtain a concise and very accurate geometric description of the cutting tool. With the resulting geometric parameters, it can be easily checked whether the tool has the required dimensions.

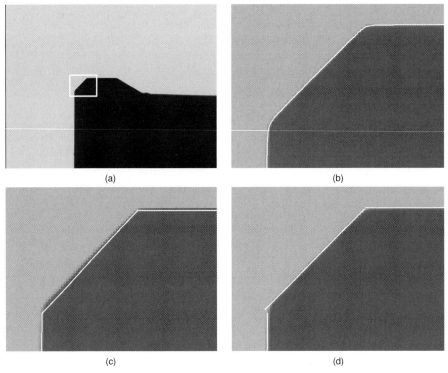

Fig. 8.91 (a) Image of a back-lit cutting tool with the part that is shown in (b)–(d) overlaid as a white rectangle. To provide a better visibility of the results the contrast of the image has been reduced in (b)–(d). (b) Subpixel-accurate edges extracted with the Lanser filter with $\alpha = 0.7$. (c) Polygons extracted with the Ramer algorithm with $d_{\max} = 2$. (d) Lines fitted robustly to the polygon segments using the Tukey weight function.

While lines are often the only geometric primitive that occurs for the objects that should be inspected, in several cases the contour must be split into several types of primitives. For example, machined tools often consist of lines and circular arcs or lines and elliptic arcs. Therefore, we will now discuss how such a segmentation of the contours can be performed.

The approaches to segment contours into lines and circles can be classified into two broad categories. The first type of algorithm tries to identify breakpoints on the contour that correspond to semantically meaningful entities. For example, if two straight lines with different angles are next to each other the tangent direction of the curve will contain a discontinuity. On the other hand, if two circular arcs with different radii meet smoothly there will be a discontinuity in the curvature of the contour. Therefore, the breakpoints typically are defined as discontinuities in the contour angle, which are equivalent to maxima of the curvature, and as discontinuities in the curvature itself. The first definition covers straight lines or circular arcs that meet with a sharp angle. The second definition covers smoothly joining circles or lines and circles [100, 83]. Since the curvature depends on the second derivative of the contour it is an unstable feature that is very prone to even small errors in the contour coordinates. Therefore, to enable these algorithms to function properly, the contour must be smoothed substantially. This, in turn, can cause the breakpoints to shift from their desired positions. Furthermore, some breakpoints may be missed. Therefore, these approaches are often followed by an additional splitting and merging stage and a refinement of the breakpoint positions [83, 17].

While the above algorithms work well for splitting contours into lines and circles, they are quite difficult to extend to lines and ellipses because ellipses do not have constant curvature like circles. In fact, the two points on the ellipse on the major axis have locally maximal curvature and consequently would be classified as break points by the above algorithms. Therefore, if we want to have a unified approach to segmenting contours into lines and circles or ellipses the second type of algorithm is more appropriate. They are characterized by initially performing a segmentation of the contour into lines only. This produces an over-segmentation in the areas of the contour that correspond to circles and ellipses since here many line segments are required to approximate the contour. Therefore, in a second phase the line segments are examined whether they can be merged into circles or ellipses [57, 77]. For example, the algorithm in [57] initially performs a polygonal approximation with the Ramer algorithm. Then, it checks each pair of adjacent line segments to see whether it can be better approximated by an ellipse (or, alternatively, a circle). This is done by fitting an ellipse to the part of the contour that corresponds to the two line segments. If the fitting error of the ellipse is smaller than the maximum error of the two lines, the two line segments are marked as candidates for merging. After examining all pairs of line segments, the pair with the smallest fitting error is merged. In the following iterations the algorithm also considers pairs of line and ellipse segments. The iterative merging is continued until there are no more segments that can be merged.

Figure 8.92 illustrates the segmentation into lines and circles. Like in the previous example, the application is the inspection of cutting tools. Figure 8.92(a) displays a cutting tool that consists of two linear parts and a circular part. The result of the initial segmentation into lines with the Ramer algorithm is shown in Fig. 8.92(b). Note that the circular arc is represented by four contour parts. The iterative merging stage of the algorithm successfully merges these four contour parts into a circular arc, as shown in Fig. 8.92(c). Finally, the angle between the linear parts of the tool is measured by fitting lines to the corresponding contour parts, while the radius of the circular arc is determined by fitting a circle to the circular arc part. Since the camera is calibrated in this application the fitting is actually performed in world coordinates. Hence, the radius of the arc is calculated in millimeters.

8.9
Template Matching

In the previous sections we have discussed various techniques that can be combined to write algorithms to find objects in an image. While these techniques can in principle be used to find any kind of object, writing a robust recognition algorithm for a particular type of objects can be quite cumbersome. Furthermore, if the objects to be recognized change frequently a new algorithm must be developed for each type of objects. Therefore, a method to find any kind of object that can be configured simply by showing the system a prototype of the class of objects to be found would be extremely useful.

The above goal can be achieved by template matching. Here, we describe the object to be found by a template image. Conceptually, the template is found in the image by computing the similarity between the template and the image for all relevant poses of the template. If the similarity is high an instance of the template has been found. Note that the term similarity is used here in a very general sense. We will see below that it can be defined in various ways, e.g., based on the gray values of the template and the image or based on the closeness of template edges to image edges.

Template matching can be used for several purposes. First, it can be used to perform completeness checks. Here, the goal is to detect the presence or absence of the object. Furthermore, template matching can be used for object discrimination, i.e., to distinguish between different types of objects. In most cases, however, we already know which type of object is present in the image.

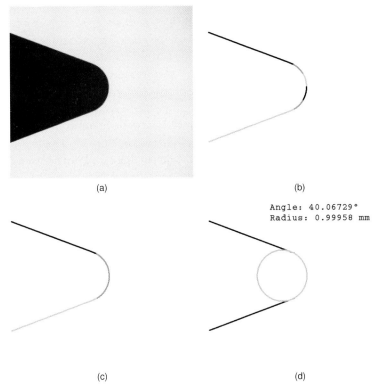

Fig. 8.92 (a) Image of a back-lit cutting tool. (b) Contour parts corresponding to the initial segmentation into lines with the Ramer algorithm. The contour parts are displayed in three different gray values. (c) Result of the merging stage of the line and circle segmentation algorithm. In this case, two lines and one circular arc are returned. (d) Geometric measurements obtained by fitting lines to the linear parts of the contour and a circle to the circular part. Because the camera was calibrated the radius is calculated in millimeters.

In these cases, template matching is used to determine the pose of the object in the image. If the orientation of the objects can be fixed mechanically the pose is described by a translation. In most applications, however, the orientation cannot be fixed completely, if at all. Therefore, often the orientation of the object, described by rotation, must also be determined. Hence, the complete pose of the object is described by a translation and a rotation. This type of transformation is called a rigid transformation. In some applications, additionally the size of the objects in the image can change. This can happen if the distance of the objects to the camera cannot be kept fixed or if the real size of the objects can change. Hence, a uniform scaling must be added to the pose in these applications. This type of pose (translation, rotation, and uniform scaling) is called a similarity transformation. If even the 3D orientation of the camera with respect to the objects can change and the objects to be recognized are planar the pose is described by a projective transformation (see Section 8.3.2). Consequently, for the purposes of this chapter we can regard the pose of the objects as a specialization of an affine or projective transformation.

In most applications a single object is present in the search image. Therefore, the goal of the template matching is to find this single instance. In some applications, more than one object is present in the image. If we know a priori how many objects are present we want to find exactly this number of objects. If we do not have this knowledge we typically must find all instances of the template in the image. In this mode, one of the goals is also to determine how many objects are present in the image.

8.9.1
Gray-Value-Based Template Matching

In this section, we will examine the simplest kind of template matching algorithms, which are based on the raw gray values in the template and the image. As mentioned above, template matching is based on computing a similarity between the template and the image. Let us formalize this notion. For the moment, we will assume that the object's pose is described by a translation. The template is specified by an image $t(r, c)$ and its corresponding ROI T. To perform the template matching, we can visualize moving the template over all positions in the image and computing a similarity measure **s** at each position. Hence, the similarity measure **s** is a function that takes the gray values in the template $t(r, c)$ and the gray values in the shifted ROI of the template at the current position in the image $f(r + u, c + v)$ and calculates a scalar value that measures the similarity based on the gray values within the respective ROI. With this approach, a similarity measure is returned for each point in the transformation space, which for translations can be regarded as an image. Hence, formally, we have:

$$s(r, c) = \mathbf{s}\{t(u, v), f(r + u, c + v); (u, v) \in T\} \quad (8.97)$$

To make this abstract notation concrete, we will discuss several possible gray-value-based similarity measures [14].

The simplest similarity measures are to sum the absolute or squared gray value differences between the template and the image (SAD and SSD). They are given by

$$sad(r, c) = \frac{1}{n} \sum_{(u,v) \in T} |t(u, v) - f(r + u, c + v)| \quad (8.98)$$

and
$$ssd(r,c) = \frac{1}{n} \sum_{(u,v) \in T} (t(u,v) - f(r+u, c+v))^2 \qquad (8.99)$$

In both cases, n is the number of points in the template ROI, i.e., $n = |T|$. Note that both similarity measures can be computed very efficiently with just two operations per pixel. These similarity measures have similar properties: if the template and the image are identical they return a similarity measure of 0. If the image and template are not identical a value greater than 0 is returned. As the dissimilarity increases the value of the similarity measure increases. Hence, in this case the similarity measure should probably better be called a dissimilarity measure. To find instances of the template in the search image, we can threshold the similarity image $sad(r,c)$ with a certain upper threshold. This typically gives us a region that contains several adjacent pixels. To obtain a unique location for the template, we must select the local minima of the similarity image within each connected component of the thresholded region.

Figure 8.93 shows a typical application for template matching. Here, the goal is to locate the position of a fiducial mark on a PCB. The ROI used for the template is displayed in Fig. 8.93(a). The similarity computed with the sum of the SAD, given by (8.98), is shown in Fig. 8.93(b). For this example, the SAD was computed with the same image from which the template was generated. If the similarity is thresholded with a threshold of 20 only a region around the position of the fiducial mark is returned (Fig. 8.93(c)). Within this region, the local minimum of the SAD must be computed (not shown) to obtain the position of the fiducial mark.

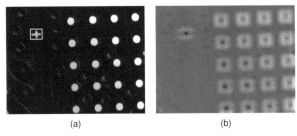

Fig. 8.93 (a) Image of a PCB with a fiducial mark, which is used as the template (indicated by the white rectangle). (b) SAD computed with the template in (a) and the image in (a). (c) Result of thresholding (b) with a threshold of 20. Only a region around the fiducial is selected.

The SAD and SSD similarity measures work very well as long as the illumination can be kept constant. If, however, the illumination can change they both return larger values, even if the same object is contained in the image, because the gray values are no longer identical. This effect is illustrated in Fig. 8.94. Here, a darker and brighter image of the fiducial mark are shown.

They were obtained by adjusting the illumination intensity. The SAD computed with the template of Fig. 8.93(a) is displayed in Figs. 8.94(b) and (e). The result of thresholding them with a threshold of 35 is shown in Figs. 8.94(c) and (f). The threshold was chosen such that the true fiducial mark is extracted in both cases. Note that because of the contrast change many extraneous instances of the template have been found.

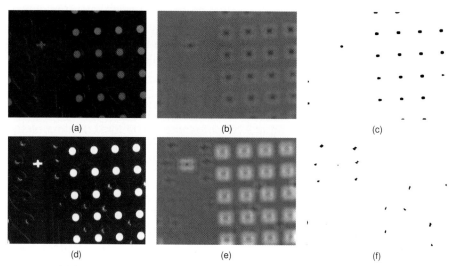

Fig. 8.94 (a) Image of a PCB with a fiducial mark with lower contrast. (b) SAD computed with the template in Fig. 8.93(a) and the image in (a). (c) Result of thresholding (b) with a threshold of 35. (d) Image of a PCB with a fiducial mark with higher contrast. (e) SAD computed with the template of Fig. 8.93(a) and the image in (d). (f) Result of thresholding (e) with a threshold of 35. In both cases, it is impossible to select a threshold that only returns the region of the fiducial.

As we can see from the above examples, the SAD and SSD similarity measures work well as long as the illumination can be kept constant. In applications where this cannot be ensured a different kind of similarity measure is required. Ideally, this similarity measure should be invariant to all linear illumination changes (see Section 8.2.1). A similarity measure that achieves this is the normalized cross correlation (NCC), given by

$$ncc(r,c) = \frac{1}{n} \sum_{(u,v) \in T} \frac{t(u,v) - m_t}{\sqrt{s_t^2}} \cdot \frac{f(r+u, c+v) - m_f(r,c)}{\sqrt{s_f^2(r,c)}} \qquad (8.100)$$

Here, m_t is the mean gray value of the template and s_t^2 is the variance of the gray values, i.e.:

$$m_t = \frac{1}{n} \sum_{(u,v) \in T} t(u,v)$$

$$s_t^2 = \frac{1}{n} \sum_{(u,v) \in T} (t(u,v) - m_t)^2 \quad (8.101)$$

Analogously, $m_f(r,c)$ and $s_f^2(r,c)$ are the mean value and variance in the image at a shifted position of the template ROI:

$$m_f(r,c) = \frac{1}{n} \sum_{(u,v) \in T} f(r+u, c+v)$$

$$s_f^2(r,c) = \frac{1}{n} \sum_{(u,v) \in T} (f(r+u, c+v) - m_f(r,c))^2 \quad (8.102)$$

The NCC has a very intuitive interpretation. First, we should note that $-1 \leq ncc(r,c) \leq 1$. Furthermore, if $ncc(r,c) = \pm 1$ the image is a linearly scaled version of the template: $ncc(r,c) = \pm 1 \Leftrightarrow f(r+u, c+v) = at(u,v) + b$. For $ncc(r,c) = 1$ we have $a > 0$, i.e., the template and the image have the same polarity, while $ncc(r,c) = -1$ implies that $a < 0$, i.e., the polarity of the template and image are reversed. Note that this property of the NCC implies the desired invariance against linear illumination changes. The invariance is achieved by explicitly subtracting the mean gray values, which cancels additive changes, and by dividing by the standard deviation of the gray values, which cancels multiplicative changes.

While the template matches the image perfectly only if $ncc(r,c) = \pm 1$, large absolute values of the NCC generally indicate that the template closely corresponds to the image part under examination, while values close to zero indicate that the template and image do not correspond well.

Figure 8.95 displays the results of computing the NCC for the template in Fig. 8.93(a) (reproduced in Fig. 8.95(a)). The NCC is shown in Fig. 8.95(b), while the result of thresholding the NCC with a threshold of 0.75 is shown in Fig. 8.95(c). This selects only a region around the fiducial mark. In this region, the local maximum of the NCC must be computed to derive the location of the fiducial mark (not shown). The results for the darker and brighter images in Fig. 8.94 are not shown because they are virtually indistinguishable from the results in Fig. 8.95(b) and (c).

In the above discussion, we have assumed that the similarity measures must be evaluated completely for every translation. This is, in fact, unnecessary since the result of calculating the similarity measure will be thresholded with a threshold t_s later on. For example, thresholding the SAD in (8.98) means that

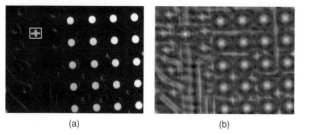

Fig. 8.95 (a) Image of a PCB with a fiducial mark, which is used as the template (indicated by the white rectangle). This is the same image as in Fig. 8.93(a). (b) NCC computed with the template in (a) and the image in (a). (c) Result of thresholding (b) with a threshold of 0.75. The results for the darker and brighter images in Fig. 8.94 are not shown because they are virtually indistinguishable from the results in (b) and (c).

we require

$$sad(r,c) = \frac{1}{n}\sum_{i=1}^{n}|t(u_i,v_i) - f(r+u_i,c+v_i)| \leq t_s \qquad (8.103)$$

Here, we have explicitly numbered the points $(u,v) \in T$ by (u_i, v_i). We can multiply both sides by n to obtain

$$sad'(r,c) = \sum_{i=1}^{n}|t(u_i,v_i) - f(r+u_i,c+v_i)| \leq nt_s \qquad (8.104)$$

Suppose we have already evaluated the first j terms in the sum in (8.104). Let us call this partial result $sad'_j(r,c)$. Then, we have

$$sad'(r,c) = sad'_j(r,c) + \underbrace{\sum_{i=j+1}^{n}|t(u_i,v_i) - f(r+u_i,c+v_i)|}_{\geq 0} \leq nt_s \qquad (8.105)$$

Hence, we can stop the evaluation as soon as $sad'_j(r,c) > nt_s$ because we are certain that we can no longer achieve the threshold. If we are looking for a maximum number of m instances of the template we can even adapt the threshold t_s based on the instance with the m-th best similarity found so far. For example, if we are looking for a single instance with $t_s = 20$ and we already have found a candidate with $sad(r,c) = 10$ we can set $t_s = 10$ for the remaining poses that need to be checked. Of course, we need to calculate the local minima of $sad(r,c)$ and use the corresponding similarity values to ensure that this approach works correctly if more than one instance should be found.

For the normalized cross correlation, there is no simple criterion to stop the evaluation of the terms. Of course, we can use the fact that the mean

m_t and standard deviation $\sqrt{s_t^2}$ of the template can be computed once offline because they are identical for every translation of the template. This leaves us with partial sums that contain the correlation of the template and the image $t(u,v) \cdot f(r+u, c+v)$ plus the mean $m_f(r,c)$ and the standard deviation $\sqrt{s_f^2(r,c)}$ of the image. It is very difficult to derive a bound for the normalized cross correlation based on these terms if all of them have been evaluated only partially. Therefore, some of the above terms must be evaluated completely before a simple stopping criterion can be derived. Since the mean and standard deviation of the image can be calculated recursively, i.e., with a constant number of operations per point, if the ROI of the template is a rectangle it is most efficient to evaluate these two terms completely [85]. Hence, we know $m_f(r,c)$ and $\sqrt{s_f^2(r,c)}$. Like for the SAD, we can multiply the correlation and the threshold by n to obtain

$$ncc'(r,c) = \sum_{i=1}^{n} \frac{t(u_i, v_i) - m_t}{\sqrt{s_t^2}} \cdot \frac{f(r+u_i, c+v_i) - m_f(r,c)}{\sqrt{s_f^2(r,c)}} \geq nt_s \qquad (8.106)$$

Let us suppose we have already evaluated the first j terms of the sum in (8.106) and call this partial sum $ncc'_j(r,c)$. Then, we have

$$ncc'(r,c) = ncc'_j(r,c)$$
$$+ \underbrace{\sum_{i=j+1}^{n} \frac{t(u_i, v_i) - m_t}{\sqrt{s_t^2}} \cdot \frac{f(r+u_i, c+v_i) - m_f(r,c)}{\sqrt{s_f^2(r,c)}}}_{\leq 1} \geq nt_s \qquad (8.107)$$

Hence, we can stop the evaluation as soon as $ncc'_j(r,c) + (n-j) < nt_s$, i.e., $ncc'_j(r,c) < (t_s - 1)n + j$. This stopping criterion, of course, works for templates with arbitrary ROIs as well. However, it can be evaluated most efficiently (recursively) for rectangular ROIs. Furthermore, analogously to the SAD we can adapt the threshold t_s based on the matches we have found so far [85].

The above stopping criteria enable us to stop the evaluation of the similarity measure as soon as we are certain that the threshold can no longer be reached. Hence, they prune unwanted parts of the space of allowed poses. It is interesting to note that improvements for the pruning of the search space are still actively being investigated. For example, in [85] further optimizations for the pruning of the search space when using the NCC are discussed. In [34] and [44] strategies for pruning the search space when using SAD or SSD are discussed. They rely on transforming the image into a representation in which a large portion of the SAD and SSD can be computed with very few evaluations so that the above stopping criteria can be reached as soon as possible.

8.9.2
Matching Using Image Pyramids

The evaluation of the similarity measures on the entire image is very time consuming, even if the stopping criteria discussed above are used. If they are not used the runtime complexity is $O(whn)$, where w and h are the width and height of the image, respectively, and n is the number of points in the template. The stopping criteria typically result in a constant factor for the speed up, but do not change the complexity. Therefore, a method to further speed up the search is necessary to be able to find the template in real time.

To derive a faster search strategy, we note that the runtime complexity of the template matching depends on the number of translations, i.e., poses, that need to be checked. This is the $O(wh)$ part of the complexity. Furthermore, it depends on the number of points in the template. This is the $O(n)$ part. Therefore, to gain a speed up, we can try to reduce the number of poses that need to be checked as well as the number of template points. Since the templates typically are large one way to do this would be to only take into account every ith point of the image and template in order to obtain an approximate pose of the template, which could later be refined by a search with a finer step size around the approximate pose. This strategy is identical to subsampling the image and template. Since subsampling can cause aliasing effects (see Section 8.3.3) this is not a very good strategy because we might miss instances of the template because of the aliasing effects. We have seen in Section 8.3.3 that we must smooth the image to avoid the aliasing effects. Furthermore, it is typically better to scale the image down multiple times by a factor of 2 than only once by a factor of $i > 2$. Scaling down the image (and template) multiple times by a factor of 2 creates a data structure that is called an image pyramid. Figure 8.96 displays why the name was chosen: we can visualize the smaller versions of the image stacked on top of each other. Since their width and height are halved in each step they form a pyramid.

When constructing the pyramid, speed is essential. Therefore, the smoothing is performed by applying a 2×2 mean filter, i.e., by averaging the gray value of each 2×2 block of pixels [90]. The smoothing could also be performed by a Gaussian filter [36]. Note, however, that in order to avoid the introduction of unwanted shifts into the image pyramid the Gaussian filter must have an even mask size. Therefore, the smallest mask size would be a 4×4 filter. Hence, using the Gaussian filter would incur a severe speed penalty in the construction of the image pyramid. Furthermore, the 2×2 mean filter does not have the frequency response problems that the larger versions of the filter have (see Section 8.2.3). In fact, it drops off smoothly toward a zero response for the highest frequencies, like the Gaussian filter. Finally, it simulates the

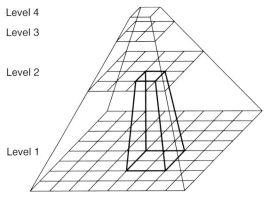

Fig. 8.96 An image pyramid is constructed by successively halving the resolution of the image and combining 2 × 2 blocks of pixels in a higher resolution into a single pixel at the next lower resolution.

effects of a perfect camera with a fill factor of 100%. Therefore, the mean filter is the preferred filter for constructing image pyramids.

Figure 8.97 displays the image pyramid levels 2–5 of the image in Fig. 8.93(a). We can see that on levels 1–4 the fiducial mark can still be discerned from the BGA pads. This is no longer the case on level 5. Therefore, if we want to find an approximate location of the template we can start the search on level 4.

The above example produces the expected result: the image is progressively smoothed and subsampled. The fiducial mark we are interested in can no longer be recognized as soon as the resolution becomes too low. Sometimes, however, creating an image pyramid can produce results that are unexpected at first glance. One example for this behavior is shown in Fig. 8.98. Here, the image pyramid levels 1–4 of an image of a PCB are shown. We can see that on the pyramid level 4 all the conductors are suddenly merged into large components with identical gray values. This happens because of the smoothing that is performed when the pyramid is constructed. Here, the neighboring thin lines start to interact with each other once the smoothing is large enough, i.e., once we reach a pyramid level that is large enough. Hence, we can see that sometimes valuable information is destroyed by the construction of the image pyramid. If we were interested in matching, say, the corners of the conductors we could only go as high as level 3 in the image pyramid.

Based on the image pyramids, we can define a hierarchical search strategy as follows: first, we calculate an image pyramid on the template and search image with an appropriate number of levels. How many levels can be used is mainly defined by the objects we are trying to find. On the highest pyramid level the relevant structures of the object must still be discernible. Then, a com-

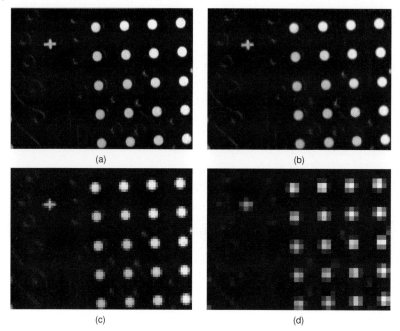

Fig. 8.97 (a)–(d) The image pyramid levels 2–5 of the image in Fig. 8.93(a). Note that in level 5 the fiducial mark can no longer be discerned from the BGA pads.

plete matching is performed on the highest pyramid level. Here, of course, we take the appropriate stopping criterion into account. What does this gain us? In each pyramid level, we reduce the number of image points and template points by a factor of 4. Hence, each pyramid level results in a speed up of a factor of 16. Therefore, if we perform the complete matching, for example, on level 4 we reduce the amount of computations by a factor of 4096.

All instances of the template that have been found on the highest pyramid level are then tracked down to the lowest pyramid level. This is done by projecting the match down to the next lower pyramid level, i.e., by multiplying the coordinates of the found match by 2. Since there is an uncertainty in the location of the match a search area is constructed around the match in the lower pyramid level, e.g., a 5×5 rectangle. Then, the matching is performed within this small ROI, i.e., the similarity measure is computed, thresholded, and the local maxima or minima are extracted. This procedure is continued until the match is lost or tracked down to the lowest level. Since the search spaces for the larger templates are very small, tracking the match down to the lowest level is very efficient.

While matching the template on the higher pyramid levels, we need to take the following effect into account: the gray values at the border of the object

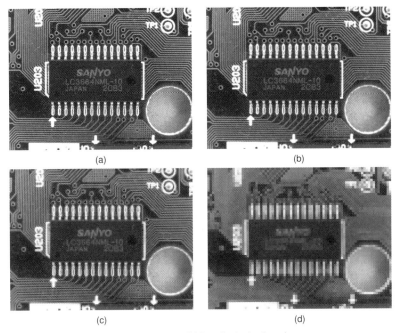

Fig. 8.98 (a)–(d) The image pyramid levels 1–4 of an image of a PCB. Note that in level 4 all the conductors are merged into large components with identical gray values because of the smoothing that is performed when the pyramid is constructed.

can change substantially on the highest pyramid level depending on where the object lies on the lowest pyramid level. This happens because a single pixel shift of the object translates to a subpixel shift on higher pyramid levels, which manifests itself as a change in the gray values on the higher pyramid levels. Therefore, on the higher pyramid levels we need to be more lenient with the matching threshold to ensure that all potential matches are being found. Hence, for the SAD and SSD similarity measures we need to use slightly higher thresholds and for the NCC similarity measure we need to use slightly lower thresholds on the higher pyramid levels.

The hierarchical search is visualized in Fig. 8.99. The template is the fiducial mark shown in Fig. 8.93(a). The template is searched in the same image from which the template was created. As discussed above, four pyramid levels are used in this case. The search starts on the level 4. Here, the ROI is the entire image. The NCC and found matches on level 4 are displayed in Fig. 8.99(a). As we can see, 12 potential matches are initially found. They are tracked down to level 3 (Fig. 8.99(b)). The ROIs created from the matches on level 4 are shown in white. For visualization purposes, the NCC is displayed for the entire image. In reality, it is, of course, only computed within the ROIs, i.e., for a total of 12 · 25 = 300 translations. Note that on this level the true match turns out to be the only viable match. It is tracked down through levels 2 and 1 (Figs. 8.99(c) and (d)). In both cases, only 25 translations need to be checked. Therefore, the match is found extremely efficiently. The zoomed part of the NCC in Figs. 8.99(b)–(d) also shows that the pose of the match is progressively refined as the match is tracked down the pyramid.

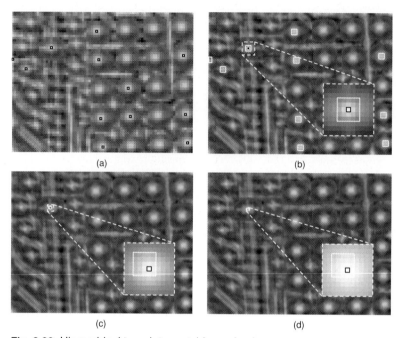

Fig. 8.99 Hierarchical template matching using image pyramids. The template is the fiducial mark shown in Fig. 8.93(a). To provide a better visualization, the NCC is shown for the entire image on each pyramid level. In reality, however, it is only calculated within the appropriate ROI on each level, shown in white. The found matches are displayed in black. (a) On the pyramid level 4, the matching is performed in the entire image. Here, 12 potential matches are found. (b) The matching is continued within the white ROIs on level 3. Only one viable match is found in the 12 ROIs. The similarity measure and ROI around the match are displayed zoomed in the lower right corner. (c)–(d) The match is tracked through the pyramid levels 2 and 1.

8.9.3
Subpixel-Accurate Gray-Value-Based Matching

So far, we have located the pose of the template with pixel precision. This has been done by extracting the local minima (SAD, SSD) or maxima (NCC) of the similarity measure. To obtain the pose of the template with higher accuracy, the local minima or maxima can be extracted with subpixel precision. This can be done in a manner that is analogous to the method we have used in edge extraction (see Section 8.7.3): we simply fit a polynomial to the similarity measure in a 3×3 neighborhood around the local minimum or maximum. Then, we extract the local minimum or maximum of the polynomial analytically. Another approach is to perform a least-squares matching of the gray values of the template and the image [92]. Since the least-squares matching of the gray values is not invariant to illumination changes, the illumination changes must be modeled explicitly, and their parameters must be determined in the least-squares fitting in order to achieve robustness to illumination changes [54].

8.9.4
Template Matching with Rotations and Scalings

Up to now, we have implicitly restricted the template matching to the case where the object must have the same orientation and scale in the template and the image, i.e., the space of possible poses was assumed to be the space of translations. The similarity measures we have discussed above can only tolerate small rotations and scalings of the object in the image. Therefore, if the object does not have the same orientation and size as the template the object will not be found. If we want to be able to handle a larger class of transformations, e.g., rigid or similarity transformations we must modify the matching approach. For simplicity, we will only discuss rotations, but the method can be extended to scalings and even more general classes of transformations in an analogous manner.

To find a rotated object, we can create the template in multiple orientations, i.e., we discretize the search space of rotations in a manner that is analogous to discretization of the translations that is imposed by the pixel grid [3]. Unlike for the translations, discretization of the orientations of the template depends on the size of the template since the similarity measures are less tolerant to small angle changes for large templates. For example, a typical value is to use an angle step size of $1°$ for templates with a radius of 100 pixels. Larger templates must use smaller angle steps, while smaller templates can use larger angle steps. To find the template, we simply match all rotations of the template with the image. Of course, this is only done on the highest pyramid level. To make the matching in the pyramid more efficient, we can also use the fact that the templates become smaller by a factor of two on each pyramid level. Consequently, the angle step size can be increased by a factor of two for each pyramid level. Hence, if an angle step size of $1°$ is used on the lowest pyramid level a step size of $8°$ can be used on the fourth pyramid level.

While tracking potential matches through the pyramid, we also need to construct a small search space for the angles in the next lower pyramid level, analogously to the small search space that we already use for the translations. Once we have tracked the match to the lowest pyramid level, we typically want to refine the pose to an accuracy that is higher than the resolution of the search space we have used. In particular, if rotations are used the pose should consist of a subpixel translation and an angle that is more accurate than the angle step size we have chosen. The techniques for subpixel-precise localization of the template described above can easily be extended for this purpose.

8.9.5
Robust Template Matching

The above template matching algorithms have served for many years as the methods of choice to find objects in machine vision applications. Recently, however, there has been an increasing demand to find objects in images even if they are occluded or disturbed in other ways so that parts of the object are missing. Furthermore, objects should be found even if there are a large number of disturbances on the object itself. These disturbances are often referred to as clutter. Finally, objects should be found if there are severe nonlinear illumination changes. The gray-value-based template matching algorithms we have discussed so far cannot handle these kinds of disturbances. Therefore, in the remainder of this section we will discuss several approaches that have been designed to find objects in the presence of occlusion, clutter, and nonlinear illumination changes.

We already have discussed a feature that is robust to nonlinear illumination changes in Section 8.7: edges are not (or at least very little) affected by illumination changes. Therefore, they are frequently used in robust matching algorithms. The only problem when using edges is the selection of a suitable threshold to segment the edges. If the threshold is chosen too low there will be many clutter edges in the image. If it is chosen too high important edges of the object will be missing. This has the same effect as if parts of the object are occluded. Since the threshold can never be chosen perfectly this is another reason why the matching must be able to handle occlusions and clutter robustly.

To match objects using edges, several strategies exist. First, we can use the raw edge points, possibly augmented with some features per edge point, for the matching (see Fig. 8.100(b)). Another strategy is to derive geometric primitives by segmenting the edges with the algorithms discussed in Section 8.8.4, and to match these to segmented geometric primitives in the image (see Fig. 8.100(c)). Finally, based on a segmentation of the edges, we can derive salient points and match them to salient points in the image (see Fig. 8.100(d)).

It should be noted that salient points can also be extracted directly from the image without extracting edges first [33, 80].

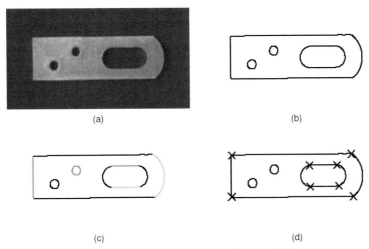

Fig. 8.100 (a) Image of a model object. (b) Edges of (a). (c) Segmentation of (b) into lines and circles. (d) Salient points derived from the segmentation in (c).

A large class of algorithms for edge matching is based on the distance of the edges in the template to the edges in the image. These algorithms typically use the raw edge points for the matching. One natural similarity measure based on this idea is to minimize the mean squared distance between the template edge points and the closest image edge points [12]. Hence, it appears that we must determine the closest image edge point for every template edge point, which would be extremely costly. Fortunately, since we are only interested in the distance to the closest edge point and not in which point is the closest point this can be done in an efficient manner by calculating the distance transform of the background of the segmented search image [12]. A model is considered as being found if the mean distance of the template edge points to the image edge points is below a threshold. Of course, to obtain a unique location of the template we must calculate the local minimum of this similarity measure. If we want to formalize this similarity measure, we can denote the edge points in the model by T and the distance transform of the background of the segmented search image (i.e., the complement of the segmented edge region in the search image) by $d(r,c)$. Hence, the mean squared edge distance (SED) for the case of translations is given by

$$\mathrm{sed}(r,c) = \frac{1}{n} \sum_{(u,v) \in T} d(r+u, c+v)^2 \qquad (8.108)$$

Note that this is very similar to the SSD similarity measure in (8.99) if we set $t(u,v) = 0$ there and use the distance transform image for $f(u,v)$. Consequently, SED matching algorithm can be implemented very easily if we already have an implementation of the SSD matching algorithm. Of course, if we use the mean distance instead of the mean squared distance, we could use an existing implementation of the SAD matching, given by (8.98), for the edge matching.

We can now ask ourselves whether the SED fulfills the above criteria for robust matching. Since it is based on edges it is robust to arbitrary illumination changes. Furthermore, since clutter, i.e., extra edges in the search image, can only decrease the distance to the closest edge in the search image it is robust to clutter. However, if edges are missing in the search image the distance of the missing template edges to the closest image edges may become very large, and consequently the model may not be found. This is illustrated in Fig. 8.101. Imagine what happens when the model in Fig. 8.101(a) is searched in a search image in which some of the edges are missing (Figs. 8.101(c) and (d)). Here, the missing edges will have a very large squared distance, which will increase the mean squared edge distance significantly. This will make it quite difficult to find the correct match.

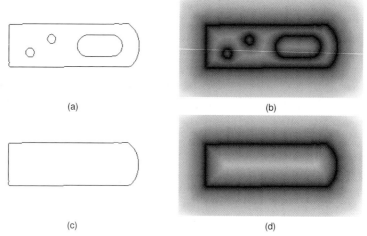

Fig. 8.101 (a) Template edges. (b) Distance transform of the background of (a). For better visualization, a square root LUT is used. (c) Search image with missing edges. (d) Distance transform of the background of (c). If the template in (a) is matched to a search image in which the edges are complete and which possibly contains more edges than the template the template will be found. If the template in (a) is matched to a search image in which template edges are missing the template may not be found because missing edge will have a large distance to the closest existing edge.

Because of the above problems of the SED, edge matching algorithms using a different distance have been proposed. They are based on the Hausdorff distance of two point sets. Let us call the edge points in the template T and the edge points in the image E. Then, the Hausdorff distance of the two point sets is given by

$$H(T, E) = \max(h(T, E), h(E, T)) \qquad (8.109)$$

where

$$h(T, E) = \max_{t \in T} \min_{e \in E} \|t - e\| \qquad (8.110)$$

and $h(E, T)$ is defined symmetrically. Hence, the Hausdorff distance consists of determining the maximum of two distances: the maximum distance of the template edges to the closest image edges and the maximum distance of the image edges to the closest template edges [78]. It is immediately clear that to achieve a low overall distance, every template edge point must be close to an image edge point and vice versa. Therefore, the Hausdorff distance is neither robust to occlusion nor to clutter. With a slight modification, however, we can achieve the desired robustness. The reason for the bad performance for occlusion and clutter is that in (8.110) the maximum distance of the template edges to the image edges is calculated. If we want to achieve robustness to occlusion, instead of computing the largest distance we can compute a distance with a different rank, e.g., the f th largest distance, where $f = 0$ denotes the largest distance. With this, the Hausdorff distance will be robust to $100 f/n\%$ occlusion, where n is the number of edge points in the template. To make the Hausdorff distance robust to clutter, we can similarly modify $h(E, T)$ to use the r th largest distance. However, normally the model covers only a small part of the search image. Consequently, there are typically many more image edge points than template edge points, and hence r would have to be chosen very large to achieve the desired robustness against clutter. Therefore, $h(E, T)$ must be modified to be calculated only within a small ROI around the template. With this, the Hausdorff distance can be made robust to $100r/m\%$ clutter, where m is the number of edge points in the ROI around the template [78]. Like the SED, the Hausdorff distance can be computed based on distance transforms: one for the edge region in the image and one for each pose (excluding translations) of the template edge region. Therefore, we must either compute a very large number of distance transforms offline, which requires an enormous amount of memory, or we must compute the distance transforms of the model during the search, which requires a large amount of computation.

As we can see, one of the drawbacks of the Hausdorff distance is the enormous computational load that is required for the matching. In [78], several possibilities are discussed to reduce the computational load, including pruning regions of the search space that cannot contain the template. Furthermore, a hierarchical subdivision of the search space is proposed. This is similar to the effect that is achieved with image pyramids. However, the method in [78] only subdivides the search space, but does not scale the template or image. Therefore, it is still very slow. A Hausdorff distance matching method using image pyramids is proposed in [53].

The major drawback of the Hausdorff distance, however, is that even with very moderate amounts of occlusion many false instances of the template will be detected in the image [70]. To reduce the false detection rate, in [70] a modification of the Hausdorff distance that takes the orientation of the edge pixels into account is proposed. Conceptually, the edge points are augmented with a third coordinate that represents the edge orientation. Then, the distance of these augmented 3D points and the corresponding augmented 3D image points is calculated as the modified Hausdorff distance. Unfortunately, this requires the calculation of a three-dimensional distance transform, which makes the algorithm too expensive for machine vision applications. A further drawback of all approaches based on the Hausdorff distance is that it is quite difficult to obtain the pose with subpixel accuracy based on the interpolation of the similarity measure.

Another algorithm to find objects that is based on the edge pixels themselves is the generalized Hough transform proposed by Ballard in [6]. The original Hough transform is a method that was designed to find straight lines in segmented edges. It was later extended to detect other shapes that can be described analytically, e.g., circles or ellipses. The principle of the generalized Hough transform can be best explained by looking at a simple case. Let us try to find circles with a known radius in an edge image. Since circles are rotationally symmetric we only need to consider translations in this case. If we want to find circles as efficiently as possible we can observe that for circles that are brighter than the background the gradient vector of the edge of the circle is perpendicular to the circle. This means that it points in the direction of the center of the circle. If the circle is darker than its background the negative gradient vector points toward the center of the circle. Therefore, as we know the radius of the circle we can theoretically determine the center of the circle from a single point on the circle. Unfortunately, we do not know which points lie on the circle (this is actually the task we would like to solve). However, we can detect the circle by observing that all points on the circle will have the property that based on the gradient vector we can construct the circle center. Therefore, we can accumulate the evidence provided by all edge points in the image to determine the circle. This can be done as follows: since we want to determine the circle center (i.e., the translation of the circle) we can set up an array that accumulates the evidence that a circle is present as a particular translation. We initialize this array with zeroes. Then, we loop through all the edge points in the image and construct the potential circle center based on the edge position, the gradient direction, and the known circle radius. With this information, we increment the accumulator array at the potential circle center by one. After we have processed all the edge points the accumulator array should contain a large evidence, i.e, a large number of votes, at the locations of the circle centers. We can then threshold the accumulator array and compute the local maxima to determine the circle centers in the image.

An example for this algorithm is shown in Fig. 8.102. Suppose we want to locate the circle on top of the capacitor in Fig. 8.102(a) and that we know that it has a radius of 39 pixels. The edges extracted with a Canny filter with $\sigma = 2$ and hysteresis thresholds of 80 and 20 are shown in Fig. 8.102(b). Furthermore, for every eighth edge point the gradient vector is shown. Note that for the circle they all point toward the circle center. The accumulator array that is obtained with the algorithm described above is displayed in Fig. 8.102(c). Note that there is only one significant peak. In fact, most of the cells in the accumulator array have received so few votes that a square root LUT had to be used to visualize that there are any votes at all in the rest of the accumulator array. If the accumulator array is thresholded and the local maxima are calculated the circle in Fig. 8.102(d) is obtained.

From the above example, we can see that we can find circles in the image extremely efficiently. If we know the polarity of the circle, i.e., whether it is brighter or darker than the background, we only need to perform a single increment of the accumulator array per edge point in the image. If we do not know the polarity of the edge we need to perform two increments per edge point. Hence, the runtime is proportional to the number of edge points in the image and not to the size of the template, i.e., the size of the circle. Ideally, we would like to find an algorithm that is equally efficient for arbitrary objects.

What can we learn from the above example? First, it is clear that for arbitrary objects the gradient direction does not necessarily point to a reference point of the object like it did for circles. Nevertheless, the gradient direction of the edge point provides a constraint where the reference point of the object can be, even for arbitrarily shaped objects. This is shown in Fig. 8.103. Suppose we have singled out the reference point o of the object. For the circle, the natural choice would be its center. For an arbitrary object, we can, for example, use the center of gravity of the edge points. Now consider an edge point e_i. We can see that the gradient vector ∇f_i and the vector r_i from e_i to o always enclose the same angle, no matter how the object is translated, rotated and scaled. For simplicity, let us only consider translations for the moment. Then, if we find an edge point in the image with a certain gradient direction or gradient angle ϕ_i, we could calculate the possible location of the template with the vector r_i and increment the accumulator array accordingly. Note that for circles the gradient vector ∇f_i has the same direction as the vector r_i. For arbitrary shapes this no longer holds. From Fig. 8.103, we can also see that the edge direction does not necessarily uniquely constrain the reference point

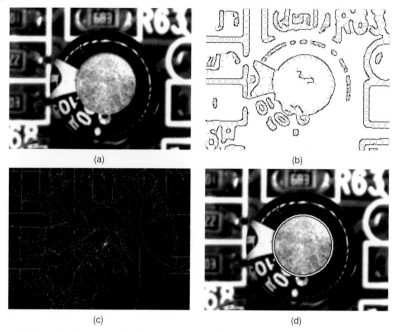

Fig. 8.102 Using the Hough transform to detect a circle. (a) Image of a PCB showing a capacitor. (b) Detected edges. For every eighth edge point the corresponding orientation is visualized by displaying the gradient vector. (c) Hough accumulator array obtained by performing the Hough transform using the edge points and orientations. A square root LUT is used to make the less populated regions of the accumulator space more visible. If a linear LUT were used only the peak would be visible. (d) Circle detected by thresholding (c) and computing the local maxima.

since there may be multiple points on the edges of the template that have the same orientation. For circles, this is not the case. For example, in the lower left part of the object in Fig. 8.103 there is a second point that has the same gradient direction as the point labeled e_i, which has a different offset vector to the reference point. Therefore, in the search we have to increment all accumulator array elements that correspond to the edge points in the template with the same edge direction. Hence, during the search we must be able to quickly determine all the offset vectors that correspond to a given edge direction in the image. This can be achieved in a preprocessing step in the template generation that is performed offline. Basically, we construct a table, called the R-table, that is indexed by the gradient angle ϕ. Each table entry contains all the offset vectors r_i of the template edges that have the gradient angle ϕ. Since the table must be discrete to enable efficient indexing the gradient angles are discretized with a certain step size $\Delta\phi$. The concept of the R-table is also shown in Fig. 8.103. With the R-table, it is very simple to find the offset vectors for incrementing the accumulator array in the search: we simply calculate the gradient angle in the image and use it as an index into the R-table. After construction of the accumulator array, we threshold the array and calculate the local maxima to find the possible locations of the object. This approach can also be extended easily to deal with rotated and scaled objects [6]. In real images, we also need to consider that there are uncertainties in the location of edges in the image and in the edge orientations. We have already seen in (8.85) that the precision of the Canny edges depends on the signal-to-noise ratio. Using similar techniques, it can be shown that the precision of the edge angle ϕ for the Canny filter is given by $\sigma_\phi^2 = \sigma_n^2/(4\sigma^2 a^2)$. These values must be used in the online phase to determine a range of cells in the accumulator array that must be incremented to ensure that the cell corresponding to the true reference point is incremented.

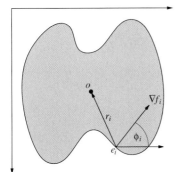

j	ϕ_j	r_i
0	0	$\{r_i \mid \phi_i = 0\}$
1	$\Delta\phi$	$\{r_i \mid \phi_i = \Delta\phi\}$
2	$2\Delta\phi$	$\{r_i \mid \phi_i = 2\Delta\phi\}$
⋮	⋮	⋮

Fig. 8.103 The principle of constructing the R-table in the GHT. The R-table on the right is constructed based on the gradient angle ϕ_i of each edge point of the model object and the vector r_i from each edge point to the reference point o of the template.

The generalized Hough transform described above is already quite efficient. On average, it increments a constant number of accumulator cells. Therefore, its runtime only depends on the number of edge points in the image. However, it is still not fast enough for machine vision applications because the accumulator space that must be searched to find the objects can quickly become very large, especially if rotations and scalings of the object are allowed. Furthermore, the accumulator uses an enormous amount of memory. For example, consider an object that should be found in a 640 × 480 image with an angle range of 360°, discretized in 1° steps. Let us suppose that two bytes are sufficient to store the accumulator array entries without overflow. Then, the accumulator array requires 640 · 480 · 360 · 2 = 221184000 bytes of memory, i.e., 211 MB. This is unacceptably large for most applications. Furthermore, it means that initializing this array alone will require a significant amount of processing time. For this reason, a hierarchical generalized Hough transform is proposed in [94]. It uses image pyramids to speed up the search and to reduce the size of the accumulator array by using matches found on higher pyramid levels to constrain the search on lower pyramid levels. The interested reader is referred to [94] for details of the implementation. With this hierarchical generalized Hough transform, objects can be found in real time even under severe occlusions, clutter, and almost arbitrary illumination changes.

The algorithms we have discussed so far were based on matching edge points directly. Another class of algorithms is based on matching geometric primitives, e.g., points, lines, and circles. These algorithms typically follow the hypothesize-and-test paradigm, i.e., they hypothesize a match, typically from a small number of primitives, and then test whether the hypothetical match has enough evidence in the image.

The biggest challenge that this type of algorithms must solve is the exponential complexity of the correspondence problem. Let us for the moment suppose that we are only using one type of geometric primitive, e.g., lines. Furthermore, let us suppose that all template primitives are visible in the image so that potentially there is a subset of primitives in the image that corresponds exactly to primitives in the template. If the template consists of m primitives and there are n primitives in the image there are $\binom{n}{m}$, i.e., $O(n^m)$, potential correspondences between the template and image primitives. If objects in the search image can be occluded the number of potential matches is even larger since we must allow that multiple primitives in the image can match a single primitive in the template because a single primitive in the template may break up into several pieces, and that some primitives in the template are not present in the search image. It is clear that even for moderately large values of m and n the cost of exhaustively checking all possible correspondences is prohibitive. Therefore, geometric constraints and strong heuristics must be used to perform the matching in an acceptable time.

One approach to perform the matching efficiently is called geometric hashing [56]. It was originally described for points as primitives, but can equally well be used with lines. Furthermore, the original description uses affine transformations as the set of allowable transformations. We will follow the original presentation and will note where modifications are necessary for other classes of transformations and lines as primitives. Geometric hashing is based on the observation that three points define an affine basis of the two-dimensional plane. Thus, once we select three points e_{00}, e_{10}, and e_{01} in general position, i.e., not collinear, we can represent every other point as a linear combination of these three points: $q = e_{00} + \alpha(e_{10} - e_{00}) + \beta(e_{01} - e_{00})$. The interesting property of this representation is that it is invariant to affine transformations, i.e., (α, β) only depend on the three basis points (the basis triplet), but not on an affine transformation, i.e., they are affine invariants. With this, the values (α, β) can be regarded as the affine coordinates of the point q. This property holds equally well for lines: three nonparallel lines can be used to define an affine basis. If we use a more restricted class of transformations fewer points are sufficient to define a basis. For example, if we restrict the transformations to similarity transformations two points are sufficient to define a basis. Note, however, that two lines are only sufficient to determine a rigid transformation.

The aim of geometric hashing is to reduce the amount of work that has to be performed to establish correspondences between the template and image points. Therefore, it constructs a hash table that enables the algorithm to quickly determine the potential matches for the template. This hash table is constructed as follows: for every combination of three noncollinear points in the template the affine coordinates (α, β) of the remaining $m - 3$ points of the template are calculated. The affine coordinates (α, β) serve as the index into the hash table. For every point, the index of the current basis triplet is stored in the hash table. If more than one template should be found, additionally the template index is stored; however, we will not consider this case further, so for our purposes only the index of the basis triplet is stored.

To find the template in the image, we randomly select three points in the image and construct the affine coordinates (α, β) of the remaining $n - 3$ points. We then use (α, β) as an index into the hash table. This returns us the index of the basis triplet. With this, we obtain a vote for the presence of a particular basis triplet in the image. If the randomly selected points do not correspond to a basis triplet of the template the votes of all the points will not agree. However, if they correspond to a basis triplet of the template many of the votes will agree and will indicate the index of the basis triplet. Therefore, if enough votes agree we have a strong indication for the presence of the model. The presence of the model is then verified as described below. Since there is a certain probability that we have selected an inappropriate basis triplet in the image, the algorithm iterates until it has reached a certain probability of having found the correct match. Here, we can make use of the fact that we only need to find one correct basis triplet to find the model. Therefore, if k of the m template points are present in the image the probability of having selected at least one correct basis triplet in t trials is approximately

$$p = 1 - \left(1 - \left(\frac{k}{n}\right)^3\right)^t \tag{8.111}$$

If similarity transforms are used, only two points are necessary to determine the affine basis. Therefore, the inner exponent will change from 3 to 2 in this case. For example, if the ratio of visible template points to image points k/n is 0.2 and we want to find the template with a probability of 99% (i.e., $p = 0.99$), 574 trials are sufficient if affine transformations are used. For similarity transformations, 113 trials would suffice. Hence, geometric hashing can be quite efficient in finding the correct correspondences, depending on how many extra features are present in the image.

After a potential match has been obtained with the algorithm described above it must be verified in the image. In [56], this is done by establishing point correspondences for the remaining template points based on the affine transformation given by the selected basis triplet. Based on these correspondences, an improved affine transformation is computed by a least-squares minimization over all corresponding points. This, in turn is used to map all the edge points of the template, i.e., not only the characteristic points that were used for the geometric hashing, to the pose of the template in the image. The transformed edges are compared to the image edges. If there is a sufficient overlap between the template and image edges the match is accepted and the corresponding points and edges are removed from the segmentation. If more than one instance of the template should be found the entire process is repeated.

The algorithm described so far works well as long as the geometric primitives can be extracted with sufficient accuracy. If there are errors in the point coordinates an erroneous affine transformation will result from the basis triplet. Therefore, all the affine coordinates (α, β) will contain errors, and hence the hashing in the online phase will access the wrong entry in the hash table. This is probably the largest drawback of the geometric hashing algorithm in practice. To circumvent this problem, the template points must be stored in multiple adjacent entries of the hash table. Which hash table entries must be used can in theory be derived through error propagation [56]. However, in practice the accuracy of the geometric primitives is seldom known. Therefore, estimates have to be used, which must be well on the safe side for the algorithm; not to miss any matches in the online phase. This, in turn, makes the algorithm slightly less efficient because more votes will have to be evaluated during the search.

The final class of algorithms we will discuss tries to match geometric primitives themselves to the image. Most of these algorithms use only line segments as the primitives [4, 37, 52]. One of the few exceptions to this rule is the approach in [95], which uses line segments and circular arcs. Furthermore, in 3D object recognition sometimes line segments and elliptic arcs are used [18]. As we discussed above, exhaustively enumerating all potential correspondences between the template and image primitives is prohibitively slow. Therefore, it is interesting to look at examples of different strategies that are employed to make the correspondence search tractable.

The approach in [4] segments the contours of the model object and the search image into line segments. Depending on the lighting conditions, the contours are obtained by thresholding or by edge detection. The ten longest line segments in the template are singled out as privileged. Furthermore, the line segments in the model are ordered by adjacency as they trace the boundary of the model object. To generate a hypothesis, a privileged template line segment is matched to a line segment in the image. Since the approach is designed to handle similarity transforms the angle, which is invariant under these transforms, to the preceding line segment in the image is compared to the angle to the preceding line segment in the template. If they are not close enough the potential match is rejected. Furthermore, the length ratio of these two segments, which also is invariant to similarity transforms, is used to check the validity of the hypothesis. The algorithm generates a certain number of hypotheses in this manner. These hypotheses are then verified by trying to match additional segments. The quality of the hypotheses, including the additionally matched segments, is then evaluated based on the ratio of the lengths of the matched segments to the length of the segments in the template. The matching is stopped once a high quality match has been found or if enough hypotheses have been evaluated. Hence, we can see that the complexity is kept manageable by using privileged segments in conjunction with their neighboring segments.

In [52], a similar method is proposed. In contrast to [4], corners (combinations of two adjacent line segments of the boundary of the template that enclose a significant angle) are matched first. To generate a matching hypothesis, two corners must be matched to the image. Geometric constraints between the corners are used to reject false matches. The algorithm then attempts to extend the hypotheses with other segments in the image. The hypotheses are evaluated based on a dissimilarity criterion. If the dissimilarity is below a threshold the match is accepted. Hence, the complexity of this approach is reduced by matching features that have distinctive geometric characteristics first.

The approach in [37] also generates matching hypotheses and tries to verify them in the image. Here, a tree of possible correspondences is generated and evaluated in a depth-first search. This search tree is called the interpretation tree. A node in the interpretation tree encodes a correspondence between a model line segment and an image line segment. Hence, the interpretation tree would exhaustively enumerate all correspondences, which would be prohibitively expensive. Therefore, the interpretation tree must be pruned as much as possible. To do this, the algorithm uses geometric constraints between the template line segments and the image line segments. Specifically, the distances and angles between pairs of line segments in the image and in the template must be consistent. This angle is checked by using normal vectors of the line segments that take the polarity of the edges into account. This consistency check prunes a large number of branches of the interpretation tree. However, since a large number of possible matchings still remain, a heuristic is used to explore the most promising hypotheses first. This is useful because the search is terminated once an acceptable match has been found. This early search termination is criticized in [51], and various strategies to speed up the search for all instances of the template in the image are discussed. The interested reader is referred to [51] for details.

To make the principles of the geometric matching algorithms clearer, let us examine a prototypical matching procedure in an example. The template to be found is shown in Fig. 8.104(a). It consists of five line segments and five circular arcs. They were segmented automatically from the image in Fig. 8.100(a) using a subpixel-precise Canny filter with $\sigma = 1$ and and by splitting the edge contours into line segments and circular arcs using the method described in Section 8.8.4. The template consists of geometric parameters of these primitives as well as the segmented contours themselves. The image in which the template should be found is shown in Fig. 8.104(b). It contains four partially occluded instances of the model along with four clutter objects. The matching starts by extracting edges in the search image and by segmenting them into line segments and circular arcs (Fig. 8.104(c)). Like for the template, the geometric parameters of the image primitives are calculated. The matching now determines possible matches for all of the primitives in the template. Of these, the largest circular arc is examined first because of a heuristic that rates moderately long circular arcs as more distinctive than even long line segments. The resulting matching hypotheses are shown in Fig. 8.104(d). Of course, the line segments could also have been examined first. Because in this case only rigid transformations are allowed the matching of the circular arcs uses the radii of circles as a matching constraint. Since the matching should be robust to occlusions the opening angle of the circular arcs is not used as a constraint. Because of this, the matched circles are not sufficient to determine a rigid transformation between the template and the image. Therefore, the algorithm tries to match an adjacent line segment (the long lower line segment in Fig. 8.104(a)) to the image primitives while using the angle of intersection between the circle and the line as a geometric constraint. The resulting matches are shown in Fig. 8.104(e). With these hypotheses, it is possible to compute a rigid transformation that transforms the template to the image. Based on this, the remaining primitives can be matched to the image based on the distances of the image primitives and the transformed template primitives. The resulting matches are shown in Fig. 8.104(f). Note that because of specular reflections sometimes multiple parallel line segments are matched to a single line segment in the template. This could be fixed by taking the polarity of the edges into account. To obtain the rigid transformation between the template and the matches in the image as accurately as possible, a least-squares optimization of the distances between the edges in the template and the edges in the image can be used. An alternative is the minimal tolerance error zone optimization described in [95]. Note that the matching has already found the four

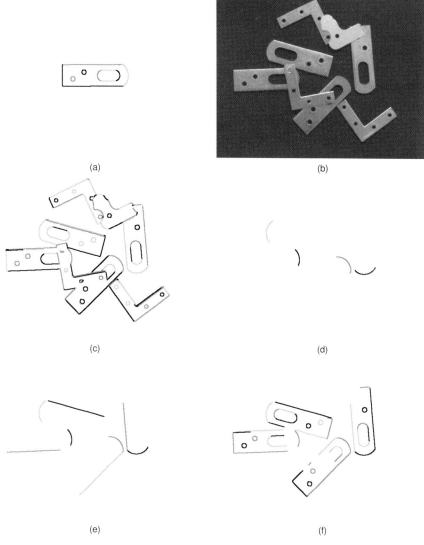

Fig. 8.104 Example of matching an object in the image using geometric primitives. (a) The template consists of five line segments and five circular arcs. The model has been generated from the image in Fig. 8.100(a). (b) The search image contains four partially occluded instances of the template along with four clutter objects. (c) Edges extracted in (b) with a Canny filter with $\sigma = 1$ and split into line segments and circular arcs. (d) The matching in this case first tries to match the largest circular arc of the model and finds four hypotheses. (e) The hypotheses are extended with the lower of the long line segments in (a). These two primitives are sufficient to estimate a rigid transform that aligns the template with the features in the image. (f) The remaining primitives of the template are matched to the image. The resulting matched primitives are displayed.

correct instances of the template. For the algorithm, the search is not finished, however, since there might be more instances of the template in the image, especially instances for which the large circular arc is occluded more than in the leftmost instance in the image. Hence, the search is continued with other primitives as the first primitives to try. In this case, however, the search does not discover new viable matches.

After having discussed different approaches for robustly finding templates in an image, the question which of these algorithms should be used in practice naturally arises. Unfortunately, no definite answer can be given since the effectiveness of a particular approach greatly depends on the shape of the template itself. Generally, the geometric matching algorithms have an advantage if the template and image contain only few, salient geometric primitives, like in the example in Fig. 8.104. Here, the combinatorics of the geometric matching algorithms work in their advantage. On the other hand, they work to their disadvantage if the template or search image contain a very large number of geometric primitives. Two examples for this are shown in Figs. 8.105 and 8.106. In Fig. 8.105, the template contains fine structures that result in 350 geometric primitives, which are not particularly salient. Consequently, the search would have to examine an extremely large number of hypotheses that could only be dismissed after examining a large number of additional primitives. Note that the model contains 35 times as many primitives as the model in Fig. 8.104, but only approximately three times as many edge points. Consequently, it could be easily found with pixel-based approaches like the generalized Hough transform.

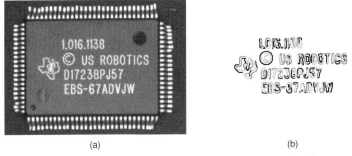

Fig. 8.105 (a) Image of a template object which is not suitable for the geometric matching algorithms. Although the segmentation of the template into line segments and circular arcs in (b) only contains approximately three times as many edge points as the template in Fig. 8.104, it contains 35 times as many geometric primitives, i.e., 350.

A difficult search image is shown in Fig. 8.106. Here, the goal is to find the circular fiducial mark. Since the contrast of the fiducial mark is very low a small segmentation threshold must be used in the edge detection to find the relevant edges of the circle. This causes a very large number of edges and broken fragments that must be examined. Again, pixel-based algorithms will have little trouble with this image.

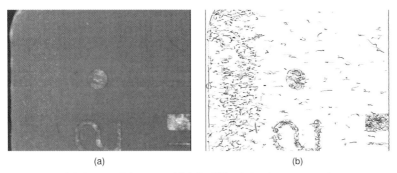

(a) (b)

Fig. 8.106 (a) A search image which is difficult for the geometric matching algorithms. Here, because of the bad contrast of the circular fiducial, the segmentation threshold must be chosen very low so that the relevant edges of the fiducial are selected. Because of this, the segmentation in (b) contains a very large number of primitives that must be examined in the search.

From the above examples, we can see that the pixel-based algorithms, e.g., the generalized Hough transform, have the advantage that they can represent arbitrarily shaped templates without problems. However, if the template and image can be represented with a small number of geometric primitives the geometric matching algorithms are preferable. Therefore, there is no clear winner.

The algorithms that are used in machine vision software packages are typically not published. Therefore, it is hard to tell which of the algorithms are currently favored in practice. From the above discussion, we can see that the basic algorithms already are fairly complex. However, typically the complexity resides in the time and effort that needs to be spent in making the algorithms very robust and fast. Consequently, these algorithms cannot be implemented easily. Therefore, wise machine vision users rely on standard software packages to provide this functionality rather than attempting to implement it themselves.

8.10
Stereo Reconstruction

In Section 8.7.4, we have seen that we can perform very accurate measurements from a single image by calibrating the camera and by determining its exterior orientation with respect to a plane in the world. We could then convert image measurements to world coordinates within the plane by intersecting optical rays with the plane. Note, however, that these measurements are still 2D measurements within the plane in the world. In fact, from a single image we cannot reconstruct the 3D geometry of the scene because we can only determine the optical ray for each point in the image. We do not know at which distance on the optical ray the point lies in the world. In the approach in Section 8.7.4, we had to assume a special geometry in the world to be able to determine the distance of a point along the optical ray. Note that this is not a true 3D reconstruction. To perform a 3D reconstruction, we must use at least two images of the same scene taken from different positions. Typically, this is done by simultaneously taking the images with two cameras. This process is called stereo reconstruction. In this section, we will examine the case of binocular stereo, i.e., we will concentrate on the two-camera case. Throughout this section, we will assume that the cameras have been calibrated, i.e, their interior orientations and relative orientation are known. This can be performed with the methods described in Chapter 4 or in [38, 59]. While uncalibrated reconstruction is also possible [27, 42], the corresponding methods have not yet been used in industrial applications.

8.10.1
Stereo Geometry

Before we can discuss the stereo reconstruction, we must examine the geometry of two cameras, as shown in Fig. 8.107. Since the cameras are assumed to be calibrated we know their interior orientations, i.e., their principal points, focal lengths (more precisely, their principal distances, i.e., the distance between the image plane and the perspective center), pixel size and aspect ratio, and distortion coefficients. In Fig. 8.107, the principal points are visualized by the points C_1 and C_2 in the first and second image, respectively. Furthermore, the perspective centers are visualized by the points O_1 and O_2. The dashed line between the perspective centers and principal points visualizes the principal distances. Note that since the image planes physically lie behind the perspective centers the image is turned upside-down. Consequently, the origin of the image coordinate system lies in the lower right corner, with the row axis pointing upward and the column axis pointing leftward. The camera coordinate system axes are defined such that the x axis points to the right, the y axis points downwards, and the z axis points forward from the image plane, i.e., along the viewing direction. The position and orientation of the two cameras with respect to each other is given by the relative orientation, which is a rigid 3D transformation specified by the rotation matrix R_r and the translation vector T_r. It can either be interpreted as the transformation of the camera coordinate system of the first camera into the camera coordinate system of the

8.10 Stereo Reconstruction

second camera or as a transformation that transforms point coordinates in the camera coordinate system of the second camera into point coordinates of the camera coordinate system of the first camera: $P_{c1} = R_r P_{c2} + T_r$. The translation vector T_r, which specifies the translation between the two perspective centers, is also called the base. With this, we can see that a point P_w in the world is mapped to a point P_1 in the first image and to a point P_2 in the second image. If there is no distortion in the lens (which we will assume for the moment) the points P_w, O_1, O_2, P_1, and P_2 all lie in a single plane.

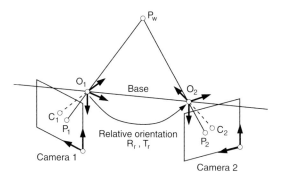

Fig. 8.107 Stereo geometry of two cameras.

To illustrate the relative orientation and the stereo calibration, Fig. 8.108 shows an image pair taken from a sequence of 15 image pairs that are used to calibrate a binocular stereo system. The calibration returns a translation vector of $(0.1534\,\text{m}, -0.0037\,\text{m}, 0.0449\,\text{m})$ between the cameras, i.e., the second camera is 15.34 cm to the right, 0.37 cm above, and 4.49 cm in front of the first camera, expressed in camera coordinates of the first camera. Furthermore, the calibration returns a rotation angle of $40.1139°$ around the axis $(-0.0035, 1.0000, 0.0008)$, i.e., almost around the vertical y axis of the camera coordinate system. Hence, the cameras are verging inwards, like in Fig. 8.107.

To reconstruct 3D points, we must find corresponding points in the two images. "Corresponding" means that the two points P_1 and P_2 in the images belong to the same point P_w in the world. At first, it might seem that, given a point P_1 in the first image, we would have to search in the entire second image for the corresponding point P_2. Fortunately, this is not the case. In Fig. 8.107 we already noted that the points P_w, O_1, O_2, P_1, and P_2 all lie in a single plane. The situation of trying to find a corresponding point for P_1 is shown in Fig. 8.109. We can note that we know P_1, O_1, and O_2. We do not know at which distance the point P_w lies on the optical ray defined by P_1 and O_1. However, we know that P_w is coplanar to the plane spanned by P_1, O_1, and O_2 (the epipolar plane). Hence, we can see that the point P_2 can only lie on the projection of the epipolar plane into the second image. Since O_2 lies

Fig. 8.108 One image pair taken from a sequence of 15 image pairs that are used to calibrate a binocular stereo system. The calibration returns a translation vector (base) of $(0.1534 \text{ m}, -0.0037 \text{ m}, 0.0449 \text{ m})$ between the cameras and a rotation angle of $40.1139°$ around the axis $(-0.0035, 1.0000, 0.0008)$, i.e., almost around the y axis of the camera coordinate system. Hence, the cameras are verging inwards.

on the epipolar plane the projection of the epipolar plane is a line called the epipolar line.

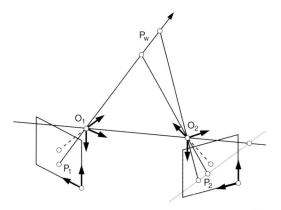

Fig. 8.109 Epipolar geometry of two cameras. Given the point P_1 in the first image, the point P_2 in the second image can only lie on the epipolar line of P_1, which is the projection of the epipolar plane spanned by P_1, O_1, and O_2 into the second image.

It is obvious that the above construction is symmetric for both images, as shown in Fig. 8.110. Hence, given a point P_2 in the second image, the corresponding point can only lie on the epipolar line in the first image. Furthermore, from Fig. 8.110 we can see that different points typically define different epipolar lines. We can also see that all epipolar lines of one image intersect in a single point called the epipole. The epipoles are the projections of the opposite projective centers into the respective image. Note that since all epipolar planes contain O_1 and O_2 the epipoles lie on the line defined by the two perspective centers (the base line).

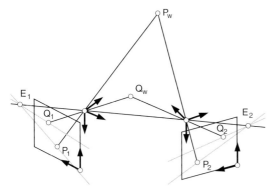

Fig. 8.110 The epipolar geometry is symmetric between the two images. Furthermore, different points typically define different epipolar lines. All epipolar lines intersect in the epipoles E_1 and E_2, which are the projections of the opposite projective centers into the respective image.

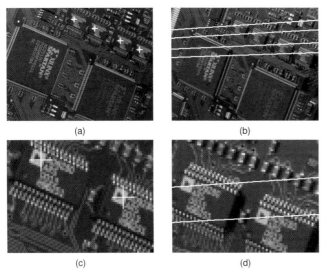

Fig. 8.111 Stereo image pair of a PCB. (a) Four points marked in the first image. (b) Corresponding epipolar lines in the second image. (c) Detail of (a). (d) Detail of (b). The four points in (a) have been selected manually at the tips of the triangles on the four small ICs. Note that the epipolar lines pass through the tips of the triangles in the second image.

Figure 8.111 shows an example for the epipolar lines. The stereo geometry is identical to Fig. 8.108. The images show a PCB. In Fig. 8.111(a), four points are marked. They have been selected manually to lie at the tips of the triangles on the four small ICs, as shown in the detail view in Fig. 8.111(c). The corresponding epipolar lines in the second image are shown in Figs. 8.111(b) and (d). Note that the epipolar lines pass through the tips of the triangles in the second image.

As noted above, we have so far assumed that the lenses have no distortions. In reality, this is very rarely true. In fact, by looking closely at Fig. 8.111(b), we can already perceive a curvature in the epipolar lines because the camera calibration has determined the radial distortion coefficients for us. If we set the displayed image part as in Fig. 8.112 we can clearly see the curvature of the epipolar lines in real images. Furthermore, we can see the epipole of the image clearly.

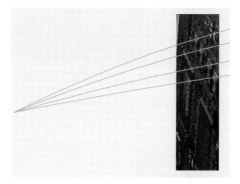

Fig. 8.112 Because of lens distortions, the epipolar lines are generally not straight. The image shows the same image as Fig. 8.111(b). The zoom has been set so that the epipole is shown in addition to the image. The aspect ratio has been chosen so that the curvature of the epipolar lines is clearly visible.

From the above discussion, we can see that the epipolar lines are different for different points. Furthermore, because of radial distortions they typically are not even straight. This means that when we try to find corresponding points we must compute a new, complicated epipolar line for each point that we are trying to match, typically for all points in the first image. The construction of the curved epipolar lines would be too time consuming for real-time applications. Hence, we can ask ourselves whether the construction of the epipolar lines can be simplified for particular stereo geometries. This is indeed the case for the stereo geometry shown in Fig. 8.113. Here, both image planes lie in the same plane and are vertically aligned. Furthermore, it is assumed that there are no lens distortions. Note that this implies that both principal distances are identical, that the principal points have the same row coordinate, that the images are rotated such that the column axis is parallel to the base, and that the relative orientation contains only a translation in the x direction and no rotation. Since the image planes are parallel to each other the epipoles lie infinitely far away on the base line. It is easy to see that this stereo geometry implies that the epipolar line for a point is simply the line that has the same row coordinate as the point, i.e., the epipolar lines are horizontal and vertically aligned. Hence, they can be computed without any overhead at all. Since almost all stereo matching algorithms assume this particular geometry, we can call it the epipolar standard geometry.

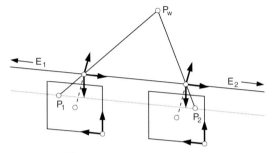

Fig. 8.113 The epipolar standard geometry is obtained if both image planes lie in the same plane and are vertically aligned. Furthermore, it is assumed that there are no lens distortions. In this geometry, the epipolar line for a point is simply the line that has the same row coordinate as the point, i.e., the epipolar lines are horizontal and vertically aligned.

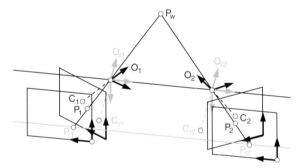

Fig. 8.114 Transformation of a stereo configuration into the epipolar standard geometry.

While the epipolar standard geometry results in very simple epipolar lines, it is extremely difficult to align real cameras into this configuration. Furthermore, it is quite difficult and expensive to obtain distortion-free lenses. Fortunately, almost any stereo configuration can be transformed into the epipolar standard geometry as indicated in Fig. 8.114 [25]. The only exceptions are if an epipole happens to lie within one of the images. This typically does not occur in practical stereo configurations. The process of transforming the images to the epipolar standard geometry is called image rectification. To rectify the images, we need to construct two new image planes that lie in the same plane. To keep the 3D geometry identical, the projective centers must remain at the same positions in space, i.e., $O_{r1} = O_1$ and $O_{r2} = O_2$. Note, however, that we need to rotate the camera coordinate systems such that their x axes become identical to the base line. Furthermore, we need to construct two new principal points C_{r1} and C_{r2}. Their connecting vector must be parallel to the base. Furthermore, the vectors from the principal points to the perspective centers must be perpendicular to the base. This leaves us two degrees of freedom. First, we must choose a common principal distance. Second, we can rotate the common plane in which the image planes lie around the base. These parameters can be chosen by requiring that the image distortion should be minimized [25]. The image dimensions are then typically chosen such that the original images are completely contained within the rectified images. Of course, we also must remove the lens distortions in the rectification.

To obtain the gray value for a pixel in the rectified image, we construct the optical ray for this pixel and intersect it with the original image plane. This is shown, e.g., for the points P_{r1} and P_1 in Fig. 8.114. Since this typically results in subpixel coordinates, the gray values must be interpolated with the techniques described in Section 8.3.3.

While it may seem that image rectification is a very time-consuming process, the entire transformation can be computed once offline and stored in a table. Hence, images can be rectified very efficiently online.

Figure 8.115 shows an example for the image rectification. The input image pair is shown in Figs. 8.115(a) and (b). The images have the same relative orientation as the images in Fig. 8.108. The principal distances of the cameras are 13.05 mm and 13.16 mm, respectively. Both images have dimensions 320×240. Their principal points are $(155.91, 126.72)$ and $(163.67, 119.20)$, i.e, they are very close to the image center. Finally, the images have a slight barrel-shaped distortion. The rectified images are shown in Figs. 8.115(c) and (d). Their relative orientation is given by the translation vector $(0.1599 \text{ m}, 0 \text{ m}, 0 \text{ m})$. As expected, the translation is solely along the x axis. Of course, the length of the translation vector is identical to Fig. 8.108 since the position of the projective centers has not changed. The new principal distance of both images is 12.27 mm. The new principal points are given by $(-88.26, 121.36)$ and $(567.38, 121.36)$. As can be expected from Fig. 8.114, they lie well outside the rectified images. Also as expected, the row coordinates of the principal points are identical. The rectified images have dimensions 336×242 and 367×242, respectively. Note that they exhibit a trapezoidal shape that is characteristic of the verging camera configuration. The barrel-shaped distortion has been removed from the images. Clearly, the epipolar lines are horizontal in both images.

Apart from the fact that rectifying the images results in a particularly simple structure for the epipolar lines, it also results in a very simple reconstruction of the depth, as shown in Fig. 8.116. In this figure, the stereo configuration is displayed as viewed along the direction of the row axis of the images, i.e., the

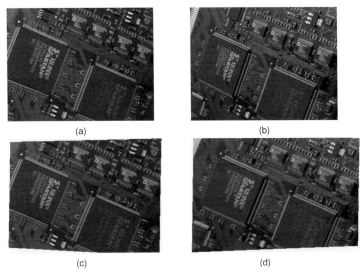

Fig. 8.115 Example for the rectification of a stereo image pair. The images in (a) and (b) have the same relative orientation as the images in Fig. 8.108. The rectified images are shown in (c) and (d). Note the trapezoidal shape of the rectified images that is caused by the verging cameras. Also note that the rectified images are slightly wider than the original images.

y axis of the camera coordinate system. Hence, the image planes are shown as the lines at the bottom of the figure. The depth of a point is quite naturally defined as its z coordinate in the camera coordinate system. By examining the similar triangles $O_1 O_2 P_w$ and $P_1 P_2 P_w$, we can see that the depth of P_w only depends on the difference of the column coordinates of the points P_1 and P_2 as follows: From the similarity of the triangles, we have $\frac{z}{b} = \frac{z+f}{d_w+b}$. Hence, the depth is given by $z = \frac{bf}{d_w}$. Here, b is the length of the base, f is the principal distance, and d_w is the sum of the signed distances of the points P_1 and P_2 to the principal points C_1 and C_2. Since the coordinates of the principal points are given in pixels, but d_w is given in world units, e.g., meters, we have to convert d_w to pixel coordinates by scaling it with the size of the pixels in the x direction: $d_p = \frac{d_w}{s_x}$. Now, we can easily see that $d_p = (c_{c1} - c_1) + (c_2 - c_{c2})$, where c_1 and c_2 denote the column coordinates of the points P_1 and P_2, while c_{c1} and c_{c2} denote the column coordinates of the principal points. Rearranging the terms, we find $d_p = (c_{c1} - c_{c2}) + (c_2 - c_1)$. Since $c_{c1} - c_{c2}$ is constant for all points and known from the calibration and rectification, we can see that the depth z only depends on the difference of the column coordinates $d = c_2 - c_1$. This difference is called the disparity. Hence, we can see that to reconstruct the depth of a point we must determine its disparity.

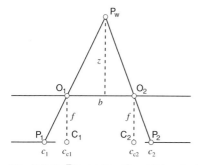

Fig. 8.116 Reconstruction of the depth z of a point only depends on the disparity $d = c_2 - c_1$ of the points, i.e., the difference of the column coordinates in the rectified images.

8.10.2 Stereo Matching

As we have seen in the previous section, the main step in the stereo reconstruction is determination of the disparity of each point in one of the images, typically the first image. Since a disparity is calculated, or at least attempted to be calculated, for each point these algorithms are called dense reconstruction algorithms. It should be noted that there is another class of algorithms that only tries to reconstruct the depth for selected features, e.g., straight lines or points. Since these algorithms require a typically expensive feature extraction they are seldom used in industrial applications. Therefore, we will concentrate on dense reconstruction algorithms. An overview of currently available algorithms is given in [79].

Since the goal of dense reconstruction is to find the disparity for each point in the image the determination of the disparity can be regarded as a template matching problem. Given a rectangular window of size $(2n + 1) \times (2n + 1)$ around the current point in the first image, we must find the most similar window along the epipolar line in the second image. Hence, we can use the techniques described in Section 8.9 to match a point. The gray value matching methods described in Section 8.9.1 are of particular interest because they do not require a costly model generation step, which would have to be performed for each point in the first image. Therefore, the gray value matching methods typically are the fastest methods for stereo reconstruction. The simplest similarity measures are the SAD and SSD measures (8.98) and (8.99). For the stereo matching problem, they are given by

$$\mathrm{sad}(r,c,d) = \frac{1}{(2n+1)^2} \sum_{j=-n}^{n} \sum_{i=-n}^{n} |g_1(r+i,c+j) - g_2(r+i,c+j+d)| \quad (8.112)$$

and

$$ssd(r,c,d) = \frac{1}{(2n+1)^2} \sum_{j=-n}^{n} \sum_{i=-n}^{n} (g_1(r+i,c+j) - g_2(r+i,c+j+d))^2 \tag{8.113}$$

As we know from Section 8.9.1, these two similarity measures can be computed very quickly. Fast implementations for stereo matching using the SAD are given in [45, 69]. Unfortunately, these similarity measures have disadvantage that they are not robust against illumination changes, which frequently happen in stereo reconstruction because of different viewing angles along the optical rays. Consequently, in some applications it may be necessary to use the normalized cross correlation (8.100) as the similarity measure. For the stereo matching problem, it is given by

$$ncc(r,c,d) = $$
$$\frac{1}{(2n+1)^2} \sum_{i=-n}^{n} \sum_{j=-n}^{n} \frac{g_1(r+i,c+j) - m_1(r+i,c+j)}{\sqrt{s_1(r+i,c+j)^2}}$$
$$\cdot \frac{g_2(r+i,c+j+d) - m_2(r+i,c+j+d)}{\sqrt{s_2(r+i,c+j+d)^2}} \tag{8.114}$$

Here, m_i and s_i ($i = 1, 2$) denote the mean and standard deviation of the window in the first and second image. They are calculated analogously to their template matching counterparts (8.101) and (8.102). The advantage of the normalized cross correlation is, of course, that it is invariant against linear illumination changes. However, it is more expensive to compute.

From the above discussion, it might appear that to match a point we will have to compute the similarity measure along the entire epipolar line in the second image. Fortunately, this is not the case. Since the disparity is inversely related to the depth of a point, and we typically know in which range of distances the objects we are interested in occur, we can restrict the disparity search space to a much smaller interval than the entire epipolar line. Hence, we have $d \in [d_{min}, d_{max}]$, where d_{min} and d_{max} can be computed from the minimum and maximum expected distance in the images. Consequently, the length of the disparity search space is given by $l = d_{max} - d_{min} + 1$.

After we have computed the similarity measure for the disparity search space for a point to be matched, we might be tempted to simply use the disparity with the minimum (SAD and SSD) or maximum (NCC) similarity measure as the match for the current point. However, this typically will lead to many false matches since some windows may not have a good match in the second image. In particular, this happens if the current point is occluded because of perspective effects in the second image. Therefore, it is necessary to threshold the similarity measure, i.e., to accept matches only if their similarity measure is below (SAD and SSD) or above (NCC) a threshold. Obviously, if we perform this thresholding some points will not have a reconstruction, and consequently the reconstruction will not be completely dense.

With the above search strategy, the matching process has a complexity of $O(whln^2)$. This is much too expensive for real-time performance. Fortunately, it can be shown that with a clever implementation the above similarity measures can be computed recursively. With this, the complexity can be made independent of the window size n and becomes (whl). With this, real-time performance becomes possible. The interested reader is referred to [26] for details.

Once we have computed the match with an accuracy of one disparity step from the extremum (minimum or maximum) of the similarity measure, the accuracy can be refined with an approach similar to the subpixel extraction of matches described in Section 8.9.3. Since the search space is one-dimensional in the stereo matching a parabola can be fitted through the three points around the extremum, and the extremum of the parabola can be extracted analytically. Obviously, this will also result in a more accurate reconstruction of the depth of the points.

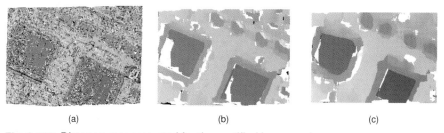

(a)   (b)   (c)

Fig. 8.117 Distance reconstructed for the rectified image pair in Figs. 8.115(c) and (d) with the NCC. (a) Window size 3×3. (b) Window size 17×17. (c) Window size 31×31. White areas correspond to the points that could not be matched because the similarity was too small.

To perform the stereo matching, we need to set one parameter; the size of the gray value windows n. It has a major influence on the result of the matching, as shown by the reconstructed depths in Fig. 8.117. Here, window sizes of 3×3, 17×17, and 31×31 have been used with the NCC as the similarity measure. We can see that if the window size is too small many erroneous results will be found, despite the fact that a threshold of 0.4 has been used to select good matches. This happens because the matching requires a sufficiently distinctive texture within the window. If the window is too small the texture is not distinctive enough, leading to erroneous matches. From Fig. 8.117(b), we see that the erroneous matches are mostly removed by the 17×17 window. However, because there is no texture in some parts of the image, especially in the lower left corners of the two large ICs, some parts of the image cannot be reconstructed. Note also that the areas of the leads around the large ICs are broader than in Fig. 8.117(a). This happens because the windows now straddle height discontinuities in a larger part of the image. Since the texture of the leads is more significant than the texture on the ICs the matching finds the best matches at the depth of the leads. To fill the gaps in the reconstruction, we could try to increase the window size further since this leads to more positions in which the windows have a significant texture. The result of setting the window size to 31×31 is shown in Fig. 8.117(c). Note that now most of the image can be reconstructed. Unfortunately, the lead area has broadened even more, which is undesirable.

From the above example, we can see that too small window sizes lead to many erroneous matches. In contrast, larger window sizes generally lead to fewer erroneous matches and a more complete reconstruction in areas with little texture. Furthermore, larger window sizes lead to a smoothing of the result, which may sometimes be desirable. However, larger window sizes lead to worse results at height discontinuities, which effectively limits the window sizes that can be used in practice.

Despite the fact that larger window sizes generally lead to fewer erroneous matches, they typically cannot be excluded completely based on the window size alone. Therefore, additional techniques are sometimes desirable to reduce the number of erroneous matches even further. Erroneous matches occur mainly for two reasons: weak texture and occlusions. Erroneous matches caused by weak texture can sometimes be eliminated based on the matching score. However, in general it is best to exclude windows with weak texture a priori from the matching. Whether a window contains a weak texture can be decided based on the output of a texture filter. Typically, the standard deviation of the gray values within the window is used as the texture filter. It has the advantage that it is computed in the NCC anyway, while it can be computed with just a few extra operations in the SAD and SSD. Therefore, to exclude windows with weak textures, we require that the standard deviation of the gray values within the window should be large.

The second reason why erroneous matches can occur are perspective occlusions, which, for example, occur at height discontinuities. To remove these errors, we can perform a consistency check that works as follows: First, we find the match from the first to the second image as usual. We then check whether matching the window around the match in the second image results in the same disparity, i.e., finds the original point in the first image. If this is implemented naively, the runtime increases by a factor of two. Fortunately, with a little extra bookkeeping the disparity consistency check can be performed with very few extra operations since most of the required data have already been computed during the matching from the first to the second image.

Figure 8.118 shows the results of different methods to increase robustness. For comparison, Fig. 8.118(a) displays the result of the standard matching from the first to the second image with a window size of 17 × 17 using the NCC. The result of applying a texture threshold of 5 is shown in Fig. 8.118(b). It mainly removes untextured areas on the two large ICs. Figure 8.118(c) shows the result of applying the disparity consistency check. Note that it mainly removes matches in the areas where occlusions occur.

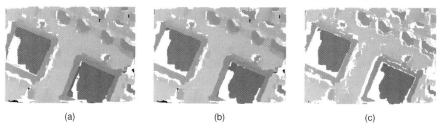

Fig. 8.118 Increasing levels of robustness of the stereo matching. (a) Standard matching from the first to the second image with a window size of 17 × 17 using the NCC. (b) Result of requiring that the standard deviations of the windows is $\geq$ 5. (c) Result of performing the check that matching from the second to the first image results in the same disparity.

8.11
Optical Character Recognition

In quite a few applications, we face the challenge of having to read characters on the object we are inspecting. For example, traceability requirements often lead to the fact that the objects to be inspected are labeled with a serial number, and that we have to read this serial number (see, for example, Figs. 8.21–8.23). In other applications, reading a serial number might be necessary to control the production flow.

Optical character recognition (OCR) is the process of reading characters in images. It consists of two tasks: segmentation of the individual characters and the classification of the segmented characters, i.e., the assignment of a symbolic label to the segmented regions. We will examine these two tasks in this section.

8.11.1
Character Segmentation

The classification of the characters requires that we have segmented the text we want to read into individual characters, i.e., each character must correspond to exactly one region.

To segment the characters, we can use all the methods that we have discussed in Section 8.4: thresholding with fixed and automatically selected thresholds, dynamic thresholding, and the extraction of connected components.

Furthermore, we might have to use the morphological operations of Section 8.6 to connect separate parts of the same character, e.g., the dot of the character "i" to its main part (see Fig. 8.40) or parts of the same character that are disconnected, e.g., because of a bad print quality. For characters on difficult surfaces, e.g., punched characters on a metal surface, gray value morphology may be necessary to segment the characters (see Fig. 8.57).

Additionally, in some applications it may be necessary to perform a geometric transformation of the image to transform the characters into a standard position, typically such that the text is horizontal. This process is called image rectification. For example, the text may have to be rotated (see Fig. 8.17), perspectively rectified (see Fig. 8.19), or rectified with a polar transformation (see Fig. 8.20).

Even though we have many segmentation strategies at our disposal, in some applications it may be difficult to segment the individual characters because the characters actually touch each other, either in reality or in the resolution we are looking at them in the image. Therefore, special methods to segment touching characters are sometimes required.

The simplest such strategy is to define a separate ROI for each character we are expecting in the image. This strategy can sometimes be used in industrial applications because the fonts typically have a fixed pitch (width) and we know a priori how many characters are present in the image, e.g., if we are trying to read serial numbers with a fixed length. The main problem with this approach is that the character ROIs must enclose the individual characters we are trying to separate. This is difficult if the position of the text can vary in the image. If this is the case, we first need to determine the pose of the text in the image based on another strategy, e.g., template matching to find a distinct feature in the vicinity of the text we are trying to read, and to use the pose of the text to either rectify the text to a standard position or to move the character ROIs to the appropriate position.

While defining separate ROIs for each character works well in some applications, it is not very flexible. A better method can be derived by realizing that the characters typically touch only with a small number of pixels. An example for this is shown in Figs. 8.119(a) and (b). To separate these characters, we can simply count the number of pixels per column in the segmented region. This is shown in Fig. 8.119(c). Since the touching part is only a narrow bridge between the characters the number of pixels in the region of the touching part only has a very small number of pixels per column. In fact, we can simply segment the characters by splitting them vertically at the position of the minimum in Fig. 8.119(c). The result is shown in Fig. 8.119(d). Note that in Fig. 8.119(c) the optimal splitting point is the global minimum of the number of pixels per column. However, in general this may not be the case. For example, if the strokes between the vertical bars of the letter "m" were slightly thinner the letter "m" might be split erroneously. Therefore, to make this algorithm more robust, it is typically necessary to define a search space for splitting of the characters based on the expected width of the characters. For example, in this application the characters are approximately 20 pixels wide. Therefore, we could restrict the search space for the optimal splitting point to a range of ± 4 pixels (20% of the expected width) around the expected width of the characters. This simple splitting method works very well in practice. Further approaches for segmenting characters are discussed in [16].

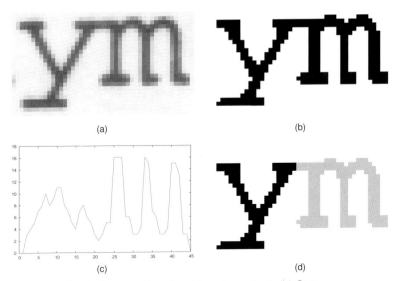

Fig. 8.119 (a) An image of two touching characters. (b) Segmented region. Note that the characters are not separated. (c) Plot of the number of pixels in each column of (b). (d) The characters have been split at the minimum of (c) at position 21.

8.11.2
Feature Extraction

As mentioned above, reading of the characters corresponds to the classification of regions, i.e., the assignment of a class ω_i to a region. For the purposes of the OCR, the classes ω_i can be thought of as the interpretation of the character, i.e., the string that represents the character. For example, if an application must read serial numbers, the classes $\{\omega_1, \ldots, \omega_{10}\}$ are simply the strings $\{0, \ldots, 9\}$. If numbers and uppercase letters must be read, the classes are $\{\omega_1, \ldots, \omega_{36}\} = \{0, \ldots, 9, A, \ldots, Z\}$. Hence, classification can be thought of as a function f that maps to the set of classes $\Omega = \{\omega_i, i = 1, \ldots, m\}$. What is the input of this function? First, to make the above mapping well-defined and easy to handle, we require that the number n of input values to the function f is constant. The input values to f are called features. They typically are real numbers. With this, the function f that performs the classification can be regarded as a mapping $f : \mathbb{R}^n \mapsto \Omega$.

For the OCR, the features that are used for the classification are features that we extract from the segmented characters. Any of the region features described in Section 8.5.1 and the gray value features described in Section 8.5.2 can be used as features. The main requirement is that the features enable us to discern different character classes. Figure 8.120 illustrates this point. The input image is shown in Fig. 8.120(a). It contains examples of lowercase letters. Suppose that we want to classify the letters based on the region features anisometry and compactness. Figures 8.120(b) and (c) show that the letters "c" and "o" as well as "i" and "j" can be distinguished easily based on these two features. In fact, they could be distinguished solely based on their compactness. As Figs. 8.120(d) and (e) show, however, these two features are not sufficient to distinguish between the classes "p" and "q" as well as "h" and "k."

From the above example, we can see that the features we use for the classification must be sufficiently powerful to enable us to classify all relevant classes correctly. The region and gray value features described in Sections 8.5.1 and 8.5.2, unfortunately, are often not powerful enough to achieve this. A set of features that is sufficiently powerful to distinguish all the classes of characters is the gray values of the image themselves. Using the gray values directly, however, is not possible because the classifier requires a constant number of input features. To achieve this, we can use the smallest enclosing rectangle around the segmented character, enlarge it slightly to include a suitable amount of background of the character in the features (e.g., by one pixel in each direction), and then can zoom the gray values within this rectangle to a standard size, e.g., 8×10 pixels. While transforming the image, we must take care to use the interpolation and smoothing techniques discussed in Sec-

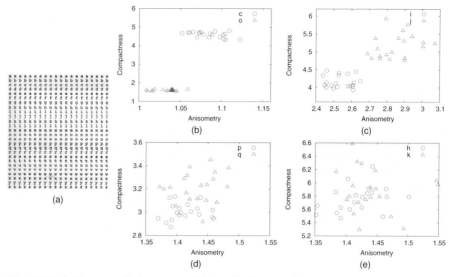

Fig. 8.120 (a) Image with lowercase letters. Features anisometry and compactness plotted for the letters "c" and "o" (b), "i" and "j" (c), "p" and "q" (d), and "h" and "k" (e). Note that the letters in (b) and (c) can be easily distinguished based on the selected features, while the letters in (d) and (e) cannot be distinguished.

tion 8.3.3. Note, however, that by zooming the image to a standard size based on the surrounding rectangle of the segmented character we lose the ability to distinguish characters like "-" (minus sign) and "l" (upper case I in fonts without serifs). The distinction can be easily done based on a single additional feature: the ratio of the width and height of the smallest surrounding rectangle of the segmented character.

Unfortunately, the gray value features defined above are not invariant to illumination changes in the image. This makes the classification very difficult. To achieve invariance to illumination changes, two options exist. The first option is to perform a robust gray value normalization of the gray values of the character, as described in Section 8.2.1, before the character is zoomed to the standard size. The second option is to convert the segmented character into a binary image before the character is zoomed to the standard size. Since the gray values generally contain more information the first strategy is preferable in most cases. The second strategy can be used whenever there is significant texture in the background of the segmented characters, which would make the classification more difficult.

Figure 8.121 displays two examples for the gray value feature extraction for the OCR. Figures 8.121(a) and (d) display two instances of the letter "5," taken from images with different contrast (Figs. 8.22(a) and (b)). Note that the characters have different sizes (14 × 21 and 13 × 20 pixels, respectively). The result of the robust contrast normalization is shown in Figs. 8.121(b) and (e). Note that both characters now have full contrast. Finally, the result of zooming the characters to a size of 8 × 10 pixels is shown in Figs. 8.121(c) and (f). Note that this feature extraction automatically makes the OCR scale-invariant because of the zooming to a standard size.

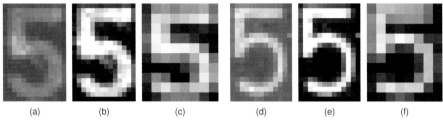

(a) (b) (c) (d) (e) (f)

Fig. 8.121 Gray value feature extraction for the OCR. (a) Image of the letter "5" taken from the second row of characters in the image in Fig. 8.22(a). (b) Robust contrast normalization of (a). (c) Result of zooming (b) to a size of 8 × 10 pixels. (d) Image of the letter "5" taken from the second row of characters in the image in Fig. 8.22(b). (e) Robust contrast normalization of (d). (f) Result of zooming (e) to a size of 8 × 10 pixels.

To conclude the discussion about the feature extraction, some words about the standard size are necessary. The discussion above has used the size 8 × 10. A large set of tests has shown that this size is a very good size to use for most industrial applications. If there are only a small number of classes to distinguish, e.g., only numbers, it may be possible to use slightly smaller sizes. For some applications involving a larger number of classes, e.g., numbers and uppercase and lowercase characters, a slightly larger size may be necessary (e.g., 10 × 12). On the other hand, using much larger sizes typically do not lead to better classification results because the features become progressively less robust against small segmentation errors if a large standard size is chosen. This happens because larger standard sizes imply that a segmentation error will lead to progressively larger position inaccuracies in the zoomed character as the standard size becomes larger. Therefore, it is best not to use a standard size that is much larger than the above recommendations. One exception to this rule is the recognition of an extremely large set of classes, e.g., ideographic characters like the Japanese Kanji characters. Here, much larger standard sizes are necessary to distinguish a large number of different characters.

8.11.3
Classification

As we saw in the previous section, classification can be regarded as a mapping from the feature space to the set of possible classes: $f : \mathbb{R}^n \mapsto \Omega$. We will now take a closer look at how the mapping can be constructed.

First, we note that the feature vector x that serves as the input to the mapping can be regarded as a random variable because of the variations that the characters exhibit. In the application, we are observing this random feature vector for each character we are trying to classify. It can be shown that to minimize the probability of erroneously classifying the feature vector, we should maximize the probability that the class ω_i occurs under the condition that we observe the feature vector x, i.e., we should maximize $P(\omega_i|x)$ over all classes ω_i, $i = 1, \ldots, m$ [91, 97]. The probability $P(\omega_i|x)$ is also called a posteriori probability because of the above property that it describes the probability of class ω_i given that we have observed the feature vector x. This decision rule is called the Bayes decision rule. It yields the best classifier if all errors have the same weight, which is a reasonable assumption for the OCR.

We now face the problem how to determine the a posteriori probability. Using Bayes's theorem, $P(\omega_i|x)$ can be computed as follows:

$$P(\omega_i|x) = \frac{P(x|\omega_i)P(\omega_i)}{P(x)} \qquad (8.115)$$

Hence, we can compute the a posteriori probability based on the a priori probability $P(x|\omega_i)$ that the feature vector x occurs given that the class of the feature vector is ω_i, the probability $P(\omega_i)$ that the class ω_i occurs, and the probability $P(x)$ that the feature vector x occurs. To simplify the calculations, we note that the Bayes decision rule only needs to maximize $P(\omega_i|x)$ and that $P(x)$ is a constant if x is given. Therefore, the Bayes decision rule can be written as:

$$x \in \omega_i \Leftrightarrow P(x|\omega_i)P(\omega_i) > P(x|\omega_j)P(\omega_j) \qquad j = 1, \ldots, m, j \neq i \qquad (8.116)$$

What does this transformation gain us? As we will see below, the probabilities $P(x|\omega_i)$ and $P(\omega_i)$ can, in principle, be determined from training samples. This enables us to evaluate $P(\omega_i|x)$, and hence to classify the feature vector x. Before we examine this point in detail, however, let us assume that the probabilities in (8.116) are known. For example, let us assume that the feature space is one-dimensional ($n = 1$) and that there are two classes ($m = 2$). Furthermore, let us assume that $P(\omega_1) = 0.3$, $P(\omega_2) = 0.7$, and that the features of the two classes have a normal distribution $N(\mu, \sigma)$ such that $P(x|\omega_1) \sim N(-3, 1.5)$ and $P(x|\omega_2) \sim N(3, 2)$. The corresponding likelihoods $P(x|\omega_i)P(\omega_i)$ are shown in Fig. 8.122. Note that features to the left of

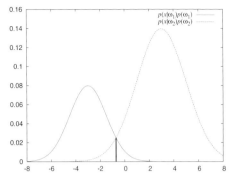

Fig. 8.122 Example for a two-class classification problem in a one-dimensional feature space in which $P(\omega_1) = 0.3$, $P(\omega_2) = 0.7$, $P(x|\omega_1) \sim N(-3, 1.5)$, and $P(x|\omega_2) \sim N(3, 2)$. Note that features to the left of $x \approx -0.7122$ are classified as belonging to ω_1, while features to the right are classified as belonging to ω_2.

$x \approx -0.7122$ are classified as belonging to ω_1, while features to the right are classified as belonging to ω_2. Hence, there is a dividing point $x \approx -0.7122$ that separates the classes from each other.

As a further example, consider a two-dimensional feature space with three classes that have normal distributions with different means and covariances, as shown in Fig. 8.123(a). Again, there are three regions in the two-dimensional feature space in which the respective class has the highest probability, as shown in Fig. 8.123(b). Note that now there are one-dimensional curves that separate the regions in the feature space from each other.

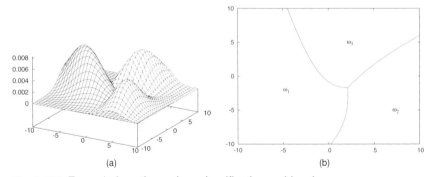

Fig. 8.123 Example for a three-class classification problem in a two-dimensional feature space in which the three classes have normal distributions with different means and covariances. (a) A posteriori probabilities of the occurrence of the three classes. (b) Regions in the two-dimensional feature space in which the respective class has the highest probability.

As the above examples suggest, the Bayes decision rule partitions the feature space into mutually disjoint regions. This is obvious from the definition

(8.116): Each region corresponds to the part of the feature space in which the class ω_i has the highest a posteriori probability. As also suggested by the above examples, the regions are separated by $n-1$ dimensional hypersurfaces (points for $n=1$ and curves for $n=2$, as in Figs. 8.122 and 8.123). The hypersurfaces that separate the regions from each other are given by the points in which two classes are equally probable, i.e., by $P(\omega_i|x) = P(\omega_j|x), i \neq j$.

Accordingly, we can identify two different types of classifiers. The first type of classifier tries to estimate a posteriori probabilities, typically via Bayes's theorem from the a priori probabilities of different classes. In contrast, the second type of classifier tries to construct the separating hypersurfaces between the classes. Below, we will examine representatives for both types of classifiers.

All classifiers require a method with which the probabilities or separating hypersurfaces are determined. To do this, a training set is required. The training set is a set of sample feature vectors x_k with corresponding class labels ω_k. For the OCR, the training set is a set of character samples, from which the corresponding feature vectors can be calculated, along with the interpretation of the respective character. The training set should be representative of the data that can be expected in the application. In particular, for the OCR the characters in the training set should contain the variations that will occur later, e.g., different character sets, stroke widths, noise, etc. Since it is often difficult to obtain a training set with all variations the image processing system must provide means to extend the training set over time with samples that are collected in the field and should optionally also provide means to artificially add variations to the training set. Furthermore, to evaluate the classifier, in particular how well it has generalized the decision rule from the training samples, it is indispensable to have a test set that is independent from the training set. This test set is essential to determine the error rate that the classifier is likely to have in the application. Without the independent test set, no meaningful statement about the quality of the classifier can be made.

The classifiers that are based on estimating probabilities, more precisely the probability densities, are called the Bayes classifiers because they try to implement the Bayes decision rule via the probability densities. The first problem they have to solve is how to obtain the probabilities $P(\omega_i)$ of the occurrence of class ω_i. There are two basic strategies for this purpose. The first strategy is to estimate $P(\omega_i)$ from the training set. Note that for this the training set must not only be representative in terms of the variations of feature vectors, but also in terms of the frequencies of the classes. Since this second requirement is often difficult to ensure an alternative strategy for the estimation of $P(\omega_i)$ is to assume that each class is equally likely to occur, and hence to use $P(\omega_i) = 1/m$. Note that in this case the Bayes decision rule reduces to the classification according to the a priori probabilities since $P(\omega_i|x) \sim P(x|\omega_i)$ should now be maximized.

The remaining problem is how to estimate $P(x|\omega_i)$. In principle, this could be done by determining the histogram of feature vectors of the training set in the feature space. To do so, we could subdivide each dimension of the feature space into b bins. Hence, the feature space would be divided into b^n bins in total. Each bin would count the number of occurrences of feature vectors in the training set that lie within this bin. If the training set and b are large enough, the histogram would be a good approximation to the probability density $P(x|\omega_i)$. Unfortunately, this approach cannot be used in practice because of the so-called curse of dimensionality: The number of bins in the histogram is b^n, i.e., its size grows exponentially with the dimension of the feature space. For example, if we use the 81 features described in the previous section and subdivide each dimension into a modest number of bins, e.g., $b = 10$, the histogram would have 10^{81} bins, which is much too large to fit into any computer memory.

To obtain a classifier that can be used in practice, we note that in the histogram approach the size of the bin is kept constant, while the number of samples in the bin varies. To get a different estimate for the probability of a feature vector, we can keep the number k of samples of class ω_i constant while varying the volume $v(x, \omega_i)$ of the region in space around the feature vector x that contains the k samples. Then, if there are t feature vectors in the training set the probability of the occurrence of class ω_i is approximately given by $P(x|\omega_i) \approx k/(tv(x, \omega_i))$. Since the volume $v(x, \omega_i)$ depends on the k nearest neighbors of class ω_i this type of density estimation is called the k nearest neighbor density estimation. In practice, this approach is often modified as follows: Instead of determining the k nearest neighbors of a particular class and computing the volume $v(x, \omega_i)$, the k nearest neighbors in the training set of any class are determined. The feature vector x is then assigned to the class that has the largest number of samples among the k nearest neighbors. This classifier is called the k nearest neighbor classifier (kNN classifier). For $k = 1$, we obtain the nearest neighbor classifier (NN classifier). It can be shown that the NN classifier has an error probability, which is at most twice as large as the error probability of the optimal Bayes classifier that uses the correct probability densities [91]: $P_B \leq P_{NN} \leq 2P_B$. Furthermore, if P_B is small we have $P_{NN} \approx 2P_B$ and $P_{3NN} \approx P_B + 3P_B^2$. Hence, the 3NN classifier is almost as good as the optimal Bayes classifier. Nevertheless, the kNN classifiers are seldom used in practice because they require that the entire training set is stored with the classifier (which can easily contain several hundred thousands of samples). Furthermore, the search for the k nearest neighbors is very time consuming.

As we have seen from the above discussion, the direct estimation of the probability density function is not practical, either because of the curse of dimensionality for the histograms or because of efficiency considerations for the kNN classifier. To obtain an algorithm that can be used in practice, we can assume that $P(x|\omega_i)$ follows a certain distribution, e.g., an n-dimensional normal distribution

$$P(x|\omega_i) = \frac{1}{(2\pi)^{n/2}|\Sigma_i|^{1/2}} \exp\left(-\frac{1}{2}(x-\mu_i)^\top \Sigma_i^{-1}(x-\mu_i)\right) \qquad (8.117)$$

With this, estimating the probability density function reduces to the estimation of parameters of the probability density function. For the normal distribution, the parameters are the mean vector μ_i and the covariance matrix Σ_i of each class. Since the covariance matrix is symmetric the normal distribution has $(n^2+3n)/2$ parameters in total. They can, for example, be estimated via the standard maximum likelihood estimators

$$\mu_i = \frac{1}{n_i}\sum_{j=1}^{n_i} x_{i,j} \qquad \Sigma_i = \frac{1}{n_i-1}\sum_{j=1}^{n_i}(x_{i,j}-\mu_i)(x_{i,j}-\mu_i)^\top \qquad (8.118)$$

Here, n_i is the number of samples for class ω_i, while $x_{i,j}$ denotes the samples for class ω_i.

While the Bayes classifier based on the normal distribution can be quite powerful, often the assumption that the classes have a normal distribution does not hold in practice. In OCR applications, this happens frequently if characters in different fonts should be recognized with the same classifier. One striking example for this is the shapes of the letters "a" and "g" in different fonts. For these letters, two basic shapes exist: "a" vs. "a" and "g" vs. "g." It is clear that a single normal distribution is insufficient to capture these variations. In these cases, each font will typically lead to a different distribution. Hence, each class consists of a mixture of l_i different densities $P(x|\omega_i,k)$, each of which occurs with probability $P_{i,k}$:

$$P(x|\omega_i) = \sum_{k=1}^{l_i} P(x|\omega_i,k)P_{i,k} \qquad (8.119)$$

Typically, the mixture densities $P(x|\omega_i,k)$ are assumed to be normally distributed. In this case, (8.119) is called a Gaussian mixture model. If we knew to which mixture density each sample belongs we could easily estimate the parameters of the normal distribution with the above maximum likelihood estimators. Unfortunately, in real applications we typically do not have this knowledge, i.e., we do not know k in (8.119). Hence, determining the parameters of the mixture model not only requires the estimation of the parameters of the mixture densities, but also the estimation of the mixture density labels k – a much harder problem, which can be solved by the expectation maximization algorithm (EM algorithm). The interested reader is referred to [91] for details. Another problem in the mixture model approach is that we need to specify how many mixture densities there are in the mixture model, i.e., we need to specify l_i in (8.119). This is quite cumbersome to do manually. Recently, algorithms that compute l_i automatically have been proposed. The interested reader is referred to [28, 96] for details.

Let us now turn our attention to classifiers that construct the separating hypersurfaces between the classes. Of all possible surfaces, the simplest ones are planes. Therefore, it is instructive to regard this special case first. Planes in the n-dimensional feature space are given by

$$w^\top x + b = 0 \tag{8.120}$$

Here, x is an n-dimensional vector that describes a point, while w is an n-dimensional vector that describes the normal vector to the plane. Note that this equation is linear. Because of this, classifiers based on separating hyperplanes are called linear classifiers.

Let us first consider the problem of classifying two classes with the plane. We can assign a feature vector to the first class ω_1 if x lies on one side of the plane, while we can assign it to the second class ω_2 if it lies on the other side of the plane. Mathematically, the test on which side of the plane a point lies is performed by looking at the sign of $w^\top x + b$. Without loss of generality, we can assign x to ω_1 if $w^\top x + b > 0$, while we assign x to ω_2 if $w^\top x + b < 0$.

For classification problems with more than two classes, we construct m separating planes (w_i, b_i) and use the following classification rule [91]:

$$x \in \omega_i \Leftrightarrow w_i^\top x + b_i > w_j^\top x + b_j, \quad j = 1, \ldots, m, j \neq i \tag{8.121}$$

Note that in this case, the separating planes do not have the same meaning as in the two-class case, where the plane actually separates the data. The interpretation of (8.121) is that the plane is chosen such that feature vectors of the correct class have the largest positive distance of all feature vectors from the plane.

Linear classifiers can also be regarded as neural networks, as shown in Fig. 8.124 for the two-class and n-class case. The neural network has processing units (neurons) that are visualized by circles. They first compute the linear combination of the feature vector x and the weights w: $w^\top x + b$. Then, a non-linear activation function f is applied. For the two-class case, the activation function simply is $\text{sgn}(w^\top x + b)$, i.e., the side of the hyperplane on which the feature vector lies. Hence, the output is mapped to its essence: -1 or 1. Note that this type of activation function essentially thresholds the input value. For the n-class case, the activation function f typically is chosen such that input values < 0 are mapped to 0, while input values ≥ 0 are mapped to 1. The goal in this approach is that a single processing unit returns the value 1, while all other units return the value 0. The index of the unit that returns 1 indicates the class of the feature vector. Note that the plane in (8.121) needs to be modified for this activation function to work since the plane is chosen such that feature vectors have the largest distance from the plane. Therefore, $w_j^\top x + b_j$ is not necessarily < 0 for all values that do not belong to the class. Nevertheless, both definitions are equivalent. Note that because the neural network has one layer of processing units, this type of neural network is also called a single-layer perceptron.

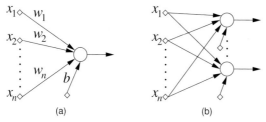

Fig. 8.124 The architecture of a linear classifier expressed as a neural network (single-layer perceptron). (a) Two-class neural network. (b) n-class neural network. In both cases, the neural network has a single layer of processing units that are visualized by circles. They first compute the linear combination of the feature vector and the weights. After this, a nonlinear activation function is computed, which maps the output to -1 or 1 (two-class neural network) or 0 or 1 (n-class neural network).

While linear classifiers are simple and easy to understand, they have very limited classification capabilities. By construction, the classes must be linearly separable, i.e., separable by a hyperplane, for the classifier to produce the correct output. Unfortunately, this is rarely the case in practice. In fact, linear classifiers are unable to represent a simple function like the *xor* function, as illustrated in Fig. 8.125, because there is no line that can separate the two classes. Furthermore, for n-class linear classifiers, there is often no separating hyperplane for each class against all the other classes although each pair of classes can be separated by a hyperplane. This happens, for example, if the samples of one class lie completely within the convex hull of all the other classes.

Fig. 8.125 A linear classifier is not able to represent the *xor* function because the two classes, corresponding to the two outputs of the *xor* function, cannot be separated by a single line.

To get a classifier that is able to construct more general separating hypersurfaces, one approach is to simply add more layers, to the neural network, as shown in Fig. 8.126. Each layer first computes the linear combination of the feature vector or the results from the previous layer

$$a_j^{(l)} = \sum_{i=1}^{n_l} w_{ji}^{(l)} x_i^{(l-1)} + b_j^{(l)} \tag{8.122}$$

Here, $x_i^{(0)}$ is simply the feature vector, while $x_i^{(l)}$, $l \geq 1$ is the result vector of layer l. The coefficients $w_{ji}^{(l)}$, $b_j^{(l)}$ are the weights of layer l. Then, the results are passed through a nonlinear activation function

$$x_j^{(l)} = f(a_j^{(l)}) \tag{8.123}$$

Let us assume for the moment that the activation function in each processing unit is the threshold function that is also used in the single-layer perceptron, i.e., the function that maps input values < 0 to 0, while mapping input values ≥ 0 to 1. Then, it can be seen that the first layer of processing units maps the feature space to the corners of the hypercube $\{0,1\}^p$, where p is the number of processing units in the first layer. Hence, the feature space is subdivided by hyperplanes into half-spaces [91]. The second layer of processing units separates the points on the hypercube by hyperplanes. This corresponds to intersections of half-spaces, i.e., convex polyhedra. Hence, the second layer is capable of constructing the boundaries of convex polyhedra as the separating hypersurfaces [91]. This is still not general enough, however, since the separating hypersurfaces might need to be more complex than this. If a third layer is added, the network can compute unions of the convex polyhedra [91]. Hence, three layers are sufficient to approximate any separating hypersurface arbitrarily closely if the threshold function is used as the activation function.

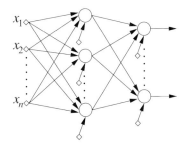

Fig. 8.126 The architecture of a multi-layer perceptron. The neural network has multiple layers of processing units that are visualized by circles. They compute the linear combination of the results of the previous layer and the network weights and then pass the results through a nonlinear activation function.

In practice, the above threshold function is rarely used because it has a discontinuity at $x = 0$, which is very detrimental for the determination of the network weights by numerical optimization. Instead, often a sigmoid activation function is used, for example the logistic function (see Fig. 8.127(a))

$$f(x) = \frac{1}{1 + e^x} \tag{8.124}$$

Similar to the hard threshold function, it maps its input to a value between 0 and 1. However, it is continuous and differentiable, which is a requirement for most numerical optimization algorithms. Another choice for the activation functions is to use the hyperbolic tangent function (see Fig. 8.127(b))

$$f(x) = \tanh(x) = \frac{e^x - e^{-x}}{e^x + e^{-x}} \tag{8.125}$$

in all layers except the output layer, in which the softmax activation function is used [9] (see Fig. 8.127(c))

$$f(x) = \frac{e^{x_i}}{\sum_{j=1}^{n} e^{x_j}} \tag{8.126}$$

The hyperbolic tangent function behaves similarly to the logistic function. The major difference is that it maps its input to values between -1 and 1. There is experimental evidence that the hyperbolic tangent function leads to a faster training of the network [9]. In the output layer, the softmax function maps the input to the range $[0, 1]$, as desired. Furthermore, it ensures that the output values sum to 1, and hence have the same properties as a probability density [9]. With any of these choices of activation functions, it can be shown that two layers are sufficient to approximate any separating hypersurface and, in fact, any output function with values in $[0, 1]$, arbitrarily closely [9]. The only requirement for this is that there is a sufficient number of processing units in the first layer (the "hidden layer").

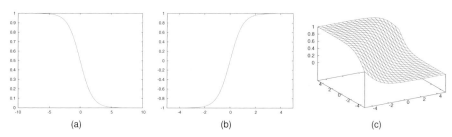

Fig. 8.127 (a) Logistic activation function (8.124). (b) Hyperbolic tangent activation function (8.125). (c) Softmax activation function (8.126) for two classes.

After having discussed the architecture of the multi-layer perceptron, we can now examine how the network is trained. Training the network means that the weights $w_{ji}^{(l)}, b_j^{(l)}, l = 1, 2$, of the network must be determined. Let us denote the number of input features by n_i, the number of hidden units (first layer units) by n_h, and the number of output units (second layer units) by n_o. Note that n_i is the dimensionality of the feature vector, while n_o is the number of classes in the classifier. Hence, the only free parameter is the number n_h of units in the hidden layer. Note that there are $(n_i + 1)n_h + (n_h + 1)n_o$ weights in total. For example, if $n_i = 81$, $n_h = 40$, and $n_o = 10$, there are 3690 weights that must be determined. It is clear that this is a very complex problem and that we can only hope to determine the weights uniquely if the number of training samples is of the same order of magnitude as the number of weights.

As described above, training of the network is performed based on a training set, which consists of the sample feature vectors x_k with the corresponding class labels ω_k, $k = 1, \ldots, m$. The sample feature vectors can be used as they are. The class labels, however, must be transformed into a representation that can be used in an optimization procedure. As described above, ideally we would like to have the multi-layer perceptron return a 1 in the output unit that corresponds to the class of the sample. Hence, a suitable representation of the classes is a target vector $y_k \in \{0, 1\}^{n_o}$, chosen such that there is a 1 at the index that corresponds to the class of the sample and 0 in all other positions. With this, we can train the network by minimizing, for example, the squared error of the outputs of the network on all the training samples [9]. In the notation of (8.123), we would like to minimize

$$\varepsilon = \sum_{k=1}^{m} \sum_{j=1}^{n_o} (x_j^{(2)} - y_{k,j})^2 \tag{8.127}$$

Here, $y_{k,j}$ is the jth element of the target vector y_k. Note that $x_j^{(2)}$ implicitly depends on all the weights $w_{ji}^{(l)}$, $b_j^{(l)}$ of the network. Hence, minimization of (8.127) determines the optimum weights. To minimize (8.127), for a long time the backpropagation algorithm was used, which successively inputs each training sample into the network, determines the output error, and derives a correction term for the weights from the error. It can be shown that this procedure corresponds to the steepest descent minimization algorithm, which is well known to converge extremely slowly [73]. Currently, the minimization of (8.127) is typically being performed by sophisticated numerical minimization algorithms, such as the conjugate gradient algorithm [73, 9] or the scaled conjugate gradient algorithm [9].

Another approach to obtain a classifier that is able to construct arbitrary separating hypersurfaces is to transform the feature vector into a space of higher dimension, in which features are linearly separable, and to use a linear classifier in the higher dimensional space. Classifiers of this type have been known for a long time as generalized linear classifiers [91]. One instance of this approach is the polynomial classifier, which transforms the feature vector by a polynomial of degree $\leq d$. For example, for $d = 2$ the transformation is

$$\Phi(x_1, \ldots, x_n) = (x_1, \ldots, x_n, x_1^2, \ldots, x_1 x_n, \ldots, x_n x_1, \ldots, x_n^2) \quad (8.128)$$

The problem with this approach is again the curse of dimensionality: The dimension of the feature space grows exponentially with the degree d of the polynomial. In fact, there are $\binom{d+n-1}{d}$ monomials of degree $= d$ alone. Hence, the dimension of the transformed feature space is

$$n' = \sum_{i=1}^{d} \binom{i+n-1}{i} = \binom{d+n}{d} - 1 \quad (8.129)$$

For example, if $n = 81$ and $d = 5$, the dimension is 34826301. Even for $d = 2$ the dimension already is 3402. Hence, transforming the features into the larger feature space seems to be infeasible, at least from an efficiency point of view. Fortunately, however, there is an elegant way to perform the classification with generalized linear classifiers that avoids the curse of dimensionality. This is achieved by support vector machine (SVM) classifiers [81].

Before we can take a look at how support vector machines avoid the curse of dimensionality, we have to take a closer look at how the optimal separating hyperplane can be constructed. Let us consider the two-class case. As described in (8.120) for linear classifiers, the separating hyperplane is given by $w^\top x + b = 0$. As noted above, the classification is performed based on the sign of $w^\top x + b$. Hence, the classification function is

$$f(x) = \mathrm{sgn}(w^\top x + b) \quad (8.130)$$

Let the training samples be denoted by x_i and their corresponding class labels by $y_i = \pm 1$. Then, a feature is classified correctly if $y_i(w^\top x + b) > 0$. However, this restriction is not sufficient to determine the hyperplane uniquely. This can be achieved by requiring that the margin between the two classes should be as large as possible. The margin is defined as the closest distance of any training sample to the separating hyperplane.

Let us look at a small example of the optimal separating hyperplane, shown in Fig. 8.128. Note that if we want to maximize the margin (shown as a dotted line) there will be samples from both classes that attain the minimum distance to the separating hyperplane defined by the margin. These samples "support" the two hyperplanes that have the margin as the distance to the separating hyperplane (shown as the dashed lines). Hence, the samples are called the support vectors.

In fact, the optimal separating hyperplane is defined entirely by the support vectors, i.e., a subset of the training samples: $w = \sum_{i=1}^{m} \alpha_i y_i x_i$, $\alpha_i \geq 0$, where

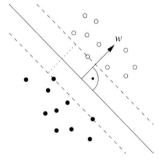

Fig. 8.128 The optimal separating hyperplane between two classes. The samples of the two classes are represented by the filled and unfilled circles. The hyperplane is visualized by the solid line. The margin is visualized by the dotted line between the two dashed lines that visualize the hyperplanes in which samples are on the margin, i.e., attain the minimum distance between the classes. The samples on the margin define the separating hyperplane. Since they "support" the margin hyperplanes they are called support vectors.

$a_i > 0$ if and only if the training sample is a support vector [81]. With this, the classification function can be written as

$$f(x) = \text{sgn}(w^\top x + b) = \text{sgn}\left(\sum_{i=1}^{m} \alpha_i y_i x_i^\top x + b\right) \quad (8.131)$$

Hence, to determine the optimal hyperplane, the coefficients α_i of the support vectors must be determined. This can be achieved by solving the following quadratic programming problem [81]: maximize

$$\sum_{i=1}^{m} \alpha_i - \frac{1}{2} \sum_{i=1}^{m} \sum_{j=1}^{m} \alpha_i \alpha_j y_i y_j x_i^\top x_j \quad (8.132)$$

subject to

$$\alpha_i \geq 0, \quad i = 1, \ldots, m \quad \text{and} \quad \sum_{i=1}^{m} \alpha_i y_i = 0 \quad (8.133)$$

Note that in both the classification function (8.131) and the optimization function (8.132) the feature vectors x, x_i, and x_j are only present in the dot product.

We now turn our attention back to the case that the feature vector x is first transformed into a higher dimensional space by a function $\Phi(x)$, e.g., by the polynomial function (8.128). Then, the only change in the above discussion is that we substitute the feature vectors x, x_i, and x_j by their transformations $\Phi(x)$, $\Phi(x_i)$, and $\Phi(x_j)$. Hence, the dot products are simply computed in the higher dimensional space. The dot products become functions of two input feature vectors: $\Phi(x)^\top \Phi(x')$. These dot products of transformed feature vectors are called kernels in the SVM literature and are denoted by $k(x,x') = \Phi(x)^\top \Phi(x')$. Hence, the decision function becomes a function of the kernel $k(x,x')$:

$$f(x) = \text{sgn}\left(\sum_{i=1}^{m} \alpha_i y_i k(x_i, x) + b\right) \qquad (8.134)$$

The same happens with the optimization function (8.132).

So far, it seems that the kernel does not gain us anything because we still have to transform the data into a feature space of a prohibitively large dimension. The ingenious trick of the SVM classification is that for a large class of kernels the kernel can be evaluated efficiently without explicitly transforming the features into the high-dimensional space, thus making the evaluation of the classification function (8.134) feasible. For example, if we transform the features by a polynomial of degree d, it can be easily shown that

$$k(x,x') = (x^\top x')^d \qquad (8.135)$$

Hence, the kernel can be evaluated solely based on the input features without going to the higher dimensional space. This kernel is called a homogeneous polynomial kernel. As another example, the transformation by a polynomial of degree $\leq d$ can simply be evaluated as

$$k(x,x') = (x^\top x' + 1)^d \qquad (8.136)$$

This kernel is called an inhomogeneous polynomial kernel. Further examples of possible kernels include the Gaussian radial basis function kernel

$$k(x,x') = \exp\left(-\frac{\|x - x'\|^2}{2\sigma^2}\right) \qquad (8.137)$$

and the sigmoid kernel

$$k(x,x') = \tanh(\kappa x^\top x' + \vartheta) \qquad (8.138)$$

Note that this is the same function that is also used in the hidden layer of the multi-layer perceptron. With any of the above four kernels, support vector machines can approximate any separating hypersurface arbitrarily closely.

Note that the above training algorithm that determines the support vectors still assumes that the classes can be separated by a hyperplane in the higher dimensional transformed feature space. This may not always be achievable. Fortunately, the training algorithm can be extended to handle overlapping classes. The reader is referred to [81] for details.

By its nature, the SVM classification can only handle two-class problems. To extend the SVM to multi-class problems, two basic approaches are possible.

The first strategy is to perform a pairwise classification of the feature vector against all pairs of classes and to use the class that obtains the most votes, i.e., is selected most often as the result of the pairwise classification. Note that this implies that $l(l-1)/2$ classifications have to be performed if there are l classes. The second strategy is to perform l classifications of one class against the union of the rest of the classes. From an efficiency point of view, the second strategy may be preferable since it depends linearly on the number of classes. Note, however, that in the second strategy there will be typically a larger number of support vectors than in the pairwise classification. Since the runtime depends linearly on the number of support vectors, the number of support vectors must grow less than quadratically for the second strategy to be faster.

After having discussed different types of classifiers, the natural question to ask is which of the classifiers should be used in practice. First, it must be noted that the quality of the classification results of all the classifiers depends to a large extend on the size and quality of the training set. Therefore, to construct a good classifier, the training set should be as large and representative as possible.

If the main criterion for comparing classifiers is the classification accuracy, i.e., the error rate on an independent test set, classifiers that construct the separating hypersurfaces, i.e., the multi-layer perceptron or the support vector machine, should be preferred. Of these two, the support vector machine is portrayed to have a slight advantage [81]. This advantage, however, is achieved by building certain invariances into the classifier that do not generalize to other classification tasks apart from OCR and a particular set of gray value features. The invariances built into the classifier in [81] are translations of the character by one pixel, rotations, and variations of the stroke width of the character. With the features in Section 8.11.2, the translation invariance is automatically achieved. The remaining invariances could also be achieved by extending the training set by systematically modified training samples. Therefore, neither the multi-layer perceptron nor the support vector machine have a definite advantage in terms of classification accuracy.

Another criterion is the training speed. Here, support vector machines have a large advantage over the multi-layer perceptron because the training times are significantly shorter. Therefore, if the speed in the training phase is important, support vector machines currently should be preferred. Unfortunately, the picture is reversed in the online phase when unknown features must be classified. As noted above, the classification time for the support vector machines depends linearly on the number of support vectors, which usually is a substantial fraction of the training samples (typically, between 10% and 40%). Hence, if the training set consists of several hundred thousands of samples, tens of thousands of support vectors must be evaluated with the kernel in the online phase. In contrast, the classification time of the multi-layer perceptron only depends on the topology of the net, i.e., the number of processing units per layer. Therefore, if the speed in the online classification phase is important the multi-layer perceptron currently should be preferred.

A final criterion is whether the classifier provides a simple means to decide whether the feature vector to be classified should be rejected because it does not belong to any of the trained classes. For example, if only digits have been trained, but the segmented character is the letter "M," it is important in some applications to be able to reject the character as being no digit. Another example is an erroneous segmentation, i.e., a segmentation of an image part that corresponds to no character at all. Here, classifiers that construct the separating hypersurfaces provide no means to tell whether the feature vector is close to a class in some sense since the only criterion is on which side of the separating hypersurface the feature lies. For the support vector machines, this behavior is obvious from the architecture of the classifier. For the multi-layer perceptron, this behavior is not immediately obvious since we have noted above that the softmax activation function (8.126) has the same properties as a probability density. Hence, one could assume that the output reflects the likelihood of each class, and can thus be used to threshold unlikely classes. In practice, however, the likelihood is very close to 1 or 0 everywhere, except in areas in which the classes overlap in the training samples, as shown by the example in Fig. 8.129. Note that all the classes have a very high likelihood for feature vectors that are far away from the training samples of each class.

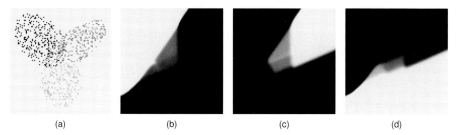

(a) (b) (c) (d)

Fig. 8.129 (a) Samples in a two-dimensional feature space for three classes, which are visualized by three gray levels. (b) Likelihood for class 1 determined in a square region of the feature space with a multi-layer perceptron with five hidden units. (c) Likelihood for class 2. (d) Likelihood for class 3. Note that all the classes have a very high likelihood for feature vectors that are far away from the training samples of each class. For example, class 3 has a very high likelihood in the lower corners of the displayed portion of the feature space. This high likelihood continues to infinity because of the architecture of the multi-layer perceptron.

As can be seen from the above discussion, the behavior of the multi-layer perceptron and the support vector machines are identical with respect to samples that lie far away from the training samples: They provide no means to evaluate the closeness of a sample to the training set. In applications where it is important that feature vectors can be rejected as not belonging to any class there are two options. First, we can train an explicit rejection class. The problem with this approach is how to obtain or construct the samples for the rejection class. Second, we can use a classifier that provides a measure of closeness to the training samples, such as the Gaussian mixture model classifier. However, in this case we have to be prepared to accept slightly higher error rates for feature vectors that are close to the training samples.

(a)

(b)

M5I8222 30
6285352
M5 I8 2 2 2   30
6 2 8 5 3 5 2

(c)

CXD1175AM
648E10E
CXD 1175AM
6 4 8 E 10 E

(d)

Fig. 8.130 (a), (b) Images of prints on ICs. (c), (d) Result of the segmentation of the characters (light gray) and the OCR (black). The images are segmented with a threshold that is selected automatically based on the gray value histogram (see Fig. 8.22). Furthermore, only the last two lines of characters are selected. Additionally, irrelevant characters like the "–" are suppressed based on the height of the characters.

We conclude this section with an example that uses the ICs that we have already used in Section 8.4. In this application, the goal is to read the characters in the last two lines of the print in the ICs. Figures 8.130(a) and (b) show images of two sample ICs. The segmentation is performed with a threshold that is selected automatically based on the gray value histogram (see Fig. 8.22). After this, the last two lines of characters are selected based on the smallest surrounding rectangle of the characters. For this, the character with the largest row coordinate is determined and characters lying within an interval above this character are selected. Furthermore, irrelevant characters like the "–" are suppressed based on the height of the characters. The characters are classified with a multi-layer perceptron that has been trained with several tens of thousands of samples of characters on electronic components, which do not include the characters on the ICs in Fig. 8.130. The result of the segmentation and classification is shown in Figs. 8.130(c) and (d). Note that all characters have been read correctly.

References

1 SUNG JOON AHN, WOLFGANG RAUH, HANS-JÜRGEN WARNECKE. *Least-squares orthogonal distances fitting of circle, sphere, ellipse, hyperbola, and parabola*. Pattern Recognition, 34(12):2283–2303, 2001.
2 SHIGERU ANDO. *Consistent gradient operators*. IEEE Transactions on Pattern Analysis and Machine Intelligence, 22(3):252–265, 2000.
3 VALERY A. ANISIMOV, NICKOLAS D. GORSKY. *Fast hierarchical matching of an arbitrarily oriented template*. Pattern Recognition Letters, 14(2):95–101, 1993.
4 NICHOLAS AYACHE, OLIVIER D. FAUGERAS. *HYPER: A new approach for the recognition and positioning of two-dimensional objects*. IEEE Transactions on Pattern Analysis and Machine Intelligence, 8(1):44–54, 1986.
5 JEAN BABAUD, ANDREW P. WITKIN, MICHEL BAUDIN, RICHARD O. DUDA. *Uniqueness of the Gaussian kernel for scale-space filtering*. IEEE Transactions on Pattern Analysis and Machine Intelligence, 8(1):26–33, 1986.
6 D. H. BALLARD. *Generalizing the Hough transform to detect arbitrary shapes*. Pattern Recognition, 13(2):111–122, 1981.
7 MARK DE BERG, MARC VAN KREVELD, MARK OVERMARS, OTFRIED SCHWARZKOPF. *Computational Geometry: Algorithms and Applications*. Springer, Berlin, 2nd edition, 2000.

8 VALDIS BERZINS. *Accuracy of Laplacian edge detectors*. Computer Vision, Graphics, and Image Processing, 27:195–210, 1984.
9 CHRISTOPHER M. BISHOP. *Neural Networks for Pattern Recognition*. Oxford University Press, Oxford, 1995.
10 D. S. BLOOMBERG. *Implementation efficiency of binary morphology*. In International Symposium on Mathematical Morphology VI, pages 209–218, 2002.
11 GUNILLA BORGEFORS. *Distance transformation in arbitrary dimensions*. Computer Vision, Graphics, and Image Processing, 27:321–345, 1984.
12 GUNILLA BORGEFORS. *Hierarchical chamfer matching: A parametric edge matching algorithm*. IEEE Transactions on Pattern Analysis and Machine Intelligence, 10(6):849–865, November 1988.
13 E. ORAN BRIGHAM. *The Fast Fourier Transform and its Applications*. Prentice-Hall, Upper Saddle River, NJ, 1988.
14 LISA GOTTESFELD BROWN. *A survey of image registration techniques*. ACM Computing Surveys, 24(4):325–376, 1992.
15 JOHN CANNY. *A computational approach to edge detection*. IEEE Transactions on Pattern Analysis and Machine Intelligence, 8(6):679–698, 1986.
16 RICHARD G. CASEY ERIC LECOLINET. *A survey of methods and strategies in character segmentation*. IEEE Transactions on Pattern Analysis and Machine Intelligence, 18(7):690–706, 1996.

17 JEN-MING CHEN, JOSE A. VENTURA, CHIH-HANG WU. *Segmentation of planar curves into circular arcs and line segments.* Image and Vision Computing, 14(1):71–83, 1996.

18 MAURO S. COSTA, LINDA G. SHAPIRO. *3D object recognition and pose with relational indexing.* Computer Vision and Image Understanding, 79(3):364–407, 2000.

19 PER-ERIK DANIELSSON. *Euclidean distance mapping.* Computer Graphics and Image Processing, 14:227–248, 1980.

20 RACHID DERICHE. *Using Canny's criteria to derive a recursively implemented optimal edge detector.* International Journal of Computer Vision, 1:167–187, 1987.

21 RACHID DERICHE. *Fast algorithms for low-level vision.* IEEE Transactions on Pattern Analysis and Machine Intelligence, 12(1):78–87, 1990.

22 RACHID DERICHE. *Recursively implementing the Gaussian and its derivatives.* Rapport de Recherche 1893, INRIA, Sophia Antipolis, April 1993.

23 MARC VAN DROOGENBROECK, HUGUES TALBOT. *Fast computation of morphological operations with arbitrary structuring elements.* Pattern Recognition Letters, 17(14):1451–1460, 1996.

24 ULRICH ECKHARDT, GERD MADERLECHNER. *Invariant thinning.* International Journal of Pattern Recognition and Artificial Intelligence, 7(5):1115–1144, 1993.

25 OLIVIER FAUGERAS. *Three-Dimensional Computer Vision: A Geometric Viewpoint.* MIT Press, Cambridge, MA, 1993.

26 OLIVIER FAUGERAS, BERNARD HOTZ, HERVÉ MATHIEU, THIERRY VIÉVILLE, ZHENGYOU ZHANG, PASCAL FUA, ERIC THÉRON, LAURENT MOLL, GÉRARD BERRY, JEAN VUILLEMIN, PATRICE BERTIN, CATHERINE PROY. *Real time correlation-based stereo: algorithm, implementations and applications.* Rapport de Recherche 2013, INRIA, Sophia-Antipolis, August 1993.

27 OLIVIER FAUGERAS, QUANG-TUAN LUONG. *The Geometry of Multiple Images: The Laws that Govern the Formation of Multiple Images of a Scene and Some of their Applications.* MIT Press, Cambridge, MA, 2001.

28 MARIO A. T. FIGUEIREDO, ANIL K. JAIN. *Unsupervised learning of finite mixture models.* IEEE Transactions on Pattern Analysis and Machine Intelligence, 24(3):381–396, 2002.

29 MARTIN A. FISCHLER, ROBERT C. BOLLES. *Random sample consensus: A paradigm for model fitting with applications to image analysis and automated cartography.* Communications of the ACM, 24(6):381–395, 1981.

30 ANDREW FITZGIBBON, MAURIZIO PILU, ROBERT B. FISHER. *Direct least square fitting of ellipses.* IEEE Transactions on Pattern Analysis and Machine Intelligence, 21(5):476–480, 1999.

31 LUC M. J. FLORACK, BART M. TER HAAR ROMENY, JAN J. KOENDERINK, MAX A. VIERGEVER. *Scale and the differential structure of images.* Image and Vision Computing, 10(6):376–388, 1992.

32 JAN FLUSSER TOMÁŠ SUK. *Pattern recognition by affine moment invariants.* Pattern Recognition, 26(1):167–174, 1993.

33 WOLFGANG FÖRSTNER. *A framework for low level feature extraction.* In Jan-Olof Eklundh, editor, Third European Conference on Computer Vision, volume 801 of Lecture Notes in Computer Science, pages 383–394, Springer, Berlin, 1994.

34 MOHAMMAD GHARAVI-ALKHANSARI. *A fast globally optimal algorithm for template matching using low-resolution pruning.* IEEE Transactions on Image Processing, 10(4):526–533, 2001.

35 JOSEPH GIL RON KIMMEL. *Efficient dilation, erosion, opening, and closing algorithms.* IEEE Transactions on Pattern Analysis and Machine Intelligence, 24(12):1606–1617, 2002.

36 F. GLAZER, G. REYNOLDS, P. ANANDAN. *Scene matching by hierarchical correlation.* In Computer Vision and Pattern Recognition, pages 432–441, 1983.

37 W. ERIC L. GRIMSON TOMÁS LOZANO-PÉREZ. *Localizing overlapping parts by searching the interpretation tree.* IEEE Transactions on Pattern Analysis and Machine Intelligence, 9(4):469–482, 1987.

38 Armin Gruen Thomas S. Huang, editors. *Calibration and Orientation of Cameras in Computer Vision.* Springer, Berlin, 2001.

39 ROBERT M. HARALICK, LINDA G. SHAPIRO. *Computer and Robot Vision*, volume I. Addison-Wesley, Reading, MA, 1992.

40. ROBERT M. HARALICK LINDA G. SHAPIRO. *Computer and Robot Vision*, volume II. Addison-Wesley Publishing Company, Reading, MA, 1993.
41. ROBERT M. HARALICK, LAYNE T. WATSON, THOMAS J. LAFFEY. The topographic primal sketch. International Journal of Robotics Research, 2(1):50–72, 1983.
42. RICHARD HARTLEY, ANDREW ZISSERMAN. *Multiple View Geometry in Computer Vision*. Cambridge University Press, Cambridge, 2nd edition, 2003.
43. JANNE HEIKKILÄ. Geometric camera calibration using circular control points. IEEE Transactions on Pattern Analysis and Machine Intelligence, 22(10):1066–1077, October 2000.
44. YACOV HEL-OR HAGIT HEL-OR. Real time pattern matching using projection kernels. In 9th International Conference on Computer Vision, volume 2, pages 1486–1493, 2003.
45. HEIKO HIRSCHMÜLLER, PETER R. INNOCENT, JON GARIBALDI. Real-time correlation-based stereo vision with reduced border errors. International Journal of Computer Vision, 47(1/2/3):229–246, 2002.
46. MING-KUEI HU. Visual pattern recognition by moment invariants. IRE Transactions on Information Theory, 8:179–187, 1962.
47. THOMAS S. HUANG, GEORGE J. YANG, GREGORY Y. TANG. A fast two-dimensional median filtering algorithm. IEEE Transactions on Acoustics, Speech, and Signal Processing, 27(1):13–18, 1979.
48. PETER J. HUBER. *Robust Statistics*. John Wiley & Sons, New York, NY, 1981.
49. RAMESH JAIN, RANGACHAR KASTURI, BRIAN G. SCHUNCK. *Machine Vision*. McGraw-Hill, New York, NY, 1995.
50. S. H. JOSEPH. Unbiased least squares fitting of circular arcs. Computer Vision, Graphics, and Image Processing: Graphical Models and Image Processing, 56(5):424–432, 1994.
51. S. H. JOSEPH. Analysing and reducing the cost of exhaustive correspondence search. Image and Vision Computing, 17:815–830, 1999.
52. MARK W. KOCH RANGASAMI L. KASHYAP. Using polygons to recognize and locate partially occluded objects. IEEE Transactions on Pattern Analysis and Machine Intelligence, 9(4):483–494, 1987.
53. OH-KYU KWON, DONG-GYU SIM, RAE-HONG PARK. Robust Hausdorff distance matching algorithms using pyramidal structures. Pattern Recognition, 34:2005–2013, 2001.
54. SHANG-HONG LAI, MING FANG. Accurate and fast pattern localization algorithm for automated visual inspection. Real-Time Imaging, 5(1):3–14, 1999.
55. LOUISA LAM, SEONG-WHAN LEE, CHING Y. SUEN. Thinning methodologies – a comprehensive survey. IEEE Transactions on Pattern Analysis and Machine Intelligence, 14(9):869–885, 1992.
56. YEHEZKEL LAMDAN, JACOB T. SCHWARTZ, HAIM J. WOLFSON. Affine invariant model-based object recognition. IEEE Transactions on Robotics and Automation, 6(5):578–589, 1990.
57. STEFAN LANSER. *Modellbasierte Lokalisation gestützt auf monokulare Videobilder*. PhD thesis, Forschungs- und Lehreinheit Informatik IX, Technische Universität München, 1997. Shaker, Aachen.
58. STEFAN LANSER, WOLFGANG ECKSTEIN. A modification of Deriche's approach to edge detection. In 11th International Conference on Pattern Recognition, volume III, pages 633–637, 1992.
59. STEFAN LANSER, CHRISTOPH ZIERL, ROLAND BEUTLHAUSER. Multibildkalibrierung einer CCD-Kamera. In G. Sagerer, S. Posch, F. Kummert, editors, Mustererkennung, Informatik aktuell, pages 481–491, Springer, Berlin, 1995.
60. REIMAR LENZ, DIETER FRITSCH. Accuracy of videometry with CCD sensors. ISPRS Journal of Photogrammetry and Remote Sensing, 45(2):90–110, 1990.
61. TONY LINDEBERG. *Scale-Space Theory in Computer Vision*. Kluwer, Dordrecht, The Netherlands, 1994.
62. ALEXANDER G. MAMISTVALOV. n-dimensional moment invariants and conceptual mathematical theory of recognition n-dimensional solids. IEEE Transactions on Pattern Analysis and Machine Intelligence, 20(8):819–831, 1998.
63. STEVE MANN RICHARD MANN. Quantigraphic imaging: Estimating the camera response and exposures from differently exposed images. In Computer Vision and Pattern

Recognition, volume I, pages 842–849, 2001.

64 DAVID M. MCKEOWN JR., STEVEN DOUGLAS COCHRAN, STEPHEN J. FORD, J. CHRIS MCGLONE, JEFFEREY A. SHUFELT, DANIEL A. YOCUM. *Fusion of HYDICE hyperspectral data with panchromatic imagery for cartographic feature extraction.* IEEE Transactions on Geoscience and Remote Sensing, 3:1261–1277, 1999.

65 JERRY M. MENDEL. *Fuzzy logic systems for engineering: A tutorial.* Proceedings of the IEEE, 83(3):345–377, 1995.

66 TOMOO MITSUNAGA SHREE K. NAYAR. *Radiometric self calibration.* In Computer Vision and Pattern Recognition, volume I, pages 374–380, 1999.

67 MADHAV MOGANTI, FIKRET ERCAL, CIHAN H. DAGLI, SHOU TSUNEKAWA. *Automatic PCB inspection algorithms: A survey.* Computer Vision and Image Understanding, 63(2):287–313, 1996.

68 FREDERICK MOSTELLER JOHN W. TUKEY. *Data Analysis and Regression.* Addison-Wesley Publishing Company, Reading, MA, 1977.

69 KARSTEN MÜHLMANN, DENNIS MAIER, JÜRGEN HESSER, REINHARD MÄNNER. *Calculating dense disparity maps from color stereo images, an efficient implementation.* International Journal of Computer Vision, 47(1/2/3):79–88, 2002.

70 CLARK F. OLSON DANIEL P. HUTTENLOCHER. *Automatic target recognition by matching oriented edge pixels.* IEEE Transactions on Image Processing, 6(1):103–113, 1997.

71 JOSEPH O'ROURKE. *Computational Geometry in C.* Cambridge University Press, Cambridge, 2nd edition, 1998.

72 ATHANASIOS PAPOULIS. *Probability, Random Variables, and Stochastic Processes.* McGraw-Hill, New York, NY, 3rd edition, 1991.

73 WILLIAM H. PRESS, SAUL A. TEUKOLSKY, WILLIAM T. VETTERLING, BRIAN P. FLANNERY. *Numerical Recipes in C: The Art of Scientific Computing.* Cambridge University Press, Cambridge, 2nd edition, 1992.

74 URS RAMER. *An iterative procedure for the polygonal approximation of plane curves.* Computer Graphics and Image Processing, 1:244–256, 1972.

75 PAUL L. ROSIN. *Techniques for assessing polygonal approximations of curves.* IEEE Transactions on Pattern Analysis and Machine Intelligence, 19(6):659–666, 1997.

76 PAUL L. ROSIN. *Assessing the behaviour of polygonal approximation algorithms.* Pattern Recognition, 36(2):508–518, 2003.

77 PAUL L. ROSIN GEOFF A. W. WEST. *Nonparametric segmentation of curves into various representations.* IEEE Transactions on Pattern Analysis and Machine Intelligence, 17(12):1140–1153, 1995.

78 WILLIAM J. RUCKLIDGE. *Efficiently locating objects using the Hausdorff distance.* International Journal of Computer Vision, 24(3):251–270, 1997.

79 DANIEL SCHARSTEIN RICHARD SZELISKI. *A taxonomy and evaluation of dense two-frame stereo correspondence algorithms.* International Journal of Computer Vision, 47(1/2/3):7–42, 2002.

80 CORDELIA SCHMID, ROGER MOHR, CHRISTIAN BAUCKHAGE. *Evaluation of interest point detectors.* International Journal of Computer Vision, 37(2):151–172, 2000.

81 BERNHARD SCHÖLKOPF ALEXANDER J. SMOLA. *Learning with Kernels – Support Vector Machines, Regularization, Optimization, and Beyond.* MIT Press, Cambridge, MA, 2002.

82 ROBERT SEDGEWICK. *Algorithms in C.* Addison-Wesley Publishing Company, Reading, MA, 1990.

83 HSIN-TENG SHEU WU-CHIH HU. *Multiprimitive segmentation of planar curves – a two-level breakpoint classification and tuning approach.* IEEE Transactions on Pattern Analysis and Machine Intelligence, 21(8):791–797, 1999.

84 PIERRE SOILLE. *Morphological Image Analysis.* Springer, Berlin, 2nd edition, 2003.

85 LUIGI DI STEFANO, STEFANO MATTOCCIA, MARTINO MOLA. *An efficient algorithm for exhaustive template matching based on normalized cross correlation.* In 12th International Conference on Image Analysis and Processing, pages 322–327, 2003.

86 CARSTEN STEGER. *On the calculation of arbitrary moments of polygons.* Technical Report FGBV–96–05, Forschungsgruppe Bildverstehen (FG BV), Informatik IX, Technische Universität München, 1996.

87 CARSTEN STEGER. *Analytical and empirical performance evaluation of subpixel line and edge detection*. In Kevin J. Bowyer P. Jonathon Phillips, editors, Empirical Evaluation Methods in Computer Vision, pages 188–210, 1998.

88 CARSTEN STEGER. *Unbiased Extraction of Curvilinear Structures from 2D and 3D Images*. PhD thesis, Fakultät für Informatik, Technische Universität München, 1998. Herbert Utz Verlag, München.

89 CARSTEN STEGER. *Subpixel-precise extraction of lines and edges*. In International Archives of Photogrammetry and Remote Sensing, volume XXXIII, part B3, pages 141–156, 2000.

90 STEVEN L. TANIMOTO. *Template matching in pyramids*. Computer Graphics and Image Processing, 16:356–369, 1981.

91 SERGIOS THEODORIDIS, KONSTANTINOS KOUTROUMBAS. *Pattern Recognition*. Academic Press, San Diego, CA, 1999.

92 QI TIAN, MICHAEL N. HUHNS. *Algorithms for subpixel registration*. Computer Vision, Graphics, and Image Processing, 35:220–233, 1986.

93 GODFRIED TOUSSAINT. *Solving geometric problems with the rotating calipers*. In Proceedings of IEEE MELECON '83, pages A10.02/1–4, Los Alamitos, CA, 1983. IEEE Press.

94 MARKUS ULRICH, CARSTEN STEGER, ALBERT BAUMGARTNER. *Real-time object recognition using a modified generalized Hough transform*. Pattern Recognition, 36(11):2557–2570, 2003.

95 JOSE A. VENTURA, WENHUA WAN. *Accurate matching of two-dimensional shapes using the minimal tolerance error zone*. Image and Vision Computing, 15:889–899, 1997.

96 HAI XIAN WANG, BIN LUO, QUAN BING ZHANG, SUI WEI. *Estimation for the number of components in a mixture model using stepwise split-and-merge EM algorithm*. Pattern Recognition Letters, 25(16):1799–1809, 2004.

97 ANDREW WEBB. *Statistical Pattern Recognition*. Arnold Publishers, London, 1999.

98 EMO WELZL. *Smallest enclosing disks (balls and ellipsoids)*. In H. Maurer, editor, New Results and Trends in Computer Science, volume 555 of Lecture Notes in Computer Science, pages 359–370, Springer, Berlin, 1991.

99 ANDREW P. WITKIN. *Scale-space filtering*. In Eigth International Joint Conference on Artificial Intelligence, volume 2, pages 1019–1022, 1983.

100 DANIEL M. WUESCHER KIM L. BOYER. *Robust contour decomposition using a constant curvature criterion*. IEEE Transactions on Pattern Analysis and Machine Intelligence, 13(1):41–51, 1991.

101 IAN T. YOUNG LUCAS J. VAN VLIET. *Recursive implementation of the Gaussian filter*. Signal Processing, 44:139–151, 1995.

9
Machine Vision in Manufacturing
Dr.-Ing. Peter Waszkewitz, Robert Bosch GmbH

9.1
Introduction

The preceding chapters have given a far-reaching and detailed view of the world of machine vision, its elements, components, tools, and methods. A lot of connections between the topics of individual chapters have already been explored, for example concerning the interactions of light, object surfaces, optics, and sensors. At this point, then, we have everything we need to create machine vision systems. But, of course, a machine vision system is not created in a void without relations to and constraints from the outer world, nor is it made up and determined by image processing components alone.

This chapter will try to illustrate the environment in which the tools, technologies and techniques presented in the previous chapters are applied and give a view of machine vision as a part of automation technology.

Whereas the universal world of physics and mathematics played a dominant role in the previous chapters, this chapter will, admittedly, show a certain bias toward (German) automotive applications; most of it should translate easily to other industrial environments, though.

Perhaps even more so than in the previous chapters, the vast variety of existing and possible applications and solutions in this field renders any claim to completeness meaningless. This chapter will therefore try to give an idea of issues that may arise in the design of a machine vision system, what they may be related to, how they have been solved under particular circumstances and what to watch out for. We will have to keep in mind, though, that since "Scientists discover what is, engineers create what has never been."[1] there will be always a point where explanations and checklists end and we will simply have to go out there and solve the problems we encounter.

1) Theodore von Karman, 1911.

Handbook of Machine Vision. Alexander Hornberg (Ed.)
Copyright © 2006 WILEY-VCH Verlag GmbH & Co. KGaA, Weinheim
ISBN: 3-527-40584-7

The Machine Vision Market

Economic data changes faster than book editions and since buying this book already proves an interest in machine vision, numbers are certainly unnecessary to prove that it is economically worthwhile and technologically important to concern oneself with machine vision.

Suffice to say that market figures, available for example at [9, 1, 2], show healthy, usually two-digit growth of the machine vision sector. Often, machine vision growth outstrips not only that of the economy as a whole, but also that of automation technology. And still experts state that only 20% of potential applications are yet implemented [9].

High labor costs in developed countries are certainly an important factor for growth in automation technology. The additional growth in machine vision is certainly due to various reasons. One factor is the increasing maturity of this still fairly young technology. Together with constantly increasing computation power, in PCs as well as compact systems, it makes more and more tasks technologically and economically viable.

And finally, but perhaps most importantly, requirements on production quality and organization are permanently increasing. Tolerances are getting tighter, cost pressures are rising, traceability has to be ensured across worldwide production setups – many production processes are hardly possible anymore without machine vision technology.

From the market data as well as from the technical development leading to increasingly powerful, versatile and cost-effective machine vision systems, it seems safe to assume that the growth of the machine vision sector will continue for quite some time. This, however, means that machine vision systems will become increasingly common elements in all kinds of production environments and that it will be necessary to integrate them into this environment as best as possible.

9.2
Application Categories

Machine vision applications can be categorized in various ways. This section will introduce some terms, which may help us to define a frame of reference to place machine vision systems in.

9.2.1
Types of Tasks

Many machine vision systems are highly specialized, designed to fulfill a unique and complicated task, and are thus special-purpose machines in themselves. Nevertheless, it is possible to identify some recurring types of applications as a first frame of reference what to expect from a new task.

The following categorization follows that in [8] and [3] with one exception: it seemed appropriate to make code recognition a category of its own, since code reading does not have much to do with actual characteristics of the work piece (like size, shape, etc.), and even more so because scanners have come into such widespread use as specialized devices, which many users probably would not even identify as image processing systems.

This leads to the following categories:

Code recognition denotes the identification of objects using markings on the objects; these are typically standardized bar codes or DataMatrix codes, but can also be custom codes. Typical applications are material flow control and logistics. Internally, methods from all areas of image processing are used, including, for example, edge detection, filtering and positioning techniques.

Object recognition denotes identification of objects using characteristic features like shape / geometry, dimensions, color, structure / topology, texture. Object identification includes the distinction of object variants and has many applications, not least as an "auxiliary science" for many other tasks. For example, position recognition or completeness checks may require prior identification of the correct objects in the scene.

Position recognition denotes determining position and orientation of an object – or a particular point of an object – in a pre-defined coordinate system, using feature computation and matching methods. Typical features are center of gravity coordinates and orientation angles. An important distinction is the dimensionality, that is whether position and orientation have to be determined in 2D or 3D. Typical applications are robot guidance, pick and place operations, insertion machines.

Completeness check denotes categorization of work pieces as correctly or incorrectly assembled; it checks whether all components are present and in the correct position, often as a pre-condition for passing the work piece on to the next assembly step or as a final check before releasing the work piece to be packed and delivered – or one step later the inspection of the package to be completely filled with products of the right type.

Shape and dimension check denotes determination of geometrical quantities with focus on precise and accurate measuring. The importance of this area increases in accordance with rising quality standards as products must meet ever tighter tolerance requirements. Applications can be found wherever work pieces or also tools have to be checked for compliance with nominal dimensions. Due to the required accuracy, these tasks typically impose high demands on sensor equipment as well as on the mechanical construction of the inspection station.

Surface inspection can be divided into *quantitative* surface inspection aiming at the determination of topographical features like roughness and *qualitative* surface inspection where the focus is on the recognition of surface defects, such as dents, scratches, pollution, or deviations from desired surface characteristics, like color or texture. Quantitative measurements of geometrical properties may be required for judging the surface quality. Typical challenges in surface inspection applications are large data sets and computationally intensive algorithms, such as gray level statistics, Fourier transformation and the like.

Of course, these categories are frequently mixed in real-world applications. Inspections of separate aspects of a part may belong to different categories – for example: reading a bar code with a serial number and performing a completeness check to ensure correct assembly; or various types of applications support and enhance each other. For example, information gained from position recognition methods can be used to adapt a dimensional check to the precise location of the work piece in the camera image.

9.2.2
Types of Production

Another important point for designing a machine vision system is the type of production the system is to work in. Three basic types of manufacturing process are [5]:

Job-shop and batch production that produces in small lots or batches. Examples are tool and die making or casting in a foundry;

Mass production typically – but not necessarily – involving machines or assembly lines that manufacture discrete units repetitively. Volumes are high, flexibility is comparatively low. Typically, assembly lines require at least partial stoppage for change-overs between different products.

Continuous production that produces in continuous flow. Examples are paper mills, refineries or extrusion processes. Product volumes are typically very high, variety and flexibility possibly even lower than with mass production.

Of course, these types rarely occur in a pure form in real-world production. Mass production frequently requires parts made in job-shop style supplied to the assembly line, raw materials for continuous flow production are prepared in batches. There are also gray areas, where for example production of very small parts is organized in the same way as continuous production although discrete units are produced.

Different types of manufacturing pose different demands on machine vision systems. Obviously there will not be a 1:1 assignment of requirements to particular production types, but some general tendencies can be noted.

9.2.2.1 Discrete Unit Production Versus Continuous Flow

As we will see in more detail in Section 9.8, it makes sense to contrast the first two categories from above as *discrete unit production* with *continuous flow production*.

In discrete unit production, it will typically be necessary to capture individual work pieces in an image. This will require some sort of trigger. In relatively slow production lines, for example using a work-piece carrier transport system, it may suffice to have the control system bring the work piece in position in front of the camera, possibly wait for some time to let remaining movement or vibrations decay, then start the machine vision system.

In faster production lines – say bottling lines running at several bottles per second – appropriate sensors, for example light barriers, will be required to determine the precise moment when the part is in the correct position for capturing. As this type of production lines approximates continuous production, in that the parts are continually in motion, additional measures, like electronic shuttering and strobe lights, may have to be taken, to avoid motion blur in the image.

In the gray area of small parts produced in an almost continuous flow (like screws on a conveyor belt) we have the typical robot vision task of picking up the pieces fast enough. Obviously, a discrete unit trigger is very difficult to implement here. Rather the image processing has to trigger itself, perform the detection of objects within its field of view fast enough to allow the robot to clear the belt completely.

There is another important distinction. Actual continuous flow production – often called endless material production which is of course a fiction, nothing man-made is ever endless–requires appropriate algorithms. For example, when looking for surface defects exceeding a certain size, we cannot arbitrarily stop the image at some line, then start acquiring a new image to have the convenience of evaluating individual rectangular images. This could lead to a defect cut in half and thus be missed. Either a strategy has to be devised to handle this with subsequent rectangular images or algorithms have to be used that can handle continuous images efficiently, that is images where continuously the first line is thrown away and a new line is added at the other end, without inefficient re-evaluation of the entire image section currently in view.

Job-Shop Production Versus Mass Production

There are mass-produced items coming in more than 1000 variants. However, new types can usually be introduced much faster in job-shop production. Mass-producing a new variant of a product will require many alterations and adjustments to a production line, a lot of test runs and so forth, so that there is considerable delay to adapt a machine vision system. Reliability and run-time diagnosis and optimization features are called for here.

The possibility to start producing new types – or even entirely new products – quite rapidly in job-shop production on the other hand puts more emphasis on the ease and speed with which a vision system can be reconfigured and adapted to perform a new task. Interactivity is much more of a concern here, so the type of production in this case may have a decisive influence on system design.

9.2.3
Types of Evaluations

We can basically distinguish between two different methods of performing image processing in a manufacturing process:

Inspection that is, checking after the production process whether the work piece is correct. This is the case for most image processing systems today.

Monitoring that is, observing during the production process whether it is carried out correctly. For example, having a camera look at a laser welding process to see whether the melt pool has the correct size during welding.

As usual, the two types can mix. For example, a machine vision system placed after the nozzle in an aluminum extrusion process does not technically look *inside* the process. The effect, however, is that of a monitoring system. Deviations in the tube profile can be used immediately to take care of a production problem instead of waiting for the tube to have cooled down and be cut into the final work pieces.

In a similar way, we can also distinguish between different goals pursued by image processing:

Recognition that is, using the results of the machine vision system to characterize the work piece (or the process); when this leads to a decision over the correctness of the work piece, we speak of *verification* (note that it is not necessarily the vision system making this decision). Verification is usually easier than pure recognition since there is typically more a-priori knowledge (namely the desired or expected outcome).

Control that is, using the results of the machine vision system to change – read: improve – or control the process; this is possible as direct feedback control or as statistical process control.

One might argue that feedback control requires monitoring as the type of execution but this is not quite true if a lag time of one work piece (in the case of unit production) is accepted. If, for example, the system notes that a component has been placed too far left on one work piece, it can feed that information back to the control system, which can use it to correct its positioning. So both inspection and monitoring systems can be used for control purposes.

On the other hand, monitoring systems can be used for verification as well as control purposes. A system guiding a robot to apply a sealing agent along a part contour can at the same time check the sealing material for interruptions or overflowing.

It goes without saying that an image processing system which produces quantitative results (as opposed to a mere yes/no decision) can be used for statistical process control like any other measuring system.

9.2.4
Value-Adding Machine Vision

Sad as it is, a machine vision system traditionally destroys value. Now, this is certainly a controversial statement in a handbook of machine vision. But it is true for every system that is used to find defects on finished or half-finished products. If it finds a defect and the part has to be scrapped, the value already added to original materials is gone. Even if the components can be salvaged, the effort spent on them is lost.

Of course, it is a little more complicated than that. Letting out a defective part to the customer will probably destroy much more value than finding and scrapping the part. Also, increasing use of inspection systems early in the process – instead of only as a final check – will filter out defective parts before additional effort is wasted on them. But it is still a game of minimizing losses instead of maximizing gains.

However, machine vision systems can be – and increasingly are – used to add value. Sometimes, they just show where the potential for process improvement lies. But they can do more than that, particularly in the control type applications mentioned in Section 9.2.3. Using machine vision systems to help machines doing their job well instead of finding parts where they didn't, to increase process quality instead of detecting process flaws, to improve precision instead of throwing out imprecise pieces – in short, enabling production to manufacture better parts instead of just finding the good ones, will add value and will certainly lead to a lot of applications in the future, which may not even be foreseeable yet.

9.3
System Categories

Similar to applications, machine vision systems can also be categorized in various ways. This section will introduce some important distinctions that affect cost as well as performance, operation and handling of the system. There are many criteria that can be used to differentiate machine vision systems, for example:

- Dimensionality: 0-, 1-, 2-, 2.5^2-, and 3D systems;
- Flexibility: from simple, hardwired sensors to systems with limited or full configurability and freely programmable systems;
- System basis: including as diverse things as common PCs, DSP-based smart cameras, custom single-purpose chips and algorithms, and large parallel computers;
- Manufacturing depth: how much of the system is built from COTS components, especially in the software area, how much is (or needs to be) custom developed.

All these categories – and certainly quite a few more – can be of importance for a system decision. This space of possible configurations far exceeds the scope of this chapter and is only partly of interest from an integration point of view. Therefore, we will restrict ourselves to some typical categories, important as well for their widespread use as for the different approaches they represent.

9.3.1
Common Types of Systems

There is little common terminology in this area. For example, some companies call a vision sensor what others call an intelligent camera; some use the term vision controller for a configurable system much like a smart camera – only without the camera built in – others for a specialized DSP-based computer, and so on. The following list therefore introduces several terms for different types of systems which we will try to use consistently in the remainder of this chapter. Those, who are used to these terms meaning different things, please bear with us:

Sensors. Small, single-purpose, hardwired systems usually giving a binary yes/no result and working in 0D or 1D, we will treat only very briefly;

---

2) commonly used for systems evaluating height maps, gray level images where the brightness encodes height.

Vision sensors. A category recently come into focus, representing a middle-ground between traditional sensors and simple image processing systems;

Compact systems for want of a better word, although there are so many terms for this type of system: intelligent camera, smart camera, smart sensor, vision sensor (not the one from above!);

Vision controllers for small, configurable, multi-purpose units; essentially small computers specialized on machine vision tasks and equipped with communication interfaces for automation purposes;

PC-based systems naming the most common hardware basis for systems built using general-purpose computers.

In the following sections, we will try to characterize these systems mainly from an automation point of view.

9.3.2
Sensors

A sensor is, from the point of view of automation, a very simple thing. It is fixed somewhere and gives a signal, indicating some state. There are all kinds of sensors using all conceivable physical quantities: force, distance, pressure, ... We are most interested, naturally, in optical sensors.

Typical optical sensors are light barriers (retro reflective or through beam) and color sensors. A light sensor can recognize the presence of an object, either by an interruption of its beam or by the reflection from the part. A color sensor can recognize the color of the light spot created on the object by the illumination of the sensor.

These sensors are essentially zero dimensional, they have a very small field of view; basically, they evaluate a single spot. It is obvious that effects such as position tolerances or variations across the part surface pose real problems to such a system. Applied appropriately, they are very useful of course, in fact, they are indispensable for high-speed image processing as trigger indicators, as we will see in Section 9.8. One has to be clear about their limitations, however.

We will not go any deeper into the subject of these sensors. Under an integration aspect, they are basically components, which are fixed and adjusted in some place and then deliver a yes/no signal, like any binary switch, or a numerical value, for example as a brightness sensor.

9.3.3
Vision Sensors

The term *Vision Sensor* has been coined for systems that use areal sensors and machine vision methods for performing specific tasks.

Thanks to the areal sensors, they have some advantages over basic sensors. They can take a larger field of view into account and thus cope, for example, with irregular reflexions created by structured surfaces (like foil packaging), and color and brightness variations across surface areas.

Together with machine vision algorithms, the areal sensors enable these systems to perform evaluations not possible for basic sensors. Many applications are feasible using such systems, from very simple tasks, like computing the average brightness over a surface area to counting, sorting, and character recognition.

The main point is that this type of system is built for a single, particular purpose. Perhaps the best-known systems of this kind are DataMatrix code scanners – although they are not typically seen as vision systems at all by end users; so we see that with the DataMatrix scanner, there is already a type of system where the underlying technology is retreating behind the application as it should be with a mature technology.

Vision sensors in this sense of the word will certainly take some market share from more complex, more expensive systems that used to be required for these applications. Even more, however, they may expand the market as a whole, making applications economically viable that have simply not been done before.

Being considerably more complex, especially in their algorithms, than simple sensors, these systems allow for – and require – more configuration than these. Nevertheless, doing an application with this kind of system is still usually less complex than with a full-fledged vision system. Being single-purpose systems makes them easier to set up and operate for end users.

From an integration point of view, they can be as simple as a basic sensor, delivering a yes/no signal according to their configuration. They may also provide numerical or string-type results, DataMatrix scanners necessarily do, but usually with little variation. Vision systems, on the other hand, being used for all kinds of applications produce very different types of results and thus may have very complex information interfaces (cf. 9.7).

To sum up: vision sensors in this sense of the word have a potential to become very cost-effective solutions to a range of applications that, so far, has been largely ignored, simple sensors not having the capabilities, vision systems being too expensive. From an automation technology point of view, they are either much simpler to integrate than vision systems or, the more versatile ones, not much different. For a system integrator, standardization will be as important as with other systems: finding a range of sensors that meet the application requirements and provide long-term solutions.

9.3.4
Compact Systems

This is, admittedly, not a common term in this field. Some companies call this type of system a Vision Sensor, others an intelligent or smart camera. The term compact system shall reflect that the entire vision system hardware, namely the camera sensor, digitization, processing and communication is integrated in a small space, typically contained in a single unit. The original compact system therefore is nothing else than a digital camera with a built-in computer.

By now, the system category has expanded, there are central units coordinating several such compact systems as well as systems with a separate camera head or an additional camera input. However, the main characteristic is unchanged: a system where the image acquisition – that is, camera sensor and digitization – is basically a unit directly connected to the processor, with built-in image processing functionality.

Compact systems come in very different forms and with different application methods. There are systems that are actually programmed using a cross compiler for the DSP used in the system and a library of machine vision routines. Everything that can be said about custom-programmed PC-based solutions of course applies here also.

Other systems use a standard CPU and operating system, for example Linux on a PowerPC or Windows CE on an Intel CPU. These systems may be conventionally programmed, the main difference being that the programmer moves in an environment that will usually provide him with a more comprehensive set of services – like file system, networking, and so on – than the environment on a DSP-based system. Or some sort of configuration program with built-in vision functionality may run on the system. They are then not much different from a PC-based solution except for size and processing power.

Then there is the kind of compact systems that is probably dominant in the market today: DSP-based systems with built-in vision functionality that is set up and configured using configuration programs, typically running on a PC and communicating with the unit over Ethernet or other suitable interfaces.

These systems are often convenient to use; sophisticated vision algorithms facilitate the solution of demanding applications while the configuration programs make these capabilities usable for nonprogrammers. Naturally, compared with a PC, the systems are restricted in some way. Processing power, RAM and permanent storage space, number of sensor connections – in some if not all of these aspects, PCs will be superior. Extensibility of hardware as well as software will also typically be inferior to PC-based systems: there are no places for extension cards, for example, to add new bus interfaces, and the software may not allow for user extensions to the vision functionality.

On the other hand, a closed system – regarding hardware as well as software – has undeniable advantages. System security is much easier to achieve than for PC-based systems (so far, no viruses for intelligent cameras have been encountered) and configuration management is also much simpler: hardware type and firmware version will usually be sufficient to reconstruct a system that will behave identically with the same vision software configuration. This can be much harder with PCs.

9.3.5
Vision Controllers

Above, we have characterized vision controllers as small, special-purpose computers with camera and communication connections. From an integration point of view, they are similar to PC-based systems, except that the units are typically smaller, thus easier to accommodate, use a 24 V DC power supply and no standard operating system, so that they do not require orderly shutdown. Otherwise, with camera, communication and power supply connections they are integrated in much the same way as PCs.

From an application point of view, they are not much different from compact systems. They usually have built-in image processing algorithms, which are set up using a configuration program, determining the sequence and parameters of the algorithms.

Vision controllers are typically less expensive than PC-based solutions and appear convenient due to their small size. They are also less flexible, for example with regard to communication, so this cost advantage has to be carefully calculated. A one-shot application of such a system will probably not be cost-effective, as it will require a lot of nonstandard expenditure; if one can use such units as a standard for a considerable number of applications, calculations will be much different, of course.

9.3.6
PC-Based Systems

In this context, a PC-based system may just as well be a Linux PowerPC. It means systems based on general purpose computers with a standard operating system; PCs, characterized by some flavor of Windows running on an Intel CPU, are just the most common type.

There are mixtures, as has already been mentioned; compact systems built upon such a CPU and running a standard operating system. Under the aspect of *how* an application is carried out, this does not make too big a difference.

However, there is a crucial difference, the extensibility of standard PCs. They can be equipped with different interface cards for various bus systems, with different frame grabbers for all kinds of cameras, with additional RAM and hard disk space and so on.

In addition, the standard operating systems on these machines offer a wide range of services, for data management, networking, display, communication and so on, and the variety of software, including development systems and machine vision software, is overwhelming. Obviously, flexibility can hardly be greater than with such a system.

This does come at a price however: the hardware is relatively expensive, at least for a single camera system; configuration management is difficult, considering that standard PC components sometimes are available for hardly more than half a year on the market and that there is such an inexhaustible variety of hardware and software components which may, or may not, work together. Also, PC systems are usually equipped with hard disks, who are not always capable of running 24/7 under production conditions. Alternatives, like flash drives which to the operating system appear just like a hard disk, are getting a foothold in the market; at viable prices, they are still considerably smaller than hard disks and especially when traceability is a requirement or the convenience to have a large back storage of images for optimizing the system, size does matter.

And of course there is the security risk posed by worms and viruses, and shutting the systems off from the network is not really a solution, considering that the networking capabilities are one of the great benefits of the PC platform. Even if the system does not really have an Internet connection – what hardly a production system would – viruses may be introduced into the system in a variety of ways that are not easy to close completely.

Among PC-based systems we can further distinguish two types of systems: library-based and application-package-based systems.

Library-Based Systems

In a library-based system, functionality from a machine vision library is in some way incorporated in an executable program that represents (the software-part of) the vision system. There are many ways to do that; most libraries today come with a rapid-prototyping tool which allows a sequence of machine vision functions to be set up rapidly and supplied with initial parameters. This sequence can be exported as source code and compiled into a larger overall program; or a complete prototype setup is loaded by an executable DLL; or the library has a built-in interpreter executing a script connecting the library functions.

The common element is that library-based systems involve a programmer in the creation of a vision system, someone who provides the necessary framework to control the execution of the vision algorithms, take care of data handling and communications, provide a user interface and so on. Since a programmer is not necessarily a specialist in lighting or optics, this will often mean an additional person in the project, hence additional cost.

The programming work to be done for such systems can be considerable, and there is the danger that after doing several applications there will be several diverse executables to be maintained. On the other hand, this approach offers the ultimate flexibility. The way the application is executed, is under your control; you can adapt to any automation environment, provide every user interface the customer wants, include any functionality you need ... we do not want to get carried away here, there will always be limitations and be it that some libraries do not work well with each other, but the fact is that with this approach, the system integrator is also the system designer.

For systems that are identically duplicated many times – for example driver-assistance systems, video surveillance systems, or standard machines – and/or that have specific requirements on performance, integration with the environment, user interface and so on, this approach is very effective, allowing for exactly the required system to be built and distributing the development cost on a large number of copies.

Application-Package-Based Systems

An application-package means a piece of software used to configure a machine vision routine which can then be loaded and executed by the same (or another, that does not really matter) piece of software *without* modifying source code or compiling anything in between. So ideally, if the machine vision functionality built into the application package suffices, no programmer is part of the application loop. The application engineer alone sets up the solution and everything that belongs to it, and deploys the solution using the application package.

The basic idea behind the application-package approach is that there are many things, outside and above the level of machine vision functionality incorporated in a library, that are identical between vision systems and that productivity and quality of program development can be improved by uncoupling programming from the application process. This can lead to rapid and cost-effective solutions for a large class of problems. For a company with a fast-changing product spectrum, this can be a very effective approach.

The downside is that some degrees of freedom are lost in comparison with the library-based approach. You, as the system integrator, are no longer master of the execution environment, of the overall application logic; depending on the architecture of the package, you may not be master of the communication taking place between the system and your (or your customer's) automation environment. This can be a problem.

Also, in many cases, some functionality may be missing from the application package, a small one to make the customer happy, or a crucial one to solve the application at all. In any case, a method to extend the package's functionality is essential. It depends on the extension interfaces provided by the application package vendor how easily and how well this is possible.

For machine vision companies making a considerable number of systems, it may prove worthwhile to create their own application package, if they cannot find one on the market meeting their needs – or simply want to avoid dependencies and keep their freedom of adaptation. When they build many basically similar systems, encapsulating that similarity in a configurable base application can be useful. Or they have such an application turnaround that there is a return on the investment to provide the resources to create such a package. There are, in fact, machine vision companies that act as turn-key system builders, but use self-developed libraries and application packages internally, which they do not sell on the market.

9.3.7
Summary

Naturally, there can be no fixed rules as to what system to use for what purpose, no single right answer to the question of the system approach. However, one thing is clear: the more powerful, and the more common, the hardware platform is, the more flexible will the system be, the wider its range of applications; at the same time, it will be more difficult to set up, more difficult to troubleshoot – nothing very surprising, actually.

Systems based on general-purpose computers give the greatest flexiblity due to their extensibility and the richness and variety of development environments, libraries, and application packages available. Vision controllers and compact systems may fare better in ease-of-use and purchasing costs, but may come up against limiting factors like computation power and flexibility, since neither hardware nor software are usually built for being extended. Special-purpose computers, like parallel DSP-systems for example, may offer the best performance and be indispensable for very demanding applications, but require highly specialized programmers. Finally, vision sensors are by definition limited in their range of applications, but within that range often hard to beat with respect to price-performance relation and ease-of-use.

Where there are several ways to solve an image processing task – for example when a PC system and a compact system are both capable of doing the inspection – the decisive factors may actually be outside the area of image processing itself: communication requirements, for example, may rule out some systems, which simply do not have the required interfaces; production organization may rule out others, which cannot cover the product spectrum and the type diversity. Cost and space constraints, as well as personnel, may impose different restrictions.

Again, there can be no single right system approach. The following sections will cover a lot of these topics outside actual machine vision. Hopefully this will give some insight into the questions to be answered when making such a decision.

9.4
Integration and Interfaces

The development of technology and market in recent years (see 9.1) clearly shows that machine vision is on its way to become a standard tool of automation technology. On the other hand, the estimate that only 20% of potential applications are yet implemented, indicates that there is still a way to go to make machine vision as common as for example a PLC.

Increasing capabilities of hardware and software will make more applications technically feasible, improving price-performance ratio will help from the economic point of view. However, the ability of integrating this technology into the production environment and the organization will probably be even more important. Together with product complexity and quality requirements, production systems become more complex also; work stations are already crammed with technology, equipped with various measuring and inspection devices in addition to the actual production processes. Production and maintenance personnel has to manage this complexity, has to operate and service all those devices.

Especially smaller companies – or smaller locations of large companies – will not be able to afford several machine vision experts, if at all; to further the use of machine vision systems, it is therefore necessary, among other things, to facilitate integration, operation, and maintenance by nonspecialists.

Take PLCs as an example. Every PLC fulfills a unique task, specialized for its production station, and it is *programmed* to do so by a specialist, a PLC programmer. However, normal production operation as well as many maintenance operations – up to for example a complete re-installation of the system after a hardware failure – can be carried out by nonprogrammers. This requires providing (nonspecialist) end-user personnel with the right information and means of intervention to operate the system, to diagnose and remedy problems, to exchange system components without breaking functionality and to adapt the system within the limits of its original specification.

However, it should not be overlooked that the physics of part surface, light, optics, and sensor introduce an additional complexity which is typically not considered in the area of PLCs. There, work usually starts with a directly usable sensor signal; making that signal usable is typically left to a specialist, unless the sensor is very simple. Here, in machine vision, it is just the point: making the system usable as a whole, including the sensor chain.

Standardization
Before we come to various interfacing requirements of vision systems, a word on standardization. Currently, there is little or no standardization in the field of machine vision beyond video signals. This lack of standardization certainly reflects the comparative youth of the field, but also the structure of the industry. The machine vision market is dominated by companies offering complete solutions, turn-key machine vision systems. Most suppliers make their own software or use a few selected products, far fewer however make their own cameras and frame grabbers, so the primary area of standardization is that of components bought by system integrators on the market.

Internally, these suppliers of course do standardize; we have already mentioned that there are system integrators who use a self-developed application package. And if you will be using machine vision systems extensively – or even buy components and make your own, be it for internal use or for sale – you may also find it worthwhile to standardize in some or all of the areas mentioned here. This will be an effort, to be sure, it may make individual systems sometimes more expensive than necessary; in the long run, however, it will certainly prove valuable, especially when considering maintenance, service and spare part costs over the lifetime of the systems – which can be surprisingly long.

Interfaces
As outlined above, a major factor for the ongoing success of machine vision is its integration into the automation technology environment; integration can be understood as interfacing with other systems; the following sections will take a look at some of the interfaces that play a role in the integration of machine vision systems:

- *mechanical* integration into the production line in Section 9.5;
- *electrical* connections in Section 9.6;
- *information* exchange in Section 9.7;
- *synchronization* aspects in Section 9.8;
- *user interfaces* in Section 9.9.

9.5
Mechanical Interfaces

It may sound trivial, but it is one of the first things to be dealt with: a machine vision system including all its components must find its place(s) inside a production station and must be mounted and fixated there.

Everywhere we find miniaturization, sensors, parts, computers become smaller; this continues in machine vision, of course; simply compare current CCD cameras with those five years ago. However, we cannot miniaturize the laws of optics. A telecentric lens for a part of certain dimensions also needs to have a certain size. Light does not change its way of propagation, working distances have to be kept to capture clear, sharp images with sufficient field of view as well as depth of focus and so on.

So machine vision has its particular demands on the mechanical setup and although a mechanical designer may have knowledge of geometrical optics, we cannot necessarily expect specific knowledge about types and characteristics of particular optical setups, lighting, cameras – much less an ingrained feel of these things – which will lead to a suitable mechanical design. So it is the duty of the machine vision specialist, as well as his advantage, to explain all these things to the mechanical designer, discussing them as early and as often as possible so that the requirements of the vision system are incorporated into the machine design.

Such information is (as usual without any claim to completeness):

- component dimensions (i.e., cameras, optics, lights, computers, always including connectors!) and means of fixation;
- working distances for lights and optics;
- position tolerances with respect to translation as well as rotation;
- forced constraints, for example for telecentric lenses and illumination (see below for an example);
- additional sensor requirements (for example the need for a light barrier as a triggering device; even if wiring and setting up such a sensor is in the responsibility of electrical engineering, the mechanical designer has to provide space and mounting for it);
- additional motion requirements;
- environmental conditions.

9.5.1
Dimensions and Fixation

As with every component to be built into a production station, the first thing the mechanical designer needs to know are the dimensions and the means of fixation.

Speaking of dimensions, it should not be forgotten that vision system components – cameras as well as lighting and processing units – usually have cables connected to them! The connectors have a certain size and the cables have certain minimum bending radii. These are factors the designer has to take into account.

The components used in imaging vary considerably in size and shape. Just look at all the various forms of LED illumination available today or at the vastly different cameras. These differences, unfortunately, continue in the means of fastening the components. Cameras may have screw threads in all possible places and may use all kinds of different screws – which may be impractically small for mounting the cameras reliably in a production environment.

Standardization is an important means of handling this variety. Restricting oneself to a minimum number of components – easier for cameras and optics than for the often highly project-specific lighting – and trying to convince the manufacturer or distributor to provide these components with practical and consistent means of fastening is a good strategy to ease the life of the mechanical designer as well as one's own and that of the maintenance personnel later on.

For example, precise, unvarying, reproducible geometrical relations (firstly between work piece and optics/camera) are of prime importance, especially for gauging applications. Software can support this, for example by providing comparison of a stored calibration image with the current image. Mechanically, accuracy can be improved in such applications by pinning the cameras to ensure a good fixation as well as precise reproducibility of the position. This requires, of course, that manufacturer or distributor equipped the camera accordingly.

9.5.2
Working Distances

Closely related to the dimensions are the working distances. Optics as well as illumination have to be in certain positions relative to the piece to be inspected. The tolerances for these working distances may be rather small, especially in the case of certain optical setups, notably telecentric lenses, or all situations where we have to work with little light and therefore little depth of focus.

The working distances and their tolerances are among the most important things to tell the mechanical designer because they strongly influence the overall design of the system and they are absolutely not obvious to a non-specialist unless he has considerable experience designing visual inspection stations.

9.5.3
Position Tolerances

Most machine vision systems have some means to handle variations of the object position or orientation in the image. However, there are limits to it. Most obviously, of course, the features to be inspected need to be completely visible in the image giving an absolute upper limit to position variations. But the practical limit may be much smaller. For example, the quality of the image may decrease toward the borders due to optical effects; lens distortion typically increases with distance from the lens center, so gauge capability may be lost or require complicated calibration methods. Or the incident angle of the light may change, rendering certain features less distinct, or …

Therefore, the effects of position variations should be estimated and an upper limit given to the mechanical designer – and it should be conservative.

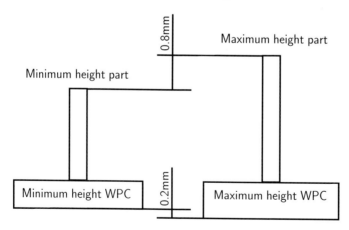

Fig. 9.1 Addition of part and setup tolerances.

Product design may be relevant here also. For example, if a part to be inspected from top has a height tolerance of 0.8 mm and the required resolution leads to an optical system with a depth of focus of just 1 mm, the vertical position tolerance for the mechanical setup is reduced to only 0.2 mm. Rarely does the chance of influencing product design for better "inspectability" present itself, but if it does, use it.

9.5.4
Forced Constraints

Sometimes there are relationships between components that are not obvious for a nonspecialist. These need to be explained and communicated.

Take the example of a system equipped with cameras with telecentric lenses and corresponding telecentric lights. As the optics can only see parallel rays and the lighting only send parallel rays, there is a forced constraint namely that lens and corresponding light always need to be parallel and preferably exactly opposite to each other. Neglecting to inform the mechanical designer about this fact can, and most probably will, result in a design where the opposite partners are not forcibly parallelized; causing a lot of unnecessary adjustment work.

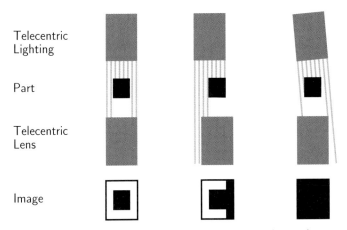

Fig. 9.2 Correctly aligned, translated and rotated telecentric components.

9.5.5
Additional Sensor Requirements

A vision system may require additional sensors, for example to detect the presence of a part in the first place – this signal might go to the PLC which in turn starts the vision system inspection. Or light barriers, used to trigger image capturing of moving objects at a precise point in time, often connected with triggering a strobe light to freeze the motion in the image.

From the mechanical point of view, these are simply additional components that the mechanical designer has to know about, including dimensions and possible restrictions to mounting and position – like maximum signal cable length.

9.5.6
Additional Motion Requirements

There are various reasons why a manufacturing station may require capabilities for additional movements because of a vision system. According to the frequency of the movement, we can distinguish among

Adjustment movements are required only rarely, often only in the first stages of an application. Not everything can be tested or calculated in advance, so some means of adjustment, for example slides or rotating joints for mounting cameras or lighting, may be required during try-out until an optimum position is found. Then the position may be fixed or the adjustment means kept in place for later changes.

Change-over movements can be required when the type of product being inspected is changed. Different part geometries may require not only changes to software procedures or parameters but also changes in, for example, distances or angles between part and optics, or part and lighting, or both. Change-over movements occur outside the area of machine vision also and can be manual or automated.

Sometimes, however, an additional set of sensor equipment (camera, optics, illumination) will be less expensive than a moving axis which requires not only purchasing, but also mounting and – in the automated case – control system programming

Process movements are carried out for every single piece; there are many reasons why process movements may be required. Here are a few examples:

- Line-scan camera capturing always requires moving (be it a translatory or rotatory movement) either the part or the camera – preferably the part to avoid tracking camera cables.
- Insertion sensors – like endoscopes – or lights have to be moved into the part cavity and back out again.
- Space constraints may require lights, optics or cameras or the part to be moved in position for inspecting and removed afterward. For example, the sensor equipment may be in the way of part transportation: if a part has to be inspected in a backlight situation in a

conveyor-belt setup, you can clearly not mount the sensor equipment over the belt in the height of the work piece, as it would have to move *through* them; mounting camera and light on both sides of the conveyor belt may lead to undesired width of the workstation. A typical solution would be to lift the part from the conveyor belt level up to the camera level for the inspection.

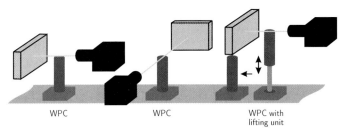

Fig. 9.3 Impossible, too wide and workable arrangement of light and camera with respect to part transportation.

- Working distances are also a frequent reason for moving pieces of equipment. For example, diffuse on-axis lighting typically works best close to the illuminated surface. There, however, it may again be in the way of part transportation.
- Mechanical changes to optical situations. for example, moving ambient light shielding in place for inspection and out of place for transportation of the parts.
- Mechanical autofocus: part tolerances may exceed the optics' depth of focus, which may be overcome by moving part or optics into the optimum position.

9.5.7
Environmental Conditions

Typical cases of environmental conditions which cause mechanical requirements are

Vibrations which may require mechanical decoupling or other measures, for example to protect sensitive computer equipment or to avoid adverse effects on image quality;

Heat which may also cause damage to equipment; possible remedies are additional cooling means or "spatial decoupling," that is putting computers further away from heat sources;

Light shielding comes in two varieties: shielding the vision system against ambient light and shielding the environment against particular light sources.

With the exception of ambient light shielding, these requirements do not differ from other systems where computer equipment is employed in a production environment, so designers should be familiar with this problem. It may be good, however, to point out the possible sensitivities of the system. Especially the effect of vibration on image quality may not be immediately obvious.

Light shielding, on the other hand, is a point which should always be taken into account when discussing the (mechanical) requirements of a vision system.

Ambient light is of course potentially a major disturbance for a machine vision system, which typically depends upon particular and stable illumination conditions. This is especially true if level, direction, or other characteristics of the ambient light vary, as may be the case in a production hall with windows letting in daylight. Light shields, however, are a major obstruction in a production station and may cause additional motion requirements if they have to be moved for transporting the part.

Protecting the environment against the vision system's illumination is basically no different from ambient light shielding as far as the mechanical requirements are concerned. Only the reaons differ. These shields are typically needed because of safety requirements, for example when using lasers or other high-intensity light sources (strobes for example). Infrared or ultraviolet lighting is also a cause of concern as our eyes do not have a natural protection strategy against infrared light like they have against visible light. So when using high-intensity or invisible radiation sources, consult a security engineer to discuss protection requirements.

9.5.8
Reproducibility

Software people sometimes tend to forget considering reproducibility since software is so easy to reproduce perfectly (although software problems sometimes are not ...), but to reproduce mechanical setups precisely, a little extra work and thinking is required.

Mechanical reproducibility is naturally a prime concern in measuring systems where accuracy often directly depends on the mechanical setup, but it is important in every machine vision system. Machines *are* disassembled sometimes, components do fail and have to be exchanged – in all these situations it must be possible to recreate the original setup of lighting and optics to achieve the same image conditions as before, and with as little effort as possible. Prefereably, recreating these conditions should be possible without the help of a machine vision specialist who may not be at hand then.

Of course this also applies to change-over movements. If these are not executed automatically but adjusted manually, appropriate limit stops and indicators are required so that various setups can be easily reconstructed. Preferably, software, either on the PLC or the vision system, should assist the user here by indicating clearly the adjustments to be made and possibly offer means of checking these adjustments – like reference images for checking the camera field of view, for example.

Standardization also helps with reproducibility. It is much easier to recreate a particular optical situation if only a few different components are involved, if these components have unvarying standard fixations and long-term availability. The difficulties of camera fixations have already been mentioned in 9.5.1. The rapid technological development does not help either, as there is a good chance that the exact same camera will not be available when an exchange or a duplicate system is needed. Standard fixations, used consistently and re-fitted on subsequent camera models, help us to meet the end-user's reasonable expectation that after replacing a camera the same image should result as before. The end user, after all, is usually not interested in the camera as such, but in the result the system delivers.

Pinning, as mentioned in Section 9.5.1, is a possible method to create a reproducible setup. It can also be used for retrofitting after finding an optimal setup in try-outs. If the system cannot be built for accurate reproducibility in such a way – for example, because adjustment possibilities must be maintained – , then position and orientation of the components may be recorded to allow for recreating the setup, or a specialized jig may be designed and manufactured as an assembly aid to assist a mechanic in re-adjusting or duplicating the setup.

The above remarks are equally valid, of course, for all parts of an optical setup. Actually, since lenses, especially telecentric lenses, are often considerably larger than modern cameras, the lens is sometimes the main object to fasten, while the camera becomes more of an attachment.

That lenses should be lockable with respect to focus and aperture is a matter of course. That does not help much, however, with exchanging lenses (or duplicating a system which is no different in principle from exchanging a lens). Again, standardization is very helpful since lens sizes and characteristics vary so much even for lenses with basically identical imaging properties.

Light sources also have to be considered. They have a limited life-span – even LEDs do – and they may change their characteristics over time, especially close to the end. Depending on the application, methods to detect this change and to compensate for it may be necessary, for example, automatic monitoring of well-defined image areas for brightness changes and adjustments to the light source power supply can help keep brightness stable. Exchanging the light source can lead to sudden, drastic changes. In addition to brightness changes, changes in wavelength are also very critical, since the exact point of focus changes with the wavelength, so the imaging characteristics may be altered considerably.

In all these cases, software tools can be very helpful to aid the user in recreating a particular optical setup. For example, the system may provide a means to store reference images for various cameras, possibly several per camera to account for different lighting situations, combined with tools to compare the current image with the stored one. Such tools may be markers identifying particular points in the image to compare the field of view, brightness computation in prescribed areas to compare lighting levels, gradient computation to estimate focus quality. Without such tools, comparing live images is basically the only – and not very precise – way of recreating a particular imaging situation.

9.5.9
Gauge Capability

Gauge capability is a difficult topic to place. It touches all areas of a machine vision system that has to perform dimensional measuring: lighting and optics, quality of sensor and signal transmission, accuracy of algorithms; all these are treated in depth in other chapters of this volume.

Of the subjects in this chapter, the most important one for gauging accuracy is the mechanics of the system. It has to guarantee the stable conditions required for precise and accurate measurements - just as with nonoptical measuring systems.

The automotive industry has long followed strict guidelines concerning the capability of measuring systems to deliver precise and accurate results[3], one of the most prominent results of this process being the Guide to the Expression of Uncertainty in Measurement [6].

We will leave the theoretical aspects of gauge capability to these guidelines as well as to works on statistical process control and the like. Suffice to say that gauge capability is a much stricter requirement than simply having sufficient resolution to measure a value with a given tolerance. And these requirements affect all parts of a machine vision system.

> 3) The distinction possibly not being obvious to non-native speakers of the English language: precision denotes the capability of a measuring system to produce stable measurement results (for the same part) within a narrow range of variation; accuracy denotes the capability of delivering a result that is close to the correct value.

9.5 Mechanical Interfaces

We can see the effect on the mechanics of a system using a simple example. Suppose we had a vision system intended to measure the diameter of a cylindrical part, a fairly easy task, or so it seems.

We will backlight the part, using all the available tools for accurate imaging, namely telecentric lighting, telecentric optics and so on. Then we could have the system measure the width of the part's shadow. However, the diameter has to be measured parallel to the cylinder's base. What, if the cylinder is not precisely horizontal? Can we simply ignore the inclination and measure horizontally anyway?

Let us assume a tolerance of ± 0.04 mm for the nominal diameter of 10 mm and an evenly distributed stochastic measuring error with an amplitude of 0.006 mm. This gauge would be capable, according to evaluation methods common in the automotive industry.

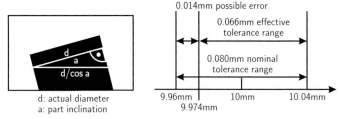

Fig. 9.4 Effect of part inclination on measurement value and tolerance range.

Now what would the effect of an inclination be. As Fig. 9.4 shows, the horizontally measured diameter is $d / \cos(\alpha)$, the actual diameter divided by the cosine of the inclination. At $3°$ of inclination this amounts to a measuring error of 0.14 %, corresponding to 14 µm on the nominal length of 10 mm. Although considerably smaller than the part tolerance of 40 µm, this error does, however, significantly reduce the *usable* tolerance range: if inclinations up to $3°$ have to be tolerated, a part measured at 9.973 mm has to be scrapped, since its actual value could be 9.959 mm, just outside of tolerance.

Although this particular error will reduce the tolerance range only on one side – as it will always result in increasing the measurement value – it still reduces the usable tolerance range from 0.08 mm to 0.066 mm, only 82.5% of its original value. An evenly distributed random inclination with an amplitude of $2°$ would cause the system to lose its gauge capability according to common evaluation methods in the automotive industry. If the construction of the work piece and the receptacle is such that the part falls with equal probability into one of two orientations, either vertical or inclined, the effect is even worse. In that constellation, an amplitude of $1.5°$ is already sufficient to make the system

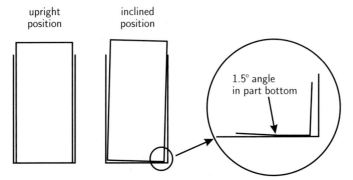

Fig. 9.5 Part with two possible inclinations, due to construction of part and receptacle.

lose its gauge capability without any other source of errors[4]. The mechanical inaccuracy therefore has a severe effect on the capability of the system.

This example clearly illustrates the importance of outside factors for the capabilities of a vision system. Especially for systems required to be gauge capable, the quality of the mechanical handling of the parts is very often a decisive factor. Expensive measuring systems with all kinds of fancy sensor equipment and software have been known to fail due to an unsuitable mechanical design.

9.6
Electrical Interfaces

Like most other automation systems of some complexity, machine vision systems require

- power supply connections for processing units and other vision system components;
- "internal" data connections between processing units and other vision system components;
- "external" data connections to the automation environment;

Note that "data" does not necessarily imply "digital" in this context.

Data connections typically have two aspects or dimensions: an informational one, describing what information is exchanged, what the timing behavior is, what the communication protocol looks like and so on; and an electrical

---

[4]) Of course, the inclination could be measured and used for compensation. However, this measurement would be subject to error also. Better, therefore, to prevent the problem in the first place.

one which is concerned with the actual wiring, voltages and the like. There is little intrinsically specific about the electrical interfaces of machine vision systems compared to other automation components, nevertheless, some points are worth mentioning.

9.6.1 Wiring and Movement

If possible, avoid moving cameras and other vision system parts; if not possible, design and wire in such a way that there is as little stress on the connectors as possible. Many of the connectors used today in vision systems are not really designed for regular mechanical strain; just think of Ethernet RJ45 connectors common on DataMatrix scanners and some intelligent cameras or of FireWire connectors. Screw-fixed connectors, as they are used for CameraLink, can certainly tolerate some movement, but they are probably not equal to the stress of – say – robot movements. Appropriate cables and mechanisms for this type of application will be a major cost factor.

This applies also to line-scan camera applications, where a relative motion between part and camera is always required. Here, also, the part should be moving – which it often does anyway, the motion just needs to be controlled to create good image capturing conditions.

9.6.2 Power Supply

Machine vision systems use the same types of power supply as other electrical devices used in automation technology. In that respect, eletrical interfaces are less varied and easier to handle than the mechanical ones. Typical power supply characteristics are 230 V AC (or 110, depending on the country) and 24 V DC.

24 V DC is usually readily available in a manufacturing station; it is less dangerous than higher voltage AC and therefore less heavily regulated. However, the power consumption of high-performance PCs of around 400 W would require impractical current ratings for the cabling. Therefore, 24 V DC is mostly limited to systems of the vision sensor or intelligent camera type, although it is also available in some types of PCs. These are typically less powerful, often fan-less, systems that, due to their lower power consumption and therefore heat-generation can be operated in closed housings or inside switch cabinets. Very demanding applications, however, will typically be outside the performance range of these systems.

Whether 24 V DC or 230 V AC, a system running a modern general purpose operating system such as Windows (2000/XP), Linux and others, in general requires a controlled system shutdown to avoid data corruption. Production stations, however, are often powered down by their mains switch without regard to specific requirements of individual devices. Therefore, these systems will usually need an uninterruptible power supply (UPS) to either help the system over a short power-down or – more frequently – supply it with power during a controlled shutdown initiated by the UPS's controlling software.

A UPS can be integrated within the system. The main disadvantage is that the electrical storage to keep the system running for some time without external supply is rather limited and difficult to exchange in the case of failure. The advantage is that control of the UPS can be encapsulated in the system and there is a very simple interface to the outside, namely the usual power supply connector.

A UPS can also be built into the switch cabinet. The advantage is that it can be larger, last longer and be more readily accessible. The disadvantage is additional cabling as the UPS requires not only a power supply connection but also a communication connection to signal to the vision system when it has to initiate shutdown due to loss of power.

Also in use are central UPSs, possibly backed by emergency generators, for an entire production line or plant, foregoing communication connections to shutdown systems and keeping all systems permanently running.

Other power consuming components are lighting and cameras. They do not require shutdown, hence they can in principle be connected like any other electrical component. They may however require higher quality of the supplied power than many other devices. This should be checked with the manufacturer.

It is also common that the central unit of the vision system supplies cameras and/or lighting with power itself, enabling the system, for example, to switch and regulate light sources.

Cameras can be supplied with power through their signal cables. This is the case with modern analog cameras using multiwire cables, often with standardized connectors, and the popular digital camera interfaces USB 2.0 and IEEE1394. The CameraLink interface of high-end cameras, however, does not have this capability, so these cameras require an additional power supply.

9.6.3
Internal Data Connections

By internal data connections we mean those between components of the vision system itself (note that "internal" is a purely logical term here, such a connection may well run over the standard factory network and thus be quite external to the vision system physically).

The first such connection that comes to mind is naturally that between processing unit and cameras (excluding, of course, systems of the smart camera type where this connection is internal and hence of no further concern for machine building). It is certainly unnecessary to elaborate on camera/computer interfaces after all that has already been said in previous chapters – at least about the informational aspect. What should be kept in mind, however, is that the electricity in these wires has to carry enormous amounts of data. A standard analog video signal carries almost 90 Mbit/s information[5] at a typical frequency of 15 MHz, close to the data rate of Fast Ethernet; the IEEE1394a bus has a maximum bandwidth of 400 Mbit/s at a strobe frequency of approximately 49 MHz, USB uses similar values; CameraLink is available in various configurations, supporting 1.2 to 3.5 Gbit/s at 85 MHz. Signal lines of such capacity are naturally susceptible to interference.

We also have to keep in mind that – in comparison to the 24 V digital I/O standard of automation systems – the signals have fairly low amplitudes: analog video has a signal amplitude of 700 mV, CameraLink (as a Low Voltage Differential Signal application) one of 350 mV. Electromagnetic noise can have a severe detrimental effect on the signals.

It is therefore not recommended to run video cables parallel to power cables, to cables carrying fast communication (like bus cables) or close to strong electrical devices, like motors. One will not often have to go so far as to put all camera cables into stainless steel tubes, but sometimes there are environments requiring such extreme shielding measures. In any case, the quality of cables, connectors, shields and the connections between them is very important for lines of such a bandwidth.

It is often necessary to control lighting during image processing. This can be as simple as switching certain lights off, others on between capturing successive images; it can also mean controlling brightness levels and LED patterns online. Traditionally – and this remains a good, simple and reliable solution – digital I/O connections have been used to switch brightness levels and/or preprogrammed lighting patterns; or lighting controllers have been used with serial interfaces. Of course these can also be connected to field buses using appropriate converters, for example bus terminals to set analog voltages for controlling brightness levels.

Recently, lighting equipment becomes available using USB or Ethernet interfaces to program and control the illumination patterns. The amount of data required is naturally much smaller than for a camera connection, posing no challenge to modern bus systems. The interesting point, however, is using the same bus for more and more devices and applications, requiring more and

---

5) Assuming 25 full frames of 768 × 576 pixels at 8 bit depth; of course, this is not directly comparable to digital data transmission.

more addresses, increasing bandwidth, and affecting realtime behavior. This is as true for Ethernet as it is for field buses.

We should certainly not forget connections between individual vision systems; this is very common for certain types of smart cameras that communicate with each other or with a specialized central unit. These connections typically use some proprietary protocol embedded in TCP/IP over Ethernet connections – again increasing the required bandwidth and number of addresses.

So as devices become more and more flexible and use of digital (bus) interfaces more and more common, requirements on these interfaces also grow, and it remains to be seen what the typical bus topology of an automated manufacturing cell with a machine vision system will look like in the future – for example with GigabitEthernet for camera signals, some kind of realtime Ethernet for control and data connections, USB for slower devices, converters for legacy devices...

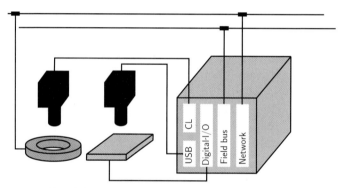

Fig. 9.6 Possible system setup with USB and CameraLink camera, Ethernet- and digital I/O controlled lighting and field bus connection.

9.6.4
External Data Connections

Finally, there are data connections between the machine vision system and its environment. Typically, these will connect the processing unit of the machine vision system with superior systems, like the station PLC or a supervisory level computer.

Traditionally, such connections have been made by digital I/O and/or serial interface and these are still used frequently. Field bus connections are becoming more and more common, barely a vision system (software or product) that does not support at least one field bus flavor. And, of course, Ethernet is making progress in this area as in all other areas of automation technology.

Actually, especially in the area of smart cameras and vision sensors, there is a growing number of devices whose *only* connection to the outside world is Ethernet-TCP/IP; parameters are set via TCP/IP, results and images are transmitted via TCP/IP.

All these connections are standard to automation technology and as far as electrics and wiring and so on are concerned no different than for any other device – including sensitivity to electromagnetic noise, of course. The interesting part are the communication protocols, the information interfaces, discussed in Section 9.7.

9.7
Information Interfaces

A vision system needs to exchange data with its environment – and be it just receiving a trigger and setting a yes/no result. Usually it is a lot more complicated. For this, a vision system needs data interfaces. In the following, some information on the two closely intertwined topics of data and interfaces is presented – unfortunately the medium of writing does allow for a sequential representation only.

There are various developments that increase the amount and importance of information exchange between vision systems and other systems. Among these is the spread of small cooperating units, for example networked smart cameras, increasing demands on the availability of actual feature values – instead of just yes/no decisions – for statistical process control, tighter integration of vision systems with the automation environment, for example with the use of common type data, traceability requirements and so on. Therefore, we will deal in some length with information exchange interfaces. What this section can only hint at, however, is the diversity – not to say, disorder – to be found in this area. Therefore, a few words on standardization first.

9.7.1
Interfaces and Standardization

Many machine vision systems or software packages offer support for various interfaces, including all those mentioned below: digital I/O, serial interface, field bus, network, file. This does not necessarily mean support for a standardized protocol or format – or any protocol at all. Often it is reduced to the system having some means of accessing corresponding hardware, like selected digital I/O or field bus boards, or operating system functions for data transfer.

This leaves protocol questions in the hands of the user. For example, the communication partner – usually a PLC – can be programmed to understand a proprietary protocol of the vision system manufacturer – causing expenses for every type of system supported. On the other hand, having the manufacturer of the system tailor its communication to the requirements of PLC and automation environment may not be free of charge either.

The fact that every machine vision system is essentially a piece of special machinery in itself, producing specific results and requiring specific parameters, will make it difficult to eliminate communication completely as a source of custom programming in vision system applications. The existence and widespread support of standard protocols would nevertheless go a long way toward that end.

9.7.2
Traceability

Many machine vision systems exist solely because of the traceability requirements on certain products; the recent rise in DataMatrix code usage – and correspondingly DataMatrix code readers – is partly due to the need or desire to be able to trace back every single part to its origins, to the place, time, shift, even the individual worker, who produced it.

Traceability may be a safety requirement on the product or it may be useful for reasons of production organization. For example, subsequent manufacturing steps may need data from previous steps, which of course has to be assignable to the individual work piece.

If the machine vision system itself stores data that needs to be connected to the individual work piece, for example, images that are subject to compulsory archiving, then it needs to do so in a traceability-aware manner. A typical method would be to use the serial number of the work piece as the (base) name for the image(s). The serial number itself is very often nowadays marked on the part in the form of a DataMatrix code. This code can be read by the machine vision system itself or it can be provided by a DataMatrix code reader. In the latter case, either the vision system needs a communication conection to the scanner or the information is handled through the control system, which keeps this essential information central to the workstation.

If the vision system itself is not capable of working in such a traceability-aware manner, then outside means have to be designed, like removing and renaming the data created by the vision system for the individual work piece – which may be inelegant and possibly expensive.

9.7.3
Types of Data and Data Transport

On one side, we can distinguish different types of data to be exchanged:
- control signals
- result/parameter data
- mass data (typically images)

On the other side, there are different ways of transporting data:
- digital I/O
- field bus (e. g. PROFIBUS, CAN-Bus, Interbus-S ...)
- serial interfaces
- network (e. g. Ethernet)
- files (images, statistical data files ...)

Note that all of these interfaces with the exception of digital I/O are actually message-based, that is, distinct data packages are serialized into bit sequences, transmitted over the signal line and deserialized again (that does also hold for files which are essentially a sequence of bytes that has to be interpreted). Field buses sometimes hide this fact from the user by presenting themselves as quasistatic I/O connections, but it should be kept in mind that they actually function in this way.

9.7.4
Control Signals

Control signals here denote signals that remote-control the run-time behavior of the vision system. Typical input and output signals (from the point of view of the control system[6]) are:

- Output signals: *Start*, (for a normal inspection run) *Change-over* (to change inspection programs), *Adjust*, *Calibrate*[7], *Reset* ...

---

6) Agreeing on a single way of looking at the data flow is very helpful when designing a communication, and as the control system is typically the master in an automation system and has to handle the data flow with a lot of peripheral systems, it is frequently easier to assume the point of view of the control system programmer.

7) As defined in [7], calibration is a "set of operations that establish ... the relationship between the values of quantities indicated by a ... measuring system ... and the corresponding values realised by standards," whereas adjustment is the "operation of bringing a measuring instrument into a state of performance suitable for its use". The term calibration is, however, frequently used to mean adjustment.

- Input signals: *Ready, Busy, Done, OK, Not OK, ...*

Of course, additional signals can be defined in a given system as circumstances required. A doubling of reciprocal signals like *OK* and *Not OK* is typically required when using interfaces – like digital I/O – without data consistency. One, and only one, of the signals must be TRUE, otherwise something is wrong, for example a wire interruption.

On message-based interfaces, these control signals can be represented in different ways. Commands, from the PLC to the vision system, may be encoded as numbers. "1," for example, may indicate a start signal, "2" a changeover command and so on. This method automatically avoids sending more than one command at a time.

The system status of the vision system, however, is frequently encoded in the form of individual bits in a status byte or word, because the system may have several status at once. For example it can at the same time have the status Done and OK, indicating that it has completed the most recent task and did so with an OK result.

9.7.5
Result and Parameter Data

Results, transferred from the vision system to the environment, and parameters, received by the vision system from the environment, are typically alphanumerical data of considerable, but not exactly overwhelming size. Results may be, for example, individual measurement values or strings identified by OCR or bar code/DataMatrix code functions. Parameters may be nominal values for measurement evaluation, search area coordinates, part identification numbers for storing traceability data and so forth.

The size of these data, be it results or parameters, varies typically between a few bytes and a few hundred bytes. In very simple cases – for example: a single small integer number – digital I/O is suitable for transferring such data. It will be rather more typical to use message-based interfaces.

It should not be overlooked that in many cases file transfer is a perfectly suitable method to exchange such data. This is especially true for result data which is to be archived for long-term traceability or statistical analysis and for type-dependent parameters.

Type-dependent parameters play an important role for efficiently adapting test programs for different types of a product. Often, the structure of the tests is completely or mostly identical for different types of the same product, but certain parameters differ. For example, nominal measurement values may vary, colors or text markings may be different and so forth. Having separate test programs for each type, differing only in a few parameter values would not be very efficient and difficult for maintenance.

A possible solution is of course to have the PLC transmit the pertaining values to the vision system in connection with a changeover to the type in question. Or the parameters may be embedded in the data describing the type, which is often available in the form of a file anyway, and have the vision system read the relevant parameters directly from the type data[8] file. An example of such a setup is presented in Section 9.10.6.

9.7.6
Mass Data

Mass data, in the case of a vision system, are typically images, be it raw camera images or "decorated" images showing the results of processing steps, like segmented objects, measuring lines, values, and the like.

Some systems use serial interfaces for this kind of data, even field bus is possible theoretically, but the typical interface is the network. Especially smart camera systems nowadays often make their live images available only through an Ethernet connection, thus saving the cost and size requirements for a VGA interface. It remains to be seen what the impact of new types of serial interfaces, like USB, will be in this area.

Note that Ethernet, even in connection with TCP/IP, can mean very different things. See the following section on network connections for various methods to access and transmit mass data over Ethernet.

9.7.7
Digital I/O

This type of interface provides a limited number of discrete signal lines; typical numbers are four for small systems, like certain types of smart cameras, 16 or 32 for PC interface cards. A high voltage level on a line is interpreted as 1 or TRUE, a low level as 0 or FALSE, so every line can carry a single bit of information. Digital I/O interfaces usually operate either in the 5 V TTL range or in the 24 V range. In automation technology, the 24 V range is usually preferred as it offers much more robust signal transmission in the electromagnetically noisy production environment.

It is in principle possible, but cumbersome and rarely done, to use digital I/O for actual data transmission. The typical use of digital I/O is for control signals. There, its strength lies in the speed of practically instantaneous transmission and in the static availability of status signals. Expensive cabling – one line per bit – and the risk of wire breaks are equally obvious weaknesses, making digital I/O less and less desirable in modern production environments.

---

8) Another common term for this data in the PLC world is "recipes," derived from its typical use in the chemical industry.

As for protocol, there is no standard assignment of signals to the digital I/O channels; this is up to the system manufacturer and/or the user to define – and to deal with, for example with custom programming on the PLC side when using a device with limited flexibility.

9.7.8
Field Bus

Field bus systems, such as PROFIBUS, Interbus-S, CAN Bus, and others, are increasingly supplanting digital I/O in the automation environment. They fill a middle ground between digital I/O and networks: they are message-based bus systems, like networks; on the other hand they have a deterministic time behavior and can often be used like large digital I/Os.

For example, the DP[9] communication profile of PROFIBUS represents the contents of a PROFIBUS message as a process image, which is automatically updated between master (usually the PLC) and slave (any other device) within each bus cycle. The bus cycle time differs between installations, depending on the data rate of the bus and the size of the process images to be exchanged, but it is fixed and guaranteed for this system – at least, as long as the number of devices using asynchronous messages does not get out of hand. The automatic update lets the process image appear to applications much like the static signal levels of digital I/O.

The possible size of the process images (small devices may use just a few bytes, but 200 bytes and more are possible) allows for the exchange of result and parameter data, while the quasi-static, deterministic behavior is well suited for control signals.

Though differing considerable in detail, field bus systems combine real-time behavior with message sizes far exceeding the width of digital I/O. Built-in data consistency and a much better ratio of bytes to wires are further advantages.

As with digital I/O, the protocol, that is, the assignment of bits and bytes in the process image to control signals and data, is the system supplier's and/or the user's to define and handle.

9.7.9
Serial Interfaces

The classical RS232 serial interface as a means of runtime communication with machine vision systems is more or less dying out. It is still used for the integration of certain sensors, like DataMatrix or bar code scanners, and for the configuration of cameras, but is gradually replaced by either field bus or net-

---

9) Decentralized peripherals.

work connections. It remains to be seen what the impact of new interfaces like USB – already well established in signal acquisition and metrology – will be in this area.

9.7.10
Network

In this context, network practically always means Ethernet as the hardware basis and TCP/IP as the protocol family.

Ethernet–TCP/IP has some distinct advantages over field bus systems, especially that it handles large amounts of data very well and is comparably very fast. Speed alone, however, can be completely useless without a defined time behavior. Therefore, many efforts have been underway for some time to impress some kind of realtime behavior upon the Ethernet. There is no space here to detail any of them; suffice to say that there is certainly the possiblity, even probability, that there will be automation systems based on Ethernet as the sole communication media.

Currently, the most prominent use of Ethernet–TCP/IP by machine vision systems is to transfer mass data, that is: images, although there is a number of systems which are completely controlled via Ethernet, mostly smart sensors, DataMatrix scanners and the like.

Various methods are used to make (live) images available over the network. Sometimes, especially with smart sensor systems, proprietary protocols are used for which the manufacturer provides decoding software – for example browser applets or ActiveX controls. FTP download is used as well as built-in web servers providing the image via HTTP directly to a standard browser.

The web server paradigm has the advantage that it allows for easy handling of different kinds of data, not only images, without having to install proprietary software and is prevalent in a lot of other devices – network printers, for example, are capable of delivering status data via HTTP as web pages. This method thus enables easy access to various systems using only a familiar web browser.

For data communication with a PLC, however, text-based HTTP is not necessarily the best option, as handling text is not a particular strength of PLCs. So for TCP/IP communication with a PLC, a (alas, proprietary) binary protocol is probably better suited.

Even so, handling images through a PLC is not very useful and could be severely detrimental to the performance of the PLC (having an HMI common to vision system and PLC display the images is quite a different matter, we will come to that in Section 9.9).

9.7.11
Files

This section may at first not seem to fit here really well. Files are not what we commonly understand by an interface. In fact, they are on a different hierarchical level, since they may, for example, be transported over an Ethernet-TCP/IP connection in the case of a file on a network share. Nevertheless, files are an important method of transferring data. We have already mentioned result and type data files for storing statistical or traceability information as well as information describing characteristics of product types.

Another very important use of files in machine vision system is of course to store images. The capability of storing "error images," that is, images of parts inspected with a "not OK" result is very helpful for the efficient optimization of machine vision systems as well as for the documentation of production problems. Error images enable us to reconstruct the chain of processing that led the vision system to the "not OK" decision so that we can judge whether the system was right or needs to be optimized.

Of course there are also systems that store not only error images but every single image they capture. This is often done in a startup phase where one wants to collect as much data as possible to have a basis for setting up the system. But it may also be necessary for safety-critical parts to store every single image and to mark the image in such a way that it can later be associated with the specific part being inspected, for example, by deriving the image file name from a part identification number or by embedding binary tags with the part identification in an image file with a suitable format to make manipulation more difficult.

9.7.12
Time and Integrity Considerations

Time and status integrity are important factors for the way data, especially control signals, is exchanged.

Take the handling of *Start* and *Ready* signals, for example. If the system is to operate with maximum speed, a practical method is to have the system react to the status of the *Start* signal and forego actual handshaking. That is, as long as the *Start* signal is on a HIGH level, the system will immediately restart itself. This method is often used in continuous motion production (see Section 9.8.2) where the actual start is given by the camera trigger. On the other hand, this gives the PLC very little control about the actual status of the vision system.

When more control over the temporal sequence is desired, for example to be able to determine whether the system actually did check every single part, then a rigid handshaking is usually employed where the system reacts on signal edges or state changes and all state transitions are signalled to and checked by the PLC. Of course, this takes some time as every handshake requires at least one PLC cycle.

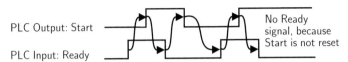

Fig. 9.7 Handshaking scheme requiring reciprocal signal changes.

As far as status data are concerned, there is an important difference between (quasi-)static (see Sections 9.7.7 and 9.7.8) interfaces on the one hand, and message-based (Ethernet, serial interface) communication on the other hand.

On a (quasi-)static interface, the PLC can get the status data at any time. A digital I/O line has its level, independent of what the vision system is currently executing; on the PROFIBUS, as an example for field bus systems, the bus logic itself takes care of the permanent exchange of status data, so again the PLC has permanent access to the status.

On a (pure) message-based interface, this is not possible. There the status has to be explicitly sent by the vision system – either automatically when the status changes, cyclically or explicitly upon request. This means that the vision system has to be able to receive a request and to communicate its status at any given time. This may seem a matter of course but is not, actually, as it requires multitasking of some kind.

A program running under the Windows operating system, for example, has at least one thread of execution, and, by default, exactly one. If this thread is busy, for example computing a filtering operation, it will not be able to react to communication requests. The typical solution is to open another thread in the program which listens permanently for communication requests and answers them. This thread will need access to status or possibly result data computed by the main thread. These data need to be locked against simultaneous access by the two threads lest one of them changes the data while the other is trying to retrieve it.

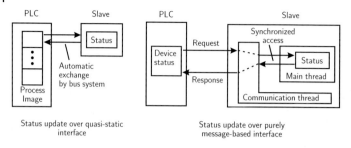

Fig. 9.8 Status data exchange methods.

Of course, these are standard problems of multithreaded programming but it is surprising how difficult it can be in practice to ensure data integrity and correct sequencing at all times. There are authorities in the field of programming who take the view that today's programming languages and methods are not capable of producing multithreaded programs that are correct under all circumstances [10], and it is definitely not a trivial task.

Other types of systems, like smart cameras, must meet the same requirements, of course, but these often have dedicated communication processors which can simplify the handling of such issues.

9.8
Temporal Interfaces

The term *temporal interfaces* is perhaps a little surprising at first. What does it mean? In Section 9.2.2 production systems are distinguished by the way in which the product moves through the production system as discrete or continous; these are essentially temporal categories. In both categories, speed can of course vary considerably: 10 units/s are quite normal for bottling systems in the beverage or pharmaceutical industry, whereas in the production of discrete, complex, products 3 s and more per unit are rather more typical. In continuous production, we have 20–30 m/s for paper production, 2-3 m/s for flat steel mill entry speed, over 100 m/s for steel rod finishing speeds. These differences naturally result in different requirements on machine vision systems and different technical solutions.

Much depends on the way production itself is carried out through time, therefore the following is organized by different types of production.

9.8.1
Discrete Motion Production

By the term "discrete motion production" we mean here types of production where individual parts are produced and – most importantly – handled individually. This can be either job-shop, batch or mass production, the salient point is that the part handling and the time constraints are such that each part can be presented separately to the camera system and typically at rest. This mode of production is often found for complicated parts that need to be assembled from other parts or sub-assemblies, for example in the automotive supplier industry or in the production of electrical devices. In contrast, simple parts, like screws or washers, are usually produced as bulk goods, not handled individually.

Cycle times for these parts can be in the area of minutes for very large parts down to seconds. Below a cycle time of somewhat less than one second there is a practical limit due to the time required for the separate handling of the parts. Below that limit we reach the area of continuous motion discussed in Section 9.8.2.

From a temporal point of view, this type of production sounds comparatively simple in contrast to parts that are in continuous, fast motion or even endless material. But it does have its own difficulties.

First of all, cycle time can be quite misleading. In a three-second production – meaning one part every 3 s – the time actually available for inspection can be much shorter. For a work-piece carrier system in which the parts need to be lifted out of the work-piece carrier for inspection and replaced afterward, the 3 s may shrink to something between 0.7 and 1.5 s, depending on how the exact part handling is designed.

Also, mechanical movements need damping periods. A part is not immediately at rest after the movement has been technically stopped. Capturing the part during that time may introduce motion blur and other image artefacts due to vibration of the part. Sometimes this is difficult to calculate or even estimate in advance putting additional time pressure on the image processing once the required lengths of the decay periods is clear.

A typical method to gain some time is to inform the environment – that is, usually, the station control system in that case – that all required images have been taken. The control system is then free to start moving the part while the vision system is still working on calculating the result. So the vision system has time to finish its task until the next part comes in, instead of just until the time when the present part has to be moved out of the inspection position.

This results in a rather typical sequence of events, which are, from the point of view of the control system:

1. Carry out movement (e.g., move work piece carrier in position, lift part from carrier, etc.);

2. Wait for stable conditions (e.g., wait for vibrations to decay);

3. Start capturing;

4. Wait for end of capturing;

5. Carry out movement (e.g., move part out of the capturing position).

Obviously, this requires more communication between the vision system and the control system than just setting a start signal and retrieving a final result. At the very least, some handshaking is required to ensure that the image capturing is complete before the part is moved.

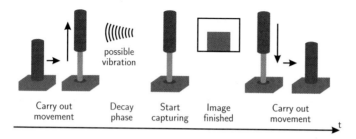

Fig. 9.9 Discrete motion production phases for image capturing.

It is impossible to list all effects that can occur and all considerations that may be required. For example, a vision system in a station with an indexing table hosting several production processes suddenly exhibited a marked increase in false alarms; investigation of error images showed apparent unfocusing, but with a preferred direction. It turned out that this was motion blur, caused by a recent cycle-time optimization of the entire station, which had placed a powerful press-in process so close in time to the image capturing that sometimes the vibration created by that process had not yet decayed when the image was captured.

This example serves to illustrate that the concentration of many, often complicated processes in the deceptively slow unit production stations can result in complicated temporal relationships; the individual stations of fast continuous motion systems typically contain less and also less complicated individual processes. Is is also a good example for the importance of being able to reconstruct inspection errors using error images. Without these, it would have been a lot more difficult to identify that particular problem.

9.8.2
Continuous Motion Production

In this type of production, discrete parts are produced – in contrast to continuous flow production of "endless" materials – but the production is too fast to stop individual parts for capturing (or it is simply not desired to do so). This means, parts have to be captured in motion. A typical example are systems in the beverage industries where we have a constant, never-stopping stream of bottles passing the camera.

Timing conditions are completely different from those of discrete motion production. On the one hand, there will never be a part at rest in front of the camera. On the other hand, the time-consuming cycle illustrated above of starting and stopping movements is eliminated. Nevertheless, continuous motion systems will usually be a lot faster than discrete motion systems.

First of all, though, there is the issue of capturing a sharp image of a single part in the first place. And there we already have the two aspects of this task: *single* part and *sharp* image.

To capture a single part, or more accurately, a particular part, namely the next to be inspected, image capturing must take place in a very narrow time window. Suppose we have a system for checking the fill level and cap presence in bottle necks. The bottles may have a diameter of 80 mm with bottle necks of 20 mm diameter. The system processes five bottles a second which means that the bottles are moving with a speed of 400 mm/s.

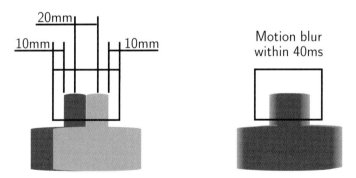

Fig. 9.10 Capturing window and motion blur for moving bottles.

As Fig. 9.10 shows, a camera field of view of 60 mm and a required distance between bottle neck and image border of at least 10 mm leave a range of 20 mm for the position of the bottle at the moment of image capturing. This corresponds to a time window of 0.05 s.

A typical solution is to use some kind of simple sensor, for example a light barrier, to detect the part, for example when it enters the field of view, and use this sensor signal to trigger the image capturing – directly on the camera or through a frame grabber. The camera should be triggered, when the bottle is at the center of the image. If the sensor detects the bottle neck when it enters the camera's field of view, the bottle has to travel 40 mm to the ideal capturing position – or 0.1 s. It will be in the prescribed capturing area for the 0.05 s time window already mentioned, so there has to be a circuitry which delays the trigger by 0.1 ± 0.025 s – or somewhat less depending on the speed with which the camera reacts to the trigger.

So triggering solves the first task, acquiring an image of a particular, single part. But of course, the movement has an additional effect: it blurs the image. Using Eq. (9.1), we can compute motion blur in mm, using Eqs. (9.2) or (9.3) in fractions of image pixels.

$$b_{mm} = v \Delta t \tag{9.1}$$

$$b_{pix} = \frac{v}{res} \Delta t \tag{9.2}$$

$$= \frac{v \cdot pix}{fov} \Delta t \tag{9.3}$$

where

b_{mm} is the motion blur in mm;

b_{pix} is the motion blur in fractions of an image pixel;

v is the velocity of the part moving past the camera in mm/s;

res is the geometrical resolution of the image, in other words, the size of a pixel in real-world coordinates in mm/pixel, in the direction of movement;

pix is the number of pixels along a given coordinate direction;

fov is the size of the field of view along the same coordinate direction in mm;

Δt is the exposure time in second;

Note that the last relation only holds if the part moves along the coordinate direction, in which the number of pixels and the size of the field of view were taken.

For a standard video camera with normal exposure time (40 ms) and image size (768 pixels horizontal) and a horizontal field of view of 60 mm, we have a motion blur of 16 mm or 204.8 pixels; this is clearly out of the question.

It is debatable, how small the motion blur should be to have no impact on the image processing and it is, in fact, dependent on the application as different algorithms will react differently to motion blur. As a first orientation, Eqs.

(9.4) and (9.5) give the exposure time that will create a motion blur exactly the size of an image pixel. They are derived from Eqs (9.2) and (9.3), respectively.

$$\exp_{pix} = \frac{res}{v} \qquad (9.4)$$

$$= \frac{fov}{v \cdot pix} \qquad (9.5)$$

With the values from above, we find that an exposure time of 0.195 ms gives a motion blur of one pixel. With a shutter speed of 1/10000, that is, an exposure time of 0.1 ms we should be fairly on the safe side for a fill level and cap presence check.

Of course, at such exposure times, little light reaches the sensor resulting in poor contrast, poor depth of focus and all the other disadvantages of low light conditions. To get a sufficient amount of light into the image during this short time span, strobe lights are typically used. The trigger signal can be used for both the camera and the strobe, possibly with different delays.

By the way, the quick light pulses that can be achieved with strobe lights makes the shuttering actually unnecessary as far as the reduction of motion blur is concerned. This is the strobe effect well known from either physical experiments or visiting discotheques: the motion appears frozen. Shutters are nevertheless frequently used to reduce the influence ambient light might have if the sensor were exposed for longer periods of time. Of course using shutter and strobe together requires increased precision of the triggering delays as the strobe must fall precisely into the shutter interval.

We see that there is considerable additional logic, circuitry and wiring expenses connected with a continuous motion system. Therefore, some companies offer ready-made "trigger boxes" where you can connect cameras, strobe lights, and also adjust delay times.

Finally, there is the question of processing speed. In our example, we have five bottles per second. The system thus has at most 200 ms for delivering the result after the bottle has entered its field of view. Actually it will probably be a little less since the superior systems may require some time to react. The available processing time is further reduced by the time required for image transmission (capturing is almost negligible when shuttered).

Figure 9.11 shows that less than 160 ms – since superior systems will need some reaction time and the vision system needs to prepare itself for the next trigger in time – are available for processing the image and computing results. Depending on what has to be checked this can be ample time, barely sufficient or simply too short.

There is one frequently used device to win some time back from the process, namely doing image capturing in parallel to the running computation. Figure 9.12 shows that now there are 200 ms for processing one image (again, a little

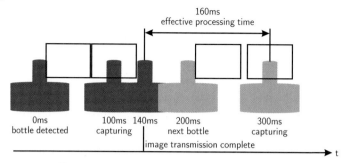

Fig. 9.11 Processing time with computation and capturing in sequence.

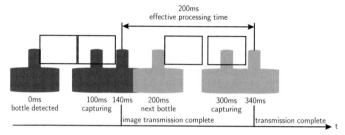

Fig. 9.12 Processing time with computation parallel to capturing.

less for overhead). In effect, the time for the image transmission has been recovered from the process since during the transmission process the system can go on processing the *previous* image. And that is important when using this method: it always processes the image of the previous part, there is a one-part offset of the result to the image trigger. This is not particularly difficult to handle, but will have to be kept in mind.

9.8.3
Line-Scan Processing

Line-scan camera systems require a relative motion between the object and the camera since the camera only captures a single line. This entails particular characteristics of the time behavior. We can distinguish three different types of line-scan processing [4]:

line-wise processing immediately evaluates the current line and makes a statement on that line; this can be seen as a kind of sophisticated light barrier, for example to monitor the width of elongated parts or continuous materials;

piece-wise processing scans the surface of a single work-piece (or a surface section of a material) into a rectangular image which is then processed as usual (apart from the fact that it is typically a lot larger than matrix camera images);

continuous processing looks at the material surface as through a sliding window: a portion of the surface is present in the image and by continuously adding new lines as they come in from the camera and throwing out old lines the image size stays the same and the image practically moves along the surface. This is the typical processing method for endless materials where it is important not to cut a possible defect (for example a knothole in the wood industry) accidentally in half as could happen with piece-wise processing.

A similar effect could be achieved by acquiring discrete rectangular images with sufficient overlap to avoid cutting objects of interest. This would allow the use of standard algorithms made for rectangular images, but it is not applicable in all situations and also requires continuous capturing and buffering capabilities.

Line-wise processing is of relatively little interest as far as time behavior goes. The system simply has to be fast enough to make the required statement before the next line is acquired and waiting to be processed. Since the evaluations – possible on a single line of pixels – are fairly limited anyway, this is typically not a problem. In fact there are vision sensors working exactly in this way as intelligent, cost-effective light barriers.

More interesting is the relationship between camera speed, speed of the required relative motion and image characteristics for the other types of processing.

Along the sensor line, the resolution can be calculated from the characteristics of the optics and the sensor in the same way as for a rectangular sensor.

Now we will assume that the direction of the relative motion is perpendicular to the sensor line which is the typical case and avoids some additional complications. What geometrical resolution do we have in the direction of motion then?

For simplicity's sake, we assume that the part in Fig. 9.13 has a circumference of 10 mm and completes a full rotation in 1 s. When the camera uses a line frequency of 1 kHz, the result is an image of 1000 lines acquired in that 1 s, covering a length of 10 mm, so every pixel corresponds to 10 μm (note that this does not say anything about the pixel resolution given by the optical setup, pixel size, field of view etc.). This may or may not be equal to the resolution in the direction of the sensor line so pixels could be square or rectangular. And that relation can easily be changed: for example doubling the line frequency results in 2000 lines, each corresponding to 5 μm of the surface

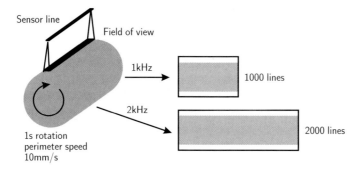

Fig. 9.13 Line-scan of rotating part.

in that direction. This is a marked difference to a rectangular sensor where the aspect ratio of the pixels is fixed by the geometrical characteristics of the sensor elements.

There is more to it, however. Camera clocks are typically more stable than the rotation of an electrical motor. The motor will have an acceleration ramp at the beginning of the movement, a deceleration ramp at its end. These we can get rid of by starting image capturing some time after starting the motor and stopping the motor only after capturing has ended. Still, variations in the motor speed during rotation will change the geometrical resolution in the direction of motion. The typical solution to this problem is using a shaft encoder measuring the actual speed of the axis and deriving from the encoder signal a line-trigger signal for the camera. Then, the camera acquisition speed will always match the speed of the motor.

Note that tying the exposure time to the line frequency will result in varying exposure – and hence brightness – when the line frequency changes due to variations in the motor speed. Exposure time will therefore have to be shorter than the typical line frequency.

Incidentally, the acceleration/deceleration ramp example again shows the importance of synchronizing various processes in the inspection station with the image capturing. The sequence here is a variation of the typical pattern for discrete motion production (cf. 9.8.1):

1. Start movement (here: accelerate motor);
2. Wait for stable conditions (here: stable rotation speed);
3. Start capturing;
4. Wait for end of capturing;
5. Start movement (here: decelerate motor).

Exploring all possible interactions of sensor, optics, speed, line frequency and exposure time affecting actual image resolution is well beyond the scope of this chapter. One effect is clear however: If the optical resolution is smaller than the geometrical resolution resulting from the relation between speed and line frequency, then the camera will not see the entire surface of the part. For example, an optical resolution of 5 µm and a geometrical increment of 10 µm will result in an undersampling effect, where the camera will see a field of 5 µm every 10 µm, that is, only half the surface.

9.9
Human–Machine Interfaces

Machine vision systems are made by humans and they are operated by humans. Therefore, they must interface with humans in different situations and at different times.

There are many differences between working with vision systems during engineering and at runtime. Environments differ, tasks differ and typically persons differ, in their goals as well as in their qualifications. Requirements on the human machine interface (HMI) differ correspondingly.

What the actual requirements are on a particular system, depends very much on what the system has to do, how complicated the task to solve is, how wide the variations expected at runtime are and so on. It does also depend on the production organization, availability of skilled staff, and many other factors, some of which will be discussed in this section.

9.9.1
Interfaces for Engineering Vision Systems

There are many human machine interface concepts – in this context more typically called user interface designs – for the task of engineering a machine vision systems. True, there are some typical, common activities in creating a vision system, like

- taking samples for setting up lighting and optics;
- arranging and configuring machine vision functionality to solve the inspection task (be they built into a vision sensor, part of an application package or a machine vision library);
- programming of required (algorithmic) extensions (provided the system can be extended);
- interfacing with the environment;

- creating a runtime HMI.

This, however, is probably the end of the common ground. For instance, how many HMIs are we talking about here? That of a single tool, which is the working environment of the machine vision engineer from taking the first images up to putting the system into operation? Or those of five specialized tools? It is not even necessarily clear which approach is superior, the all-encompassing general tool or the specialized ones.

An integrated working environment is very convenient by eliminating the need to switch between programs and by presenting a unified user interface, a single one, to get used to. For a company, for example, whose product spectrum varies considerably – and can undergo extensions at any moment due to customer demands – the rapid deployment and reconfiguration possible with integrated packages can be a decisive factor. If such a company employs a vision specialist at each location to create vision systems for their own production, this person will usually be available for fine-tuning the system, working directly with the development interface. There will be no great need for a sophisticated runtime interface.

However, for a manufacturer of standard machines, built in series, all equipped with the same machine vision system for the same purpose, the investment in a specially programmed single-purpose solution which is then duplicated again and again may well be worthwhile. There will probably not be any specialists for the vision system at the customers' sites, so a well-designed runtime HMI can be a major success factor. A custom-programmed system gives the manufacturer the freedom to create exactly the solution his customers need. In that case, the HMI for the engineering phase is actually a standard program development environment.

Too different are the needs and circumstances of various users for a single right answer as to how a user interface, or the overall package, for the development of machine vision systems should be designed. And since this chapter is supposed to be about machine vision in *manufacturing* not in the vision laboratory, we will turn to runtime HMIs in a moment.

There is one thing, however, which can be said with a high degree of certainty: just as installed machine vision systems will be integrated more and more with their automation environment, so will the engineering work on machine vision systems be integrated more and more with other engineering processes in automation technology. We are currently seeing the emergence of engineering frameworks, facilitated by progress in software interoperability and interprocess communication; initial versions of such engineering frameworks by various major automation companies are hitting the market. Sooner or later, machine vision will be expected to become part of the engineering processes carried out within such frameworks. Activities occurring at the interfaces between vision system and automation world will be affected first by this development, but there is no saying where it may end.

As a simple example, instead of the tedious and error-prone procedure of defining type data (see Sections 9.7.5 and 9.10.6) structures on each system individually and tracking changes on each system individually, type data structures would be defined in the framework and used alike by all systems involved, machine vision, PLC, and SCADA. The framework becomes the central authority for this information, avoiding redundancy and thus reducing the required work and the possibilities for errors.

9.9.2
Runtime Interface

This section is concerned with the human machine interface of a vision system which has successfully undergone its engineering phase and its deployment and is now running in production. We will call this an *installed* vision system.

For the installed system, the degree of variation of *how* to perform required activities through the HMI is usually smaller than for the engineering environment; on the other hand, the variation of *what* to do may actually be quite large, at least when comparing different companies. The fundamental question remains: what tasks have to be performed using the (runtime) HMI, where is the line to be drawn between tasks to be done through the runtime HMI and tasks which require a specialist to delve into the depths of the system and deal with the full complexity of the test program.

The answer to that question may differ considerably between production organizations. Every organization has its particular ideas about the tasks to be performed by production personnel, about the borders between production, maintenance, technical assistance, and specialist staff. And so activities that are in one company performed by maintenance may in others be the duty of the production personnel directly – or, the other way round, of a specialist.

Therefore, the requirements what the HMI should allow the user to do, down to what level of the system it lets him reach, and what level of assistance it gives, vary greatly. Nevertheless, we can list some typical activities that may have to be performed on or with an installed vision system. They are roughly sorted according to increasing involvement and interaction with the system.

Monitoring , that is, watching the vision system do its work; of course this is not an activity pursued for lengthy periods of time. But watching the screen display of a vision system may lead to discovering trends in the production, emerging anomalies. For this to be effective, the system display has to convey a clear idea of what is going on in the system, where it stopped in the case of an error and also, if possible, why – especially when the system can be used to analyze stored error images.

Changing over, that is, changing the system setup for inspecting different types of products is usually an automatic process, initiated by a command from the PLC and carried out by loading a different test program or a different parameter set; nevertheless, confirmation of certain steps by the operator may be required, for example when change-over involves mechanical changes, and this may be done either on the PLC or on the vision system whichever seems more appropriate.

Adjustment and calibration are essential tasks, at least for gauging systems, which may have to be performed periodically using specially manufactured calibration pieces under a "testing device administration regime," and of course every time a change occurs on the system that may affect the relationship between camera and world coordinates, for example a mechanical change-over. Often, calibration is a special mode of operation on the PLC which automatically initiates calibration on the vision system. As with change-over, operator-input may nevertheless be required, either on the PLC or the vision system. And, obviously, in PLC-less systems, the operator will have to initiate the calibration himself.

Troubleshooting refers to the task of finding out why the vision system is not functioning as expected and removing the cause of the problem. For this task, meaningful messages are very helpful, in particular, messages personnel can relate to observable circumstances. For example a message like "Image of top-view camera is too dark" is probably more helpful than "No object in image 2," as it will induce the operator to check a particular camera and the corresponding illumination. Admittedly, thinking up and managing messages for all possible combinations of errors is a tall order, especially for low-cost systems. But the fundamental idea holds that the system should give a clear indication of what went wrong and possibly why.

Also useful are easy-to-use test functions, starting with a live image function to check the operation of the cameras, and not ending with functions to execute communication processes for testing interfaces.

Tweaking that is, making small changes to parameters during production in order to compensate for batch variations in the appearance of parts, for example. It is essential that there are sanity checks for the parameter changes – that is, the operator is limited in what he can change – and that there is some method to check that parameter changes do not break existing inspections.

Re-teaching, like optimizing, is a borderline case. Re-teaching means adapting the test program to a new type of product, usually involving more than just changes to numerical parameters. The typical example is that of recognizing a new type in a RobotVision application, which is a position recognition type of application usually based on contour-/edge-features; the distinctive feature here is the shape of the part which has to be made known to the system.

Re-teaching may necessitate changes to a test program that go deep into the engineering area; ideally, however, the system should be designed in such a way that plant personnel can initiate a process by which the system "learns" the characteristic features of the new part and stores them as a new type. This process should not require plant personnel to deal with the full internal complexity of the machine vision program, be it source code or the configuration mode of an application package.

Optimizing as opposed to *tweaking* is typically an offline activity, where stacks of images – for example, error images, of parts considered faulty by the system – are re-evaluated to find out why the system evaluated the part the way it did and if it was correct in doing so. Here it is important that the system is capable of storing and loading data in such a way that an inspection cycle for a part can be completely reconstructed. This, though, is less a requirement on the HMI than on the overall system architecture. The changes needed to optimize the system may require reverting to the engineering environment, but the runtime HMI can offer means of analyzing and changing the system behavior using stored images. So this is clearly an activity on the borderline between runtime and engineering.

So what general requirements can we distill from these typical activities?

Obviously, an HMI must be capable of conveying an idea of the current system state and how that state was reached, using display of images as well as results and protocols. Images can either be "naked," that is pure camera images, or "decorated," that is containing visible results of the image processing, like detected objects, measuring lines and so on. This decoration should enable the user, together with results and protocols, to understand why a particular situation occurred.

Meaningful messages are an important requirement. It is increasingly required and becoming the state-of-the-art in the PLC world to have these messages in local language.

The ability to get and set parameters of the test program and/or the vision system itself is required for tweaking and optimizing. Ideally, visual feedback should be given on the effects of parameter changes, this can be quite difficult however. More important is that it is not foreseeable which parameter may be useful to change in a given system and situation. Therefore, the system *itself* should not impose restrictions on the parameters that can be set at runtime. This restriction should be in the responsibility of the application engineer, production lead technician, plant vision specialist, whoever is responsible for maintaining the system.

Incidentally, these requirements are often easier met with library-based approaches than with application packages, since there you can program everything just the way you need it. Note, however, that easier does not necessarily mean simpler or cheaper.

Using the PLC HMI for Machine Vision
Today's production stations are highly complicated affairs. Sometimes they are so densely packed with technology that a screw falling into the station from top would never reach the ground. Many different systems, manufacturing processes as well as measuring and inspection devices, are built into the stations, each with its own user interface, its own operating philosophy, data formats, displays, and so on. This diversity is hard to handle for the production personnel. They cannot possibly be intimately familiar with all these devices, and therefore, measures must be taken to make the operation of the station as easy and streamlined as possible.

One way of doing this is to enrich the data flow between the station's subsystems and the PLC of the station such that normal production operation can be carried out completely from the PLC. Automating change-over and calibration is a step in that direction. Remote setting of parameters for tweaking the vision system could be another. Troubleshooting is a very important point. Production staff is typically used to error messages on the PLC clearly indicating what device is malfunctioning in what way, like jammed stop gates, failed end switches and so on. This can also be approximated only by a rich information flow between vision system and PLC.

Somewhere in the area of optimizing and re-teaching is probably where the line for further integration is currently to be drawn. We forgot, however, the monitoring aspect.

In many setups, HMI and PLC are actually separate systems, though in the case of PC-based PLCs they may run on the same hardware. Sufficiently powerful information channels provided (that means hardware, i.e., bandwidth, as well as software, i.e., protocol), there is no fundamental obstacle to using the same HMI for PLC and machine vision. The HMI could then display not only numerical but also iconic results of the image processing system – perhaps not permanently since screen real estate is always at a premium – but switchable, whenever the user feels the need to see the output of the vision system.

Using the same display for PLC and vision system has a number of advantages; it saves a monitor, it saves the space and construction effort of stowing the additional monitor; it saves the user the trouble to watch two systems.

Naturally, there is a strong case for a separate display for the vision system. Performance, especially for large images, and the possibility of watching the system permanently at work are definitive advantages. Even then, however, the ability to display iconic results on the PLC makes sense.

Moving many operating steps, like setting parameters for tweaking the vision system, to the PLC is useful because the user interface of the PLC is familiar to the production personnel, and integrating devices in this way gives some degree of uniformity to the handling of various systems. Also, these activities can then be integrated into a user authorization management on the station and an activity logging which will not have to be duplicated on the vision system side. But many parameters of a vision system require a visual feedback to check their correct setting. Displaying iconic results on the PLC would allow that and thus further the transfer of operating activities and responsibility to the PLC as the central system.

Integrating the vision system runtime more and more with the PLC in this way is a certain trend which is visible in the offerings of large automation system and component suppliers. This may result in a new drive for standardization in this area.

9.9.3
Remote Maintenance

Modern production is a global undertaking. All around the world, we expect to be able to manufacture the same products, meeting the same quality standards everywhere. Inevitably, machine vision systems of all kinds and degrees of complexity are also proliferating throughout the world – including locations where no vision specialist may be available or affordable. The tasks, production or maintenance personnel can carry out, are always limited, by their skills in such a specialized area as well as by the interface a particular system provides for them in the production environment.

Transporting vision specialists around the globe to set up systems or solve problems is sometimes necessary, but usually only the second best choice. Ideally, any work on a vision system which can be done without actually having to touch the system hardware, should be possible from everywhere in the world – and sometimes even more if you have someone on site who can do the touching for you.

This requirement is easily fulfilled with smart sensors that are configured over the network anyway. There is no essential difference as to the location of the system, apart from a possibly inconvenient lack of speed when working on locations with less highly developed infrastructure. Clearly, any system that requires other means than an Ethernet–TCP/IP connection to be configured – like for example a serial connection or a memory card – is at a clear disadvantage here.

Also at a disadvantage, perhaps surprisingly, are PC-based systems (or other systems on general-purpose computers). The machine vision software on these systems is rarely designed in such a way as to allow every operation over a remote connection also – and even then, there may be limits due to necessary interactions with the operating system. The only way to work remotely with such a system, then, is to remote-control the entire operating system using appropriate software. Since this software must at least transmit changes to the system display, it requires considerable network bandwidth and may be unwieldy over slow connections. Also it may be considered a security risk by network administrators, blocked by firewalls, or downright forbidden.

The very power and versatility of general purpose computers running a modern operating system is actually the problem here; single-purpose systems designed for a particular task are much easier to handle remotely. They have a limited set of possible actions, hence they can be controlled by a limited set of commands.

Safety precaution: no movements

A word of caution with regard to remote maintenance: Machine vision systems seldom do cause movements in a machine, but there are cases where a vision system does initiate axis or cylinder movements, for example by digital I/O signals. It is a matter of course in control system programming but may not be so well known or deeply ingrained in a machine vision programmer: *Never ever* initiate a movement on a machine that you do not have directly in your sight. You, yourself, personally, may be responsible for severe injuries.

9.9.4
Offline Setup

The ability to set up a vision system (software) offline, that is, without actually using the target system, is particularly useful in connection with remote maintenance; consider a lengthy optimization job, looking at loads of error images, fine-tuning parameters, etc. Doing that online, on the production system, is most likely an unpopular method, as it would cause a lengthy production stop and possibly further disturbances.

Therefore, it is highly desirable to be able to carry out any configuration of a system also offline, without direct access to the target system. In a maintenance or optimization situation, this means configuring – or even creating – a complete test routine which can then be downloaded to the target system and without further changes executed there in automatic operation mode.

Obviously, this requires that the system software can work from stored images as well as from camera images since the required production parts will not be available at the maintenance location nor are the necessary lighting and imaging setup, and optimization typically done on error images anyway. PC-based application packages or libraries are usually capable of this (specific library-based programs only if it was considered during their development, however); with vision sensor setup programs it depends on the overall architecture of the program: whether it is capable of working without the sensor hardware or, otherwise, whether the sensor hardware is capable of storing images or receiving stored images for processing. This is an area where PC-based systems are usually superior.

There is another not so obvious requirement, however: the problem of hardware-related parameters. Some settings cannot really be made without at least knowing what hardware there is in the system. For example, how do you select the correct camera for image capturing in a particular test subroutine if you have no idea how many cameras there are in the system? How do you set up image section sizes if the size of the camera image is not available? How do you configure digital I/O handshakes without the information how many input/output lines are available?

Now it is obviously economically – and also often technically – infeasible to have duplicates of all target systems likely to be in need of remote configuration; even having a complete selection of all possible hardware would not help much as configurations would have to be changed back and forth permanently which is clearly impractical if not actually impossible.

To solve this problem, a system would have to be able to completely simulate all machine-vision related hardware possibly available in the target system. Here, smart sensors are at an advantage again. They *are* a specialized, stand-alone hardware and the setup program is typically dedicated to that hardware. Therefore, the manufacturer has complete control over the entire system and can provide his configuration program with the ability to simulate the image capturing, processing, and communication abilities of his hardware. Such systems actually do exist and are very convenient to use especially in distributed environments.

Hardware simulation is much more difficult for a PC-based system due to the variety of possible hardware setups. There are helpful approaches in some software packages, but a system capable of overall hardware-simulation is currently unknown (which does not necessarily mean that it does not exist, of course).

9.10
Industrial Case Studies

This section will present several actual industrial applications. Despite the necessary discretion with respect to product – and even more, nowadays, production – information, they will hopefully raise a few interesting ideas, reinforce some of the points made in this chapter and lead the reader to some novel solutions and applications in her/his own environment.

At the end of each application description, a brief description of equipment and important algorithms is given. The equipment description is most detailed in Section 9.10.1, in the following descriptions only a brief overview is given.

9.10.1
Glue Check under UV Light

By kind permission of Robert Bosch GmbH, Eisenach plant, Germany.

Task
The application presented here falls into the *completeness check* category, and it is a typical verification inspection, that is the part is checked for correctness after some manufacturing process has been carried out.

The part in question (see Fig. 9.14) is an automotive sensor element, the process is the application of glue on the part's lid. The part has to meet stringent quality requirements with respect to sealing and resistance against chemical aggressive media. The casing therefore has to be tightly and securely closed. This is achieved by using special glue.

To ensure strength and sealing properties of the joint, an image processing system is used to verify the amount and shape of glue before connecting the pieces. The vision system also checks whether there is glue outside the prescribed spots.

9.10.1.1 Solution
The most interesting part of this application is the illumination. In visible light, the glue exhibits almost no contrast to the part. Under UV light, however, the glue is fluorescent, resulting in an image with excellent contrast properties (see Fig. 9.15).

Equipment
Except for the UV illumination, the application was mostly solved with standard components and methods. A standard CCD matrix camera with a 50 mm lens and a UV filter to avoid saturating the sensor with reflection of the UV

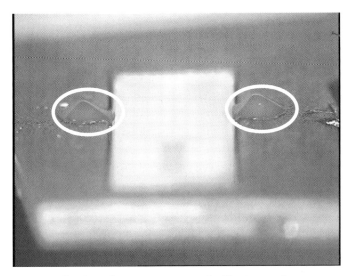

Fig. 9.14 Section of sensor component with glue spots (marked).

 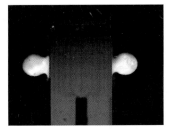

Fig. 9.15 Glue spots of different quality under UV illumination.

illumination from the component was used to capture the image. The speed of the production line did not require specific triggering mechanisms, image capture is done directly after the start signal from the control system has been received.

On the computer side, a custom-built fan-less PC was used with a 24 V DC power supply, a standard frame grabber, a PROFIBUS board for control system communication and an Ethernet connection for remote maintenance. Since 24 V is a standard voltage in automation technology systems, also used for digital I/O connections or to supply lighting equipment, using such a power supply makes for easy cabling (and makes the mandatory CE certification process less costly in terms of work as well as money).

As for the software, a configurable Windows-based image processing system was used running under Windows 2000 Professional. Interfacing to the control system was done with a configurable protocol on top of PROFIBUS

DP. In addition to constantly available control signals (like Start, Change Program, Ready, Done, etc.) this protocol uses the PROFIBUS DP process image for the transmission of user-definable data structures which can contain practically any information generated by an image processing system. It would therefore be possible at any time to enhance the system not only to evaluate the quality of the glue spots, but also to transmit actual measurement values to the control system. These could be used, for example, to adapt dispenser settings based on statistical process control results.

Algorithms

Due to the excellent contrast of the fluorescent glue under the UV illumination, the system can rely mostly on thresholding for segmentation. Template matching is also used, mainly for position determination. Of course, various feature computation algorithms are used to describe characteristics of the glue spots and thus determine their correctness.

Key Points

This application holds some points of interest: first, it proves once more that finding a suitable illumination is the key point in most vision applications. The rest is basically simple – which does not mean that it is not a lot of effort and far from trivial. And we see what makes a suitable illumination: one that is capable of clearly and consistently emphasizing the salient features of the inspection task.

The second is very much related to the first. It shows the importance of the knowledge, the help, and the commitment of the people from the product development and manufacturing side for the vision system engineer. The solution to use UV illumination to make the glue visible can only be found by either receiving the information *that* the glue used is fluorescent or by influencing production to use such glue – or, in some cases, to have one developed. Only by talking to the people who know about the product and about the process, by asking questions and making suggestions, by using all available knowledge, good solutions can be found.

9.10.2
Completeness Check

By kind permission of Robert Bosch GmbH, Nuremberg plant, Germany.

Task

The application presented here is a *completeness check* of the verification inspection type, meaning that the part is checked for correctness after execution of a particular manufacturing step.

The part in question is an automotive control component. For protection against the environmental conditions prevalent in the engine space, the component is encased in injection molded plastic. Before that process, the presence and correct position of various components are checked – which are then encased in layers of plastic. Afterward, several features of the finished molding are checked. Figure 9.16 shows an image of the part in background illumination.

Fig. 9.16 Backlit image of controller component before and after molding.

Solution
The parts are captured by standard cameras in front of diffuse infrared backlights. In the clear-cut contour image, a number of prescribed areas are checked for being covered by the required components; other areas are checked for being free of obstacles. This ensures not only the completeness of the required components but also that no other material is present inside the injection molding volume and that the components are actually in their required positions (because inacceptable deviations from these positions would result in nominally free areas to be at least partially covered). The position checks are augmented by edge-based measurements to verify certain geometrical features with the required accuracy.

Key Point: Mechanical Setup
A very interesting feature of the application is the mechanical setup. A four-position indexing table is used to transport the parts. The parts are checked in three of the four positions. Figure 9.17 shows a sketch of the indexing table and the setup of cameras and illumination. Note that nowhere a camera is placed directly beside any of the lighting components providing the illumination for a different camera to avoid problems with stray light.

Also interesting is the complexity of the check sequence and data management on the control system for the round-table check. Since a single computer is responsible for the checks at the loading and unloading position, it receives several start signals per indexing table step. Each time a different test program is executed and the control system has to assign the correct results to various parts in different positions.

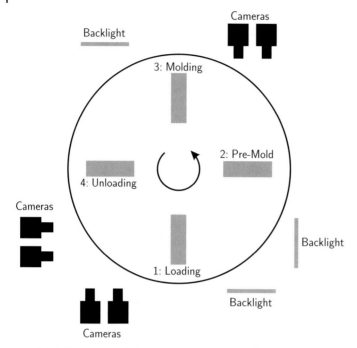

Fig. 9.17 Camera and illumination arrangement on indexing table.

Equipment

The system uses standard format CCD cameras, one 24 V industrial PC for the loading and unloading position, one for the pre-molding position, both equipped with standard frame grabbers for up to four cameras and PROFIBUS boards for communication with the control system.

All illuminations are diffuse LED area lights. Since there are no particular requirements on gauge capability in this application, standard 35 mm and 50 mm lenses are used.

As for the software, a configurable Windows based image processing system running under Windows 2000 Professional was used in combination with a custom-developed standard protocol for control system communication.

Algorithms

The excellent contrast properties of the backlit images allow the application to rely almost completely on thresholding segmentation. For the dimensional checks, where particular component positions are checked, edge detection is used for increased accuracy.

9.10.3
Multiple Position and Completeness Check

By kind permission of Robert Bosch Espana Fca. Madrid S.A., this application has been implemented in cooperation with the technical functions department of Robert Bosch Espana, RBEM/TEF3, and the vision group of the special machinery department of Robert Bosch GmbH, PA-ATMO1/EES22.

Task

The application presented here is a *completeness check* of the verification inspection type, that is, a part or in this case 42 parts, are checked for presence and correctness, combined with position recognition.

The actual part in question is a small hybrid circuit. Several components on this circuit have to be checked for the presence and position; in addition a type has to be verified by reading and checking a character. So far, so basic. The interesting thing about the application is that production logistics puts up to 42 of these circuits on a single foil, and still the resolution has to be sufficient to check distances between individual components with a considerable accuracy. Figure 9.18 shows such a foil.

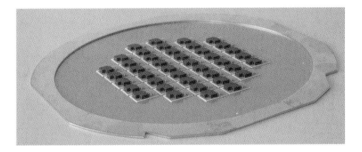

Fig. 9.18 Fully occupied foil with 42 hybrid circuits.

Solution

The size vs. resolution obstacle can be overcome in different ways; one would be to use a sufficient number of cameras – large enough cameras at that – or a line-scan camera to capture a single high-resolution image of the entire foil. In the end, neither of these two options came into play.

Increasing the number of cameras ran into a cost barrier, led to a difficult – or space-consuming – mechanical design which had to find room for all the cameras and would have complicated the application insofar that it would have been necessary to check images for overlap; otherwise, components might have been checked twice or entirely overlooked[10].

The large line-scan camera image would have required, naturally for a line-scan camera, a mechanical setup capable of smoothly moving either foil or camera. In addition, it would have resulted in a very large image and advances in computer hardware notwithstanding, it is always nicer to work with images of "normal" size.

So in the end, it was decided to use a single camera and do a separate check for each of the 42 components. This made the image processing programs simpler – the same procedure every time for 42 hybrids – moving some of the complexity toward the interface between vision system and control system. The test procedure was realized in the following way:

- For a new foil, the control system gives a reset message to the vision system, indicating precisely that: delete all current result, position and statistics data, we are starting a new foil.

- For each hybrid, the control system proceeds as follows:

    – Bring the camera into position;

    – Transmit the position number to the vision system;

    – Start the evaluation.

The vision system carries out the following steps:

    – Capture an image with backlighting through the foil;

    – Execute the evaluation of the distances of the circuit to its neighbors;

    – Capture an image with IR top-lighting;

    – Capture an image with red top-lighting;

    – Execute the evaluation of the components on the circuit;

    – Visualize the result at the appropriate overall position (this is the main purpose of the reset signal, clearing the visualization);

    – Transmit the results for the given position to the control system (including the position itself as a cross check).

Key Point: Cycle Time
Incidentally, this procedure solved another requirement, namely that a foil need not be fully occupied, that is, any position on a given foil could be empty.

10) Of course, current image-processing packages provide support for stitching images, but it is usually easier to be able to do without.

Only the control system would have this knowledge, and could thus simply skip the position without any additional communication overhead or logic on the vision system side.

A note on the result itself: the specification called for a detailed result, not merely yes and no but a yes/no on a number of properties, namely the presence and position of several individual components as well as the presence and correctness of the type mark. This communication was also very easy to handle using the above procedure.

Obviously, the procedure is easily scalable – in principle – but as always in life, there is no free lunch. Of course, the separate positioning and execution for each individual hybrid takes time. For each hybrid, three high-resolution images are captured. Together with switching the LED illumination, this takes approx. 0.7 s. Then the camera has to be moved from one position to the next, fortunately these are only about 20 mm apart. Since even the tiniest movement by the most simple axis requires an acceleration and a braking phase, there is clearly a limit on the achievable cycle time.

On the current system, there is ample time for this procedure. If the timeframe tightens, there are two measures that can immediately be taken to help with keeping the cycle time:

- Immediately after acquiring the images, the vision system could set a handshake signal on the PROFIBUS, indicating to the control system that the camera can be moved to the next position. So the control system does not have to wait for the actual evaluation result to move the camera. Of course, it does have to wait for the evaluation to be finished before giving the next start signal.

- Assume that one row of circuits has been processed from left to right; then it is of course faster to move the camera merely to the next row and start the row from right to left than to move the camera the entire way back to the beginning of the line. Thanks to the transmission of the position number to the vision system, it can nevertheless assign and visualize its results correctly.

Equipment

The system uses a high-resolution CCD camera and an appropriate frame grabber. For the three images, the system switches between an infrared diffuse light below the foil and an infrared directed light above the foil and a red directed light above the foil. The light beneath the foil gives a good backlit image of the circuit in question and parts of its neighbors, so distances between a circuit and its neighbors can be measured with a high accuracy. The directed lights above the foil enhance different features to be checked on the circuits.

The system was implemented using a configurable vision software package running under Windows 2000 on an 19" IPC. The rather product-specific requirements on the management of results (that is, visualizing and transmitting the separate feature results for the individual circuits) made some custom programming necessary; the control system interface itself, however, is a standardized protocol on top of PROFIBUS DP that allows for the transmission of configurable data structures in addition to the usual control signals.

Algorithms
The system has to check distances between each circuit and its neighbors. With a single circuit at the center of the image and just the borders of its neighbors visible (for maximum resolution), these distances necessarily reach far into the border areas of the image and thus into the areas of maximum lens distortion. An algorithm for calibrating the measured lengths according to the distance from the center proved to be helpful for reaching gauge capability here.

Also, sub-pixel edge detection techniques and corresponding distance and angle measuring algorithms played an important role. For the distinction between the type characters, template matching is used. Template matching is usually better at finding things than at distinguishing them, but with only two clearly distinct characters, this is no problem at all.

9.10.4
Pin Type Verification

By kind permission of Robert Bosch GmbH, Bamberg plant.

Task
The purpose of this application is to check the type of contact fed into the station before actually attaching it to the product, in itself a fairly simple *object recognition*. It is interesting because constraints imposed by the mechanical setup of the machine prevented the optimal solution from being implemented, so an indirect solution had to be found.

Solution
The contact pins are distinguished by the following main characteristics:

- length and width

- angle to the horizontal and bending points

Figure 9.19 shows some of different types of pins. Obviously, the width can only be seen from top (or bottom at that), whereas the angle and bending points are best determined in a side view. The length could be determined from both aspects. Both images would ideally be taken in backlight. Unfortunately, the geometry of the machine did not allow for such a lighting and capturing setup. A side view could not be achieved at all and for the top view, a back light was not possible.

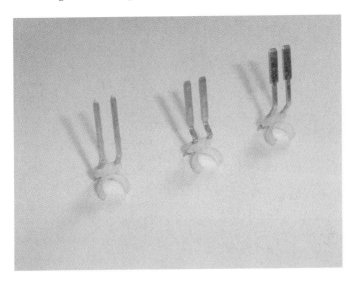

Fig. 9.19 Different types of pins.

It was impossible, therefore, to measure all the distinguishing features directly. An indirect method had to be found. Figure 9.20 shows its principle. The parts are illuminated from top at such an angle that the flat surfaces reflect maximum light into the camera whereas the bend reflects the light in a completely different direction. This makes it possible to deduce the position of the bends from the position of dark areas interrupting the reflexion of the pins. Of course in reality, light and lens are much larger in relation to the pin than in the sketch and there is also diffuse reflexion on the pin so that the entire length of the pin will appear bright to the camera.

In this way, it was possible to distinguish between all required types of pins. Figure 9.21 shows some different types. The check then becomes relatively easy: at certain, type-dependent positions in the image, the presence of dark and light areas is checked.

Key Point: Self-Test
There are, of course, limits to this setup. It is very difficult if not impossible, to distinguish pins whose only difference is their angle to the horizontal. If the bends are not in different places, this is practically not visible. More importantly, however, is that changes to the position of the checked regions would

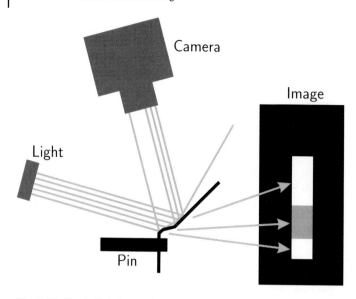

Fig. 9.20 Basic lighting and camera setup.

Fig. 9.21 Different types of pins under top lighting.

lead to a high proportion of false alarms and could, in principle, lead to different types being confused – which is very unlikely with the existing types, but nevertheless possible.

Therefore, the system was equipped with a "self-test-capability". This self-test uses stored reference images of each type to check whether a test program performs as expected after a change to the program. To this end, all the images are loaded into the test program one by one – instead of camera images – and the result is determined and compared to the expected result for the image. Only for images of the correct type should the program yield OK, for all other images not OK. Thus it can be easily seen whether a particular change led to the program accepting the wrong type or rejecting the correct type.

Equipment
The system uses a standard CCD matrix camera and a 50-mm lens to capture the image. The camera is connected to a high-precision frame grabber in a custom-built fan-less 24V IPC with a PROFIBUS board for control system communication and an Ethernet connection for remote maintenance.

As for the software, a configurable Windows-based image processing system was used running under Windows 2000 Professional. Interfacing to the control system was done with a configurable protocol on top of PROFIBUS DP. The protocol would allow at any time to enhance the system not only to evaluate the type of the pins internally but also to transmit actual measurement values to the control system.

Algorithms
Due to the excellent contrast of the illuminated parts of the pins, the system can rely mostly on thresholding for segmentation to check the required bright and dark areas of the pins. For measuring the width and distance of the pins, additional edge-detection algorithms are used.

Note that this application is not a measuring application by nature but a type verification with considerable distances between the nominal measurements of different types of pins. Therefore, the system has not been designed for gauge capability requirements.

9.10.5
Robot Guidance

By kind permission of Robert Bosch GmbH.

Task
The goal of this application is a *position recognition* with the added requirement of transmitting this position so that a robot can take the component and automatically join it with its counterpart.

The part in question is a plastic lid which is glued onto a metal casing. The robot has to take the lid from the feeder, gripping it at the correct points, turn and move it into the required end-position and press it onto the glue-covered casing to fix it securely and precisely into place.

Solution

To this end, the original position of the part coming from the feeder must be recognized with sufficient accuracy that the robot gripper can take hold of it and can turn and move it precisely into the required end position. The feeder places the lid on an LED backlight as shown in Fig. 9.22 where it is captured by a high-resolution camera mounted at a fixed position inside the manufacturing cell (which means that we have a fixed camera coordinate system as well as a fixed robot coordinate system).

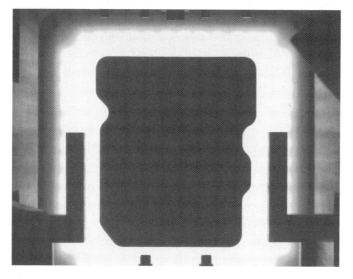

Fig. 9.22 Plastic lid on LED backlight.

Using a geometrical matching algorithm, the position and orientation of the lid can be accurately determined. The frame of reference for these coordinates is of course the camera-coordinate system. Subsequently, a coordinate transformation into the robot-coordinate system is carried out. The transformed coordinates are then transmitted to the robot control system.

A geometrical matching algorithm – in contrast to a gray-level template matching method – correlates the position of geometrical features (like corners, straight edges, curves etc.) found in the image with those in the matching model. It can thus achieve a high degree of accuracy and can also be made independent from orientation and scaling of the part in the image.

Key Point: Calibration

A calibration is required to enable the system to transform between the two coordinate systems. Due to the geometrical and optical characteristics of the system, no image rectification is needed to achieve the required accuracy. Therefore, a three-point-calibration is sufficient.

Calibration is requested by the operator through the control system. The operator has to insert a calibration part, that is in this case a part similar to the actual production parts, but made with particular high accuracy and – for durability – of metal. This part is then presented to the vision system in three different positions, involving a rotation of the part (which should therefore not be symmetrical). In each position, a handshaking takes place, that is, the robot control system transmits a message to the vision system that it has reached the position, the vision system captures the image of the part and in turn transmits a message to the robot control system to indicate that it has captured the image and the part can be moved again.

The robot coordinates of this procedure are pre-determined and are transmitted to the vision system. The vision system compares the coordinates of the part in the three images with the robot coordinates and determines and stores the transformation parameters.

The coordinates thus determined refer to the tool center point of the robot, because that is what the pre-determined coordinates refer to. Suppose we had different types of lids with different sizes so that the tool center point is at a different distance from the border of the parts. In that case, the robot control system needs a corrective offset in order to place all types at the correct position. This corrective factor could be fixed in the robot control program but it has been incorporated into the vision system programs for two reasons:

- changing the calibration of the vision system may affect the correction factors;

- the vision system will need different matching models for different lids since their geometrical features will be different; it is thus quite natural, to place the correction factors together with different models. The robot can then – in principle for simple movements, it may not be practical in all circumstances – use the same program for all types.

The correction factors are determined by a similar procedure as the calibration. Instead of a calibration part, however, a part of the pertaining type is used to determine the correction factor.

Key Point: Communication

Finally a word on communication in this application. There are three components talking to each other in this setup:

1. The station control system which is the overall master; it controls the feeder as well as all the glue dispensers, the safety equipment and so on;

2. The robot control system which is slave to the station control system but is in itself the master for the vision system;

Fig. 9.23 Communication partners in robot vision application.

3. The vision system as slave to the robot control system

Figure 9.23 shows the relationship between the three components. The vision system receives its start commands, handshake signals and other information (like type numbers and calibration coordinates) from the robot control system and sends its handshake responses and results to the robot control system. This entire communication is carried out over Ethernet in the TCP/IP protocol. The robot control system in turn is controlled also via Ethernet from the station control system.

Equipment
The system uses a high-resolution (1392*1040) camera on an appropriate frame grabber in a 24V industrial PC. A large LED area backlight is used to render the contours of the lid for the positioning algorithm.

Algorithms
The most prominent algorithm here is the geometrical pattern matching, which enables the system to find different part types reliably and accurately. Also important is the three-point calibration and the offset adjustment algorithm for processing different types of lids.

Another key factor is the close interaction between vision system and robot control carried out through the TCP/IP-based protocol.

9.10.6
Type and Result Data Management

By kind permission of Robert Bosch GmbH, Bamberg plant.

Task
Modern spark plugs are specifically optimized for particular gasoline engines. Therefore, a practically endless variety of different spark plug types is produced nowadays, potentially more than 1000 in a single plant. Figure 9.24 shows a few of them. Accordingly, there is a large number of features to check. Some of these checks are quite basic, others are extremely difficult, but that is not our main concern here. The data management for such a number of variants is a challenge in itself.

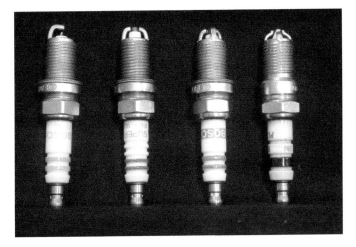

Fig. 9.24 Various types of spark plugs.

Solution
It is clearly not possible to have a completely separate check routine for every single type. Assume that all these routines have some common procedure and parameters. Now someone develops an optimization of these parameters. Imagine the work and cost required to upgrade all of the check routines. And the necessary configuration management!

Fortunately, the spark plugs are made from several basic building blocks which in themselves occur in several variants but not near as much than the overall spark plug. Its variants are to a large extent created by assembling different varieties of the basic components, so clearly, the way to go is using separate sub-routines for these basic components and reconfigure those according to the type currently manufactured.

Key Point: Type Data
The problem of variety concerns parameters as well as nominal values and result data. Obviously, when there are spark plugs with different dimensions that have to be checked, there will be different nominal values and tolerances for those dimensions. At the same time, however, basic parameters of an algorithm may depend on this dimension. Take for example a high-precision edge detection as a basis for a geometrical check. This edge detection needs to take place in the correct location for the given type of spark plug. Thus, region of interest, search direction as well as parameters of the edge model may change with the type of spark plug being checked.

To meet these requirements, the software has been designed using sub-routines corresponding to various checks on the individual building blocks of the spark plugs. Each of these sub-routines has been implemented in a generic way so that its detailed behavior is actually controlled by parameters set externally.

The parameters, on the other hand, are managed in a type database, which contains data describing the type of product currently manufactured and inspected. Inspection parameters thus become part of the characteristics of the part type being produced.

Here, type data is centrally managed on a supervisory level server. From this server the type data relevant for the type currently produced is handed down to the inspection stations. During initialization or change-over, the inspection programs read the type data and adjust their parameters for the type to be inspected. This scheme would also naturally allow for central optimization of the parameters of various inspection stations doing the same type of inspection.

Key Point: Result Data

Another interesting point in this inspection system is also related to the management of data, in this case result data. For each inspected work-piece, a considerable amount of results is produced. What we are typically used to is that the vision system itself draws the conclusion whether the part is good or not and submits its decision to the control system. This decision has been broken down here to the level of individual inspection features. For some features, the actual value is transmitted, for most a good/faulty decision of the vision system – but referring only to this feature, not the overall part.

The reason for this is that different things may be done with the part subsequently. For example, there may be defects on a part that can be remedied by re-work; or a particular fault occurs with a higher than normal frequency so that the production management deems it appropriate to investigate this type of fault by sending work pieces exhibiting that fault to parts analysis. What parts to send for re-work and parts analysis and what parts to actually scrap is a question of manufacturing strategy and therefore not necessarily appropriate to be decided by a machine vision system.

It is of course always a question of overall production organization and strategy, how to distribute responsibilities between various systems. In this case, the decision about how to handle these "borderline parts" was placed in the hands of the control system, for several reasons; one being that this is the tool the line personnel is most familiar with. Also, the actions required to steer the parts to the appropriate places are carried out by the control system anyway. Most importantly, however, because it was perceived as the appropriate way to assign the responsibility: the vision system delivers results, that is: data, like any other measuring system; the control system controls the handling of the parts and what is done with them.

Equipment

There are various image processing systems for this product. All use high-performance industrial PCs to meet the cycle-time requirements. The different algorithms run inside a shell program providing communication to the PLC via PROFIBUS, loading of type data provided by the PLC via Ethernet and visualization and statistics for the operator.

The PROFIBUS communication is used for control signals (Start, Ready, etc.) as well as for communicating result data as described above in a configurable protocol enabling the transmission of practically arbitrary data structures through the PROFIBUS process image.

9.11
Constraints and Conditions

The previous sections have given a little information on various kinds of machine vision applications you may encounter in practice, different types of systems, aspects of their integration into the automation technology environment. This section will explore, given the background from this chapter, some aspects of approaching machine vision applications.

An idea of how various constraints and conditions influence the operation of a machine vision system is important for determining the technology to be used. Having a clear view of the task in question, the environment in which it is to be carried out, the requirements and constraints under which it is to be performed, is the first step of a strategy to design a machine vision system.

There are various constraints which are of importance for making a system decision. Obvious constraints are speed, resolution, and costs. Equally important however may be the technical and organizational structure of the production, the personnel as well as the technical environment.

9.11.1
Inspection Task Requirements

The actual requirements on the inspection to be performed are obviously the starting point for designing a vision system. There are many requirements which can be systematically checked and several check lists have been designed for that purpose. Typical requirements are, for example, resolution, type spectrum, cycle time or speed of motion; all the things that can be quantified or categorized.

In an ideal world, this would describe the problem completely and there would be sufficient samples – from actual production of course, not pre-series parts from model making – to test the entire range of variation the parts' appearance may have. Alas, this is not an ideal world and so we frequently have to deal with small sets of samples that are not at all representative of series production on the one hand, incomplete and sometimes undescribable requirements on the other hand.

Often, machine vision systems are installed to eliminate human visual inspection. Frequently, these human inspectors work using a "defect catalog," examples of what constitutes a good part, what a bad part. They do not typically use a lot of numerical, or objective, criteria. Sometimes, the defects may not even be known, at least not completely. This is frequently the case with surface inspection tasks. Often, it is quite well known what a good surface looks like, but no one knows all the defects that could possibly occur.

It may well be that the very vividness, the close relationship to humans' most important sense, is what makes it so difficult to specify the requirements on a vision system. A sentence frequently heard in this area is "But you can easily see that". Yes, a human can, with a visual apparatus and a processing unit formed by million years of evolution. A machine cannot.

So the task of finding out what the system actually has to do is not as trivial as it may seem at first. After the obvious, quantitative, requirements mentioned above have been covered, it will be necessary to keep close contact with the customer, mechanical designers, all people involved to make sure that you really build the system that is required.

It is important to get some understanding of the manufacturing process, which produces the part to be inspected. This will give you a better idea of what defects can be there – or not – and also perhaps of better, indirect ways to check a feature that is not immediately obvious. The process engineer, the one responsible for the manufacturing process whose outcome has to be checked, is a very important contact for the vision engineer. He usually knows much more about the part and the possible problems than there will ever be on paper.

9.11.2
Circumstantial Requirements

Even when all the direct requirements are known, when it is clear, how lighting and optics have to be set up to render the salient features visible, when the algorithms have been found, that can solve the task, there is still considerable freedom of design.

It may be equally feasible, technically, to implement the solution in a smart camera or a PC-based system. You may have a choice between using a distributed system of less powerful, networked units or a single, powerful machine. What factors affect such design decisions if the fundamental feasibility does not?

An area we will not cover here are legal requirements. Of course, a vision system, like any other part of a machine underlies various legal requirements, ruling all kinds of things, but this is no different from other areas of machine building.

Cost
Cost is a very obvious factor but it is not always obvious what has to be taken into account. Cost is more than the money required to buy the system. It has to be installed and integrated and that may cost more with a less flexible low-cost unit than with a – at first glance more expensive – PC-based system. This, of course, depends very much on the overall way of building machines and integrating subsystems.

But neither task nor expenses are finished after buying the system and getting it up and running. It will in many cases be necessary to optimize the system after initial start of operation, for example because the actual range of appearance of the series parts becomes clear only after the systems starts operation. A system that can store and re-evaluate error images is clearly superior in this respect than one that cannot. The cost factor of an efficient way of reducing false alarm rates should not be underestimated. A lower rate of false alarms can easily achieve amortization of a higher system price within half a year.

Another important cost factor are spare parts. Using many different systems requires managing and holding available a corresponding number of spare parts. Therefore, using a particular system just once because of its price, will most certainly not pay off in the long run. Although possibly more expensive at first glance, a standard system that has the flexibility to be used frequently in many applications will probably win out in the long run. In every company, such considerations depend, of course, on internal calculation conditions and therefore cannot be generalized.

These are but a few examples that the price of purchase is not the only measure of the cost of a system. Ease-of-use, hard as it is to measure, reliability, flexibility, extensibility, standardization to reduce training as well as spare parts cost, may prove more important in the long run.

Automation Environment
Not every system can be integrated with the same ease into every environment and every machine builder typically has his particular style of building workstations. Here are just a few examples, as we have covered various interfaces between machine vision systems and their environment already in some depth.

On the geometrical side, for example, when switch cabinets are commonly used, which do not accommodate 19" PCs, then it may be difficult to find space for such a system and it will cause extra costs. On the other hand, smart cameras are typically larger than normal cameras, so if space is at a premium inside the station, only systems with normal cameras – or smart cameras with separate head, there are a few of those – can be used.

On the informational side, if, for example, PROFIBUS is the communication media of choice in the facilities, then systems with other interfaces will cause additional cost for converters. Control system programmers are used to adapt their systems to proprietary protocols of device manufacturers since flexibility is typically higher on the control system side; if, however, the communication protocol of the device is not powerful enough to transmit the desired information, then a compromise in functionality may be necessary.

In the same vein, the automation style of a machine builder may require master-slave behavior from a vision system (as an "inferior" device with respect to the control system), and not every vision system may be equally well-suited for that. For example, some vision systems initiate communication by themselves at – from the control system's point of view – arbitrary moments in time: whenever they have a result available, they send it out. This can lead to considerable effort on the PLC side to handle the communication.

We could go on like this for a while, but the general idea should be clear: the physical as well as informational interfaces of the system, which we already have covered in some detail previously, must fit the automation environment. Depending on whether the vision system is mostly compiled from off-the-shelf components or programmed from scratch, this is either a requirement on component selection or on system development or both.

Organizational Environment

The environment that the organization provides for the vision system to work in, covers a wide range of topics, reaching from the type of production to the departmental and personnel structure of the company.

Some questions that can arise in this area are:

- What is the type of production; we have covered that in some detail in Sections 9.2.2 and 9.8.

- What is the product spectrum of the company; are there few similar products or very different ones; are product life cycles long or do products change frequently. This question is most interesting when a decision about a standardized machine vision system, applied throughout the company, has to be made, because it indicates, for example, whether there is a focus on rapid reconfiguration capabilities or not. A system integrator, building machine vision systems for other companies, will have to ask similar questions about his customer base.

- How is production and maintenance personnel organized, what are their tasks; for example, it is a big difference, whether there is a technical function or maintenance group, which will take care of the vision systems, or whether this is part of the responsibilities of the direct production personnel, who will have much less time to devote to these tasks.

- What are the skills of the personnel working with the system; this refers to the people responsible for tuning and troubleshooting the system as well as to those who only operate it – or even less, in whose production area the system is operating. Design of visualization, message texts, available parameters and how to set them, all these things can be affected by the question, *who* are the people working with the system and what is their way to work.

  It can be an extremely enlightening experience for a machine vision programmer to work in production for some time, for example putting into operation one of his own systems and answering questions of production personnel. This may lead to completely new ways of looking at software features and user interfaces.

- For a large organization, a question may be, how much optimization and maintenance work is to be done by people on-site, how much by a centralized vision specialist group. For an independent builder of machine vision systems, the question is similar (from the technical point of view, the economical aspect may be quite different): how much service work is to be provided by the system builder, to what extent can (or should) the customer be enabled to do this work himself.

  This is less a technical question than one of business model, but it has technical consequences. If the provider of the system (be it an independent company or an internal vision specialist group) also provides service and optimization work, his employees will either have to travel a lot or powerful means of remote maintenance and setup will be necessary – or both.

We could prolong this list more or less indefinitely. The main point is that a technically feasible – even excellent – solution may be impractical under the conditions at a particular manufacturing location.

9.11.3
Limits and Prospects

As a conclusion, a few words shall try to summarize the many aspects of machine vision systems' life within automation technology covered in this chapter.

Market data shows a remarkable growth in the application of machine vision systems, and there is no sign of this trend to end soon. On the contrary, the increasing performance of hardware as well as software creates progress at all levels: the improving cost-performance relationship of small systems allows more and more applications to be implemented, which used to be too expensive to do; small, easy-to-use systems conquer more and more areas formerly the domain of PC-based solutions; at the same time, far from shrinking, the PC segment moves upstream to increasingly demanding applications.

But there are, of course, limits to what the technology can do and also what people will want to (or be able to) do with the technology.

It will be a long time, if ever, before machine vision comes close to the capabilities of the human visual system and the human information processing. Of course there are areas, certain applications and aspects where machine vision systems do things humans would not be capable of, but the ease with which humans recognize and evaluate an overwhelming diversity of things will elude machine vision systems for some time to come.

Weakly specified tasks, tasks based on examples rather than quantifiable criteria, will remain a difficult field, especially tasks where the range of possible appearances is not known in advance. It is much harder to design a system for detecting "anything" – unspecified deviations from a desired state, for example – than "something".

This often touches a nontechnical limit: customer acceptance. A system may be functioning quite well; if it is not possible to explain its operation to the customer, to make it plausible that and why it works and delivers results, it may still fail in the eyes of the customer. Advanced algorithms and methods are often faced with this problem, which may be an effect of a conservatism on the part of the customer. Although perhaps sometimes exaggerated, this conservatism is quite understandable, considering that the customer (of the vision system supplier) is a manufacturer at the same time, responsible for the quality of his products, and wants to be able to explain to *his* customer, how he guarantees product quality and why the inspection system is safe and reliable.

It is certainly not our intention to advocate not to use modern methods and algorithms, or not to watch closely what is happening in machine vision (and other) research and introduce such methods into the practice. But on the other hand, as engineers who have to build systems, which may at some time or other be safety-critical, we should also not succumb to the temptation of technology for technology's sake. Powerful, general algorithms, capable of recognizing anything anywhere, are an important research topic, which we will all benefit from, but at least for the time being, the task of the engineer is to incorporate as much a-priori-knowledge into the system as possible. In the manufacturing industry, we often have the luxury of knowing comparatively well, what the vision system has to look for and where, and we should make use of that knowledge.

It follows that the task of the system designer – or rather, the tool designer – is to facilitate this as far and as easily as possible. This is basically what "integration" of vision systems into manufacturing is all about: making machine vision methods easier to use, vision systems easier to operate, bringing technology closer to the customer, enabling OEM users as well as end users to make better, easier use of the technology – and thus paving the way for the use of machine vision in more and more application areas.

The approach demonstrated by Siemens AG on the SPS/IPC/Drives trade show in 2004 to operate a vision sensor through an ActiveX control embedded in the HMI of the control system is an expression of the general effort to make the vision system a part of automation technology that is no longer recognized as something special, a matter-of-course. This can be seen as an indicator of the increasing maturity of machine vision: in general, the more mature a technology is, the less visible it is, the less thought is required to use it. How many car drivers have the least idea of how its engine works, how many cell phone users know anything about the myriad lines of code making up the software that enables them to play games, surf the Internet, watch movies, and, yes, make phone calls?

So, to sum up, the following trends appear fairly certain to continue at least for the foreseeable future of machine vision:

- Solid growth, increasing coverage in low-cost areas as well as in demanding applications;
- Development of new fields of use in addition to the traditional quality control;
- Continuing improvement of technology as well as usability;
- Increasing integration into automation technology.

And perhaps growing maturity – as well as cost-pressure – may actually at some time lead to widespread standards, used over company limits.

References

1 *http://www.machinevisiononline.org* Automated Imaging Assocation.
2 *http://www.emva.org* European Machine Vision Association.
3 Norbert Bauer *Leitfaden zur Industriellen Bildverarbeitung*. (in German: Industrial Image Processing Manual). Fraunhofer Allianz Vision, Erlangen, 2001.
4 Christian Demant, Bernd Streicher-Abel, Peter Waszkewitz (ed) *Industrial Image Processing*. Springer, Berlin, Heidelberg, New York, 1998.

5 Richard C. Dorf (eds.) *The Engineering Handbook*. CRC Press, Boca Raton, 1996.
6 *Guide to the Expression of Uncertainty in Measurement*. BIPM, IEC, IFCC, ISO, IUPAC, IUPAP, and OIML, 1995.
7 *International Vocabulary of Basic and General Terms in Metrology*. International Organisation for Standardisation (ISO), 1995.
8 Jähne, B., Massen, R., Nickolay, B., Scharfenberg, H. (1995): *Technische Bildverarbeitung – Maschinelles Sehen*. (in German; Technical Image Processing – Machine Vision). Springer, Berlin, Heidelberg, New York, 1995.
9 Patrick Schwarzkopf. *The Machine Vision Market in Germany 2004*. VDMA, Frankfurt a. Main, Germany, 2004. Online summary available at http://www.emva.org.
10 Herb Sutter *The Concurrency Revolution* C/C++ Users' Journal, 2005/02.

Appendix

Checklist for designing a machine vision system

General

Date _____
Company _____
Contact person _____
Department _____
City _____
Zip code _____
Street _____
Phone _____
E-mail _____

Task

Description of task and benefit
_____
_____

Size of the smallest feature to be detected ____
Required accuracy ____
100% inspection ____   Random control ____
Offline inspection ____   In-line inspection ____
Retrofit ____   New design ____

Parts

Description of parts

Discrete parts ___   Endless material ___
Dimensions (min, max)   to
    to
    to
Color   Surface finish

Corrosion, adhesives
Changes due to handling

Number of part types
Difference of parts: _____
Batch production ___
Can production change be addressed ___

Part Presentation

Indexed positioning ___   Time of non-movement ___
Manual positioning ___   Time of non-movement ___
Continuous positioning ___   Speed ___
Tolerances in positioning x
    y
    z
    Rotation about x
    Rotation about y
    Rotation about z
number of parts in view
Overlapping parts ___   Touching parts ___

Time Requirements

Maximum processing time ___
Processing time is variable in tolerances ___

Information Interfaces

Triggering manual ___ automatic ___
What information has to be passed by interfaces?
_____

What interfaces are necessary?
Digital I/O ___
TCP/IP ___
Fieldbus ___
RS-232 ___
Requirements for user interface
_____

Miscellaneous

Installation space
    x
    y
    z
Maximum distance between camera and PC
Ambient light ___
Protection class ___
Dirt or dust ___
Shock or vibration ___
Variations in temperature ___
Electromagnetic influences ___
Availability of power supply

Index

$V(\lambda)$-curve, 85
$\cos^4$-law, 323
f-number, 240
1/f noise, 422
1394 Trade Association, 446
3D reconstruction, 653–666
64-bit PCI, 428

a posteriori probability, 672
a priori probability, 672, 674
A/D conversion, 400
aberration, 212
aberrations, 316
absorbance, 110, 113, 127
absorbing filter, 141
absorption, 113
accommodation, 3, 5, 29
accumulator array, 642–645
accuracy, 605–606, 695, 711, 718, 720, 759, 764, 765
    contour moments, 566
    edges, 520, 607–612
    gray value features, 520
    gray value moments, 562–563, 566
    hardware requirements, 611–612
    region moments, 562–563
    subpixel-precise threshold, 520
achromatic axis, 12
acquisition, 481
action potentials, 18
active pixel sensor, 376
ActiveX, 477
adaptation, 13, 14, 17, 20, 29
    contrast, 29
    ligth/dark, 29
    motion, 29
    tilt, 29
    to optical aberrations, 30
    to scaling, 29
adaptive lighting, 134, 195
additive mixed illumination colors, 117
additive noise, 309
additivity, 288
adjustment, 713, 714, 717, 727
adjustment vs. calibration, 727

AFE, 398
affine transformation, 537–538, 559, 626, 646
afocal system, 262
aging, 102
AIA, 437
Airy disc, 278, 280
algebraic distance, 619
algebraic error, 619
aliasing, 542–543, 632
aliasing effect, 305, 307
aliasing function, 315
aliasing potential, 315
aliasing ratio, 315
alignment, 537
amacrine cells, 19, 21
amplitude distribution in the image plane, 276
analog, 428
analog front end, 398
analog processing, 396
analog video signal, 429
analyzer, 123
Ando filter, 598
angel
    kappa, 3
anisometry, 558, 562, 564, 669
anterior chamber, 3
anti-reflection coating, 323
anti-reflective coating, 143
AOI, 406
aperture, 235, 521
aperture correction, 392
aperture stop, 235, 236
API, 475, 476
Application Programming Interface, 476
application-package, *see* systems, application-package
applications
    code recognition, 695
    completeness check, 695, 752, 754, 757
    object recognition, 695, 760
    position recognition, 695, 763
    shape and dimension check, 695
    surface inspection, 696

Handbook of Machine Vision. Alexander Hornberg (Ed.)
Copyright © 2006 WILEY-VCH Verlag GmbH & Co. KGaA, Weinheim
ISBN: 3-527-40584-7

types, 696
AR, 143
area, 556, 562, 564, 565
area MT, 26
area of interest, 406
artificial astigmatism, 140
aspect ratio, 742
astigmatism, 5, 12
asynchronous, 411, 443, 446
asynchronous reset, 499
asynchronous shutter, 366
attention, 25, 26
auto shutter/gain, 400
axial magnification ratio, 229, 266
axons, 3, 13, 18, 19, 21

b/w, 382
Back Porch, 430
back porch, 431, 432
backlight, 199
backlighting, 158
band-limited system, 307
bandpass filter, 142
bandwidth, 138, 410, 415, 417, 425, *see* signal, bandwidth
bar codes, *see* codes, bar codes
batch production, *see* production, batch
Bayer, 446, 454, 480, 489, 490
BAYER pattern, 396
Bayer pattern, 389
Bayes decision rule, 672, 673
Bayes's theorem, 672
beam converging lens, 222
beam diverging lens, 222
Bessel function, 278
Big Endian, 419
bilateral telecentric systems, 262
binarization, 502
binary image, 513, 552, 556, 566, 567, 569, 573, 579
binning, 382, 385, 386, 406
binocular stereo reconstruction, 653–666
bipolar cells, 18
    OFF, 15
    ON, 15
bit depth, 486
Bitmap, 491
black light, 122
blemish pixel, 378
blobs, 23, 26
blocks, 484
blooming, 372
BMP, 491
boundary, 553, 560, 573–574
boundary condition, 273
bounding box, 559–560, 565, 669
BP, 142

brightfield illumination, 158
brightness, 235
brightness behavior, 104
brightness control, 189
brightness perception , 86
buffer, 481, 501
buffering, 471, 500
bundle adjustment, 346
Bus_Id, 418, 419

C, 476
C-mount, 362
cables, 457
calcium, 15
calibration, 727, 746, 748, 764–766
    geometric, 610, 611, 654
        camera coordinate system, 654
        distortion coefficients, 654, 658
        exterior orientation, 611
        focal length, 654
        interior orientation, 611, 654
        pixel aspect ratio, 654
        pixel size, 654
        principal distance, 654
        principal point, 654
        relative orientation, 654
    radiometric, 519–524, 610
        calibration target, 520
        chart based, 520–521
        chart-less, 521–524
        defining equation for chart-less calibration, 522
        discretization of inverse response function, 522–523
        gamma response function, 520, 524
        inverse response function, 522
        normalization of inverse response function, 523
        polynomial inverse response function, 524
        response function, 520, 521, 524
        smoothness constraint, 523–524
calibration vs. adjustment, 727
camera, 511–512
    fill factor, 563, 608, 611
    gamma response function, 520
    gray value response, 519–520
        linear, 519, 610, 611
        nonlinear, 520, 610
    inverse response function, 522
    response function, 520, 521, 524
    shutter time, 521
camera bus, 427, 428
camera calibration, 333
camera constant, 209
Camera Link, 383, 426, 437, 439, 455–458
candela, 125

Canny filter, 594–595, 599, 606
    edge accuracy, 607, 608
    edge precision, 606
cardinal elements, 219
Cat5, 458
CCD, 361, 364
CCIR, 379, 429, 456, 457
CCTV, 362
CDS, 388
center of gravity, 557–558, 562, 565, 614, 615
center wavelength, 96, 138
central disc, 278, 283
central moments, 557–558, 562, 565, 613
central perspective, 338
central projection, 208, 251
centre of perspective, 208
CFA, 489
chamfer-3-4 distance, 581
change-over, 714, 746, 748
channel, 496
channel capacity, 12, 18
characteristic function, 513, 561, 562, 583
charge, 373, 384
chessboard distance, 580, 581
chief ray, 236
chip size, 382
Chroma, 392, 430–432
chromatic aberration, 5, 9, 11
    longitudinal, 11
    transverse, 11
circle fitting, 617–619
    outlier suppression, 617–618
    robust, 617–618
circle of confusion, 241
    permissible size, 249
circles of confusion, 245
circular aperture, 278
city-block distance, 580, 581
classification, 669, 672–688
    a posteriori probability, 672
    a priori probability, 672, 674
    Bayes classifier, 674–677
    classification accuracy, 685
    classifier types, 674
    curse of dimensionality, 675, 682
    decision theory, 672–674
        Bayes decision rule, 672, 673
        Bayes's theorem, 672
    error rate, 674
    expectation maximization algorithm, 676
    features, 669
    Gaussian mixture model classifier, 676–677, 687
    generalized linear classifier, 682
    k nearest neighbor classifier, 675
    linear classifier, 677–678

    neural network, 677–681, 685–687
        hyperbolic tangent activation function, 680
        logistic activation function, 680
        multi-layer perceptron, 678–681, 685–687
        multi-layer perceptron training, 680–681
        sigmoid activation function, 680
        single-layer perceptron, 677–678
        softmax activation function, 680
        threshold activation function, 677, 679
        universal approximator, 679, 680
    nonlinear classifier, 678–687
    polynomial classifier, 682
    rejection, 686–687
    support vector machine, 681–687
        Gaussian radial basis function kernel, 684
        homogeneous polynomial kernel, 684
        inhomogeneous polynomial kernel, 684
        kernel, 684
        margin, 682
        separating hyperplane, 682–683
        sigmoid kernel, 684
        universal approximator, 684
    test set, 674
    training set, 674, 681
    training speed, 685–686
closed circuit television, 362
closing, 577–578, 584–586
cloudy day illumination, 171
clutter, 638, 641, 646
CMOS, 362, 373
coaxial diffuse light, 160
coaxial directed light, 162
coaxial telecentric light, 163
code recognition, *see* codes, recognition
coded light, 166
codes
    bar codes, 695
    DataMatrix, 695, 702, 726, 731
    recognition, 695
coherent light, 99
cold light source, 90
color, 382
color alias, 405
color burst, 431
color constancy, 27, 28
color contrast, 27, 28
color correction, 406
color difference, 392
color filter, 148, 397
color filter array, 489

color information, 404
color interpolation, 404
color perception, 85
coma, 12
    horizontal, 5
combined lighting technique, 184
compact systems, *see* systems, compact systems
compact vision system, 427
compactness, 560, 669
CompactPCI, 464, 465
CompactPCI Express, 466
complement, 567
complementary color, 116
complementary colour filter, 389
completeness check, *see* applications, completeness check
complex cells, 25
component labeling, 552
composite video format, 430
computer bus, 427, 428
cone pedicle, 18
cones, 7, 14, 17, 18, 27, 29, 30
configuration management, 704, 705, 767
connected components, 551–552, 667
connectivity, 551–552, 567, 574, 580
constancy of luminance, 128
continuous production, *see* production, continuous
contour, 515–516
contour feature, *see* features, contour
contour length, 560
contour segmentation, 620–624, 638
    lines, 620–622
    lines and circles, 623–624
    lines and ellipses, 623–624
contrast, 134, 504
contrast adaptation, 12
contrast enhancement, 517–519
contrast normalization, 517–519
    robust, 517–519, 671
contrast sensitivity function, 19, 20
contrast threshold, 19
control, *see* evaluation, control
control signals, *see* data, control signals
controlled lighting, 190
convex hull, 560, 565
convexity, 560
convolution, 530–531
    kernel, 530
convolution integral, 292
convolution theorem, 295, 304
coordinates
    homogeneous, 537, 538
    inhomogeneous, 537, 538
    polar, 544
cornea, 1, 3, 5

correct perspective viewing distance, 243
corrected image, 403
correction data, 403
correlated double sampling, 387
CPU, 480
CRC, 410
cross talk, 307
crystalline lens, 1
cumulative histogram, 519, 561
cutoff spatial frequency, 7
cutoff wavelength , 138
cuton wavelength, 138
cycle start packet, 411
cycle time, *see* production, cycle time

dark current noise, 421
dark field lighting, 173
dark noise, 17
dark signal non uniformity, 378
darkfield illumination, 159
data
    control signals, 727, 732
    images, 729, 731
    type data, 729, 767–769
data acquisition, 497
data packets, 410
data payload size, 417
data rate, *see* signal, data rate
data structures
    images, 511
    regions, 513–515
    subpixel-precise contours, 515–516
Data Valid, 439
DataMatrix code, *see* codes, DataMatrix
daylight suppression filter, 144
DCAM, 406
decision theory, 672–674
    a posteriori probability, 672
    a priori probability, 672, 674
    Bayes decision rule, 672, 673
    Bayes's theorem, 672
deferred image transport, 404
definition, 486
defocused image plane, 282
deighborhood, 551–552
delay time, 190
depth of field, 244, 246, 267
depth of field T, 246
depth of focus, *see* focus, depth of
depth of focus T, 247
depth-first search, 552
Deriche filter, 595, 607
    edge accuracy, 607, 609
    edge precision, 607
deriche filter, 599
derivative
    directional, 589

first, 587, 591
   gradient, 589
   Laplacian, 590
   partial, 589, 598, 599
   second, 588, 591
determinism, 459
deterministic system, 288
deviation from telecentricity, 325
dichromatic, 27
difference, 567
diffraction, 114, 268
diffraction integral, 269, 270
diffraction limit, 5
diffraction limited, 275
diffraction limited depth of focus, 282
diffraction-limited MTF, 5
diffuse area lighting, 161
diffuse bright field incident light, 160
diffuse bright field transmitted lighting, 177
diffuse dark field incident light, 173
diffuse directed partial bright field incident light, 169
diffuse lighting, 155
diffuse on axis light, 160
diffuse transmitted dark field lighting, 184
digital camera specification, 418
digital I/O, *see* interface, digital I/O
dilation, 568–570, 573–574, 583, 604
dimension check, *see* applications, dimension check
DIN 1335, 233
DIN 19040, 231
diplopia, 29
Dirac comb, 304, 307
direct linear transformation, 353
Direct Memory Access, 476
directed bright field incident light, 162
directed bright field transmitted lighting, 178
directed dark field incident light, 173
directed lighting, 155
directed on axis light, 162
directed reflection, 108
directed transmission, 112
directed transmitted dark field lighting, 184
DirectFirePackage, 420
direction of light, 220
directional properties of the light, 155
DirectX/DirectShow, 420
discharging lamp, 88, 91
discrete Fourier transform, 309
disk, 485
disparity, 29, 31, 660–661
dispersion, 118
dispersion of light, 210

display, 481
   image, 492
distance
   chamfer-3-4, 581
   chessboard, 580, 581
   city-block, 580, 581
   Euclidean, 581
distance transform, 580–582, 639, 641
distortion, 320, 336
distribution of illuminance, 131
DMA, 410, 476, 480
DNL, 401
doise
   variance, 592, 606
don-maximum suppression, 596
dormalized moments, 557, 562, 565
dorsal stream, 25, 26, 29
downshift, 493
drift, 90, 98, 102
driver, 480
driver software, 427, 429
DSNU, 379
duality
   dilation–erosion, 573, 584
   hit-or-miss transform, 575
   opening–closing, 578, 584
DV, 445
dynamic range, 374, 423
dynamic thresholding, 548–550, 586, 667

edge
   amplitude, 589, 598
   definition
      1D, 587–589
      2D, 589–590
   gradient magnitude, 589, 598
   gradient vector, 589
   Laplacian, 590, 604–605
   nonmaximum suppression, 588, 596, 600–601
   polarity, 589
edge extraction, 587–612, 638
   1D, 591–597
      Canny filter, 594–595
      Deriche filter, 595
      derivative, 587, 588, 591
      gray value profile, 591–593
      nonmaximum suppression, 596
      subpixel-accurate, 596–597
   2D, 597–605
      Ando filter, 598
      Canny filter, 599, 606
      Deriche filter, 599, 607
      Frei filter, 598
      gradient, 589
      hysteresis thresholding, 601–602
      Lanser filter, 600, 607

Laplacian, 590, 604–605
    nonmaximum suppression, 600–601
    Prewitt filter, 598
    Sobel filter, 598
    subpixel-accurate, 602–605
edge filter, 593–595
    Ando, 598
    Canny, 594–595, 599, 606
        edge accuracy, 607, 608
        edge precision, 606
    Deriche, 595, 599, 607
        edge accuracy, 607, 609
        edge precision, 607
    Frei, 598
    Lanser, 600, 607
        edge accuracy, 607
        edge precision, 607
    optimal, 594–595, 599–600
    Prewitt, 598
    Sobel, 598
edge spread function, 317
effective f-number, 247
EIA (RS170), 379
Eikonal equation, 206
EISA, 428, 462
electromagnetic wave, 205
electronic shutter, 368
electronic shutter , 365
ellipse fitting, 619–620
    algebraic error, 619
    geometric error, 619
    outlier suppression, 619–620
    robust, 619–620
ellipse parameters, 557–558, 560, 562, 564, 565, 614, 615, 619
embedded vision systems, 479
EMC, 93
emmetropia, 3
enclosing circle, 559–560, 565
enclosing rectangle, 559–560, 565, 669
encoder, 497
engineering framework, 744
entocentric perspective, 253
entrance pupil, 227, 238
entrance window, 239
epipolar image rectification, 659–660
epipolar line, 656
epipolar plane, 655
epipolar standard geometry, 658–659
epipole, 656
erosion, 571–574, 583–584
error of second order, 182
errors of the first order, 182
ESF, 317
Ethernet, 450, 459, *see* interface, Ethernet
Euclidean distance, 581

evaluation
    control, 698, 699, 763
    inspection, 698, 699, 752, 754, 757
    monitoring, 698, 699
    recognition, 698
    verification, 698, 699, 752, 754, 757, 760
event, 410
exit pupil, 227, 239
exit window, 239
exposure, 131, 136
exposure time, 136
extensibility, 705, 707, 771
exterior orientation, 335, 611
extraneous light, 200
EXView HAD, 381
eye lens, 1
eye movement, 11, 13, 31
eye protection, 121

face-distortion aftereffect, 30
facet model, 602
far cells, 29
far point, 245, 256
feature extraction, 555–566
features
    contour, 556, 565–566
        area, 565
        center of gravity, 565
        central moments, 565
        contour length, 565
        ellipse parameters, 565
        major axis, 565
        minor axis, 565
        moments, 565
        normalized moments, 565
        orientation, 565
        smallest enclosing circle, 565
        smallest enclosing rectangle, 565
    gray value, 556, 560–564
        α-quantile, 561
        anisometry, 562, 564
        area, 562, 564
        center of gravity, 562
        central moments, 562
        ellipse parameters, 562, 564
        major axis, 562, 564
        maximum, 517, 560
        mean, 561
        median, 561
        minimum, 517, 560
        minor axis, 562, 564
        moments, 561–564
        normalized moments, 562
        orientation, 562
        standard deviation, 561
        variance, 561

region, 556–560
  anisometry, 558, 669
  area, 556, 562
  center of gravity, 557–558, 562
  central moments, 557–558
  compactness, 560, 669
  contour length, 560
  convexity, 560
  ellipse parameters, 557–558, 560
  major axis, 557–558
  minor axis, 557–558
  moments, 557–559
  normalized moments, 557
  orientation, 557–558, 560
  smallest enclosing circle, 559–560
  smallest enclosing rectangle, 559–560, 669
Fermat's principle, 206
field angle
  object side, 231
field bus, *see* interface, field bus
field integration, 367, 368
field programmable gate array, 393
field readout, 391
field stop, 236
fields, 430
FIFO, 404
files, *see* interface, files
fill factor, 365, 563, 608, 611
filter
  anisotropic, 532, 600
  border treatment, 528–529
  convolution, 530–531
    kernel, 530
  definition, 530
  edge, 593–595
    Ando, 598
    Canny, 594–595, 599, 606
    Deriche, 595, 599, 607
    Frei, 598
    Lanser, 600, 607
    optimal, 594–595, 599–600
    Prewitt, 598
    Sobel, 598
  Gaussian, 532–534, 543, 548–550, 595, 599, 632
    frequency response, 533
  isotropic, 533, 600
  linear, 530–531, 591
  mask, 530
  maximum, *see* Morphology, gray value, dilation
  mean, 527–530, 534, 543, 549, 550, 592, 632
    frequency response, 531–532, 632
  median, 534–536, 549, 550

  minimum, *see* Morphology, gray value, erosion
  nonlinear, 534–536, 584
  rank, 536, 584
  recursive, 530, 531, 533, 584
  runtime complexity, 529–530
  separable, 529, 531, 533
  smoothing, 525–536
    optimal, 532–533, 595, 599
    spatial averaging, 527–530
    temporal averaging, 526–527
filter combination, 149
filter factor, 139
finite extension of ray pencils, 234
Fire4Linux, 420
FirePackage, 420
FireWire, 383, *see* interface, FireWire
first Bessel function, 20
fitting
  circles, 617–619
    outlier suppression, 617–618
    robust, 617–618
  ellipses, 619–620
    algebraic error, 619
    geometric error, 619
    outlier suppression, 619–620
    robust, 619–620
  lines, 613–617
    outlier suppression, 614–617
    robust, 614–617
fixation, 710, 711, 717
fixed pattern noise, 422
fixed-pattern noise, 377
flash, 367
flash duration, 93
flash light, 98
flash lighting, 190
flash mode, 190
flash repeating frequency, 190
flash time, 193
fluorescent lamp, 93
focal length
  image side, 219
  object side, 219
focal point
  image side, 216
  object side, 216
focus
  depth of, 712, 739
focussing plane, 241, 244
Fourier transform, 273, 531
fovea, 3, 11
FPGA, 393
FPN, 422
fps, 418
frame, 430
frame grabber, 388, 428, 432, 475, 498, 500

frame integration, 367
frame memory, 404
frame rate, 457
frame rates, 415
frame readout, 369
frame transfer, 365, 366
Frame Valid, 439
framework, *see* engineering framework
Fraunhofer approximation, 274
Frei filter, 598
frequency, 210, *see* signal, frequency
Fresnel's approximation, 273
FTP, *see* interface, network
full dynamic, 493
full frame, 365
full well capacity, 424
fuzzy membership, 562–564
fuzzy set, 562–564

gain, 400
gamma correction, 502, 504
gamma function, 401
gamma LUT, 401, 402
gamma response function, 520, 524
gamut, 390
ganglion cells, 13, 19, 21, 23, 24, 27, 31
gauge capability, 712, 718, 719, 760
Gaussian filter, 532–632
gaussian filter, 534
    frequency response, 533
Gaussian optics, 212, 268
generalized Hough transform, 642–646
    accumulator array, 642–645
    R-table, 644
geometric camera calibration, 610, 611, 654
    exterior orientation, 611
    interior orientation, 611, 654
        camera coordinate system, 654
        distortion coefficients, 654, 658
        focal length, 654
        pixel aspect ratio, 654
        pixel size, 654
        principal distance, 654
        principal point, 654
    relative orientation, 654
        base, 655
        base line, 656
geometric error, 619
geometric hashing, 646–648
geometric matching, 646–652
geometrical optics, 205, 206
geometrical path, 206
Gigabit Ethernet, 383, 426, 428, 450, 457, 475
given range, 493
global shutter, 377
glutamate, 15

grab, 482
gradient, 589
    amplitude, 589, 598
    angle, 589, 643
    direction, 589, 642, 643
    length, 589
    magnitude, 589, 598
    morphological, 586
gradient algorithm, 135
gradient index lens, 5
grating acuity, 9, 20
gray value, 512
    1D histogram, 517–519, 546–548, 561
        cumulative, 519, 561
        maximum, 546–548
        minimum, 546–548
        peak, 546–548
    2D histogram, 522–523
    α-quantile, 561
    camera response, 519–520
        linear, 519, 610, 611
        nonlinear, 520, 610
    feature, *see* features, gray value
    maximum, 517, 560
    mean, 561
    median, 561
    minimum, 517, 560
    normalization, 517–519
        robust, 517–519, 671
    profile, 592
    robust normalization, 670
    scaling, 517
    standard deviation, 561
    transformation, 516–519, 561, 564
    variance, 561
gray value difference, 134
grey filter, 146

half-power points, 138
half-width, 96
halogen lamp, 89
handling, *see* production, part handling
handshaking, *see* interface, handshaking
hardware trigger, 408
harmonic wave, 295
Hausdorff distance, 640–642
Helmholtz equation, 270
Hessian normal form, 613
HF-ballast, 94
high speed imaging, 375
high-speed inspection, 191
higher order aberrations, 5
HiRose, 388, 414
histogram
    1D, 517–519, 546–548, 561
        cumulative, 519, 561
        maximum, 546–548

minimum, 546–548
peak, 546–548
2D, 522–523
hit-or-miss opening, 576
hit-or-miss transform, 574–575, 579
HMI, *see* human machine interface
hole accumulation Diode, 381
homocentric pencil, 207, 211
homogeneity, 288
homogeneous coordinates, 537, 538
homogenous lighting, 171
horizon line, 256
horizontal mirror, 403
horizontal synchronization, 431
host, 428, 449
HSL, 432, 486
Hsync, 431–434
HTTP, *see* interface, network
Huber weight function, 615
human machine interface, 731, 743–745, 747, 748, 775
human perception, 135
hyperacuity, 29
hypercentric perspective, 258
hypercolumn, 26
hyperfocal depth
near limit, 248
hyperfocal distance, 248
hyperoparization, 15
hyperpolarization, 15, 17
hypothesize-and-test paradigm, 646
hysteresis thresholding, 601–602

ideal lens transformation, 276
idealized impulse, 289
IEEE 1394, 402, 420, 428, 442, 443, 446, 447, 449, 454, 456, 457, 475
IEEE 1394a, 425
IEEE 1394b, 412, 425, 443
IEEE-1394 plug, 412
IEEE1394, *see* interface, IEEE1394
IIDC, 399, 406, 407, 409, 410, 418, 446, 447, 454
illuminance, 125, 127, 128, 192
at image sensor, 130
illuminations component, 78
image, 481, 511
binary, 513, 552, 556, 566, 567, 569, 573, 579
bit depth, 512
complement, 584
domain, *see* region of interest
enhancement, 516–536
function, 512–513
gray value, 512
gray value normalization, 517–519
robust, 517–519, 671

gray value scaling, 517
gray value transformation, 516–519
label, 513, 552
multi-channel, 512
noise, *see* Noise
pyramid, 632–636
real, 214
RGB, 512
segmentation, *see* segmentation
single-channel, 512
smoothing, 525–536
spatial averaging, 527–530
temporal averaging, 526–527
transformation, 539–544
virtual, 214
image circle, 292
image circle diameter, 241
image construction
graphical, 223
image orientation, 213
image quality, 316
image reconstruction, 496
image side focal length, 219
image space, 218, 229
images, *see* data, images
imaging equation
general, 227
imaging equations, 224
Newtonian, 226
imaging optics, 153
impulse response, 272, 275, 276
incandescent emission, 88
incandescent lamp, 89
incident light, 171, 198
incident lighting, 157
incoherent imaging, 277
incoherent light, 157
incoherent transmission chain, 304
incoming light, 107
industrial lighting, 78
information theoretical aspect, 288
inhomogeneous coordinates, 537, 538
INL, 401
input, 414
input voltage, 414
inspection, *see* evaluation, inspection
integration time, 15, 17, 19
intelligent camera, *see* systems, compact systems
intensity distribution near focus, 281
interface
digital I/O, 723, 724, 728–730
Ethernet, 723–725, 729, 731, 733, 750, 766
field bus, 723, 724, 730, 731
files, 732
handshaking, 732, 733, 736, 765

multithreading, 734
network, 731
serial, 729
USB, 722–724, 729, 731
interface card, 427, 428
interface device, 427
interference filter, 142
interior orientation, 335, 611, 654
camera coordinate system, 654
distortion coefficients, 654, 658
focal length, 654
pixel aspect ratio, 654
pixel size, 654
principal distance, 654
principal point, 654
interlaced, 369, 430
interlaced scan, 365, 367
interline transfer, 365–367, 370
intermediate pupil, 239
International System of Units, 125
interpolation, 405
bilinear, 541–542, 553, 593
nearest neighbor, 539–541, 593
interrupt, 410, 480
interrupt service routine, 480
intersection, 567
invariant moments, 558
IP, 451
IR, 82, 367, 386
IR cut filter, 386, 395
IR suppression filter, 146
irregular diffuse reflection, 108
irregular diffuse transmission, 112
ISA, 428, 462
ISO9001:2000, 81
isochronous, 410, 411, 441, 443, 444, 446, 450
isoplanasie condition, 291
isoplanatic region, 292
ISR, 480

jitter, 409
job-shop production, *see* production, job-shop
Joint Photographic Experts Group, 491
JPEG, 491

Keplerian telescope, 262
kernel mode, 475
knee point, 375
koniocellular pathway, 23, 28
kT/C noise, 387

L cone, 11, 27, 28, 30
label image, 513, 552
labeling, 552
LabVIEW, 476, 477

Lambert radiator, 132, 155
Lanser filter, 600, 607
edge accuracy, 607
edge precision, 607
Laplacian, 590, 604–605
laser, 88, 99
laser protection classes, 99
latency, 409
law of reflection, 208
law of refraction, 208, 210
LCD shutter, 365
LED, 88, 95
lens
aperture, 521
vignetting, 520
lens aberrations
astigmatism, 610
chromatic, 610
coma, 610
LGN, *see* lateral genicalate nucleus, *see* lateral genicalate nucleus
libdc1394, 420
lifetime, 89–91, 93, 94, 96, 98, 100, 401
light, 82, 205
infrared, 82
ultraviolett, 82
visible, 82
light color, 116
light deflection, 199
light distribution, 131
light emitting diode, 95
light filter, 138
light perception, 84
light propagation, 83
light ray, 83, 205
light shielding, 715, 716
light source, 88
life-span, 717
wavelength, 718
light-slit method, 167
lighting control, 186
lighting systematic, 154
lighting technique, 150, 154
lighting with through-camera view, 170
limit switch, 497
line
Hessian normal form, 613
line fitting, 613–617
outlier suppression, 614–617
robust, 614–617
line frequency, 741–743
line scan, 436, 458
line spread function, 301, 317
Line Valid, 439
line-scan processing, 740
linea camera model, 243
linear, 289

linear system, 288
linearity, 401
Linux, 479
Lommel, 278
Lommel functions, 280
long-wavelength pass filter, 142
look-up table, 401, 502, 517, 522
low pass filter, 297
LSF, 317
Luminance, 126, 432
luminance channel, 23, 27
luminance distribution, 132
luminance indicatrix, 132
luminance of the object, 128
luminescence radiator, 88
luminous efficiency, 89
luminous intensity, 125
LUT, 401, 502, see Look-up table
LVDS, 428, 436
LWP, 142

M cone, 11, 27, 30
machine vision, 362
machine vision hardware, 152
machine vision software, 152
magnification ratio, 214
magnocellular pathway, 23, 27
major axis, 557–558, 562, 564, 565, 614, 615
manufacturing tolerances, 325
marginal ray, 236
market growth, 694
mass production, see production, mass
material flow, 695
mathematical operator, 288
maximum filter, see Morphology, gray value, dilation
maximum likelihood estimator, 676
mean filter, 527–530, 534, 543, 549, 550, 592, 632
    frequency response, 531–532, 632
mean squared edge distance, 639–640
measuring error, 182
mechanical shutter, 365
median filter, 534–536, 549, 550
memory, 485
memory handling, 500
metal vapor lamp, 91
microlens, 381
microlenses, 381
minimum filter, see Morphology, gray value, erosion
Minkowski addition, 568–569, 573, 583
Minkowski subtraction, 570–571, 573, 583
minor axis, 557–558, 562, 564, 565, 614
mirror, 200
missing codes, 401
modulation transfer function, 6, 19, 321

moiré pattern, 7
moments, 557–559, 561–564
    invariant, 558
monitoring, see evaluation, monitoring
monochromatic aberration, 9
monochromatic light, 96, 99, 116
morphological adaptation, 9
morphology, 566–586, 667
    duality
        dilation–erosion, 573, 584
        hit-or-miss transform, 575
        opening–closing, 578, 584
    gray value
        closing, 584–586
        complement, 584
        dilation, 583
        erosion, 583–584
        gradient, 586
        Minkowski addition, 583
        Minkowski subtraction, 583
        opening, 584–586
        range, 586
    region, 566–582
        boundary, 573–574
        closing, 577–578
        complement, 567
        difference, 567
        dilation, 568–570, 573–574, 604
        distance transform, 580–582, 639, 641
        erosion, 571–574
        hit-or-miss opening, 576
        hit-or-miss transform, 574–575, 579
        intersection, 567
        Minkowski addition, 568–569, 573
        Minkowski subtraction, 570–571, 573
        opening, 575–577
        skeleton, 578–580
        translation, 567
        transposition, 567
        union, 566–567
    structuring element, 567–568, 574, 583
mosaic, 454
motion blur, 697, 735–739
    computation, 738, 739
motion sensitivity, 3, 23
motional blurring, 137, 192
movements, 714, 721, 735, 737, 750, 765
    adjustment, 714
    change-over, 714, 717
    process, 714
MTF, 321
    semiconductor imaging device, 302
multi coated lens, 144
multi-drop, 464

multithreading, *see* interface,
        multithreading
myopia, 3

natural light, 122
natural vignetting, 129, 323
near cells, 29
near limit, 248
near point, 245
negative principal points, 227
neighborhood, 574, 580
neural network, 677–681, 685–687
        activation function
            hyperbolic tangent, 680
            logistic, 680
            sigmoid, 680
            softmax, 680
            threshold, 677, 679
        multi-layer perceptron, 678–681, 685–687
            training, 680–681
        single-layer perceptron, 677–678
        universal approximator, 679, 680
neutral density filter, 143, 146
neutral filter, 146
Node_Id, 418, 419
noise, 420, 525–526
        suppression, 526–536
        variance, 525, 527, 528, 533–534
noise floor, 423
nondestructive overlays, 492, 494
nonintegrating sensors, 374
nonmaximum suppression, 588, 600–601
normal distribution, 676
normalized cross correlation, 628–631, 663
normalized moments, 557–615
NTSC, 429, 456, 457
null direction, 21
numerical aperture, 240
        image side, 262
Nyquist band pass, 309
Nyquist frequency, 315, 321
Nyquist limit, 8
Nyquist sampling theorem, 307

object
        virtual, 215
object field angle, 231
object field stop, 239
object plane, 158
object recognition, *see* applications
object side field angle, 231, 241
object side focal length, 219
object side telecentric perspective, 257
object side telecentric system, 260
object space, 218, 229
obliquity factor, 271

occlusion, 638, 641, 646
OCR, *see* optical character recognition
ocular dominance column, 26
ON and OFF channels, 21
onboard memory, 501
one shot, 409
one-chip, 396
one-dimensional representation, 300
opening, 575–577, 584–586
optic nerve, 11, 13, 30
optic nerve head, 3
optical axis, 3, 212
        eye, 3
optical character recognition, 519, 537, 538, 555, 556, 666–688
        character segmentation, 667–668
            touching characters, 667–668
        classification, *see* Classification
        features, 669–671
        image rectification, 541, 543, 667
optical density, 139
optical path difference, 274
optical path length, 140
optical transfer function, 296
orientation, 334, 557–558, 560, 562, 565
        exterior, 611
        interior, 611, 654
            camera coordinate system, 654
            distortion coefficients, 654, 658
            focal length, 654
            pixel aspect ratio, 654
            pixel size, 654
            principal distance, 654
            principal point, 654
        relative, 654
            base, 655
            base line, 656
orientation column, 26
OTF, 296
        multiplication rule, 315
outlier, 613, 614
outlier suppression, 614–620
        Huber weight function, 615
        random sample consensus, 617
        RANSAC, 617
        Tukey weight function, 616
output, 414
overtemperature, 103

PAL, 429, 456, 457
palettes, 494
parallax, 29
parallel capturing, 739
parallel lighting, 156
parallel offset, 140
parameter data, *see* data, parameters
parasitic sensitivity, 378

paraxial region, 212
part handling, *see* production, part handling
partial bright field illumination, 159
parvocellular pathway, 23, 27
PC-based systems, *see* systems, PC-based
PCI, 428, 464, 470
PCI Express, 428, 465, 466, 470
PCI-X, 428, 466, 470
performance, 479
permissible defocusing tolerance, 281
permissible size for the circle of confusion, 249
perspective, 251
perspective transformation, 538–539, 543, 667
pharoid ray, 237
Phase Lock Loop, 432
photo response nonuniformity, 422
photodiodes, 374
photogrammetry, 333
photometric inverse square law, 127
photometric quantity, 125
photon, 15, 83
photon counter, 27
photon noise, 17, 21, 421
photoreceptor, 1, 3, 6, 8, 13, 14, 27, 29, 30
phototransduction, 14, 30
photovoltaic effect, 373
pinhole camera, 208, 252
pinning
    cameras, 711
pipelined global shutter, 377
pixel, 302, 305, 364, 370–377, 379, 381, 382, 384, 386–389, 392, 396–398, 400–403, 405–407, 410, 421, 511–512
pixel sensitivity function, 302
plane wave, 270
platform, 479
PLC, 708, 726, 728–733, 746, 748, 749, 769
PLL, 388, 432
PNG, 491
point spread extension, 284
point spread function, 291
Poisson statistics, 15
polar coordinates, 544
polar transformation, 544, 667
polarization, 122
polarization filter, 142, 147
polarization phenomena, 269
polarized light, 122
    circular, 122
    elliptical, 122
    linear, 122
    unpolarized, 122
polarizer, 123

polling, 480
polygonal approximation, 620–622
    Ramer algorithm, 621–622
Portable Network Graphics, 491
pose, 536–537, 625
position recognition, *see* applications, position recognition
position variation, 712
power, 414
power supply, 721, 722, 753
    uninterruptible, 722
precision, 605–606, 718
    edge angle, 645
    edges, 606–607
    hardware requirements, 607
preferred direction, 21, 24, 31
presynaptic, 15
Prewitt filter, 598
primary colors, 370, 396
principal plane, 217
    image side, 217
    object side, 217
principal point, 340, 654
principal ray, 236
principle of triangulation, 166
prism, 199
PRNU, 422
processing, 482
processing time, 135
production
    batch, 696
    continuous, 696, 697, 734, 737
    continuous motion, 732, 737
    cycle time, 735, 758, 759
    discrete motion, 735, 742
    discrete unit, 697, 699, 736
    job-shop, 698
    mass, 696, 698
    part handling, 735
progressive scan, 365, 367, 369–371, 383, 431
projection, 182
projection centre, 208, 209, 243, 253
projective transformation, 208, 538–539, 543, 626, 667
PSF, 291
pulse mode, 189
pulse-duty-factor, 96, 104
pupil, 5, 11, 14, 31
pupil function, 278
pupil magnification ratio, 229, 242, 267
PXI, 464–466

quadrature encoder, 499
quantization noise, 423

r-table, 644

radial spatial frequency, 301
radiant intensity, 125
Radii of curvature, 220
radiometric camera calibration, 519–524, 610
    calibration target, 520
    chart based, 520–521
    chart-less, 521–524
        defining equation, 522
        gamma response function, 520, 524
        inverse response function, 522
            discretization, 522–523
            normalization, 523
            polynomial, 524
            smoothness constraint, 523–524
    response function, 520, 521, 524
radiometric quantity, 125
Ramer algorithm, 621–622
rank filter, 536, 584
rapid-prototyping, 705
RBG to YUV, 406
real object
    real, 214
receptive field, 20, 21, 23, 24, 31
recipes, see data, type data
reciprocity equation, 225
recognition, see evaluation, recognition
reduced coordinates, 279
reflectance, 107, 110, 127
reflection, 107, 205
    directed, 108
    irregular diffuse, 108
    regular diffuse , 108
reflection law, 107
refraction, 115, 205
refraction law, 83
refractive index, 210
region, 513–515
    as binary image, 513, 552, 556, 566, 567, 569, 573, 579
    boundary, 560, 573–574
    characteristic function, 513, 561, 562, 583
    complement, 567
    connected components, 551–552, 667
    convex hull, 560
    definition, 513
    difference, 567
    feature, see features, region
    intersection, 567
    run-length representation, 514–515, 552, 556, 557, 566, 567, 569, 579
    translation, 567
    transposition, 567
    union, 566–567
region of interest, 504, 513, 537, 539, 570, 582, 604, 626

regular diffuse reflection, 108
relative illumination, 323
relative irradiance, 323
relative orientation, 654
    base, 655
    base line, 656
remote maintenance, 750, 773
removal of surface reflexes, 123
representation, 294
representation space, 294
reproducibility, 711, 716, 717
reset noise, 422
Resolution, 455
resolution, 382, 485
resolution limit, 321
result data, see data, results
retina, 3, 6, 12, 13, 21, 30, 31
RGB, 432, 486
rhodopsin, 17
rigid transformation, 625, 637, 647
ring, 483
ring light, 169
rod photoreceptor, 14
rods, 7, 14, 17, 18, 21, 30
ROI, 504, see region of interest
rolling curtain, 376
rotation, 538, 541, 559, 625, 637, 667
RS-170, 429, 456, 457
RS-232, 414
RS-422, 428, 436
RS-644, 436
RS232, see interface, serial
run-length encoding, 514–515, 552, 556, 557, 566, 567, 569, 579

S cone, 11, 27
S-Video, 433
sampling interval, 6
sampling theorem, 7
saturation, 14, 30, 135
scaling, 538, 542–543, 559, 625, 637
scattering, 107
secondary wavelet, 268
security, see systems, security
segmentation, 544–555
    connected components, 551–552, 667
    dynamic thresholding, 548–550, 586, 667
    hysteresis thresholding, 601–602
    subpixel-precise thresholding, 553–555, 604
    thresholding, 545–548, 667
        automatic threshold selection, 546–548, 667
sensitivity, 365, 382, 423
sensor, 511–512
sequence, 483

SG pulse, 384
shading correction, 402, 507
shading images, 403
Shannon, 7
shape check, *see* applications, shape check
shift theorem, 313
shifting property, 289
short-wavelength pass filter, 142
shutter efficiency, 378
shutter speed, 739
shutter time, 136, 193, 521
SI, 125
signal
    bandwidth, 723, 724, 748, 750
    frequency, 723, 741–743, 768
signal to noise ratio, 401, 424
signal-to-noise ratio, 594, 606, 607, 645
similarity measure, 626–631
    normalized cross correlation, 628–631, 663
    sum of absolute gray value differences, 626–630, 662
    sum of squared gray value differences, 626–628, 662
similarity transformation, 626, 637, 647
simple cells, 24
skeleton, 578–580
skew, 538
slant, 538
smallest enclosing circle, 559–560, 565
smallest enclosing rectangle, 559–560, 565, 669
smart camera, 427, *see* systems, compact systems
smart cameras, 479
smear, 365, 372
smoothing filter, 525–536
    Gaussian, 532–534, 543, 548–550, 595, 599, 632
        frequency response, 533
    mean, 527–530, 534, 543, 549, 550, 592, 632
    median, 534–536, 549, 550
    optimal, 532–533, 595, 599
    spatial averaging, 527–530
    temporal averaging, 526–527
snap, 481
SNR, 401
Sobel filter, 598
solid angle, 126
solid state material, 96
space variant nature of aliasing, 312
space-invariance, 291
spatial averaging, 527–530
spatial depth, 251
spatial frequencies, 274
spatial frequency, 5–7, 12, 31, 296

spatial frequency components, 298
SPC, *see* statistical process control
speckle, 157
speckle pattern, 99
spectral band, 91
spectral opponency, 23
spectral response, 84, 139
    Gaussian filter, 533
    mean filter, 531–532, 632
    sensor, 512
spectral sensitivity, 386, 390, 394, 398
spectral transmittance, 323
spectrum, 90
spectrum of light, 82
speed, 382
spherical aberration, 5
spherical wave, 270
spikes, 18, 19
spontaneous spike activity, 19
spurious resolution, 7
square pixel, 371
stabilization, 90
standardization, 702, 709, 717, 725, 771
standardized viewing distance , 250
static lighting, 190
statistical process control, 699
stereo geometry, 654–661
    corresponding points, 655
    disparity, 660–661
    epipolar line, 656
    epipolar plane, 655
    epipolar standard geometry, 658–659
    epipole, 656
    image rectification, 659–660
stereo matching, 662–666
    robust, 665–666
        disparity consistency check, 665
        excluding weakly textured areas, 665
    similarity measure
        normalized cross correlation, 663
        sum of absolute gray value differences, 662
        sum of squared gray value differences, 662
    subpixel accurate, 664
    window size, 664–665
stereo reconstruction, 653–666
stereopsis, 29, 31
stochastic process, 525, 527
    ergodic, 527
    stationary, 525
streaking light, 174
Strehl ratio, 281
strobe, 370
strobe light, 497
strongly translucent part, 184

structured bright field incident light, 165
structured light, 166
structured lighting, 157
structuring element, 567–568, 574, 583
sub sampling, 369
sub-pixel region, 302
sub-sampling, 382
subpixel-precise contour, 515–516
    convex hull, 565
    features, see features, contour
subpixel-precise thresholding, 553–555, 604
sum of absolute gray value differences, 626–630, 662
sum of squared gray value differences, 626–628, 662
superdiffuse ring lights and shadow-free lighting, 171
SuperHAD, 381
superposition integral, 271
support vector machine, 681–687
    kernel, 684
        Gaussian radial basis function, 684
        homogeneous polynomial, 684
        inhomogeneous polynomial, 684
        sigmoid, 684
    margin, 682
    separating hyperplane, 682–683
    universal approximator, 684
surface inspection, see applications, surface inspection
SVGA, 379
SWP, 142
synaptic plasticity, 30
sync bit, 410
sync pulse, 388
synchronization, 194
synchronization of flash lighting, 193
system of lenses, 231
systems
    application-package, 705, 706
    compact systems, 703, 704, 707, 723–725, 729
    library-based, 705
    PC-based, 703–705, 750, 751
    security, 704, 705, 750
    types of, 700–708
    vision controllers, 704
    vision sensors, 702, 707, 725, 741

Tagged Image File Format, 491
tangential spatial frequency, 301
tap, 496
TCP, 451
TCP/IP, 724, 725, 729, 731, 750, 766
telecentric bright field incident light, 163
telecentric bright field transmitted lighting, 180
telecentric lighting, 156, 180
telecentric objective, 156
telecentric on axis light, 163
telecentric perspective, 260
telecentric system, 251
    object side , 260
telescope magnification, 263
temperature compensation, 189
temperature radiator, 88
template matching, 624–653, 667
    clutter, 638, 641, 646
    erosion, 572
    generalized Hough transform, 642–646
        accumulator array, 642–645
        R-table, 644
    geometric hashing, 646–648
    geometric matching, 646–652
    Hausdorff distance, 640–642
    hierarchical search, 633–636
    hit-or-miss transform, 574
    hypothesize-and-test paradigm, 646
    image pyramid, 632–636
    linear illumination changes, 628
    matching geometric primitives, 648–652
    mean squared edge distance, 639–640
    nonlinear illumination changes, 638, 646
    occlusion, 638, 641, 646
    opening, 576
    robust, 638–653
    rotation, 637–638
    scaling, 637–638
    similarity measure, 626–631
        normalized cross correlation, 628–631, 663
        sum of absolute gray value differences, 626–630, 662
        sum of squared gray value differences, 626–628, 662
    stopping criterion, 629–631
        normalized cross correlation, 630–631
        sum of absolute gray value differences, 629–630
    subpixel accurate, 636–637
    translation, 626–636
temporal averaging, 526–527
temporal control, 189
test object, 105, 152
thermal noise, 17
thickness of the lens, 220
thin films, 143
thin lens, 221
three field readout, 369
thresholding, 545–548, 667

automatic threshold selection, 546–548, 667
subpixel-precise, 553–555, 604
throughput, 469
TIFF, 491
tilt rule, 216
time response, 409
time-invariant system, 291
timing, 497
topography, 168
total reflection, 111
traceability, 694, 726, 728, 732
transfer function, 295
transfer function of the pixel, 304
transformation
  affine, 537–538, 559, 626, 646
  geometric, 536–544, 667
  gray value, 516–519
  image, 539–544
  perspective, 538–539, 543, 667
  polar, 544, 667
  projective, 538–539, 543, 626, 667
  rigid, 625, 637, 647
  rotation, 538, 541, 559, 625, 637, 667
  scaling, 538, 542–543, 559, 625, 637
  similarity, 626, 637, 647
  skew, 538
  slant, 538
  translation, 538, 558, 567, 625, 637
translation, 538, 558, 567, 625, 637
transmission, 110, 112, 138, 139
  diffuse, 112
  directed, 112
  irregular diffuse, 112
transmission
  through the lens, 129
transmission chain, 304
transmittance, 112, 127
transmitted lighting, 158
transposition, 567
trichromatic, 27
trigger, 407, 459, 697, 701, 713, 732, 738–740, 742
triggering, 497
triggering pulse, 190
tristimulus theory, 396
tukey weight function, 616
two field readout, 369
type data, *see* data, type data

UART, 439
UDP, 451
ultraviolet light, 121
uninterruptible power supply, *see* power supply, uninterruptible
union, 566–567

unit production, *see* production, discrete unit
Universal Serial Bus, 428
unpolarized light, 122
UPS, *see* power supply, uninterruptible
USB, 383, 440, 441, 457, 475, *see* interface, USB
USB2.0, 425
user interface, *see* human machine interface
user mode, 475
UV, 82, 121, 386
UV blocking filter, 144

variable height acquisition, 499
VB.NET, 477
veiling glare, 325
velocity of light, 206
ventral stream, 25
verification, *see* evaluation, verification
vertical synchronization, 431
vertices, 220
VGA, 379
VHA, 499
vibrations, 736
video data format, 410
video formats, 406
video signal, 305
video standard, 379
vignetting, 132, 520
VIS, 82, 116, 118
visible light, 116, 118
vision controllers, *see* systems, vision controllers
vision sensors, *see* systems, vision sensors
Visual Basic, 476
visual cortex, 13, 23, 25, 26, 29
vitreous chamber, 3
Vsync, 431–434
VxWorks, 479

wave equation, 205
wave nature of light, 268
wave number, 270
wave–particle dualism, 83
wavefront
  plane, 207
  spherical, 207
wavefront aberration, 281
waveguides, 8
wavelength, 210, *see* light source, wavelength
WDM, 420
Web server, *see* interface, network
weight function
  Huber, 615
  Tukey, 616
white balance, 399

white balancing, 490
white light, 118
Windows, 479
windows, 239
wiring, 725, 739
wood effect, 120
working distance, 140, 710–712
world coordinates
    from single image, 611
    from stereo reconstruction, 653–666

Xenon flash lamp, 92
Xenon lamp, 92

YUV, 480

zero crossing, 588, 590, 604

vision
WE MAKE THE COMPONENTS
control

MACHINE VISION COMPONENTS

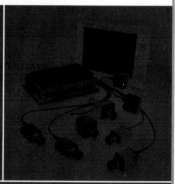

[we make the components]

Vision & Control GmbH
Pfuetschbergstrasse 14
98527 Suhl · Germany

Phone +49 [0] 36 81 / 79 74 - 0
Fax +49 [0] 36 81 / 79 74 - 44
infopoint@vision-control.com

www.vision-control.com

LIGHTING – OPTICS – VISION SYSTEMS

**Schneider knows how the world
of industrial optics is changing.**

We're changing it.

We recently introduced the world's first 12k line scan 5-micron capable lens. Virtually indestructible compact c-mounts. Bilateral telecentric lenses, and so much more. For industry-standard and custom image processing solutions of tomorrow, anywhere in the world, contact a Schneider optical expert today.

In the USA: +1 631 761-5000
Outside the USA: +49.671.601.387
www.schneiderindustrialoptics.com

Making vision technology work.

Digital cameras for industrial vision systems

BE AHEAD
with Baumer vision components.

- ✔ Digital cameras, interface boards and software drivers from a single source
- ✔ 32 camera models – CCD and CMOS – in 8 / 12 bit
- ✔ VGA up to 4 million pixel color and monochrome
- ✔ Outstanding image quality
- ✔ Multi-camera real-time systems, up to 8 cameras / PC
- ✔ Standard PCI, Compact PCI and *PCI Express*™
- ✔ Customized OEM solutions
- ✔ Cameras also available with FireWire™ interface

Baumer Optronic GmbH
Badstraße 30, 01454 Radeberg, Germany
Phone +49 (0)3528-4386-0, Fax +49 (0)3528-4386-86
Email: sales@baumeroptronic.com

www.baumeroptronic.com

CTMV
Consulting Team Machine Vision
Ihr Partner für Industrielle Bildverarbeitung & Automation

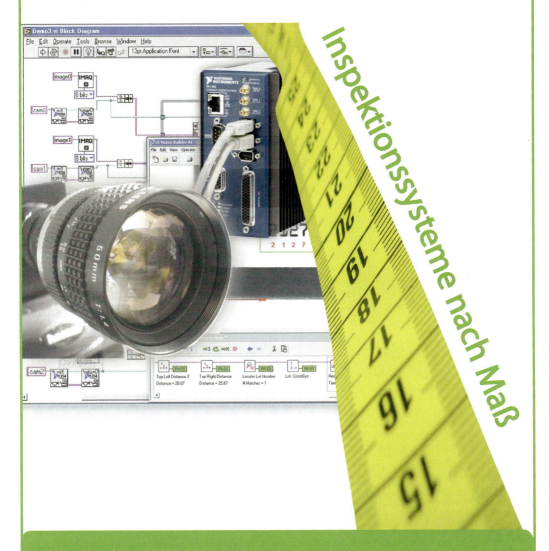

Inspektionssysteme nach Maß

www.ctmv.de

CTMV GmbH & Co. KG
Kronprinzenstr. 38 | 75177 Pforzheim
Tel.: +49 7231 5853-259 | info@ctmv.de

HALCON

the Power of Machine Vision

Matches your requirements
with a comprehensive library for all machine vision applications

Includes advanced algorithms
for matching, identification, positioning, and 1D, 2D, and 3D metrology

Speeds up your development
with powerful tools and IDE for machine vision

Secures your investment
by compatibility with Linux/UNIX and Windows (incl. x64)

Runs your application faster
by instant support of multi-processor computers

Try HALCON for free. Request a free product CD or use our free application evaluation service at: www.mvtec.com/halcon/now

MVTec Software GmbH   *Building Vision for Business*

Positioniersysteme & Optische Systeme

Lineartische
Drehtische
Goniometer
Multiaxes/ Hexapod
Steuerungen
Optische Bänke
Laser- Schulungssysteme
Kundenspezifische Lösungen

Contact:
MICOS GmbH
Freiburger Str. 30
79427 Eschbach, Germany
Phone +49/7634/5057-231
Fax   +49/7634/5057-393
sales@micos-online.com
www.micos.ws

---

6 Volume Set

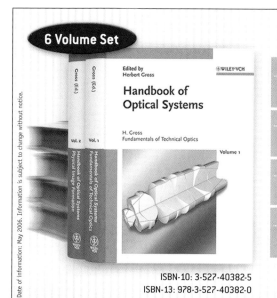

Edited by Herbert Gross
Handbook of Optical Systems
H. Gross
Fundamentals of Technical Optics
Volume 1

ISBN-10: 3-527-40382-5
ISBN-13: 978-3-527-40382-0

- gives a unique overview for both newcomers and professionals in academia and industry
- balances comprehensive introduction with latest research results in a uniform style
- features over 3,000 color illustrations that facilitate access to complex problems
- written by experts at the world's leading manufacturer of optical systems

WILEY-VCH
Wiley-VCH • Tel.: +49 (0) 6201 - 606 400 • Fax: +49 (0) 6201 - 606 184
e-Mail: service@wiley-vch.de • www.wiley-vch.de

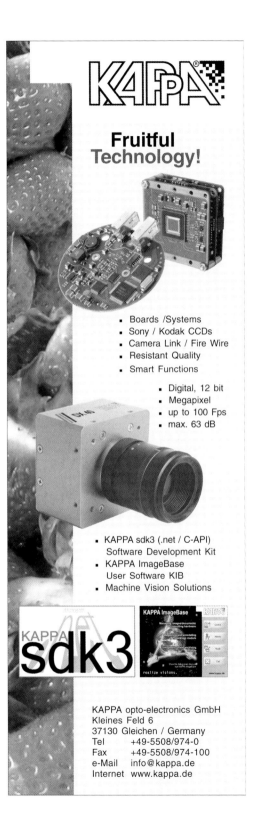

Fruitful Technology!

- Boards /Systems
- Sony / Kodak CCDs
- Camera Link / Fire Wire
- Resistant Quality
- Smart Functions

- Digital, 12 bit
- Megapixel
- up to 100 Fps
- max. 63 dB

- KAPPA sdk3 (.net / C-API) Software Development Kit
- KAPPA ImageBase User Software KIB
- Machine Vision Solutions

KAPPA opto-electronics GmbH
Kleines Feld 6
37130 Gleichen / Germany
Tel   +49-5508/974-0
Fax   +49-5508/974-100
e-Mail info@kappa.de
Internet www.kappa.de

all-inclusive sozusagen ...

FISBA OPTIK entwickelt und produziert für Sie individuelle Optiklösungen nach Ihrem spezifischen Anforderungsprofil. Profitieren Sie dabei von unser jahrzehntelangen Erfahrung in der Optikbranche.

Know How in Design & Engineering
- damit wir Ihnen innovative Lösungen nach Ihrem Anforderungsprofil schnell und effizient entwickeln können

Projekt Management
- damit wir Ihnen einen reibungslosen Ablauf von der Konzeptphase bis zum fertigen Produkt gewährleisten können

QM Standards / Sauberkeit
- damit Ihre optischen Systeme direkt und störungsfrei im täglichen Betrieb eingesetzt werden können

In-Haus Produktion
- damit unserproduktionstechnisches Wissen bereits von Beginn weg zu Ihrer Verfügung steht

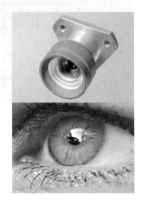

 FISBA OPTIK

FISBA OPTIK AG
Rorschacher Str. 268
CH-9016 St. Gallen
T. +41 (0)71 282 31 31
F. +41 (0)71 282 31 30
www.fisba.ch